The Biology, Structure, and Systematics of the Cycadales

Proceedings of CYCAD 90, the Second International Conference on Cycad Biology

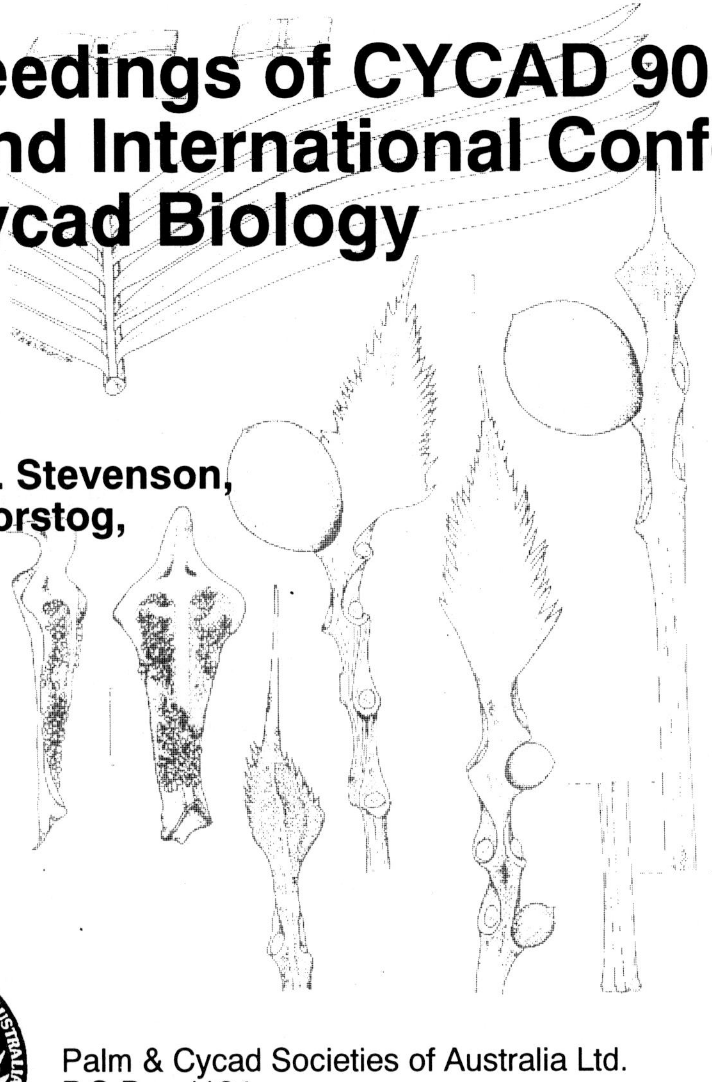

Dennis W. Stevenson,
Knut J. Norstog,
Editors

Palm & Cycad Societies of Australia Ltd.
P.O.Box 1134,
Milton, Queensland 4064
AUSTRALIA

Published by the: Palm & Cycad Societies of Australia Ltd.
June, 1993

Cover: Background motif of *Cycas media* reproductive and leaf parts by courtesy of Royal Botanic Gardens Sydney, Australia. M. Westmacott, del.

Published by:

Palm & Cycad Societies of Australia Ltd. Publication Fund
P.O. Box 1134,
Milton, Queensland, 4064
Australia

The Biology, and systematics of Cycadales : Proceedings of Cycad 90, the second international conference on cycad biology, Townsville, Queensland, Australia, July 22 - 28, 1990. / Dennis W. Stevenson and Knut J. Norstog, editors.

Includes bibliographical references.
ISBN 0-9587-931-6-6
1. Cycadales—Congresses. I. Stevenson, Dennis W., II Norstog, Knut J.

International Standard Book Number 0-9587-931-6-6

Page publishing by: Page People Pty. Ltd., 5 Kilnatoora St., The Gap, Queensland 4061
Printed by: A. Webb & Sons, Baxter Street, Fortitude Valley, Queensland 4006

In Memoriam

Sergio Sabato

30-4-1941 30-4-1991

Table of Contents

Part One - Introduction

Part Two - Toxicity

Part Three - Population Ecology

Table of Contents (cont'd)

Part Four - Physiology

Part Five - Cycad Diversity

Table of Contents

Part Six - Conservation

Part Seven - Abstracts

Table of Contents (cont'd)

List of Authors

Ackery, Phillip R. , Robert J. Nash, E. Arthur Bell, Knut Norstog:

Beaton, John M.

Beck, Wendy

Bell, E. A.

Butt, L.P.

Byrne, M. and James, S.H.

Caputo, Paolo, Cecilia Marquis, T. Wurtzel, Dennis W. Stevenson, Eleanore T. Wurtzel

Chadwick, C.E.

Chakavarika, Charles

Chirgwin, S.

Connell, S.W., P.G. Ladd

Dehgan, Bijan, Bart Schutzman, and Fe Almira

Dehgan, Bijan and Fe Almira

Delevoryas, T.

Donaldson, John S.

Duncan, Mark W.

Fawcett, Priscilla K.S., Knut J. Norstog

Gaudio, L. and Esposito, E., Siniscalco Gigliano, G., Caputo, P., Muoio, R., and Moretti, A.

Giddy, Cynthia

Goode, Douglas

Grilli Caiola, Maria and Antonella Canini

Grobbelaar, N.

Grobbelaar, N. and Marshall, J.

Guo-fan, Liu , Deng Ting-xiu, Zhou Lin and Yang Si-Yuan

Hawley, Grant

Hill, K.D.

Hill, Chris. R.

Hoffmann, George R.

Kurland, Leonard T.

Ladd, P.G., S.W. Connell, B. Harrison

Loconte, Henry

Moretti, A., Caputo, P., and Cozzolino, S., Gaudio, L.

Norstog, Knut J.

Norstog, Knut J., Priscilla K.S. Fawcett, Andrew P. Vovides, Mark W. Dunn

Nunn, Peter B. and Malarvizhi Ponnusamy

Ornduff, Robert

Osborne, Roy

Osborne, R. & J.J. Nair

Osborne, R., N. Grobbelaar, P.L.D. Vincent

Pant, Divya Darshan and Kabita Das

Pant, D.D. and Rita Singh

Pate, John S.

Resslar, P.M.

Roach, JF and John, C

Sanchez-Tinoco, Maria Ydelia, Maria Vazquez-Torres and Jorge Alberto Alejandre-Rosas

Schutzman, Bart and Bijan Dehgan

Seawright, A., A.W. Brown, C.C. Nolan and J.B. Cavanagh

Siniscalco Gigliani, Gesualdo

Siniscalco Gigliano, G., De Luca, P., Caputo, P., Moretti, A., Cozzolino, S., Gaudio, L., and Tat, R.

Spencer, Peter S.

Staff, Ian A. and Charmian P. Ahern

Stevenson, D.

List of Authors (cont'd)

Stevenson, Dennis W.M. & Roy Osborne

Tang, William

Turner, Ian S.

Vazquez-Torres, M.

Vorster, P.

Vovides, Andrew P.

Whitelock, L.M.

Wilson, Alan F.

Part One

Introduction

Introduction

It was with considerable trepidation that the organisers of Cycad 90 undertook the planning of this symposium, principally because the planning committee members resided in such distant places as the United States of America, South Africa, Italy and Australia. That they succeeded in presenting a diversified program of high quality was generally recognised as the conference proceeded an then drew to its closing. What, perhaps, was not always apparent to those in attendance was the enormous amount of background labour in Townsville by Mr and Mrs David Coutts and their able assistants. They not only established the logistics for the program but also in large measure funded the conference and its events. The Organising Committee and, indeed, all of those attending Cycad 90, owe them a great debt.

We also wish to acknowledge our indebtedness to Mr Jean-Pierre Sclavo who initiated these conferences with the very successful First International Conference on Cycad Biology, Cycad 87, held in July 1987 in Nice, France. The fact that we now have experienced the second such conference and are planning for Cycad 93 in South Africa and even beyond to Cycad 96 in China and Cycad 99 in the United States of America, testifies to the importance of the first conference. In a very real sense, these conferences were conceived and initiated by Jean-Pierre and we thank him for his far sighted vision.

For those of us who in one way or another been involved with cycads, the past decades have been both heartening and discouraging. Heartening because we see so much evidence of growing interest in the world's cycads, their study and preservation; discouraging because the cycads are sustaining such heavy pressure from over-collecting and habitat destruction. Consequently, for many rare species there is no assurance that they can survive much longer in the wild.

In this conference, new information is presented on the ecology, biogeography, physiology, propagation, toxicology, ethnobotany, paleobotany, and conservation of cycads. Since Cycad 87 in Nice, for example, a totally unsuspected dependency of cycads upon insect pollinators has surfaced, and new information is presented relative to the impact of cycad toxins on such cycad predators as humans, livestock, and certain insects.

The reader of these proceedings will be in no doubt that the papers presented herein reflect a rather wide variety of experiences with cycads. This was not, stictly speaking, a typical scientific symposium in which professional research workers present their most recent findings along rather conventional lines.

Rather, we have included papers from nurserymen, museum botanists, medical researchers, agriculturalists, university faculty, talented hobbyists, and those of other interests. All, however, are tied together by a common and shared interests in those unique and fascinating, living fossils, the cycads.

We wish to express our thanks to all our participants who individually and collectively helped to make the conference a success. We also thank the Planning Committee, Pat Coutts, Aldo Moretti, Knut Norstog, Roy Osborne and Dennis Stevenson for their contributiuons to the conference. We are grateful to The Palm and Cycads Societies of Australia Ltd for their support and to its offices, especially so to John Dowe, the editor of its magazine *Palms and Cycads*. Finally, our thanks are extended to all the organisations and friends of cycads in Australia and elsewhere throughout the world who supported Cycad 90.

We dedicate this volume to the memory of Professor Sergio Sabato. Sergio was an outstanding cycadologist who had already made many significant contributions to cycad biology. He understood Central American cycads better than any of his predecessors. Sergio was much more than just a cycad biologist, he was also a scholar. His knowledge of botany, science in general, art, literature, and music was both thorough and fundamental. He was a unique colleague who touched the lives of many of us very deeply. As testimony to our respect and with great sadness at the loss of our dear friend, this volume is dedicated to his memory

Part Two

Toxicity

Sequestration of MAM-Glycosides by Insects

Phillip R. Ackery
The Natural History Museum
Cromwell Road, London SW7 5BD, U.K.

Robert J. Nash, E. Arthur Bell, Knut Norstog
Fairchild Tropical Garden
Miami, Florida 33156, U.S.A.

Abstract

Cycasin was detected in dried museum specimens of the cycad-feeding butterflies *Taenaris butleri*, *T. catops*, *T. onolaus* and *Luthrodes cleotas* but was absent from the more cryptic species *Theclinesthes onycha*. It is suggested that in New Guinea the toxic *Taenaris* species are models for a mimicry complex that involves various butterfly species, including other *Taenaris*. The snout weevils *Rhopalotria slossoni* and *R.mollis* were also found to contain cycasin and macrozamin respectively.

Introduction

The Cycadales are characterised by their ability to synthesise and accumulate glycosides of methylazoxymethanol (MAM), (Fig. 1), a potent toxin and carcinogen which is released enzymatically from the glycoside on ingestion by animals (Spatz, McDaniel and Laqueur, 1966). Sequestration of cycasin, the β-D-glucoside of MAM, has been shown to occur in two species of Lepidoptera, *Seirarctia echo*, a Tiger Moth *Teas, 1967), and the Lycaenid butterfly *Eumaeus atala florida* (Rothschild, Nash and Bell, 1986). Both feed on *Zamia floridana* and can be considered to exhibit warning attributes. The former was shown to avoid MAM intoxication by glycosylating or reglycosylating MAM released by its gut β-glucosidase activity. Cycasin so formed is stored by the larva in the haemolymph. There are also other Lepidoptera known to feed on *Cycas*, namely *Theclinesthes onycha* Hewitson (Lycaenidae) in Australia, and *Luthrodes cleotas* Guerin (Lycaenidae) and some *Taenaris* species (Nymphalidae) in New Guinea (Parsons, 1984). The latter genus is part of a mimicry complex involving several other nymphalid genera and a papilinoid species. The possible sequestration of MAM-glycosides by the *Cycas* feeding butterflies and two snout weevils which feed on *Zamia floridana* cones was studied.

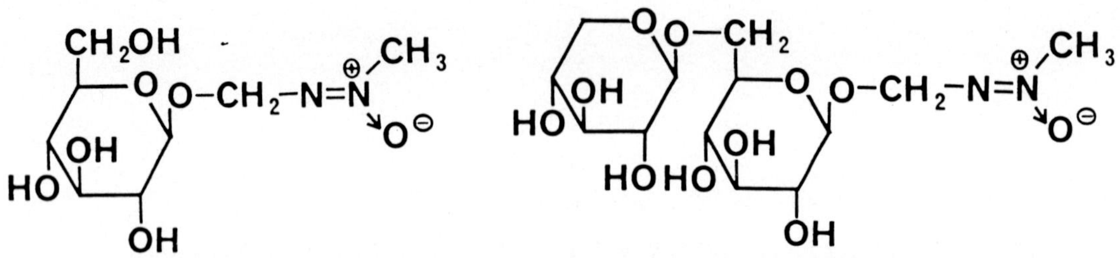

I

II

Figure 1: Structures of MAM glycosides cycasin (I) and Macrozamin (II)

Materials and Methods

Insect Material

Dead, dried butterfly specimens were obtained from the collections at The Natural History Museum, London. Knut Norstog and Priscilla Fawcett collected *Rhopalotria slossoni* pupae and adult beetles from male cones of *Zamia pumila* and *Rhopalotria mollis* larvae, pupae and adults from *Zamia furfuracea* male cones. The beetles were stored in methanol.

Gas Chromatographic Analysis of Cycasin and Macrozamin

The analysis of 80% aqueous methanol extracts for cycasin was based on that described by Wells et al., (1968). The trimethysilyl-derivative of cycasin in the extracts was co-chromatographed with the pure compound and the concentration determined. The addition of almond β-glucosidase to the extracts in water resulted in the total disappearance of the cycasin peak.

Macrozamin was determined by scraping the band corresponding with authentic macrozamin from silica gel plates run in n-BuOH-Me2Co-H2O (4:1:1) by resorcinol - HC1 (Nishida, Kobayashi and Nagahama, 1955).

Results

Obtaining representative living material of *Taenaris* is extremely difficult but it was fortunately possible to analyse whole or parts of specimens from The Natural History Museum in London and detect cycasin in some specimens which were 70 years old (Table 1). Three of the *Taenaris* species reported to feed on cycads, namely *T. butleri, T. onolaus* and *T. catops*, were found to contain cycasin. Specimens of *T. phorcas* and *T. myops* did not contain MAM-glycosides. *Luthrodes cleotas gades* contained high levels of cycasin (Table 1).

Of the mimics in the *Taenaris*-like complex in New Guinea, it was possible to analyse males and females of *Hypolimnas deois divina* Fruhstorfer, *Elymnias agondas muscosa* Fruhstorfer, *Mycalesis drusillodes* Oberthur and *Papilio aegeus ormenus* Guerin. All were negative for cycasin, as were two specimens of the cryptic *Theclinesthes onycha onycha* (Hewiston) kindly provided by Max Moulds of the Australian museum.

Cycasin was detected in pupae and adults of *Rhopalotria slossoni* but the levels were extremely low (0.1 μg/adult beetle n = 64). In the pupae/larvae extracted from the pupal case the level was calculated to be 0.24μg/pupa (n - 13). The pupal case was negative. *Rhopalotria mollis* larvae, pupae and adults gave negative results for cycasin but were all positive for macrozamin; the adults contained 1.7 μg per

Table 1
Concentration of Cycasin in Adult *Taenaris* and *Luthrodes*

Specimen and where Collected	Collector or Collection	Part of Body Analysed	Cycasin Content
Taenaris butleri (male) Oberthur, Lower Aroa River 1939 New Guinea	A.S. Meek (Rothschild)	Abdomen	0.08%
Taenaris butleri (female) Port Moresby, New Guinea	(Rothschild)	Abdomen	0.03%
Taenaris catops adriana (male) Fruhstorfer 1934	Joicey Bequest	Abdomen	0.08%
Taenaris catops adriana (female) 1912	Adams Bequest	Abdomen	0
Taenaris onolaus (male) Kirsch Momi Coast, Arfak Pen. 1928	Dr. E. Mayr	Abdomen	0
Taenaris onalaus (female) Wanggar/S.Geelvink Bay New Guinea 1920	C.F. and J. Pratt	Abdomen	0.12%
Taenaris phorcas anableps (male) Vollenhoven Wahnes	N. Ireland (Rothschild) 1939)	Abdomen	0
Taenaris phorcas anableps (female)	N.Ireland	Abdomen	0
Taenaris myops (male) Felder & Felder	Australian Museum Sydney	Whole	0
Luthrodes cleotas gades (male) Nissan I., Solomons 1924	A.F. Eichhorn	Whole	0.27%
Luthrodes cleotas gades (female)	(Rothschild)	Whole	0.35%

insect (*n* - *63). It was not possible to calculate the weight of the insects because of their storage in methanol.*

Discussion

Larvae of *Taenaris* species are known to feed on a number of monocotyledons, as is common among other Amathusiinae (Ackery, 1988). For example, *T.catops* is known from *Musa, Areca, Caryota, Cordyline* and *Phaius* species, *T. phorcas* from *Cordyline* and *T. myops* from *Tapeinochilus*. *Taenaris butleri* and *T. onolaus* larvae are known to feed on cycads and, unlike the larvae of the former monocotyledon-feeding *Taenaris* species, they are wine-red. The presence of cycasin in adults of *T. butleri* and *T. onolaus* and the similarity in coloration of the larvae of these species to those of a totally unrelated New World species which is nonetheless undoubtedly aposematic (*Eumaeus atala*), strongly suggests cycasin has a protective function in these insects. In view of the age of the specimens the absence of cycasin in the male of *T. onolaus* is probably not reliable. The discovery of cycasin in one adult of *T. catops* would seem surprising if it were not for the observation by Parsons (1984) that the adults of both *T. catops* and *T.onolaus* are fond of feeding on cycad juices from fermenting fruits and damaged leaves and *T. catops* has also been seen probing the fresh green frass of larvae of a chrysomelid beetle (*Crioceris clarkii* B. Baly) feeding on cycads. There are about 25 species of *Taenaris*, 18 of which occur in New Guinea (Brooks, 1950). There is a remarkable similarity in the markings of the majority of the species, particularly in the females. The colour range is confined to yellow, shades of brown, to black and white, with varying shades of grey; blue occurring only in the eye spots or ocelli. There is extreme variability in many of the races and certain species are directly separable in geographically localised light and dark forms. The light forms are highly visible in the shady lowlands which they usually frequent. *T. catops* is the most widely spread and commonest species found in New Guinea and shows a great variation in its wing pattern and colours.

With the presence of cycasin determined in some *Taenaris* species, they clearly could be the models for the mimicry complex. *Taenaris myops, T. catops* and the related amathusine *Hyantis hodeva* Hewitson, are extremely alike where they occur together and *Mycalesis drusillodes* is thought to be a mimic of *H. hodeva* (not available for analysis). Of the other *Taenaris*-like mimics, *Elymnias agondas* is a good mimic of *T. bioculatus* Guerin and *T. catops*, and *Hypolimnas deois* resembles *T. onolaus* in the Bulolo Valley and perhaps also *E. agondas* (Parsons, 1984; Vane-Wright, Ackery and Smiles, 1977). Females of *Papilio aegeus* around Bulolo resemble *T. catops* and *T.onolaus*. A more comprehensive study of *Taenaris* species is planned.

Luthrodes cleotas is not part of the *Taenaris*-like complex but in view of its cycasin content, the large patches of orange on the upper and underside of its hindwings can be considered as possible aposematic characteristics. The cryptic *Theclinesthes onycha* did not appear to sequester cycasin. The role of MAM-glycosides in *Rhopalotria* species is unclear.

References

***Ackery, P.R.* 1988.** Hostplants and classification: a review of nymphalid butterflies. Biol. J. Linn. Soc. 33: 95 - 203.

***Brooks, C.J.* 1950.** A revision of the genus *Taenaris* Hubner (Lepidoptera: Amathusiidae). Trans. R. Ent. Soc. Lond. 101: 170 -238.

***Nishida, K., Kobayashi, A.* and *Nagahama, T.* 1955.** Cycasin, a new toxic glycoside of *Cycas revoluta* Thunb. 1. Isolation Bull. Agric. Chem. Soc. Japan 19: 77 - 84.

***Parsons, M.* 1984.** Life histories of *Taenaris* (Nymphalidae) from Papua New Guinea. J. Lepidopterists' Soc. 38: 69 - 84.

Rothschild, M., Nash, R.J.and Bell, E.A. 1986. Cycasin in the endangered butterfly *Eumaeus atala florida*. Phytochemistry 25: 1853 - 1854.

***Spatz, M., McDaniel, E.G.* and *Laqueur, G.L.* 1966.** Cycasin excretion in conventional and germ-free rats. Proc. Soc. Exp. Biol. Med. 121: 417 - 422.

***Teas, H.J.* 1967.** Cycasin synthesis in *Seirarctia echo* (Lepidoptera) larvae fed methylazoxymethanol. Biochem. Biophys. Res. Commun. 26: 686 - 690.

***Vane-Wright, R.I., Ackery, P.R., and Smiles, R.L.* 1977.** The polymorphism, mimicry, and host plant relationships of Hypolimnas butterflies. Biol. J. Linn. Soc. 9: 285 - 297.

***Wells, W.W., Yang, M.G., Bolzer, W.,* and *Mickelson, O.* 1968.** Gas-liquid chromatographic analysis of cycasin in cycad flour. Analyt. Biochem. 25: 325 - 329.

Cycas Use by Australian Aboriginal People

Wendy Beck
Department of Archaeology and Palaeoanthropology
University of New England
NSW 2351 Australia

Abstract

Despite the well-documented toxicity of cycad seeds to humans, processed *Cycas* seeds are a traditional food eaten by Aboriginal people in northern Australia. This paper investigates the different types of Aboriginal *Cycas* processing technology, the possible history of this technology, and examines the cycasin content of various food products.

Introduction

Cycads are an extremely old component of the Australian landscape, but the origins of cycad use generally, in Asia and the Pacific has received little attention from researchers. In order to understand the history of this technology it is first necessary to understand its variability and its biological implications.

This paper has two specific aims: (1) to document the range of *Cycas* processing techniques used by the Australian Aboriginal people in the last 200 years and up to the present and (2) to provide the results of experiments designed to explore the toxicity of untreated *Cycas angulata* seeds and the effects of Aboriginal processing on cycasin, as observed in the Northern Territory. The field observations reported here were undertaken in collaboration with N.G. White of La Trobe University.

Cycas plants occur in the northern parts of the continent, areas which have been the last to be colonised by white's large-scale agriculture. Fig. 1 shows the approximate distribution of *Cycas* species in Australian (after Johnson, 1959). Many Aboriginal people live here in isolated communities on their traditional homelands in small towns and bush settlements. As bush foods are still known and used by these people, information regarding *Cycas* food processing can still be obtained first-hand. The other important source of information regarding recent methods of Aboriginal food processing is the anthropological and historical records which have been written over the last one hundred years or so (for example: Palmer, 1983; Lumholz, 1889; Roth, 1901; Tindale, 1925; Love, 1936; Harvey, 1945; Thomson, 1945; Specht, 1958; Hart & Pilling, 1960; Levitt, 1981; Scarlett & White, 1981). The records of different types of *Cycas* processing have been collated and are shown in Fig. 1. Processing

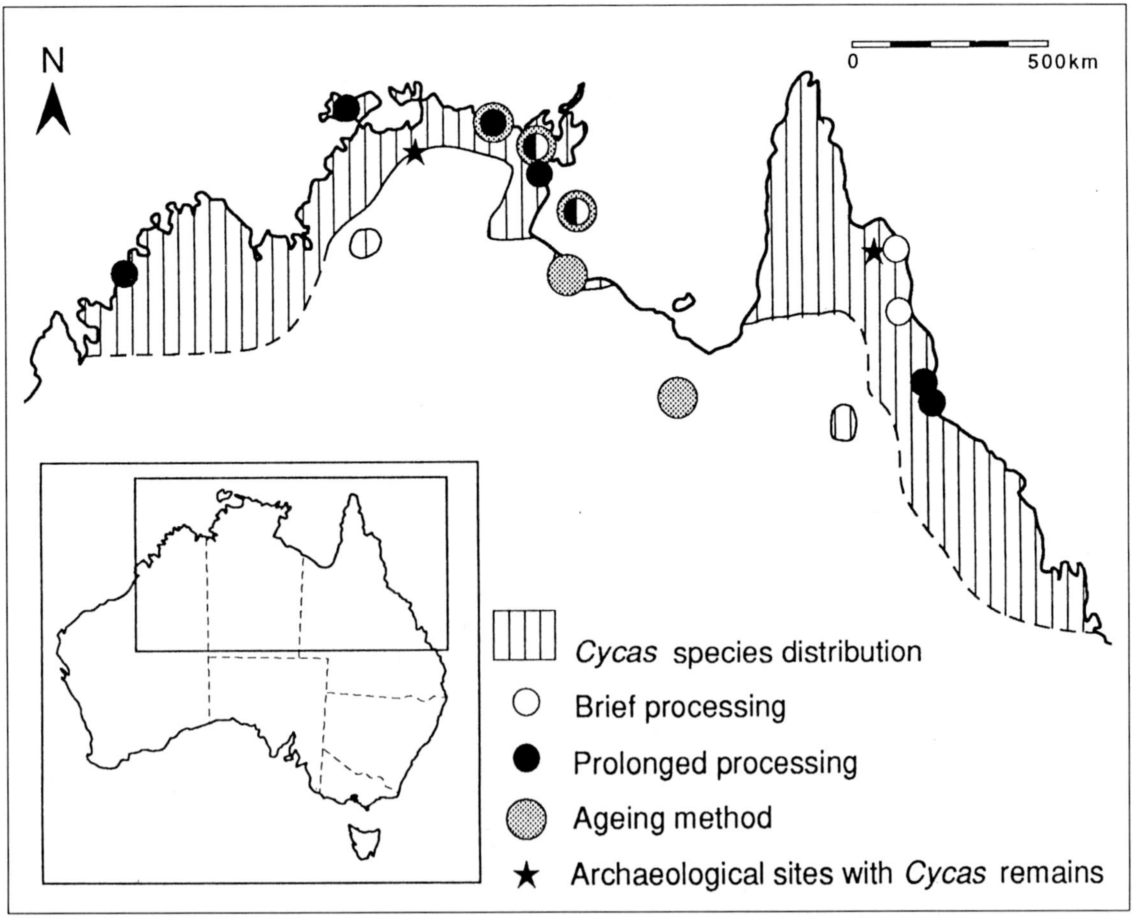

Figure 1: Ethnographic accounts of different processing types and archaeological evidence for *Cycas* processing in Australia.

can take three general forms: (1) *brief leaching*, where the seeds are harvested directly from the plants and immediately cooked, followed by overnight leaching in running water; (2) *prolonged leaching*, where the starting material is either windfall seeds or those directly harvested, dried, pounded, then soaked in water for 3-7 days and eaten raw or roasted; (3) *ageing*, which may be artificially controlled by drying or burying the seeds for several months, or may be the harvesting of naturally aged windfall seeds, eaten raw or roasted. Brief processing is found in Northern Queensland and the north-eastern parts of the Northern Territory, the prolonged processing method is the most commonly reported and most widely distributed and the ageing method is found in the central area of the range of *Cycas* species occurrence.

There is a great deal of regional diversity in Aboriginal societies and I suspect that the distribution of the three types of processing reflects cultural preference and the type of food product required, rather than any strictly environmental factors such as the availability of water for leaching (as suggested by Beaton, 1982). It is likely that in most places more than one processing technique would be known and the distribution map reflects most closely the distribution of

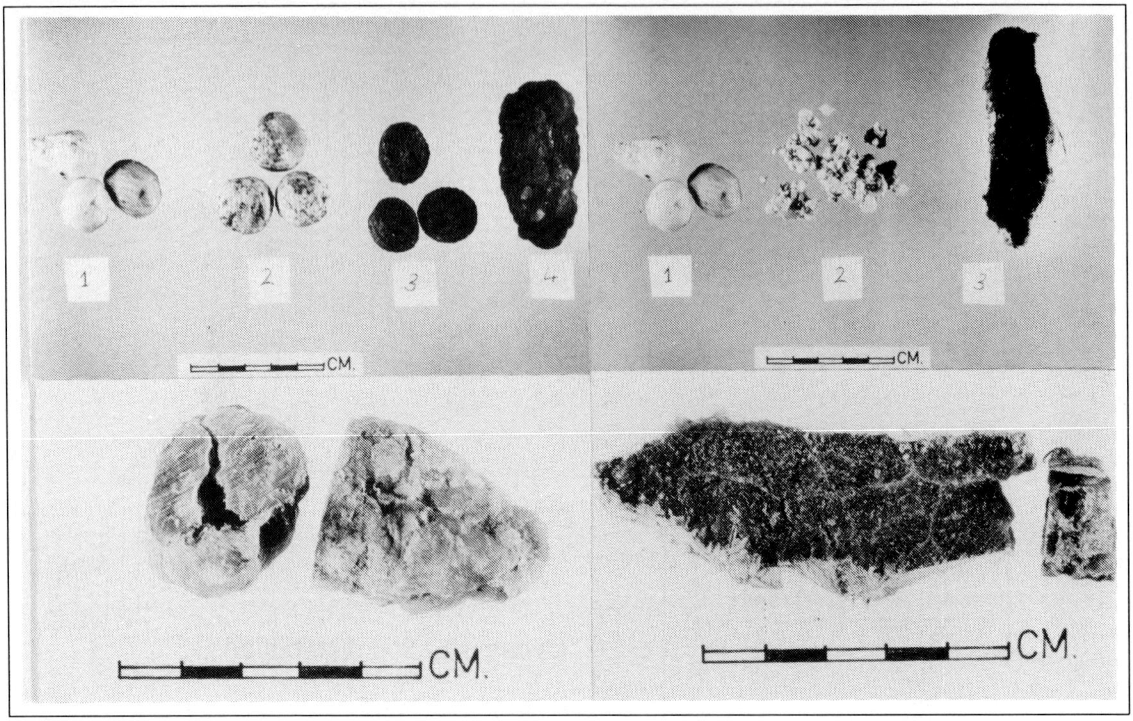

Figures 2-5: *Cycad Bread Preparations*

Figure 2: *Rarrak-rarrak* bread preparation, an example of prolonged leaching; (1) seeds after extraction; (2) seeds after preliminary cooking in hot sand and ash; (3) seeds after drying and rolling in charcoal; (4) prepared bread after leaching for 4 days.

Figure 3: Section through prepared bread.

Figure 4: *Ngathu* bread preparation, an example of prolonged leaching; (1) seeds after extraction; (2) crushed seeds after leaching for 1 week; (3) final bread product after cooking.

Figure 5: Section through finished bread.

written records (and their relative completeness), rather than real geographic boundaries between processing types.

The final source of evidence for *Cycas* processing is archaeology. The sclerotestae of *Cycas* seeds are robust and survive relatively intact in rockshelter archaeological deposits. I know of two sites which contain such remains (Schire, 1982; M. Morwood, University of New England pers. comm) and both of these date to whin the last five thousand years, and it is also likely that more will be discovered in the future. It is possible to distinguish between *Cycas* seeds which may have been incorporated into sites by humans and those which are derived from animal activities, by examination of the fracture patterns on the hard seed coats, and it may also be possible to distinguish between leaching and ageing processing techniques (Beck, White and Fullagar, 1988).

So, *Cycas* processing by Aboriginal people took a number of forms and was known throughout the range of *Cycas* species in northern Australia.

Archaeological evidence suggests that people have been successfully eating *Cycas* seeds for at least five thousand years in Australia.

Methods

Field observations were based on work carried out at Donydji outstation in north-east Arnhem Land in 1981 - 83, and cycad samples were collected in the early dry season (May - June). In this community, Aboriginal people, from two dialect units, *ritharrngu* and *wagilak*, lived in their traditional range and to a large extent, followed traditionally-orientated life-styles. Flour, tea, sugar and other groceries were purchased sporadically (White, 1978, 1984).

Cycasin assays were carried out using (1) the method of Wells et al., 1968 and (2) HPLC using a 15cm x 3mm Lichrosorb NH2 column with 80% acetonitrile: 20% water as the solvent. The results were quantified by reference to known amounts of authentic cycasin and macrozamin samples provided by N.V. Riggs, then Department of Chemistry, University of New England.

Results and Discussions

Processing Technology

At Donydji, naturally ageing the seeds of *Cycas angulata* was the most common technique applied (*munbuwa* bread), rather than the prolonged or brief leaching methods. This technique requires considerable skill and know-how rather than material hardware. Nonetheless at Donydji,, subtle variations within leaching technology can be seen. Figs. 2-5 illustrate two of the cycad breads produced by variants of the prolonged leaching of *Cycas angulata* seeds observed by N.G. White (Department of Genetics and Human Variation, La Trove University) at Donydji outstation in north-east Arnhem Land, Northern territory in 1981 and 1983. Fig. 2 illustrates the steps in the process of *rarrak-rarrak* bread preparation, where the seeds were cracked and the gametophytes (kernels) were removed whole form the outer sclerotesta. The kernels were than rolled in hot sand mixed with charcoal and placed in a bag with more charcoal. This bag was dried in the sun for several days before leaching in running water. After 4 days the kernels were removed from the water and pounded into a long cake before being roasted in ashes. Fig. 3 illustrates the steps in the preparation of *ngathu* bread.

The kernels are similarly extracted from the sarcotesta, then roughly crushed and placed in a bag made form *Pandanus* fibre. The crushed material was leached in running water for about a week, or in still water for 5-8 days, then ground into a past, wrapped in paperbark and roasted in ashes for 30 minutes to one hour (depending on the size of the damper). Fig. 4 shows a detail of *rarrak-rarrak* break and Fig. 5 shows a close-up of *ngathu* slice. These two different methods produce breads with different tastes and textures, but the latter method was by far the most commonly observed for leached seeds at Donydji.

Cycasin Analyses of Seeds

The field specimens of *Cycas angulata* seeds collected from Donydji in 1981 were assayed for cycasin content using the gas-liquid chromatography method of Wells et al (1968). The mean value for seeds harvested directly from the tree was 0.65g/100g of fresh gametophyte, which is very similar to the value of 0.72g/100g obtained by Moretti et al (1981) for another Northern Territory species. On the basis of comparative data in the toxicological literature the levels of cycasin experimentally determined for fresh seeds would probably be acutely toxic and lethal to small animals in doses of 1-5 seeds. The direct implications of these findings for human toxicity are difficult to establish, since it is not possible to extrapolate between animal species, but 1-3 seeds have been shown to be toxic in accidental poisonings of humans in Australia (Beaton, 1982).

Cycas Assays of Bush Foods

Leaching experiments were also carried out. In a sample of crushed seeds which were left in still water for eighteen hours the cycasin level dropped to 0.013g/100g fresh seeds. Leaching in running water resulted in a more rapid loss with cycasin declining from 0.6g/100g fresh seeds to below the detectible

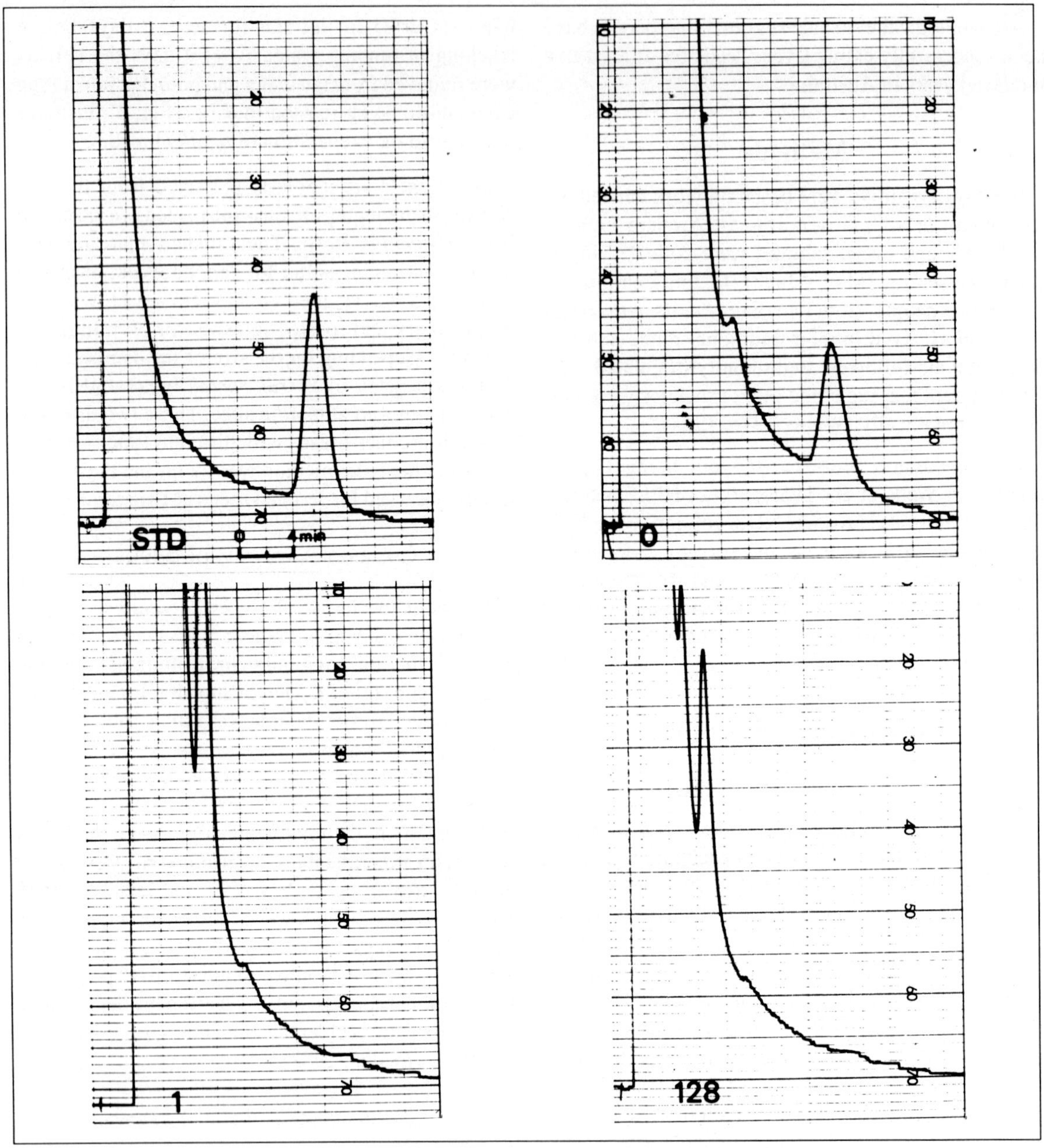

***Figures 6-9**: GLC Chromatogram traces for Leaching Experiment with Cycas angulata*

Figure 6: Standard curve for cycasin.

Figure 7: Unleached sample.

Figure 8: After 1 hour leaching.

Figure 9: After 128 hours leaching. Note disappearance of peak due to cycasin.

limit (about 0.007g/100g). The actual chromatographic traces for the second leaching experiment are shown in Figs. 6-9. The peak representing cycasin has disappeared after one hour's leaching in running water. (Fig. 8). The peaks between the injection and the cycasin peak were not identified, but are almost certainly due to excess solvent and isomers of glucose (Wells et al., 1968).

Two batches of aged seeds (*munbuwa*) were analysed from collections made by Aboriginal people in the bush near Donydji in 1982. The batch which was rejected during bread-making as being inedible had a cycasin content of 1.67g/100g fresh wt. (determined by HPLC) which is very similar to that obtained for fresh seeds (Fig. 10). This suggests that some of the seeds collected from the bush had only been aged for a very short period. The mean cycasin content of the edible batch was below the detectable limit (Fig. 11). Natural ageing was effective in reducing the cycasin content of some fresh seeds to a low level, and Aboriginal detection techniques could discriminate between these seeds and those with a high cycasin content (Beck et al., 1988).

Conclusions

Human populations including the Australian Aborigines have developed effective means of removing cycasin from *Cycas* seeds before eating. All the samples of Aboriginal foods which were assayed had no detectable cycasin. Leaching in running or still water was demonstrated to remove the cycasin present in experimental simulations, although running water was the more rapid technique of the two. The mechanism by which the water-soluble cycasin is leached out is still unknown, however, the presence of gas bubbles in the leaching water suggests the enzymic hydrolysis followed by spontaneous breakdown into gaseous products is most likely occurring.

Ageing the seeds, that is, the seeds being exposed to natural weathering while lying on the ground, can also reduce the cycasin content to an undetectable level. This is probably due to a combination of enzyme actions, leaching, drying and fungal infection. However, not all the seeds are equally affected and some

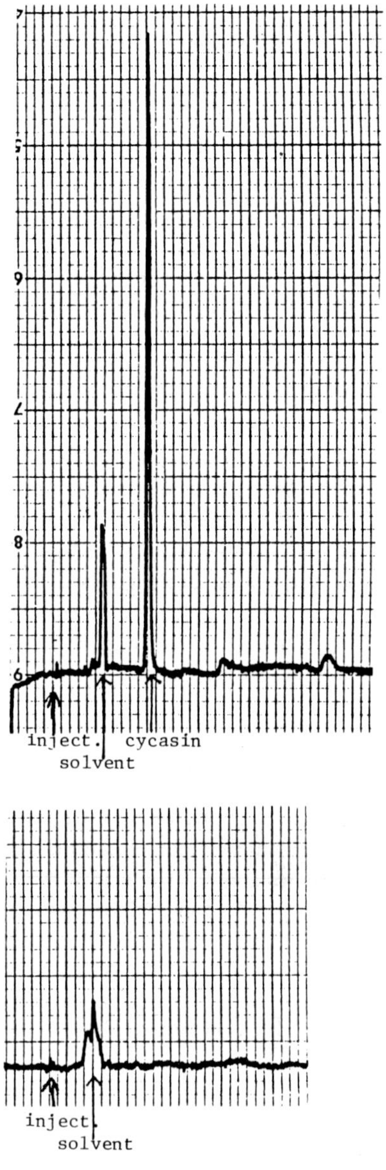

Figures 10-11: HPLC Chromatogram traces of Cycasin content of aged seeds prepared by Aboriginal people in North Australia (Cycas angulata)

Figure 10: Seeds rejected during selection.

Figure 11: Seeds made into *munbuwa* bread.

still retain high levels of cycasin. Skilled people can distinguish 'cycasin-free' kernels from others by a combination of visual, olfactory and tactile cues.

This group of technologies represents a sophisticated adaptation to *Cycas* toxicity, which although time-consuming and probably energy-consuming also, provided a rich and predictable starch resource, not only to meet the nutritional needs of Aboriginal people, but social and ceremonial needs as well. The alternative techniques allowed for the preparation of food products with different storage and handling techniques. *Cycas* preparation was primarily a women's subsistence activity.

The antiquity of *Cycas* processing in Australia is currently being investigated through archaeology, and most sites indicate the presence of cycad seeds from 5,000 years ago to the present. No information is yet at hand regarding the chronology or origins of different processing types, although a method for distinguishing between aged and leached seeds has been developed. The potential certainly exists for further research into the prehistoric use of *Cycas.*

Acknowledgment

This research was supported by a postgraduate research fellowship from the (then) Australian Institute of Aboriginal Studies and La Trove University. I wish to thank the Aboriginal people of Donydji, and my supervisors: Drs. D. Frankel and N.G. White.

References

***Beaton, J.* 1982.** Fire and Water: aspects of Australian Aboriginal management of cycads. Arch. in Oceania 17(1): 51 - 58.

***Beck, W., Fullagar, R., White, N.G.* 1988.** Archaeology with ethnography: the Aboriginal use of cycad as an example. Pages 137 - 147 *in* B. Meehan and R. Jones, eds. Archaeology with ethnography: an Australian perspective. Department of Prehistory, Research School of Pacific Studies, Australian National University, Canberra.

***Harvey, A.* 1945.** Food preservation in Australian tribes. Mankind 3: 191 - 192.

***Johnson, L.A.S.* 1959.** The families of cycads and the Zamiaceae of Australia. Proc. Linn. Soc. N.S.W. 84: 64 - 117.

***Levitt, D.* 1981.** Plants and People, Aboriginal Uses of Plants on Groote Eylandt. Australian Institute of Aboriginal Studies, Canberra.

***Love, D.R.B.* 1936.** Stone-age Bushmen of Today. Blackie and Son, London.

***Lumholtz, M.A.* 1989.** Among Cannibals. An account of Four Years in Australia. John Murray, London.

***Moretti, A., Sabato, S., Siniscalco, G.C.* 1981.** Distribution of macrozamin in Australian cycads. Phytochem. 20(6): 1415 - 1416.

***Palmer, E.* 1983.** On plants used by the natives of Northern Queensland, Flinders Ranges and Mitchell Rivers, for food, medicine, etc. J. and Proc. Roy. Soc. N.S.W. 17: 93 - 113.

***Roth, W.E.* 1901.** Food, its search, capture and preparation. North Queensland Ethnography Bulletin No. 3.

***Thomson, D.F.* 1949.** Economic Structure and the Ceremonial Exchange Cycle in Arnhem Land. Melbourne University Press, Melbourne.

***Specht, R.L.* 1958.** An introduction to the ethno-botany of Arnhem Land. Pages 479 - 504 *in* R.L. Specht and C.P. Mountford, eds. Records of the American-Australian Scientific Expedition to Arnhem Land, Vol. 3, Botany and Plant Ecology. Melbourne University Press, Melbourne.

***Scarlett, N., White, N.G.* 1981.** List of names and uses of Aboriginal food plants, Human Ecology project, north east Arnhem Land: Note Books 1979 - Australian Institute of Aboriginal Studies, Manuscript A3 236, No. 86-1.

***Schire, C.* 1982.** The Alligator Rivers, prehistory and ecology in western Arnhem Land. Terra Australis 7, Department of Prehistory, Research School of Pacific Studies, Australian National University.

***Wells W.W., Yang, M.G., Bolzer, W., Mickelsen, O.* 1968.** Gas-liquid chromatographic analysis of cycasin in cycad flour. An. Biochem. 25: 325 - 329.

***White, N.G.* 1978.** A human ecology research project in the Arnhem Land region: an outline. Aust. Inst. of Aboriginal Studies Newsletter 9: 39 - 52.

***White, N.G.* 1984.** Sex differences in Australian Aboriginal subsistence: possible implications for the biology of hunter-gatherers. Pages 323 - 361 in Newcombe, R., Ghesquire, J., and Martin, R. Eds. Sexual dimorphism. Symposium of the Society for the study of Human Biology. vol. 24. Taylor and Francis, London.

From Lathyrus to *Cycas*: A Search for Dietary Neurotoxins

E.A. Bell
Department of Biochemistry, King's College London
Strand, London WC2R 2LS

Lathyrism

A survey of ninhydrin-reacting compounds present in the seeds of 53 species of *Lathyrus* showed that L-amino-oxalylaminopropanoic acid (ODAP or BOAA), the neurotoxin first isolated from seeds of *L. sativus* (Murti, Seshadri and Venkitasubramanian, 1964; Rao, Adiga and Sarma, 1964), was the only toxic component detectable in eleven species. These eleven species included all three (*L. sativus, L. cicera* and *L. clymenum*) which are used as food and have been implicated as causes of the motor neurone disease lathyrism (Bell, 1964).

Cycad Toxicity

The similarity of lathyrism found in Indian and African populations, to the amyotrophic lateral sclerosis-Parkinsonism dementia complex (ALS-PD) found among the Chamorro people of Guam and other Mariana Islands (see Kurland, 1988) led Dr. Marjorie Whiting to suggest to me that the seeds of *Cycas circinalis*, a suspected cause of ALS-PD, might also contain amino-oxalylaminopropanoic acid. An analysis of these failed to reveal the presence of the oxalylaminopropanoic acid. It did, however, reveal (Vega and Bell, 1967) the presence of L-amino-methylaminopropanoic acid (MeDAPA, or BMAA) a previously unknown amino acid which was acutely neurotoxic in chicks, rats and mice (Bell, Vega and Nunn, 1967; Polsky, Nunn and Bell, 1972). The subcutaneous administration of one third of the convulsant dose of amino-melthylamino-propanoic acid to rats over a period of 78 days failed to induce any signs of neurotoxicity however. This failure still left in doubt the questions as to whether the ingestion of food made from seeds of *C. circinalis* played any part in the development of ALS-PD and if it did whether amino-methylamino-propanoic acid was the toxin responsible. The concentrations of free amino-methylaminopropanoic acid present in the seeds certainly seemed too low to create neurotoxic problems in the short-term.

In the years that have elapsed since the discovery of amino-methylaminopropanoic acid a number of alternative theories as to cause of ALS-PD have been proposed and rejected (see Bell and Nunn, 1988) and attention has once more become focussed on the possible presence of slow-acting toxin in *C. circinalis*. The reasons for this renewed interest may be summarised as follows. Firstly, the abandonment of traditional life styles and diets in Guam has been accompanied by a reduction in the incidence of ALS-PD (Reed and Brady, 1975; Garruto, Yanagihara and Gajdusek, 1985). Secondly, post-mortem studies have revealed that some Guamanians who were apparently free of neurological disease and died from unrelated causes nevertheless showed evidence of characteristic but as yet subclinical damage to the central nervous system (Anderson, et al., 1975). Thirdly, Chamorro people who have migrated to the continental United States and Filipino migrants to Guam, both suffer a greater risk of developing ALS-PD than individuals in the U.S. population as a whole (Garruto et al., 1985). The likelihood that amino-methylaminopropanoic acid in the diet may be involved has also been increased by the observation that primates are more sensitive to the compound than the rodents and chicks used in earlier experiments. Macaques developed a motor systems disease when amino-methylaminopropanoic acid was administered orally for several weeks and the mask-like features and tremors observed in these animals were relieved by the administration of L-dihydroxyphenylalanine (L-DOPA) which is used as a drug in the treatment of Parkinson's Disease (Spencer, et al., 1987). Conflicting evidence has however been provided by a recent study (Duncan, et al., 1990) that has shown that 90 - 99% of free amino-methylaminopropanoic acid may be removed from cycad flour during processing and the final product may contain as little as 0.005% by weight of the neurotoxin. The authors suggest that it is unlikely that concentrations as low as this will cause ALS-PD even if ingested for prolonged periods. It may be unlikely, but it cannot be ruled out until we know the minimum concentration of toxin that affects the cells of the nervous system and whether the effects are cumulative.

Another aspect of the problem which has still to be addressed is the potential toxicity of precursors and derivatives of amino-melthylaminopropanoic acid. These could be toxic in their own right or broken down to yield amino-methylaminopropanoic acid after ingestion. They might moreover be less easily removed during the processing of *C. circinalis* flour than the free amino acids.

It is not yet possible to say whether amino-methylaminopropanoic acid, cycasin or any other toxin present in seeds of *C. circinalis* is responsible for ALS-PD. The solution of the problem clearly requires the continuing collaboration of botanists, chemists and medical scientists in a whole range of disciplines. Such collaboration may have significance far beyond Guam however if it provides a greater appreciation and understanding of the way in which glutamic acid analogues and related compounds synthesized by plants can act as neurotoxins when included in the human diet.

References

***Anderson, F.H., Richardson, E.P., Okazaki, H., Brody, J.A.* 1979.** Neurofibrillary degeneration on Guam: frequency in Chamorros and non-Chamorros with no known neurological disease. Brain. 102: 65 - 77.

***Bell, E.A.* 1964.** Relevance of biochemical taxonomy to the problem of lathyrism. Nature. 203: 378 - 380.

***Bell, E.A., Nunn, P.B.* 1988.** Neurological diseases in man - are plants to blame? Biologist. 35: 39 - 43.

***Bell, E.A., Vega, A. Nunn, P.B.* 1967.** A neurotoxic amino acid in seeds of *Cycas circinalis*. Proc. 5th conf. Toxicity of cycads. Miami. xi: 1 - 7.

***Duncan, M.W., Steele, J.C. Kopin, I.J., Markey, S.P.* 1990.** amino-(melthylamino)-propanoic and (BMAA) in cycad flour: An unlikely cause of amyotrophic lateral sclerosis and parkinsonism-dementia of Guam. Neurology. 40: 767 - 772.

***Garruto, R.M., Yanagihara, R., Gajdusek, D.C.* 1985.** Disappearance of high-incidence amyotrophic lateral sclerosis and parkinsonism-dementia on Guam. Neurology. 35: 193 - 198.

***Kurland, L.T.* 1988.** Amyotrophic lateral sclerosis and parkinson's disease complex on Guam linked to environmental neurotoxin Trends in Neurosci. 11: 51 - 53.

***Murti, V.V.S., Seshadri, T.R., Venkitasubramanian, T.A.* 1964.** Neurotoxic compounds in the seeds of *Lathyrus sativus*. Phytochemistry. 3: 73 - 78.

***Polsky, F.I., Nunn, P.B., Bell, E.A.* 1972.** Distribution and toxicity of amino-methylaminopropionic acid. Fed. Proc. 31: 1473 - 1475.

***Rao, S.L.N., Adiga, P.R., Sarma, P.S.* 1964.** The isolation of β-N-oxalyl-a, β-diaminopropionic acid; a neurotoxin from seeds of *Lathyrus sativus*. Biochemistry., 3: 432 - 436.

***Reed, D.M., Brody, J.A.* 1975.** Amyotrophic lateral sclerosis and parkinsonism-dementia on Guam, 1945 - 1972. Am. J. Epidemiol. 101: 287 - 301.

***Spencer, P.S., Nunn, P.B. Hugon, J., Ludolph, A.C., Ross, S.M., Roy, D.N., Robertson., R.C.* 1987.** Guam amyotrophic lateral sclerosis-parkinsonism dementia linked to a plant excitant neurotoxin. Science 237: 517 - 522.

***Vega, A., Bell, E.A.* 1967.** amino-methylamino-propionic acid, a new amino acid from seeds of *Cycas circinalis*. Phytochemistry. 7: 1885 - 1887.

Cycads as a Possible Cause of Neurological Disorders in the Western Pacific

Mark W. Duncan
Biomedical Mass Spectrometry Unit
Faculty of Medicine
University of New South Wales
Kensington, NSW 2033 Australia

Abstract

In the early 1950s a relatively rare neurological disorder, amyotrophic lateral sclerosis (ALS), occurring with high incidence on the Mariana island of Guam, was linked to the reliance of the indigenes on the seeds of *Cycas circinalis* L. as a food source. However, thirty years later no component had been identified that could account for the onset of the specific symptoms of ALS. In 1987, Spencer and colleagues re-examined one cycad constituent, the non-protein amino acid β-methylamino-L-alanine (BMAA), reporting the induction of neurological symptoms in primates when fed large quantities of this compound (≥100mg/kg/day, 3 months). We examined BMAA as a possible cause of the ALS of the western Pacific region, measuring the BMAA content of cycad seeds and cycad-derived foods. These data were correlated with the known toxic dose BMAA. We also calculated the BMAA dose associated with the application of cycad poultices. These calculations indicate that neither diet nor medicinal use of cycads provide humans with toxic doses of BMAA. Tests of cycad extracts in neuronal cell culture systems indicate no correlation between toxicity and BMAA concentrations. These data lead us to conclude that there is insufficient evidence to link the neurological disorders of the western Pacific to BMAA.

Cycads are majestic plants with a fascination all of their own, but for some of us the central interest is the key role these plants may play in our attempts to understand the mechanisms of the aging of the human brain. The association between cycads and the human brain is far from obvious, but some believe that the cycad contains specific slow-acting toxins that can destroy part of the brain and result in the onset of neurological conditions similar to those that plague our aging population. If this is true, and if the toxic constituents can be identified and their mechanism of action determined, then cycads may help us in gaining a better understanding of neurodegeneration and the neurodegenerative disorders that occur worldwide.

The association between cycads and neurodegenerative disorders dates back over thirty years.

Beginning in the 1950s it was noted that a variant of amyotrophic lateral sclerosis (ALS) (Kurland and Mulder, 1954), and a parkinsonism dementia complex both occurred with a high incidence amongst the Chamorros of the island of Guam (Hirano et al., 1961 a; Hirano et al., 1961 b). These disorders sometimes occurred together in the one individual, or they could exist independently. Neuropathological studies indicated that the brains of the ALS patients showed neurofibrillary tangles - a classical hallmark of Alzheimer-type dementia (Hirano et al., 1966). Subsequently, two other high incident foci for this same disorder were identified: one in the Kii peninsula of Japan (Kimura et al., 1963), and the other in the southern lowlands of west New Guinea (Gajdusek, 1963).

The exceedingly high incidence of these disorders in Guam prompted extensive studies to identify the cause. Early efforts concentrated on three possible causative factors: a genetic origin, infectious agents and environmental toxins. After thirty years of research it is now generally agreed that infectious agents are not involved: all attempts to transmit the disorder to non-human primates have been unsuccessful, and attempts to isolate an infectious agent have failed (Gibbs and Gajdusek, 1982). Furthermore, the disorder lacks any patterns characteristic of Mendelian genetics, and hence a hereditary cause is considered unlikely (Garruto et al., 1985; Kurland and Molgaard, 1982). It is now agreed that an environmental factor (or toxin) is responsible, but its identify is unknown (Garruto and Yase, 1986; Spencer et al., 1987 a). Strength is given to this claim because the annual incidence rates for these disorders have been dropping, a finding consistent with a disappearing environmental factor (Garruto et al., 1985).

Early investigations were focussed on the cycad plant as a possible environmental factor. Dr. Marjorie Whiting was the first to suggest that the ingestion of cycads (i.e., *Cycas circinalis*) was responsible for this disorder (Whiting, 1963). Whiting provided compelling evidence in support of this association after spending time in Guam living in the villages of Umatac and Yigo. She noted that cycads are acutely toxic, and the consumption of raw seeds (and other plant parts) can result in violent gastrointestinal disturbances and even death. Whiting suggested that the onset of neurological disorders may be linked to incomplete detoxification of the seeds, or the use of raw cycads - sometimes employed as medicinals. During the Second World War cycad starch became a dietary staple when rice was unavailable, and the Chamorros assumed that the seeds were fit for consumption once the emetic agents had been removed. Whiting suggested that this may not be the case, and pointed out that consumption of plant parts causes ataxia in cattle - a finding consistent with the putative link between the plant and neurological disorders (Whiting, 1963). Furthermore, comparisons were made between ALS on Guam and lathyrism - a disorder that was shown to be linked to diet (Selye, 1957). Thus the stage was set for a thorough investigation of cycads, and in particular, an intensive research effort to isolate and identify their toxic components.

Soon two compounds were isolated and characterised: the non-protein amino acid β-methylamino-L-alanine (BMAA) (Vega and Bell, 1967), and the azoxyglycoside cycasin (Nishida et al., 1955). These two compounds were thoroughly investigated over the next decade, but failure to induce experimental animal models of cycad-induced ALS-like disorders damped enthusiasm for the cycad hypothesis (see reports cited in: Whiting, 1988). Despite more than a decade of detailed scientific work, much of which was orchestrated by Kurland, the cycad hypothesis had all but died completely by 1978. In fact, attention had been redirected from the issue of neurotoxicity when it was found that cycasin was a potent hepatotoxin and carcinogen and this became the central focus of research in the late 1970s.

Then, in 1987, there was a dramatic change in the emphasis of research. Dr. Peter Spencer and his colleagues reported the induction of neurological symptoms in primates when fed large quantities of the cycad component BMAA (Spencer et al., 1987 a). This exciting finding provided new impetus for researchers investigating the link between cycads and ALS, but more importantly, a specific cycad component was now targeted as the putative etiologic agent. As a consequence it became possible to design experiments to gain support for this association.

Our studies began soon after the report of Spencer and colleagues, and were aimed at examining the hypothesis that cycads are the causative agents of ALS-PD in the western Pacific, and that toxicity is mediated by BMAA. We began by developing an assay for BMAA based on combined gas chromatography-mass spectrometry (GC/MS) (Duncan et al., 1989). This approach gave us a specific, sensitive and precise assay. Subsequent studies quantified BMAA in the gametophyte tissue of the seeds of various members of the genus *Cycas*; notably *C. circinalis, C. media and C. revoluta* (Duncan et al., 1989). The highest BMAA content (approx 1mg/g or 0.1% wt/wt) was present in the seeds of *C. circinalis* and closely related species. Although these levels are relatively high, they have little relevance when discussing the use of cycads as a food force. Cycad seeds are 'processed' by washing prior to consumption, therefore the crucial issue is the residual BMAA content in the flour that is used to prepare food. Because BMAA is a water-soluble amino acid, a significant proportion of the total BMAA content could be expected to be leached out during processing. To test this hypothesis we collected samples of processed cycad flour prepared on Guam, and assayed them for residual BMAA content (Duncan et al., 1990). We found a value of 126μ/g of BMAA for a sample of flour washed for only 24 h with no change of rinse water. Based on a total BMAA content of raw cycad gametophyte tissue of 1,140 μg/g, this indicates that about 90% of the total BMAA content is readily removed from the seeds. We found that more than 85% of the BMAA is removed by processing in all the flour samples we tested. In more than half of the samples 99% or more of the BMAA had been extracted by the leaching process.

These data led us to question the original hypothesis. Doses of greater than 100 mg/kg/day were used in the original studies of Spencer and colleagues (Spencer et al., 1987 a). Based on a BMAA content of 100 μg/g, a 70 kg human would have to consume 70 kg of processed flour in one sitting to obtain an equivalent dose. Realistically, doses about two orders of magnitude lower would arise from dietary use of cycads. Although it could be argued that a lower dose over longer time intervals might also be neurotoxic, there is no experimental evidence to support this claim. These data suggests that diet would be unlikely to ever provide man with toxic levels of BMAA.

There are other important components of the cycad hypothesis that we have examined. For example, consumption of cycad flour was not common to all three foci for ALS-Pd, leading Spencer and collaborators to suggest that exposure to raw cycads (notably the unprocessed seed gametophyte tissue) is the common etiologic factor in ALS-Pd. In west New Guinea, for example, the application of cycad seed poultices to open wounds was proposed to be the principal route for absorption of neurotoxin (Spencer et al., 1987 b). However, calculations indicate that the maximum daily does of BMAA that would be available from the application of a cycad seed poultice is in the order of 1mg/kg (Duncan et al., 1989; Duncan, unpublished observations). Nevertheless, it could be argued that the oral bioavailability of BMAA is exceedingly low, and therefore large doses are required to achieve high circulating levels of the neurotoxin. If this were the case, significantly lower doses of BMAA absorbed directly through the skin might result in higher circulating levels. To check this possibility we have determined the oral bioavailability of the BMAA in the primate. At a dose of 2mg/kg we estimated the oral bioavailability was effectively 100% (Duncan, unpublished observations). Thus, topical application or any other route of administration offers no meaningful advantage regarding the appearance of BMAA in the circulation. The small amount of BMAA available through the use of cycad seed poultices would result in circulating levels of BMAA that are orders of magnitude lower than those demonstrated to be toxic to primates.

Blood-brain barrier (BBB) permeability is another important consideration in assessing the possible role of BMAA in the etiology of ALS-PD in the Western Pacific region. *In vitro* cell culture systems indicate that relatively high concentrations of BMAA (i.e. 1-3 mM) are required to demonstrate toxicity (Ross et al., 1987; Spencer et al., 1987 a; Weiss and Choi, 1988) Hence, if enough BMAA is to reach the central nervous system to result in neuronal destruction, a relatively high BBB permeability would be desirable. Our *in vivo* studies indicate that BMAA is rapidly cleared from the circulation, poorly transported into the brain and that uptake and long-term storage does not take place (Duncan et al., 1991). We have determined a blood-brain barrier permeability surface area constant for BMAA of 2.3 ± 0.3 x 10^{-5} mL/s/g, a value less than that of any of the naturally occurring protein amino acids. Furthermore, preferential uptake of BMAA does not occur in those regions that appear most susceptible to this agent (i.e. the motor cortex and spinal cord). Thus it is not likely that BMAA concentrations could be sufficient to kill a significant portion of brain cells in such a way as to produce an acute lesion which, for unknown reasons, would flare into a cascade after 10 - 20 years of dormancy.

In collaboration with Dr. Ann Marini, National Institute of Neurological Disorders and Stroke, rational Institutes of Health, we have used neuronal cell culture systems to determine the toxicity of various cycad extracts. In contrast to agents such as α-methyl-4-phenyl-1,2,3,6 tetrahydrophyridine (MPTP) and L-β-oxalyl-α,β-diaminopropionic acid (BOAA), BMAA toxicity is only expressed when high concentrations are present in the cell culture medium (i.e., millimolar concentrations). Furthermore, when cycad extracts were tested in our neuronal cell culture system it was clear that toxicity was not mediated by BMAA. There was no correlation between the concentration of BMAA in the medium and the toxicity of that extract. In fact, some of the most toxic seed extracts had the lowest BMAA concentrations, and the concentration of BMAA in the culture medium was always significantly lower than that determined to be toxic when the pure compound was tested.

Our studies on the possible role of BMAA, and other cycad toxins, in the onset of neurodegenerative disorders in the western Pacific are continuing. While our studies so far do not rule against an association between BMAA and ALS-PD on Guam, it would seem unlikely that foods prepared from cycads, or cycad seed poultices, can provide neurotoxic levels of BMAA. If BMAA is involved, then exposure to raw seeds, either through ingestion or topical application, would be necessary to obtain significant quantities of this putative neurotoxin.

In conclusion, high circulating concentrations of BMAA are unlikely to arise from either the use of processed cycad flour or the application of cycad poultices. BMAA does not appear to be particularly neurotoxic, and at this time there is no evidence to suggest that the low circulating concentration of this compound that might arise from occasional cycad use is harmful. BMAA blood-brain barrier permeability is low, and in cell culture studies show there is not correlation between toxicity and BMAA concentration. All these data lead us to conclude that if cycads are involved in the etiology of neurodegenerative disorders in the western Pacific, then other cycad constituents are likely to be involved. Work in our laboratory, and in other centers, is now focussed on other constituents, including the azoxyglycosides. Although there may be a causal link between exposure to cycads and neurological disorders, we are yet to identify the responsible agent.

References

Duncan, M.W., Kopin, I.J., Crowley, J.S., Jones, S.M., and Markey, S.P. **1989.** Quantification of the putative neurotoxin amino-(methylamino)-propanoic acid (BMAA) in cycadales: analysis of the seeds of some members of the family Cycadaceae. J. Anal. Toxicol. 13: 169 - 175.

Duncan, M.W., Steele, J.C., Kopin I.J., and Markey, S.P. **1990** amino-(methylamino)-propanoic acid (BMAA) in cycad flour: an unlikely cause of amyotrophic lateral sclerosis and parkinsonism-dementia of Guam. Neurology: 40: 767 - 72.

Duncan, M.W., Villacreses, N.E.,Pearson, P.G., Wyatt, L., Rapoport, S.I., Kopin, I.J., Markey, S.P., and Smith, Q.R. **1991.** amino-(methylamino)-propanoic acid (BMAA) pharmacokinetics and blood-brain barrier permeability in the rat. J. Pharm Exp. Ther. In Press.

Gajdusek, D.C. **1963.** Motor-neuron disease in natives of New Guinea, N. Engl. J. Med. 268: 474 - 476.

Garruto, R. and Yase, Y. **1986.** Neurodegenerative disorders of the western Pacific: the search for mechanisms of pathogenesis. Trends Neurosci. 9: 368 - 374.

Garruto, R.M., Yanagihara, R. and Gajdusek, D.C. **1985.** Disappearance of high-incidence amyotrophic lateral sclerosis and parkinsonism-dementia on Guam. Neurology. 35: 193 - 8.

***Gibbs, C.J.J. and Gajdusek, D.C.* 1982** An update on long term in vivo and in vitro studies designed to identify a virus as the cause of amyotrophic lateral sclerosis, parkinsonism-dementia, and Parkinson's disease. In Human motor neuron diseases, ed. by L.P. Rowland, pp. 343 - 353, Raven Press, New York, 1982.

***Hirano, A., Kurland, L.T., Krooth, R.S. and Lessell, S.* 1961.** a. Parkinsonism-Dementia complex, an endemic disease on the island of Guam. I. Clinical features. Brain. 84: 642 - 661.

***Hirano, A., Malamud, N. and Kurland, L.T.* 1961** b. Parkinsonism-dementia complex, an endemic disease on the island of Guam. II. Pathological features. Brain. 84: 662 - 679.

***Hirano, A., Malamud, N. and Kurland, L.T.* 1966.** Parkinsonism-dementia complex of Guam. Neurological revaluation. Arch. Neurol. 14: 356 - 368.

***Kimura, K., Yase, Y., Higashi, Y., Uno, s., Yamamoto, Y., Iwasaki, M., Tsumoto, I., Sugiura, M., Yoshimura, S., Namikawa, K., Kumura, J., Iwamato, S. and Yamamoto I.* 1963.** Epidemiological and geomedical studies on amyotrophic lateral sclerosis. Dis. Nerv. Syst. 24: 155 - 159.

***Kurland, L.T. and Molgaard, C.A.* 1982.** Guamanian ALS: hereditary or acquired? Adv Neurol. 36: 165 - 71.

***Kurland, L.T. and Mulder, D.W.* 1954.** Epidemiological investigations on amyotrophic lateral sclerosis. Preliminary report on geographic distribution with special reference to the marinara islands, including clinical and pathological observations. Neurology. 4: 355 - 378 and 438 - 448.

***Nishida, K., Kobayashi, A. and Nagahama, T.* 1955.** Studies on cycasin, a new toxic glycoside, of *Cycas revoluta* Thunb. Bul. Agr. Chem. Soc. 19: 77 - 83.

***Ross, S.M., Seelig, M. and Spencer, P.S.* 1987.** Specific antagonism of excitotoxic action of 'uncommon' amino acids assayed in organotypic mouse cortical cultures. Brain Res. 425: 120 - 7.

***Selye, H.* 1957.** Lathyrism. Rev. Can. Biol. 16: 1 - 82.

***Spencer, P.S., Nunn, P.B., Hugon, J., Ludolph, A.C., Ross, S.M., Roy, D.N. and Robertson, R.C.* 1987** a. Guam amyotrophic lateral sclerosis-parkinsonism-dementia linked to a plant excitant neurotoxin. Science. 237: 517 - 22.

***Vega, A. and Bell, E.A.* 1967.** Alpha-amino-beta-methylaminopropionic acid, a new amino acid from seeds of *Cycas circinalis*. Phytochem. 6: 759 - 762.

***Weiss, J.H. and Choi, D.W.* 1988.** Beta-N-methylamino-L-alanine neurotoxicity: requirement for bicarbonate as a cofactor. Science. 241: 973 - 5.

***Whiting, M.G.* 1963.** Toxicity of cycads. Econ Bot. 17: 270 - 302.

Mutagens Produced by Cycads

George R. Hoffmann
Department of Biology, College of the Holy Cross
Worcester, MA 01610, USA

Abstract

The mutagenicity and carcinogenicity of cycad azoxyglycosides depend on cleavage of their sugar linkage, releasing methylazoxymethanol (MAM). MAM is active in assays for DNA damage, gene mutations, chromosomal alterations, and carcinogenicity. This paper reviews the genetic toxicology of cycad azoxyglycosides and the concern about azoxyglycosides as potential mutagens and carcinogens in the diet.

Introduction

The azoxyglycosides are cycad metabolites that consist of the highly reactive compound MAM linked to a sugar. They have been found in *Bowenia, Ceratozamia, Cycas, Dioon, Encephalartos, Lepidozamia, Macrozamia, Microcycas, Stangeria,* and *Zamia* (Moretti, Sabato, and Siniscalco Gigliano, 1983). Cycasin, the β-D-glucoside of MAM, is the most thoroughly studied azoxyglycoside. It can serve as a model for the mutagenicity and carcinogenicity of other azoxyglycosides, such as macrozamin and the neocycasins, because they also have MAM as their reactive component (Morgan and Hoffmann, 1983).

Scope of Genetic Toxicology

Genetic toxicology concerns mutagens and their effects on DNA and on mechanisms of inheritance in cells and organisms (Hoffmann, 1991). A mutagen is any agent that causes mutations or chromosomal alterations. Mutagens are of interest in toxicology because mutations induced in human germ cells can cause an increase in genetic disease in future generations and because mutations in somatic cells can contribute to various disorders, most notably, cancer.

Three principal classes of genetic damage are gene mutations, chromosome aberrations, and aneuploidy (Hoffmann, 1991).

A gene mutation, also called a point mutation, is a change in the DNA sequence within a gene. Gene mutations include base-pair substitutions, in which one base pair in DNA (e.g., guanine-cytosine) is replaced by another (e.g., adenine-thymine) and frameshift mutations, in which one or a few base pairs are added to the normal sequence of DNA or deleted from it.

Laboratory tests for gene mutations have been developed in viruses, bacteria, fungi, plants, insects, mammalian cell cultures, and mammals (Hoffmann, 1991). Mutation assays are designed to compensate for the fact that mutation is a rare event. In higher organisms, this typically entails detecting mutations on the basis of unequivocal phenotypes and several or many genes. In contrast, assays in microorganisms and cell cultures often involve a positive selection for mutations in a single gene. For example, the Ames test is based on strains of *Salmonella* that require histidine in order to grow. Plating in petri dishes that lack histidine selects for bacteria that have experienced a mutation that enables them to grow without it. The induction of mutations is detected as an increase in the number of colonies with increasing doses of a mutagenic chemical.

Methods for detecting genetic damage at the chromosomal level differ from gene-mutation assays, in that the alterations are observed by microscopy rather than relying on the observation of altered phenotypes (Hoffmann, 1991). Chromosome breaks and various rearrangements can be detected in appropriately prepared dividing cells from plants or animals. Most commonly, the assays are conducted in mammalian-cell cultures or in cells from mouse bone marrow. Besides chromosome aberrations, other microscopically observable events, such as sister-chromatid exchanges (SCE) and micronuclei, can serve as indicators of genetic damage.

Reactions Involving Cycasin and MAM

The mutagenicity of an azoxyglycoside depends on the cleavage of its sugar linkage, releasing MAM (Morgan and Hoffmann, 1983). The bond between glucose and MAM in cycasin is cleaved by β-glucosidase, which occurs in various microorganisms and tissues of plants and neonatal mammals, but not in tissues of mature mammals. In mature mammals, the glycoside bond is cleaved by the bacterial flora of the gut, so oral exposure to cycasin leads to MAM exposure.

The efficiencies with which different azoxyglycosides are deglycosylated vary (Yagi and Tadera, 1987), so metabolic differences among the compounds must be considered if one attempts to predict effects of other glycosides from cycasin.

The mutagenicity of MAM involves the methylation of DNA by highly reactive ions derived from MAM (Morgan and Hoffmann, 1983). Methyl groups are added to various sites in the DNA bases. The highly nucleophilic nitrogen atom at the 7-position of guanine is most susceptible to alkylation, but other sites tend to be more important for mutagenesis. For example, 0^6-methylguanine is particularly apt to mispair when DNA replicates (Hoffmann, 1991). The ratio of 0^6 to N7 methylation by MAM is 0.18 (Kumari, et a., 1985), which means that MAM forms more 0^6-alkylguanine than many alkylating agents. This distribution of alkylation makes MAM a potent mutagen, causing base-pair-substitution mutations by base mispairing.

MAM gives rise to methylcarbonium ions and methyldiazonium ions spontaneously, so enzymatic metabolism is not required for methylation of DNA and mutagenesis. Enzymatic reactions, however, can facilitate this process (Morgan and Hoffmann, 1983; Fiala, Sohn and Hamilton, 1987). MAM can be oxidized by alcohol dehydrogenase to an aldehyde called methylazoxyformaldehyde or MAMAL. This aldehyde is highly reactive and generates carbonium ions even more readily than MAM does. Alcohol dehydrogenase may therefore potentiate the mutagenic activity of MAM even though it is not obligately required.

Mutagenicity of Cycasin and MAM

Cycasin is mutagenic in assays with β-glucosidase activity but not in assays that lack this activity. The mutagenicity of cycasin is therefore ascribable to MAM. Detailed information on the genetic toxicology of cycasin and MAM is available in previous reviews (Morgan and Hoffmann, 1983; Hoffmann and Morgan, 1984). These reviews are summarized here and updated on the basis of recent literature.

Cycasin induces base-pair-substitution mutations in *Salmonella* when its sugar linkage is cleaved (Morgan and Hoffmann, 1983). *Salmonella* does not have the enzymatic activity for this cleavage, but the assay responds positively when exogenous β-glucosidase is provided.

The enzymatic cleavage can also be performed by fecalase, which is an extract of human feces that contains the glucosidase activity of the microbial flora of the gut. Like cycasin, neocycasin is mutagenic when metabolised by fecalase. Macrozamin was not mutagenic in this assay, perhaps because of less-effective cleavage of its primeverose sugar. Further studies comparing cycasin and macrozamin would be of interest.

A host-mediated assay combines a microbial test for mutation with *in-vivo* mammalian metabolism. An indicator organism, such as *Salmonella*, is inserted into the animal (e.g., injected into the peritoneal cavity), and the chemical is administered to the animal by another route. The bacteria are later removed from the animal and assayed for mutation, having been exposed to both the chemical and its metabolites. In this assay, cycasin is mutagenic in *Salmonella* when given orally but not when given by intraperitoneal, intramuscular, or intravenous routes. This result reflects the requirement for deglucosylation by the gut flora. The findings that the mutagenicity of orally administered cycasin is prevented by ampicillin, which eliminates the microbial flora, and that MAM is mutagenic whether given orally or by other routes support the same interpretation (Morgan and Hoffmann, 1983). The mutagenicity of orally administered cycasin in a host-mediated assay has been confirmed in a test for DNA damage in *Escherichia coli* (Knasmuller and Mohn, 1986).

Cycasin has been tested in diverse assays in eukaryotes (Morgan and Hoffmann, 1983). It induces gene mutations and mitotic recombination in yeasts in the presence of rat-gut flora (Mayer and Goin, 1983) and both gene mutations and chromosome aberrations in plants that have β-glucosidase activity. *Zamia integrifolia* root tips are among the plant tissues in which chromosome aberrations have been reported; one may speculate that cellular compartmentalization protects *Zamia* from its own azoxyglycosides. Cycasin is nonmutagenic in the *Drosophila* sex-linked recessive lethal (SLRL) test; this negative result can be explained by lack of cleavage of the glycoside bond, because MAM is mutagenic in the same assay. DNA strand breakage and unscheduled DNA synthesis (UDS) in mammalian cells are other indicators of genetic damage induced by cycasin under conditions that provide for its deglucosylation.

Unlike cycasin, MAM is mutagenic without β-glucosidase activity(Morgan and Hoffmann, 1983).

Results in such bacterial strains as *hisG46* and TA1535 of *Salmonella* and WP2 of *E. coli* indicate that it induces base-pair substitutions. The effectiveness of mutagenic treatment varies with experimental conditions, such as preincubation assays relative to standard plate tests, and occasional negative results are encountered because of experimental difficulties associated with the instability of MAM. Nevertheless, the evidence for the induction of base-pair substitutions by MAM is unequivocal. Results for the induction of frameshift mutations are less clear, in that positive responses in the presence of liver microsomal enzymes were observed in some studies but not in others. Therefore, weak frameshift mutagenesis remains a possible secondary mechanism of mutagenesis by MAM. Tests for forward mutations in *Salmonella* (Dorado and Pueyo, 1989), induction of the SOS system in *E. coli* (Quillardet, de Bellecombe, and Hofnung, 1985) and repairable DNA damage in *E. coli* and *Bacillus* further indicate that MAM is genetically active in diverse bacterial assays (Morgan and Hoffmann, 1983).

MAM is mutagenic in many assays in eukaryotes (Morgan and Hoffmann, 1983). It induces gene mutations in yeast and in the SLRL test in fruit flies. Tests for mutations at the *tk* locus in cultured mouse lymphoma cells (Caspary, et al., 1988) and at the *hprt* and ouabain-resistance loci in Chinese hamster cells were also positive. Chromosome aberrations were detected after treatment of mammalian cells both *in vitro* and *in vivo* (Hoffmann and Morgan, 1984). Other effects of MAM include the induction of DNA strand breakage (Kumari, et al., 1985), SCE, and UDS in mammalian cells and mitotic recombination in yeast.

The extent of UDS (Mori, et al., 1984) and DNA strand breakage (Morgan and Hoffmann, 1983) was greater in the colon than in the jejunum, perhaps because the colon, but not the jejunum, converts MAM into MAMAL. The enhanced generation of carbonium ions when MAM is converted to MAMAL may also contribute to liver being more susceptible to DNA strand breakage than kidney, colon, and lung, because the liver is an active site of alcohol dehydrogenase.

An exception among the preponderance of positive results with MAM is a negative test for the induction of morphological abnormalities in mouse sperm. Although this assay can show that a chemical affects germ cells, its genetic basis and sensitivity are uncertain. Therefore, the negative result does not indicate that MAM is ineffective in germ cells.

It has not been tested in a well-characterized mammalian germ-cell assay, but it is mutagenic in many other assays, including the SLRL test in *Drosophila* germ cells. Moreover, many other alkylating agents are mutagenic in mammalian germ cells.

Carcinogenicity of Cycasin and MAM

The carcinogenicity of cycasin and MAM has been covered in previous reviews (Laqueur and Spatz, 1975; Laqueur, 1977; Morgan and Hoffmann, 1983; Hirono, 1986), a brief synopsis is presented here. Cycasin and MAM are carcinogenic in diverse species, including; fish, mice, rats, hamsters, guinea pigs, rabbits, and monkeys. Of these, the laboratory rat has received the most study. The principal sites of tumors induced in rats are the liver, kidneys, colon, lungs, brain, and duodenum. Treatment conditions affect the responses; for example, liver tumors are the predominant effect when cycasin or MAM is given by chronic feeding, while acute exposures have a greater effect on the kidneys. Besides the purified compounds, crude cycad meal and cycad husk (i.e. seed coats) are carcinogenic in rats when ingested; unlike these unrefined cycad materials, prepared Cycas flour was not carcinogenic. Transplacental exposure to cycasin and MAM causes tumors in the offspring of treated rats. Exposure to MAM *in utero* also causes neurotoxic and teratogenic effects.

The carcinogenicity of cycasin, like its mutagenicity, requires deglucosylation. The requirement for cleavage of the sugar linkage explains the findings that orally administered cycasin is carcinogenic in rats with a normal gut flora but not in germfree rats, whereas MAM is carcinogenic in both. These results indicate that microbial metabolism is a critical step in the carcinogenicity of cycasin. Unlike mature germfree rats, newborn germfree rats have β-glucosidase activity and are susceptible to the carcinogenic effects of cycasin.

Mutagenic and Carcinogenic Risk for Humans

Mutagenicity in diverse assays suggests that MAM can pose a mutagenic risk for humans. Azoxyglycosides can similarly pose a mutagenic risk if ingested, because the gut flora can deglycosylate them. Interpretations about human carcinogenicity must rest on the laboratory findings that cycasin and MAM are carcinogenic in several organs and in diverse species, because epidemiologic data are lacking.

However, the genetic toxicology data and the carcinogenicity data are consistent and provide a strong basis for suspecting that cycasin and MAM would be mutagenic and carcinogenic in humans who received sufficient exposure.

Means for removing toxic compounds from cycads have been developed by peoples who have traditionally used cycads as foods (Whiting, 1963; Hirono, 1987). Despite the evidence that carcinogens are removed in the processing of cycads, there is insufficient information on the degree to which various methods are effective in doing so. The noncarcinogenicity of some samples of cycad flour does not preclude the presence of carcinogens in other samples or in other cycad foods.

It is not feasible to test many cycad samples for carcinogenicity because of the great expense of the assays. To circumvent this problem, simple mutagenicity tests can be used as surrogates for carcinogenicity tests; doing so seems more appropriate for cycasin and MAM than for many other agents, because their mutagenic and carcinogenic responses are consistent. A mutagenicity assay, used in combination with chemical analysis, can provide a practical approach for testing cycad products prepared by various methods.

The risk posed by cycad products may be significant or negligible, depending on the extent of their use and the concentrations of azoxyglycosides in them. The most obvious means of human exposure is through foods prepared from cycads, but medicinal uses should not be overlooked. Even in populations that use cycads as foods, exposure to azoxyglycosides may be slight if adequate precautions are taken in food preparation. If cycads are improperly processed or eaten without preparation, as is reported by cycad husk in Guam (Hirono, 1987), significant exposures can occur.

References

***Caspary, W.J., Daston, D.S. Myhr, B.C., Mitchell, A.D., Rudd, C.J., Lee, P.S.* 1988.** Evaluation of the L5178Y mouse lymphoma cell mutagenesis assay: Interlaboratory reproducibility and assessment. Environ. Molec. Mutagenesis 12 (Supp). 13: 195 - 229.

***Dorado, G., Pueyo, C.* 1988.** L-Arabinose resistance test with *Salmonella typhimurium* as a primary tool for carcinogen screening. Cancer Res. 48: 907 - 912.

***Fiala, E.S., Sohn, O.S., Puz, C., Czerniak, R.* 1987.** Differential effects of 4-lodopyrazole and 3-methylprazole on the metabolic activation of methylazoxymethanol to a DNA methylating species by rat liver and rat colon mucosa in vivo. J. Cancer Res. Clin. Oncol. 113: 145 - 150.

***Hirono, I.* 1987.** Cycasin. *In* I. Hirono (ed.), Naturally occurring carcinogens of plant origin: Toxicology, pathology and biochemistry, pp. 3-24. Kodansha, Tokyo/Elsevier, Amsterdam.

***Hoffmann, G.R.* 1991.** Genetic Toxicology. *In* M.O. Amdur, J. Doull, and C.D. Klaassen (eds.), Casarett and Doull's Toxicology: The Basic Science of Poisons, ed. 4. Pergamon Press, Elmsford, New York, (in press).

***Hoffmann, G.R., Morgan, R.W.* 1984.** Review: Putative mutagens and carcinogens in foods. V. cycad azoxyglycosides. Environ. Mutagenesis 6: 103 - 116.

***Knasmuller, S., Mohn, G.R.* 1986.** On the distribution of genotoxic factors in various organs of mice treated with cycasin. Chem.-Biol. Interactions 58: 109 - 116.

***Kumari, M.L., Kamat, P.L., D'Ambrosio, S.M., Witiak, D.T., Milo, G.E.* 1985.** A comparative study of dimethylhydrazine regio-isomers and the methylazoxymethanol metabolite of 1,1 and 1,2-dimethlhydrazine in relation to transformation in human fibroblasts. Cancer Lett. 29: 265 - 275.

***Laqueur, G.L.* 1977.** Oncogenicity of cycads and its implications. Adv. Mod. Toxicol. Environ. Cancer 3: 231 - 261.

***Laqueur, G.L., Spatz, M.* 1975.** Oncogenicity of cycasin and methylazoxymethanol. Gann Monogr. Cancer Res. 17: 189 - 204.

***Mayer, V.W., Goin, C.J.* 1983.** Activation of cycasin to a mutagen for *Saccharomyces cerevisiae* by rat intestinal flora. Appl. Environ. Microbial. 45: 651 - 657.

***Morgan, R.W., Hoffmann, G.R.* 1983.** Cycasin and its mutagenic metabolites. Mutat. Res. 114: 19 - 58.

Moretti, S., Sabato, S., Siniscalco Gigliano, G. 1983. Taxonomic significance of methylazoxymethanol glycosides in the cycads. Phytochemistry 22: 115 - 117.

***Mori, H., Ohbayashi, F., Sugie, S., Tanaka, T., Takahashi, M., Mori, N.* 1984.** Measurement of unscheduled DNA synthesis on alkylating carcinogens in rat intestinal mucosal cells from different sites. J. Toxicol. Sci. 9: 23 - 28.

***Quillardet, P., de Bellecombe, C. Hofnung, M.* 1985.** The SOS Chromotest, a colorimetric bacterial assay for genotoxins: validation study with 83 compounds. Mutat. Res. 147: 79 - 95.

***Whiting, M.G.* 1963.** Toxicity of cycads. Econ. Bot. 17: 270 - 329.

***Yagi, F., Tadera, K.* 1987.** Azoxyglycoside contents in seeds of several cycad species and various parts of Japanese cycad. Agric. Biol. Chem. 51: 1719 - 1721.

Cycas circinalis as an Etiologic Risk Factor in Amyotrophic Lateral Sclerosis and Other Neurodegenerative Diseases on Guam

Leonard T. Kurland
Senior consultant and Professor of Epidemiology
Mayo Clinic, Rochester, Minnesota 55905, U.S.A.
Supported in part by grants AG 08802 and NS 17750
from the National Institutes of Health, Bethesda, Maryland

Introduction

This report deals with the possible role of cycads in amyotrophic lateral sclerosis (ALS) and Parkinsonism-Dementia Complex (PDC) in the indigenous Chamorro population of Guam. ALS is a progressive and fatal disease of adults characterized by wasting and weakness of skeletal muscles due to motor neurone degeneration at spinal level (progressive spinal muscular atrophy) or the bulbar level (progressive bulbar palsy). Hyperactive reflexes are usually present and are due to degeneration of the neurones of the Betz cells of the motor cortex.

On the basis of national mortality studies (Bharucha, et al., 1984), and the medical records survey for the population of Rochester, Minnesota (Yoshida, et al., 1984), the disease newly effects about 5,000 persons per year from about age 20 years onward. The rate increases with advancing age and has a median age of onset of 65 years. Worldwide there may be 100,000 new cases (and deaths) each year. The male to female ratio is about 1.6:1 and the average duration is about three years. As a result of studies on Guam and at the Mayo Clinic (Kurland and Mulder, 1954, 1955), at least three major forms of ALS are now recognised: (1) the classical or sporadic form which accounts for about 90% to 95% of all cases in the continental U.S. and other developed countries; (2) the familial form which accounts for about 5% to 10% of the remaining cases in these locales; and, (3) the Marianas or Western Pacific form of ALS which has been prevalent in several Pacific foci (Guam, the Kii Peninsula of Japan, and the Irian Jaya are of New Guinea).

ALS was first described by Charcot in Paris in the 1860's although Aran, also in Paris, had described progressive muscular atrophy about 15 years earlier. However, ALS, as a prevalent disease and major cause of death on Guam, has been traced back even earlier to about 1815. It is known to have occurred long before modern technology, industry and pollution, or epidemics of poliomyelitis arrived on this island.

Although diagnoses of ALS or *paralytico* or *lytico*, as it is known by the Chamorros, were being recorded in the earliest available death certificates on the island in 1902, ALS was not recognised as epidemic until a series of studies began at the end of World War II by medical personnel of the U.S. Navy Hospital on Guam. As a Public Health Service officer with an interest in neuroepidemiology, I arranged with Dr. Donald Mulder of Mayo Clinic for formal and systematic surveys in 1953 which led to the organisation of a research centre on Guam in 1956 from which studies were carried out (with some brief interruptions) until the centre was closed in 1985.

On this, the 35th anniversary of the founding of the NINDB Research Centre on Guam, we recognize its many contributions to neurology since it has served as a model of collaboration in international, private and governmental research. The centre was formally established soon after we began our population surveys in the Mariana and Caroline Islands. These surveys, over the next 35 years, were to generate a series of studies and more than 200 publications in an effort to explain the high prevalence of ALS and a clinical variant known as Parkinsonism-Dementia Complex (PDC) found in the same Western Pacific foci. Clarification of these foci may hold a key to the understanding of Parkinson's and Alzheimer's diseases as well as ALS in other parts of the world.

The research began among the Chamorro people of the Marianas and those who had migrated to California and subsequently spread to involve the population in the Kii Peninsula of Japan and the Auyu and Jakai people in Irian Jaya, Western New Guinea. At the Guam Research Centre, intensive efforts by scientists from many nations representing many disciplines, were concentrated on ALS and PDC and the feature common to both; namely, neurofibrillary tangles in the degenerating central nervous system. About 1962, Hirano described, in great detail, the unusual pathological features of ALS and PDC on Guam; many of these same features were subsequently recognised in the pathology of Alzheimer's disease and other populations, (Hirano et al., 1961 a; Hirano, Malamud and Kurland 1961 b). The large number of collaborators who made significant contributions to this research over the past 35 years are listed in one of the more recent reviews of this research effort (Kurland and Mulder, 1988).

The possible role of cycad as a toxic cause of ALS is found in some of my correspondence with Dr Margaret Whiting in the 1950's. We were sufficiently concerned over the role of cycad toxin so that six international conferences on cycad were organized between 1962 and 1972 to explore this possibility. Many of the papers presented at those meetings provided an in-depth analysis of cycasin as a hepatotoxic and carcinogenic agent; there was also some indication of neurological disease in a few animals (Bharucha, et al., 1984 and Yoshida et al., 1984). But it was not until the report in 1987 by Spencer et al., (1987) that BMAA, a neurotoxic amino acid derived form *Cycas circinalis*, provided the first relevant animal model with motor neurone disease and possibly a Parkinsonism and dementia as well.

The more recent studies by Spencer and others at this conference that concern cycad and neurodegenerative diseases are found elsewhere in this volume. At this point, it may be more appropriate to review the general epidemiological features of the 'Guam' disease and emphasize those issues, some still controversial, which may contribute to a better understanding of the geographic distribution, frequency, trend and population selectivity of ALS and PDC in the Western pacific.

With respect to the role of a possible toxin in cycad, the proceeding of the Third and Sixth conferences are especially useful (Bharucha, et al., 1984 and Yoshida et al., 1984). However, by the time of the Third Cycad Conference;

i. Proceedings of the Third Conference on the Toxicity of Cycads, April 17, 1964, Chicago, Illinois, Fed. Proc. 23(1) : 133 - 783.

ii. Proceedings of the Sixth International Cycad Conference, April 17, 1972, Chicago, Illinois. Fed. Proc. 1972: 31(5) : 1465 - 1546.

In 1964 (that is, within ten years of the initiation of the continuing program on Guam), the following neurological, pathological and epidemiological studies had been completed.

1. Confirmation of the earlier reports of Zimmerman (1945), Arnold et al., (1953), Koerner (1952), and

Tillema and Wynberg (1953), of an increased prevalence of ALS (known as lytico or paralytico) on Guam. The disease was also prevalent among the Chamorros on the nearby island of Rota and among those who had migrated to California in the previous ten years (Torres, Iriarte and Kurland, 1957). It was somewhat less among the Chamorros of Saipan and it had not been observed among the Carolineans in the islands to the south of Guam.

2. As noted earlier, an increased incidence of ALS among the Chamorro population of Guam was traced back as early as 1815. In the first survey by Mulder and myself on Guam in 1953, we encountered patients with what we initially regarded as post-encephalitic parkinsonism; but it soon became apparent that what we were seeing was a new entity that we referred to as parkinsonism-dementia complex or PDC which was well recognised by the native population as bodig or raput, an inappropriate term for slowness or laziness. The intensive clinical, pathological, and familial studies that were carried out soon made it clear that PDC was a clinical variant of the local form of ALS (Hirano et al., 1961 a, b).

3. In 1957, numerous neurofibrillary tangles and granulovacuolar bodies were found in the hippocampus and other subcortical areas in both ALS and PDC. These findings led to the realisation that the ALS on Guam was different than ALS in the continental U.S. (Kurland, Choi and Sayre, 1957). PDC had similar neurofibrillary tangles but in PDC there was also cortical degeneration and loss of pigment in the substantial nigra and locus ceruleus. The latter changes are far less pronounced, if present at all, in the patients with ALS only. Among those with PDC, however, lower motor neurone changes of ALS ultimately developed clinically in about one out of six cases although EMG changes consistent with lower motor neurone diseases were said to be present in two-thirds of the PDC cases [personal communication, Prof. Andrew Eisen, University of British Colombia, 1988]. Virtually all patients with ALS or PDC had upper motor neurone disease. A small proportion of the patients with PDC also appeared to have difficulties with upward gaze but these had not been regarded as a key feature of PDC in the earlier years of the study. Steele subsequently called attention to his and its similarity to progressive supranuclear palsy (Steel, Richardson and Olszewski, 1964). Thus, ALS and PDC are considered variants of a single disease entity on Guam in which the neurofibrillary tangles are the 'common denominator'. This concept was supported by later observations by Anderson et al., (1979) that there was premature occurrence of neurofibrillary tangles in many adult Chamorros dying of non neurological causes, indicating that a subclinical form of the neurological disease was also prevalent in the same population.

4. Although familial ALS was recognised as occurring in other populations, mounting evidence pointed to an environmental cause for the disease in the Western Pacific foci. Intensive efforts to isolate an infectious agent to account for these diseases have been negative (Gibbs and Gajdusek, 1981).

5. In 1955, in the search for risk factors, Dr. Margaret Whiting, a nutritionist, and Dr. F. Raymond Fosberg, a botanist, had identified cycad as a staple food among the Chamorros for which there was a particular concern over a known toxic component that had to be removed by repeated soaking over a period of several days. The variability in soaking procedures and the uncertainty of toxin removal led us to consider *Cycas circinalis* as a prime candidate in the etiology of ALS and PDC (Kurland, 1972).

The seeds of *Cycas circinalis* have been known to be toxic for humans since ancient times although some populations had learned how to process the seeds for food so that acute toxicity could be avoided. Among the Chamorros, the ritual for long soaking developed and during times of famine, especially following typhoons and times of conflict such as World War II, cycad became an important source of carbohydrate in the diet. The husk of the nut was sometimes used as a chew and the freshly ground cycad seeds were also used as a medicinal, particularly in the form of a poultice applied to ulcers and other lesions of the skin. The toxin known at that time was cycasin which is a glycoside component and for humans and other mammals, is perhaps the most hepatotoxic and carcinogenic compound discovered in nature.

6. However, by 1962, the studies on Guam had also disclosed three features of experimental lathyrism or odoratism associated with *Lathyrus odoratus*. These were the neurological disease, multiple exostoses, and collagen abnormalities (Fullmer et al., 1960; Kurland, 1832). A search for a lathyrogenic agent was deemed an appropriate line of study in the 1960's. It was the identification of the amino acid α-amino-β-methylaminopropionic acid or β-N-methylamino-L-alanine (BMAA) by Bell and his co-workers in the late 1960's (Vega and Bell, 1967) as a neurotoxic agent in *Cycas circinalis* that led to Spencer's recent work suggested BMAA, possibly with cycasin, played a crucial role in the etiology of ALS and PDC on Guam (Spencer et al., 1987).

The long delay between the observations at the cycad conferences and the report by Spencer deserves explanation. Laqueur (1964) and Mickelsen et al., (1964) of the National Institutes of Health in the late 1950's showed that ingestion of the cycasin glycoside in large doses could produce acute yellow atrophy of the liver. Small doses ingested or following application to eroded skin produced cancer of the liver, kidneys, and other organs after a latent period of a year or more. The aglycone of cycasin, methylasoxymethanol (MAM), was found to be the carcinogenic agent with a methylating mechanism similar to that of dimethylnitrosamine. These early efforts failed to produce a neurological disease corresponding to ALS or PDC in animals in any consistent manner. It was not until 1964 at the Third Cycad Conference that Datsun (1964), of Bombay, India reported that one of several monkeys fed washed cycad (presumably free of cycasin) appeared to develop a motor neurone disease. This, unfortunately, was not replicated by other investigators at the time nor could the neurological disorder be induced in other laboratory animals such as rodents.

In 1967, shortly after the publication of the Third Cycad Conference, Vega and Bell (1967), at the University of Texas, described the isolation of a proprionitrile from *Cycas circinalis* which had lathyrogenic features. They had also found this compound (BMAA) occurred in free or bound form in other members of the Cycas genus. Toxicity experiments continued and only the L-isomer form of BMAA was found to be toxic in the concentrations tested but 'convulsions and paralysis' were produced in mice and rats, although only with higher doses. No convulsions or paralysis were produced when the compound was injected at low concentrations or over a long period of time. Unfortunately, detailed neuropathology was not carried out and, with Dr. Bell's return to London, the work on this compound appears to have lapsed until it was reactivated by Spencer.

Dr Spencer's interest in lathyrism and another neurotoxic amino acid, β-N-oxalylamino-L-alanine (BOAA), led him to reopen the question of a lathyrogenic mechanism in the Guam disease (Spencer et al., 1987). This time, with knowledge and experience derived form his studies of lathyrism and BOAA, he embarked on a new research effort to demonstrate neurotoxic effects of BMAA. In his experiments on moneys, he was able to produce an illness with features of human ALS and possibly Parkinson's disease. Spencer subsequently showed that cycad, as a medicinal or a poultice, might also be the prime etiologic factor in the cases which had occurred in the Kii Peninsula and in Western New Guinea (Kimura, Yase and Handa, 1965; Spencer et al., 1990).

In the early years of our studies (1953 - 1955), there were, among the 40,000 Chamorros, about 60 patients with ALS and 20 with PDC on Guam. Reed and Brody (1975) noted a decrease in the incidence rate for both disorders by the mid 1960s and Garutto, Yanagihara and Gajdusek reported a marked decrease in both disorders in 1985. Unfortunately, the impressions by these latter observers that ALS and PDC incidence had decreased substantially were not based on intense house-to-house surveys such as those conducted in the early 1950s. It is possible that many patients were not spontaneously being referred to the NIH research facility for diagnosis in the 1980s. Under the auspices of the NINCDS, a house-to-house survey of ALS and PDC was carried out in 1987 - 1988 in Umatac, Merizo and Inarajan, three villages in southern Guam. That study suggested ALS was still about 50 times more prevalent than in Rochester, Minnesota and parkinsonism and dementia were also highly prevalent in comparison to the estimated rates for Parkinson's disease and Alzheimer's disease in the continental U.S. (Lavine et al., 1991).

A shift towards the later age of onset had also been noted for both ALS and PDC on Guam (Garruto et al., 1985; Lavine et al., 1991). The median age of onset of ALS in 1988 was now about 55 years versus the 44 years which had been reported in 1954; PDC onset was now occurring at about 62 years on average versus 54 years as reported in the 1960's. There was also a high frequency of dementia in the population 50 years and over; many of these had features of parkinsonism but the determination of whether these were PDC or a form of Alzheimer's disease must await ongoing autopsy studies.

Other recent studies (Doty et al., in press) have shown that odour identification, which appears to be lost in early Parkinson's disease and Alzheimer's disease in the continental U.S., is also lost in early PDC on Guam. This has led to the hypothesis that a portal of entry of a toxin may be through the nasal mucosa and the olfactory nerve since the olfactory neurones are seriously depleted in the PDC cases. For that reason, I suggested that various cycad pollens be analysed chemically. Whether the pollen of *Cycas circinalis* contains cycasin, BMAA or other toxins and enzymes that may aid their penetration of nasal mucosa remains to be determined. Whether such toxins are differentially present in the pollen of *Cycas circinalis* compared to other cycads should also be studied. Whether such toxins, if present, could enter the olfactory nerve fibres, be passed to other neurones, and have such a devastating effect on the central nervous system is a topic that would require a major pathophysiologic effort. However, the initial chemical studies on various cycad pollens are currently under way thanks to the efforts of several investigators attending the Cycad 90 conference (Seawright, Orriell, Spencer, Nunn, Giddy, Duncan, Norstog and also Hubbuch at the Fairchild Gardens in Miami and Mabry, at the University of Texas). The extent that additional research on this topic can be justified will depend on the results of chemical analysis of several pollens which have been collected.

The earlier conferences and Cycad 90 provide an important reference for those interested in the role of plants in disease. Out of this interest may well come the precise understanding of the mechanism of action of cycasin on many tissues and of BMAA on motor and other nerve cells in the brain and spinal cord. The research of BMAA can be expected to stimulate nutritional biochemistry, pharmacology, and botany to locate BMAA and similar toxins in food, medicinals, and other products that we use. Such studies may help to explain the cause and pathogenesis not only of ALS as it occurs around the world, but also of other neurodegenerative diseases such as Parkinson's, progressive supranuclear palsy, and Alzheimer's dementia. If BMAA is indeed the cause of such disorders, the age and intensity of exposure may have influenced the clinical picture. It has been suggested that heavy exposure at an early age on Guam could lead to ALS; that later and/or less exposure could result in PDC or Marianas dementia and that minimal exposure may result in the subclinical form of the disease. It has also been suggested that neuritic or senile plagues which are a component of Alzheimer's disease, may now be more numerous than noted in the earlier years of the study on Guam, especially in those 60 years and older. If this is so, it would suggest that the evolutionary process over the last half century may be a remarkably rapid transition from ALS to PDC and now possibly to a state in which dementia predominates. The interaction of a specific toxin, dose, and genetic susceptibility may be the determinates for a subsequent disease process such as ALS and the parkinsonism-dementia complex which resembles both Parkinson's disease and Alzheimer's dementia.

One additional feature of interest followed Steele's observation on gaze palsy in the PDC cases. Cox, an ophthalmologist at the University of British Colombia, noticed that about half of the patients with ALS/PDC had retinal 'tracks' (Cox et al., 1989). These resembled *ophthalmomyiasis interna*, as described by Gass and Lewis (1976) which presumably is due to a larval migrans. However, in some 60 cases followed on Guam, no larva has been found. The first eyes with such retinal tracks examined at autopsy are those with ALS/PDC from Guam. A focal degenerative retinal pigment epitheliopathy has been described; but there are no inflammatory changes nor any evidence of a parasite in these cases [personal communication, R. Jean Campbell, Mayo Clinic, 1991].

Epidemiologic studies are now under way on Guam and other islands in the region to determine the frequency and distribution of the retinopathy and further autopsy studies are also planned. It is also of interest that a high proportion of the small Japanese fish (medaka) *Oryzias latipes* in 4 to 20 weeks after a single brief exposure to methylazoxymethanol acetate from cycasin developed a retinal malignant neoplasm diagnosed as medulloepithelioma (Hawkins et al., 1986).

The mechanism of tumorigenesis may be quite different than the degenerative process that was described in the eyes of the ALS/PDC patients. But these observations should lead to further investigations of the diversity of possible effects of cycad toxins.

Although the emphasis in this presentation has been on cycad, we must also take into consideration the role of calcium, magnesium, and aluminium, independently or in association with cycad, as emphasised in the work of Yase (1980), Chen et al., (1981), Gajdusek, Garruto and Yangihara et al., (1980, 1982 a, b).

If the cycad has a role in the etiology of ALS and PDC on Guam, it would be fitting for this report to be dedicated to the people of Guam, to whom *Cycas circinalis* was so important for survival; but this plant may also have carried the agents responsible for the premature deaths of a large proportion of that same population. If the work of Spencer is confirmed, the people of Guam may have some consolation that the knowledge obtained from cycad research may, in the near future, eradicate the 'Guam disease' and that we are on the scientific threshold of solving the long-standing puzzle on the cause and pathogenesis of ALS and PDC. This particular research may also disclose the means of prevention and cure of ALS elsewhere in the world and provide knowledge of the pathogenesis of other neurodegenerative diseases such as Parkinson's, progressive supranuclear palsy, and Alzheimer's disease. Lastly, if the solution is indeed at hand, it is indicative of the interrelationship of epidemiology, laboratory sciences including botany, and clinical practice and is an illustration of the tremendous opportunities afforded by the recognition of disease occurrence in geographically isolated populations.

Acknowledgments

The author gratefully acknowledges the assistance of Donald Mulder, M.D., John Steele, M.D., and David Williams, M.D. in the development of this review. The author also extends his appreciation to Mrs Marlee Boyd for preparation of the manuscript.

References

Anderson, F.H., Richardson, E.P. Jr. Okazaki, H., Brody, J.A. 1979. Brain. 102(1): 65 - 77.

***Arnold, A., Edgren D.C., Palladino, V.S.* 1953.** N. Nerve. Ment. Dis. 117(2): 135 - 139.

***Bharucha, N.E. Schoenberg, B.S. Raven, R.H., Pickre, L.W. and Byar, D.P.* 1984.** Geographic Distribution of Motor Neurone Disease in the U.S. and Correlation with Possible Aetiologic Factors. *In*: Motor Neurone Disease, M. Gourie-Devi (Ed.), Proceedings of the International Symposium on Motor Neurone Disease, Bangalore, India, October 29 - 30.

***Chen, K-M, Mukai, E., Gajdusek, D.C., Gibbs, C.J. Jr., and Chase, T.N.* 1981.** *In* Abstracts of the Twelfth World Congress of Neurology, Kyoto, Japan, September 20 - 25, 1981 (International Congress Series No. 548). Excertpa Medica, p. 16, Amsterdam.

***Choi, N.W.* and *Sayre, G.P.* 1969.** Motor Neuron Diseases, Norris FH Jr and Kurland L.T. (Eds.), pp. 28 - 50. Grune & Stratton, New York.

***Cox, T.A., McDarby, J.V., Lavine, L., Steele, J.C. and Calne, D.B.* 1989.** A Retinopathy on Guam with High Prevalence in Lytico-Bodig. Ophthalmology. 96: 1731 - 1735.

***Dastur, D.K.* 1964.** Fed. Proc. 23(1): 1368 - 1369.

***Doty, R.L., Perl, D.P., Steele, J.C., Chen K-M, Pierce, J.D. Jr., Reyes, P. and Kurland, L.T.* (in press).** The Odour Identification Deficit of the Parkinsonism-Dementia Complex of Guam is Equivalent to that of Alzheimer's and Parkinson's Disease. Ann. Neurol.

***Fullmer, H.M., Siedler, H.D., Krooth, R.S.,* and *Kurland, L.T.* 1960.** Neurology. 10(8): 717 - 724.

***Gajdusek, D.C.* 1963.** N. Engl. J. Med. 268(9): 474 - 476.

***Garruto, R.M. and Yanagihara, R.* 1980.** *In* Abstracts of the Tenth International Congress of Tropical Medicine and Malaria, Manila, Philippines. November 9 - 15, 1980. 383.

***Garutto, R.M., Yanagihara, R., and Gajdusek, D.C.* 1985.** Neurology. 35: 193 - 198.

***Gass, J.D.M. and Lewis, R.A.* 1976.** Subretinal Tracks in Ophthalmomyiasis. Arch. Ophthalmol. 94: 1500 - 1505.

***Gibbs, C.J. Jr. and Gajdusek, D.C.* 1981.** Advances in Neurology, Vol. 36: Human Motor Neuron Diseases, Raven Press, New York. 36: 343 - 353.

***Gibbs, C.J. Jr. and Gajdusek, D.C.* 1972.** An Appraisal of the Neurotoxicity of Cycad and the Etiology of Amyotrophic Lateral Sclerosis on Guam. (6th International Cycad Conference). Fed. Proc, 31(5): 1540 - 1542.

***Hawkins, W.E., Fournie, J.W., Overstreet, R.M., and Walker, W.W.* 1986.** Intraocular Neoplasms Induced by Methylazoxymethanol Acetate in Japanese Medaka (*Oryzias latipes*). JNCI. 76(3): 453 - 465.

***Hirano, A., Kurland, L.T. Krooth, R.S., and Lessell, S.* 1961** a. Brain. 84: 642 - 661.

***Kimura, K., Yase, Y., and Handa, Y.* 1965.** Proceedings of the Eighth International Congress of Neurology, Vienna, Austria, September 5 - 10, 1965 (International Congress Series No. 94), Excertpa Medica, Amsterdam, 1965: 199 - 205.

***Koerner, D.R.* 1952.** Ann. Intern. Med. 37(6): 1204 - 1220.

***Krooth, R.S., Macklin, M.T. and Hilbish, T.E.* 1961.** Am. J. Human. Genet. 13(3): 340 - 347.

***Kurland, L.T. and Mulder, D.W.* 1954.** Epidemiologic Investigations of Amyotrophic Lateral Sclerosis. (1) Preliminary Report on Geographic Distribution, with Special Reference to the Mariana Islands, including Clinical and Pathologic Observations. Neurology 4 (5 - 6): 355 - 378.

***Kurland, L.T. and Mulder, D.W.* 1955.** Epidemiologic Investigations of Amyotrophic Lateral Sclerosis. (2) Familial Aggregations Indicative of Dominant Inheritance (Part I). Neurology 5 (3 - 4): 182 -268.

***Laqueur, G.L.* 1964.** Fed. Proc. 23(1): 1386 - 1388.

***Lavine, L., Steele, J.C., Wolfe, N., Calne, D.B., Obrien, P.C., Williams, D.B., Kurland, L.T., and Schoenberg, B.S.* 1991.** Amyotrophic Lateral Sclerosis/Parkinsonism-Dementia Complex in Southern Guam: Is It Disappearing? *In* Amyotrophic Lateral Sclerosis and Motor Neuron Diseases: Research Progress, Rowland, L.P., (Ed.). New York: Raven Press, (in press).

***Malamud, N., and Kurland, L.T.* 1961** b. Brain. 84: 662 - 679.

***Michelsen, O., Campbell, E., Yang, M, Mugera, G. and Whitehair, C.K.* 1964.** Fed. Proc. 23(1): 1363 - 1365.

***Mulder, D.W.* 1988.** Recent Epidemiologic Developments in the Context of Earlier Observations on ALS: Geographic Isolates in the Marianas and Other Islands of the Western Pacific: *In* Amyotrophic Lateral Sclerosis, Tsubaki, T. and Yase, Y. (Eds.). Elsevier Science Publishers, B.V. (Biochemical Division).

***Perl, D.P., Gajdusek, D.C., Garruto, R.M., Yanagihara, R.T., and Gibbs, C.J. Jr.* 1982** a. Abstracts of the Ninth International Congress on Neuropathology, Vienna, Austria, September 5 - 10, 1982. 31.

***Perl, D.P., Gajdusek, D.C., Garruto, R.M., Yanagihara, R.T., and Gibbs, C.J. Jr.* 1982** b. Science. 217: 1053 - 1055.

Reed, D.M., Brody, J.A. *1975. Am. J. Epidemiol. 101: 287 - 301.*

***Ross, S.M., Kisby, G. and Roy, D.N.* 1990.** Western Pacific Amyotrophic Lateral Sclerosis: Putative Role of cycad Toxins. *In* Amyotrophic Lateral Sclerosis: Concepts in Pathogenesis and Etiology, p. 263. Hudson A.J. (Ed.). University of Toronto Press, Toronto Buffalo London.

***Spencer, P.S., Nunn, P.B., Hugon, J., Ludolph, A.C., Ross, S.M., Roy D.N. and Robertson, R.C.* 1987.** Science. 237: 517 - 522.

***Steele, J.C., Richardson, J.E., Olszewski, J.* 1964.** *Progressive Supranuclear Palsy, A Heterogenous Degeneration Involving Brain Stem Basal Ganglia and Cerebellar with Vertical Gaze and Pseudobulbar Palsy, Nuchal Dystonia and Dementia. Arch. Neurol. 10: 333 - 359.*

***Tillema, S., and Wynberg, C.J.* 1953.** Documents de Medi Geographics et Tropica. 5(4): 366 -370.

***Torres, J.M., Iriarte, L.L.G., Kurland, L.T.* 1957.** Calif Med. 86: 385 - 388.

***Vega, A., Bell, E.A.* 1967.** Phytochemistry, 6: 759.

***Yase, Y.* 1980.** Neurotoxicology. 1(4): 101 - 109.

***Yoshida, S., Mulder, D.W., Kurland, L.T., Chu C-P., Okazaki, H.* 1986.** Follow-up Study on Amyotrophic Lateral Sclerosis in Rochester, Minn., 1925 through 1984. Neuroepidemiology. 5: 61 - 70.

***Zimmerman, H.M.* 1945.** Monthly Report to Medical Officer in Command, U.S. Naval Medical Research Unit No. 2, June 1, 1945.

Seed Toxicity in *Macrozamia riedlei*

P.G. Ladd
School of Biological & Environmental Sciences, Murdoch University
Murdoch, W.A. 6150 Australia

S.W. Connell
Department of Agriculture,Barron Bay Court
Como, W.A. 6151 Australia

B. Harrison
Department of Chemistry, University of Western Australia
Nedlands, W.A. 6009 Australia

Abstract

Macrozamia seeds, with their colourful, fleshy covering (sarcotesta) and very hard stoney gametophyte protection (sclerotesta) are well adapted to animal dispersal. Despite earlier claims that the flesh is nontoxic, the sarcotesta of *M. riedlei* contains a relatively high level of the poison macrozamin, but this does not seem to be a problem for native animals which have coexisted with *Macrozamia* species for a long time. The red flesh of newly shed seeds is almost always removed from the sclerotesta in the wild. The phenomenon of resistance of fauna to macrozamin is similar to the situation of other fauna and coextensive toxic plants such as the fluoroacetate-containing *Gastrolobium* spp. in Western Australia. The mechanisms for resistance are not known, but may involve the animals lacking an enzyme to hydrolyse macrozamin or the possession of a gut flora which detoxifies the toxic breakdown products of macrozamin.

Introduction

The seeds of many cycads are distinguished by a brilliant red or orange seed coat (sarcotesta). From one point of view it could be considered that this is a warning sign such as is found in some animals (e.g., poison arrow frog of the Amazon) that the organism is poisonous and therefore dangerous. On the other hand it could be considered that this bright colourisation is to attract animals to eat the seeds and in so doing aid dispersal of the species (cf. Willson, 1983).

The study of the toxic compounds in cycads has been very extensive (de Luca, et al., 1980; Moretti, Sabato and Siniscalco Gigliano, 1981, and many others) because of the importance of domestic animal poisoning associated with cycads and the use by some human groups of cycad seeds as food. In almost all published cases it is only vegetative parts or the kernels (gametophyte with embryo) of the seed which has been studied. In addition any investigation of the food value of cycads has been to do with the kernel component of the seed because this has been used by many human races, from the Australian Aborigines (Beaton, 1982), to the inhabitants of Guam in the Pacific (Kurland, Williams and Steele, 1990). The implications from archaeological and ethnographic evidence are that the seeds were an important food source for Australian Aborigines, possibly for over 10,000 years, in the areas where cycads grew (Beck, 1990).

The study is concerned principally with the composition of the sarcotesta of the seed because there has been little attention paid to this part of the seed, despite the fact that this is a most important element in the species' biology.

Macrozamia riedlei (Ficher ex. Gaudich) Gardner is a species restricted to the southwest of Western Australia. Females periodically produce cones, which may be quite large (weighing up to 14 kg), particularly after fire. Over 150 seeds may be produced per cone and this represents a considerable investment of photosynthate for the plant and a resource to be exploited by animals.

Materials and Methods

The kernel and sarcotesta used for the macrozamin determination where from *M. riedlei* collected at Wanneroo (a suburb of Perth, Western Australia), while for the food value analysis seeds came from the Murdoch University campus (also in Perth).

Pure macrozamin 19.9 mg was dissolved in water and the solution was diluted to 10ml to provide the standard solution for HPLC analysis. The standard solution gave a retention time of 2.0 minutes when run on a 15cm C18 column with 7% acetonitrile/water as the mobile phase at a flow rate of 2.5ml/minute. The UV detector was monitoring absorbency at 280nm.

Sarcotesta (50g) of fresh seeds of *M. riedlei* was minced and extracted with ether exhaustively (Soxhlet). The orange ether extract, which contained some droplets of water was extracted with a small amount of water, then dried and evaporated whereupon a dark orange oil (11.75g) with a characteristic odour was obtained. This was not examined further. The aqueous washings were evaporated at < 40ºC, then the residue was taken up in water, filtered and diluted to 10 ml. This solution was analysed for macrozamin by HPLC and found to contain 150 mg (i.e. 0.30% freshweight).

The ether was evaporated from the flesh under reduced pressure. This yielded 33.9g of dried plant material which was extracted with water under N_2. The extract was squeezed through muslin and then filtered through paper under gravity. The cloudy solution was boiled under N_2, cooled quickly under N_2 and then filtered. The filtrate (125ml) was diluted with ethanol (185ml) then allowed to stand overnight. The clear supernatant pale yellow solution was decanted from the gum and evaporated at < 50ºC. The residue was taken up in water, diluted to 25ml and analysed for macrozamin by HPLC. This solution contained 1,790mg of macrozamin (i.e. 3.58% freshweight).

Seeds were analysed for oil, crude protein, sugars and polysaccharides by the Western Australian Government Chemical Laboratory. The oil was extracted with hexane, crude protein calculated as nitrogen content times 6.25, sugars were extracted by hot 80% ethanol and calculated as glucose, while polysaccharides were calculated as glucose.

Results

Table 1 shows the macrozamin content (based on fresh weight) of various cycad species (Moretti et. al.,, 1981) together with the values for *M. riedlei* sarcotesta (present work) and kernel (Cannon et. al., 1980).

The sarcotesta has a macrozamin content almost 10 times greater than in the kernel tissue. The amount of macrozamin in the *M. riedlei* kernel is rather low in

Table 1	
Macrozamin content of Ripe Seeds in selected Cycads	
Cycad	**% Fresh Weight**
Cycas spp.	0.2 — 0.45
Stangeria eriopus	4.7
Dioon spp.	0.62 - 0.68
Encephalartos spp.	2.09 - 2.86
Lepidozamia peroffskyana	1.11
Macrozamia diplomera	2.41
M.fawcettii	2.49
M.heteromera	2.59
M.miquelii	3.88
M.moorei	3.72
M.riedlei	Sarcotesta 3.88, kernel 0.4
Bowenia spp.	4.83 - 5.04
Ceratozamia spp.	1.01 - 1.06
Zamia spp.	1.01 - 1.25
Microcycas calocoma	0.13

comparison with other *Macrozamia* species. However the sarcotesta value is comparable with that found in the kernels of *M. riedlei* and *M. moorei*, higher than in other *Macrozamia* species and somewhat lower than for *Bowenia*, which has the highest macrozamin values of any cycad.

The sarcotesta comprises about 16% of the fresh weight of seeds and averages about 4.4g in weight. For a cone which produces 158 seeds (around the order of seeds produced in a cone of a large plant) this means that there is about 0.7kg of sarcotesta produced per cone. In contrast the kernel comprises 67.6% of the seed weight and averages 18.3g.

The organic chemical composition (food value) of *M. riedlei* seeds is shown in Table 2. The sarcotesta has a slightly higher water and much higher oil and 'other material' content than the kernel, while the polysaccharide level of the kernel is much greater than in the sarcotesta. A crude iodine test showed that the majority of the kernel is starch. In comparison Beaton (1977) reported 44% sugars and polysaccharides in *Cycas* kernel.

Discussion

There has been a diversity of views about the toxicity of individual cycad taxa. It has been claimed (Giddy, 1974) that in *Encephalartos* the kernel of the seed is the only toxic part of the plant, while it is acknowledged that virtually all parts of *Macrozamia* species contain the toxic methylazoxymethanol glycoside.

A number of native animals are recorded as consuming the sarcotesta of the seeds (Burbidge and Whelan, 1982) and it is claimed in that paper that the 'outer integument' is non-poisonous.

Table 2		
Composition of Kernel and Sarcotesta in M.riedlei		
	Sarcotesta %	**Kernel %**
Water (& volatiles)	53	42
Oil	14 (28+)	1
Protein (crude)	3	8
Sugars	3	3
Polysaccharides	11	43
Other Material	16	3

from Thieret (1958) Economic Botany of Cycads.

Everist (1974) reported that cattle which consumed the ripe seeds of *M. moorei*, but spat out the intact sclerotesta, later developed symptoms of poisoning. However, this is the only record which demonstrates toxicity in the sarcotesta of a *Macrozamia* species. In this study the sarcotesta of *M.riedlei* is shown to have a high content of macrozamin and it can reasonably be inferred that this is the case in other *Macrozamia* species. Despite this poisonous nature of the sarcotesta the seeds are almost always cleaned of the red flesh soon after the cones open to expose the seeds. The sarcotesta of *M. riedlei* has a moderate food value (25% oil and polysaccharides) compared with the predominantly starchy content of the kernel.

In almost all cases of natural utilisation by animals it is the sarcotesta which is eaten rather than the kernel, which is protected by the hard sclerotesta. The combination of bright colouration, easily accessed food value of the outer covering, combined with the very hard sclerotesta, indicate an animal seed dispersal syndrome. Bauman and Yokoyama (1976) have suggested that the syndrome in cycads was originally a coevolution with birds which have a visual system sensitive to red and yellow. However, mammals have readily adapted to it and are now probably more effective visitors than are birds.

Most animals do not pierce the sclerotesta, the published exception being in New South Wales where *Rattus fuscipes* is an important seed predator of *M. communis* (Ballardie and Whelan, 1986). In Western Australia damage to the sclerotesta which would cause seed death was caused by birds (probably black cockatoos, *Calyptorhynchus* sp.) but the damage was incidental to removal of the kernel.

In contrast, most human utilisation of cycads is of the starch present in the seed kernels. References to rare exceptions are given in Thieret (1958). This correlates with the greater resource of the kernel compared with the sarcotesta and the greater recovery of edible components after pre-treatment to remove the toxic principle. The tissue of the kernel is compact and the starch is relatively insoluble in water and will remain through the soaking and leaching process. However the sarcotesta flesh is soft and spongy at maturity and the oil components of the sarcotesta would be quickly lost in any process which disrupted the tissue, leaving 'other material' - principally waxes, cell walls and cuticle of little food value.

Clearly the sarcotesta contains the poisonous macrozamin, so it is surprising that some animals consume it so voraciously with no apparent ill-effects. During the seeding session it is common to find emu droppings containing several to many stripped seeds in areas of coning macrozamias. The large size and avian metabolism of emus could mean that they can ingest moderate amounts of sarcotesta without ill-effect. However, much smaller mammals such as the chudich (*Dasyurus geoffroyi*) and especially the brush-tailed possum (*Trichosurus vulpecula*) have also been recorded as cleaning the seeds (Burbidge and Whelan, 1982), and they are of much smaller size and have a mammalian metabolism.

In Western Australia there is a parallel example of tolerance to particular plant poisons of native animals which have 'evolved' with poisonous plants. Marsupials which come from areas where *Gastrolobium* spp. (Fabaceae) grow are tolerant of levels of the normally acutely poisonous sodium monofluoroacetate (1080 poison) which are much higher than can be survived by eutherians or marsupials from areas where the gastrolobiums do not grow (King et al., 1989) Although the toxic principal is quite different from macrozamin the same situation is likely to apply to resistance to macrozamin. The actual mechanism of resistance to macrozamin is unknown, but it has been suggested (J. Cannon, pers. comm.) that either the animals lack a β glucosidase which is necessary to hydrolyse macrozamin to the very toxic (methyl-ONN-azoxy) methanol or that the microorganisms in the gut of the native animals might be able to reduce the azoxy compound to something less toxic. Given that the resistance to the toxin occurs over several groups of animals, ie. birds, eutherians (*R. fuscipes*) and marsupials, the gut flora theory may be the most likely explanation.

References

***Ballardie, R.T. and Whelan, R.J.* 1986.** Masting, seed dispersal and seed predation in the cycad *Macrozamia communis*. Oecologia 70: 100 - 105.

***Bauman, A.J. and Yokoyama, H.* 1970.** Seed coat carotenoids of the cycad genera *Dioon, Encephalartos, Macrozamia* and *Zamia*: Evolutionary significance. Biochem. Syst. Ecol. 4: 73 - 4.

***Beaton, J.M.* 1977.** Dangerous Harvest. Unpublished PhD thesis, Australian National University, Canberra.

***Beaton, J.M.* 1982.** Fire and water: Aspects of Australian Aboriginal management of cycads. Archaeology in Oceania 17: 51 - 58.

***Beck, W.* 1990.** Aboriginal preparation of *Cycas* seeds in Australia. Conference Abstracts, Cycad '90.

***Burbidge, A.H. and Whelan, R.J.* 1982.** Seed dispersal in a cycad, *Macrozamia riedlei* Aust. J. Ecol. 7: 63 - 7.

***Cannon, J.R., Raston, C.L., Toia, R.F. and White, A.H.* 1980.** Crystal structures of (Z) - O - β - D Xylo - pyranosyl (1->6) - β - D glucopyranosyloxy - *NNO* azoxymethane (macrozamin) and 5 -) Methyl - *myo* - inositol (sequoyitol). Aust. J. Chem. 33: 2229 - 36.

***De Luca, P., Moretti, A., Sabato, S., and Siniscalco Gigliano, G.* 1980.** The ubiquity of cycasin in cycads. Phytochem. 19: 2230 - 31.

***Everist, S.L.* 1974.** Poisonous Plants of Australia. Angus and Robertson.

***Giddy, C.* 1974.** Cycads of South Africa. Purnell, Cape Town.

***King, D.R., Twigg, L.E., and Gardner, J.L.* 1989.** Tolerance to sodium monofluoroacetate in dasyurids from Western Australia. Aust. Wildl. Res. 16: 131 - 40.

***Kurland, L.T., Williams, D.B., and Steele, J.C.* 1990.** Does cycad affect the clinical, pathologic and epidemiologic characteristics of neurodegenerative diseases in the Mariana Islands. Conference Abstract, Cycad '90.

***Moretti, A., Sabato, S., and Siniscalco Gigliano, G.* 1981.** Distribution of macrozamin in Australasian cycads. Phytochem. 20: 1415 - 1416.

***Thieret, J.W.* 1958.** Economic botany of cycads. Economic Botany 12: 3 - 41.

***Willson, M.F.* 1983.** Plant Reproductive Ecology. Wiley - Interscience, N.Y.

Toxicity and Insect-Cycad Symbiosis in *Zamia furfuracea*

Knut J. Norstog, Priscilla K.S. Fawcett
Fairchild Tropical Garden
11935 Old Cutler Road, Miami, Florida 33156, U.S.A.

Andrew P. Vovides
Instituto de Ecologia, Xalapa, Mexico

Mark W. Dunn
Faculty of Medicine
University of New South Wales, Sydney, Australia

Abstract

The Mexican cycad *Zamia furfuracea*, is pollinated by the host-specific snout weevil, *Rhopalotria mollis*, which swarms upon male cones, where mating, feeding, and egg-laying occur. Larvae feed exclusively upon tissues of male cones and all stages of metamorphosis occur within the cones. A neurotoxin α-amino-β-(methylamino) propanoic acid (BMAA) is sequestered in male-cone idioblasts but appears to be released from idioblasts into female cone tissues. Sporophylls of male cones contain much starch; those of female cones have little starch, and weevils feed upon male cones and are visitors to but not feeders upon or within female cones. Weevils transport pollen to female cones but neither adults nor larvae feed on pollen and they do not damage pollen sacs or pollen. Near the end of the breeding season of the weevil (and the cycad), some larvae become dormant in thick-walled pupal cases and remain so until the next reproductive season of the cycad.

Introduction

The pollination of cycads has been of considerable interest at Fairchild Tropical garden (FTG) because most of the exotic cycad genera and species growing there require artificial pollination in order to produce viable seeds. In nearly all cases, male and female plants of the same species are growing at distances no greater than a few meters from each other, sometimes separated only by just a meter or two. Since there still exists a fairly common assumption that cycads are wind-pollinated plants, as are most other gymnosperms (Norstog and Stevenson, 1980; Niklas and Norstog, 1984;

Norstog, Stevenson and Niklas, 1985; Norstog, 1987), one wonders why healthy and cone-producing transplants are not wind-pollinated in cultivation as they are supposed to be in nature. After all, wind is wind! During the past decade, we have accumulated evidence that one cycad, *Zamia furfuracea* L. fil., in cultivation is effectively pollinated by a small snout weevil, *Rhopalotria mollis* Sharp (Fig. 1), which completes at least one phase of its life cycle in pollen cones of that cycad (Norstog and Stevenson, 1980; Norstog, Stevenson and Niklas, 1985; Norstog and Fawcett, 1989). Field studies of pollination of *Z. furfuracea* are in progress and one of us (APV) observes that *R. mollis* is present in wild plants in Veracruz, Mexico and parasitizes male cones of this cycad in much the same way as related here.

At the time of pollination, it is observed that the ground tissue (parenchyma) of male cones is rich in starch, whereas female-cone parenchyma is poor in starch (Norstog and Fawcett, 1989). In addition, a type of storage cell (idioblast) is intact in male-cone scales (microsporophylls) but appears to lose some content in female-cone scales (megasporophylls). Several toxic substances, including the glycosides cycasin and macrozamin and an amino acid α-amino-β(methylamino) propanoic acid (BMAA), have been found in cycad tissues (Vega and Bell, 1967; Moretti, Sabato and Siniscalco Gigliano, 1983; Roschild, Nash and Bell 1986; Spencer et al., 1987; Duncan et al., 1989). Norstog and Fawcett propose that idioblasts and *Z. furfuracea* contain stored toxins in male-cone tissues, but in female-cone tissues toxins are liberated from idioblasts and dispersed in the parenchyma. In addition, a reward, in the form of starch in male cones may account for use of male cones as feeding and breeding sites and absence of predation on female cones which are relatively poor in starch.

Methods

During the reproductive season of the *Zamia furfuracea* (and the weevil, *Rhopalotria mollis*) we collected specimens daily and examined partially and wholly dissected cones and their inhabitants. Respective samples of cones and weevils were fixed in formalin-acetic acid (FAA) for sectioning and study or were sent by overnight express mail to the Institute of Neurological Disorders and Stroke, National Institutes of Health, Bethesda, MD, where one of us (MWD) extracted and identified the cycad neurotoxin BMAA from the samples by appropriate means described in Duncan et al. (1989).

For microscopy, sections of cones and weevils were stained in PAS (periodic acid-Schiff, for carbohydrate), NIN (ninhydrin-Schiff for amino acids), I_2KI-erythrosine (weevils only, for starch and protein) or I_2KI-fast green (cones, for starch and cellulose) and Sudan IV for lipid (Jensen, 1962).

Idioblasts were extracted by grinding chopped sporophylls from fresh cones with a mortar and pestle and a little sand and filtering the product through cheesecloth. Microscopic observation of the crude idioblast fraction assured us that the majority of idioblasts were whole and intact and they then were analysed for BMAA content as outlined above for other samples.

When a female cone of *Z. furfuracea* is pollen receptive, a single, vertical crack measuring about 2cm in open area, appears between adjacent megasporophylls and provides the only access to the cone's interior. Because this narrow access seems inconsistent with entry of airborne pollen, we (Norstog and Fawcett, 1989) earlier decided to monitor the vicinity of four pollen-receptive female cones for airborne pollen by placing greased microscope slides next to them. After 4 days (the appropriate time cones remain open) they were examined microscopically for trapped pollen and used to estimate airborne transport of pollen to receptive areas of female cones.

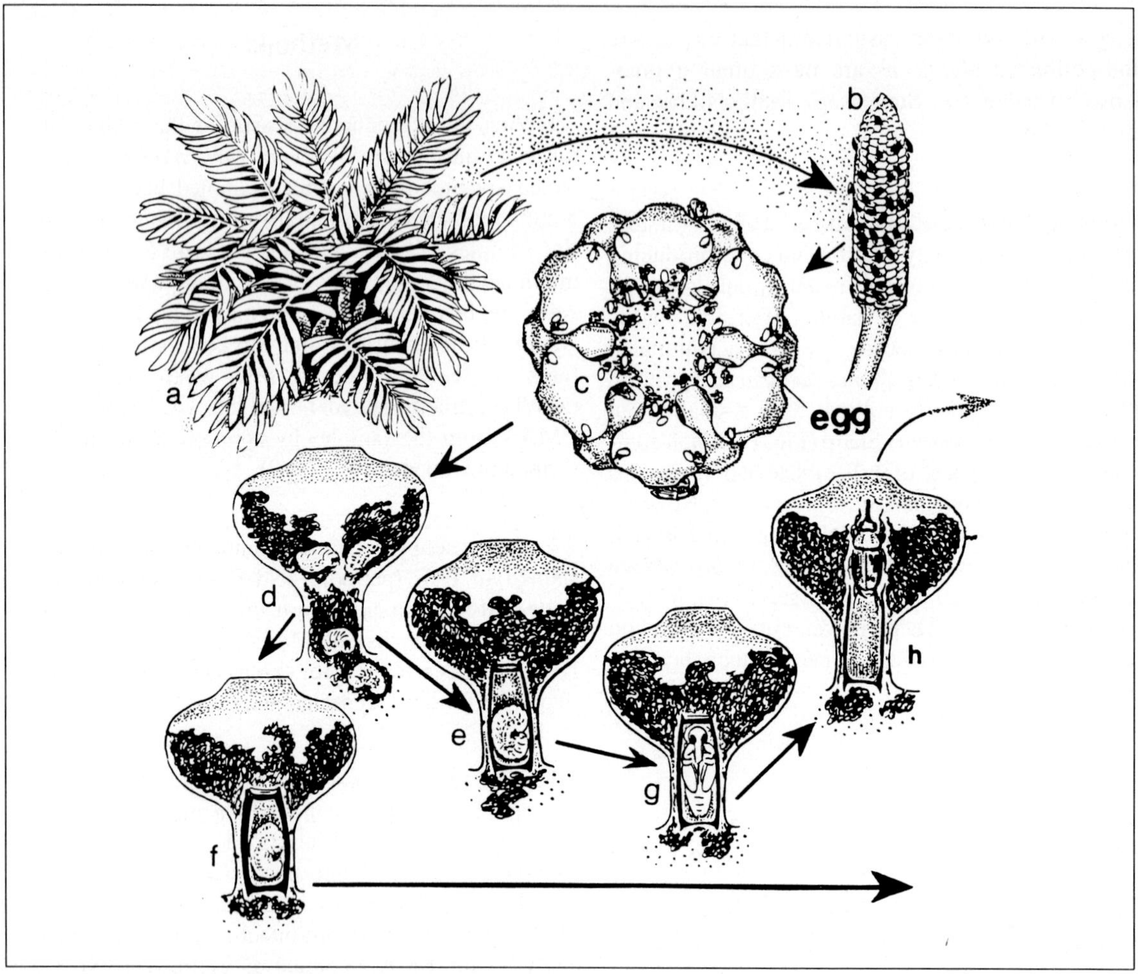

Figure 1. Breeding cycle of *Rhopalotria mollis*:

(a) male plant of *Z. furfuracea*;

(b) weevil swarm on male cone;

(c) interior of male cone showing weevil eggs (small arrows);

(d) larvae excavating parenchyma in microsporophyll, near the end of the season larvae avoid tissue tissue at the tips of the microsporophylls but earlier the microsporophyll is entirely excavated;

(e-h) pupation and metamorphosis of weevil in thin-walled pupal case;

(f) late-season dormant larva in thick-walled case.

Observations and Discussions

Rhopalotria mollis is an excellent flyer and during the early part of the reproductive season of *Z. furfuracea*, males and females are attracted to male cones where they feed upon sporophyll tissue by making small holes in the ends of the sporophylls through which they probe for food. The swarming weevils exhibit cone preferences, selecting only one or two cones from among as many as 15 - 20 at various stages of pollen maturation, and remaining upon the chosen cone 2 - 3 days before abandoning it (Fig. 1b). Although we are not sure where the specific weevils that start the annual breeding cycle on *R.mollis* come from, we suspect they emerge from long-dormant larvae that "winter" in debris at the base of the cycads they later infest. Their presence in such debris as well as in male cones may, in fact, account for their inadvertent introduction with specimens imported from Mexico.

Mating takes place at the time of swarming and feeding upon male cones approaching maturity. The female weevils lay eggs in feeding holes just excavated (Fig. 1c); three large eggs per female is the usual number. This apparently is a general mode of oviposition among female snout weevils (Comstock, 1936). As a result ends of cone scales (microsporophylls) often become rather hollowed out and as many as six eggs can be present in one microsporophyll (Fig. 1c, d). Eggs hatch within a day or so of oviposition, producing larvae which eat their way inwardly from the outer end of a microsporophyll into its stalk, and consume much of the parenchyma (Fig. 1d).

Weevils prefer Male Cones - Food Content of Male and Female Cones

Male-cone tissue is soft and rather mushy and stained intensely with I_2KI, indicating high starch content, but only lightly with Sudan IV indicating low lipid content (i.e. fats and/or oils). Microsporophyll sections prepared just before, during, and just after egg-laying activity, show parenchyma cells loaded with starch grains (Figs. 2, 4). The larvae feed almost exclusively on parenchyma; the gut is packed with material which stains lightly with I_2KI but intensely with PAS, indicating low starch content but high content of other carbohydrates (Fig. 6).

As is typical of weevils, the larvae are legless grubs and within the confines of the sporophylls are quite active. Up to six have been observed in one microsporophyll and they eat away the sporophyll tissue until they meet, then one eats the other (Fig. 1d, e). Subsequently each sporophyll will contain only one large larva the others being represented only by shrivelled bits of cuticle.

In the meantime, pollen attains maturity and weevils enter newly open male cones, sheltering and feeding, and become covered with pollen. It is important to note that we have never seen weevils feeding upon pollen nor found pollen in the guts of either adult or larval weevils. A single cone at this stage may contain more than 100 weevils, and it is at this stage that visitation to female cones takes place. However, female cones are not fed upon, although interior surfaces sometimes exhibit superficial wounds which may have been made by weevils. Megasporophyll sections stained less intensely with I_2KI than did the microsporophylls and sections show a low number of starch grains (Fig. 3). Post-pollination female cones contain idioblasts that are considerably larger than those in pre-pollination cones or in male cones.

Metamorphosis of Weevils

Within about 3 - 4 days after hatching, a larva constructs a chamber in the sporophyll stalk within which it pupates (Fig. 1e). The pupa nearly always is orientated with its head toward the outer end of the microsporophyll and its abdomen toward the base of the cone; female pupae often are found to contain up to three developing eggs. The pupal case is constructed from fecal matter, which, upon hardening, lines the hollowed-out sporophyll stalk, so that the larva and later pupa is housed within a tough, cylindrical capsule.

When metamorphosis is complete, the adult emerges, usually by boring out through the tip of the sporophyll, leaving a distinctive exit hole (Fig. 1h). Our observations suggest that the entire process is completed in 7 - 9 days. Thus it is possible that over

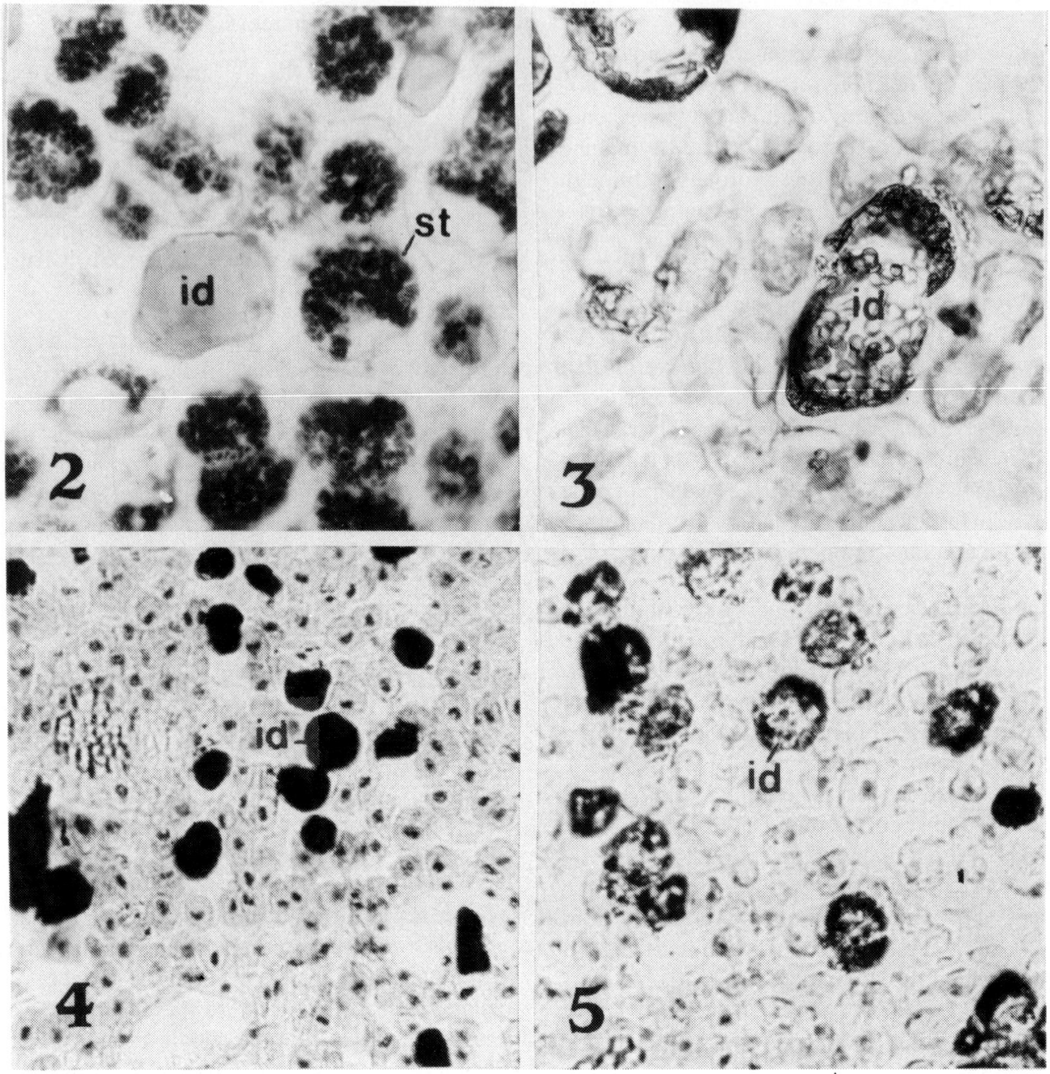

Figures 2 - 5: Idioblasts of *Z. furfuracea*.

Figure 2: Microsporophyll section stained with I_2KI for starch, cells have dark contents mainly composed of iodine-stained starch (st), idioblasts, unstained idioblasts (id), X400.

Figure 3: Megasporophyll section stained with I_2KI to show lack of starch in cells and modified idioblasts (id), X400.

Figure 4: Microsporophyll section stained with NIN, showing NIN+ idioblasts (id), X200.

Figure 5: Megasporophyll section stained with NIN showing relatively 'empty' idioblasts (id), a few NIN+ idioblasts are present (darkly stained), X200.

Table 1

α-Amino-β-(methylamino) propanoic acid (BMAA) content of tissues of *Zamia furfuracea* and of larvae and pupal cases of *Rhopalotria mollis*, a faithful pollinator of *Z.furfuracea*

Sample Name	Relative BMAA Content
Male cone, sample 1	++
Male cone, sample 2	+
Female cone	++
R.mollis, pooled larvae	++
R.mollis, pooled pupae cases	++++
Idioblasts of pre-pollinated Cones	
Male cone	+++
Male cone	+++
Female cone	++++

+ = *approximately 5 μg/g of sample*

the 6 week period, or so, of the annual coning cycle of the cycad, as many as six generations of weevils may have been produced from the initial mating.

Idioblasts contain Toxins

In both male and female-cone tissues there are scattered storage cells which we term idioblasts in accordance with Foster (1956) and Greguss (1968). Those of *Z. furfuracea* contain substance(s) that do not stain intensively with I_2KI, but do stain variously with Sudan IV, PAS and NIN indicating that they contain a complex of substances (Figs. 2 - 5). In male tissue they are amber-coloured before staining and do not stain with Sudan IV, I_2KI, and PAS, but they do stain intensely with NIN (Fig. 4). Because NIN stains BMAA a purple-brown colour (Vega and Bell, 1967), and produces a similar colour in male-cone idioblasts we think they may contain BMAA and this is bourne out by analyses of idioblasts extracted from male cones (Table 1). Those of female-cone tissue stain intensely with Sudan IV, and PAS, but variably with NIN (Figs. 3, 5). This may mean they contain some lipid and variable amounts of BMAA.

Functions of Idioblasts

A great variety of idioblasts is found in plants (Foster, 1956), most of them of unknown function. However, in the elder (*Sambucus*) release of contents of idioblasts is associated with protection from insect predation (Zobel, 1986) and it is probable that they have a similar function in cycads. Norstog and Fawcett (1989) speculate that whereas idioblasts are ingested by weevils, they are not digested but pass through the gut and are excreted (e.g. Fig. 6). Now, we find that pupal cases of *R. mollis*, which are constructed with dung excreted by the larvae, contain relatively high concentrations of BMAA (Table 1). Therefore, intact idioblasts in microsporophylls, by storing toxin may, in effect, detoxify parenchyma tissue for feeding by adult and larval weevils. On the other hand, tissues of post-pollination female cones, collected on 8 August and stained with I_2KI, or with NIN (Figs. 3, 5), show both low amounts of starch, and enlarged idioblasts, many of which are unstained with NIN. The latter may constitute evidence that contents (BMAA?) continue to diffuse from idioblasts into surrounding tissue even after pollination. Thus the toxin could continue to protect the maturing cone and seeds from predation by insects and/or other animals.

Pollination

Male and female cones of *Z. furfuracea* are of similar form, which may account for visitations to female cones by pollen-carrying adult weevils (Norstog, Stevenson and Niklas, 1986). Earlier exclusion experiments (Norstog et al., 1986) provided good evidence that *R. mollis* is an effective pollinator of *Z. furfuracea*, but did not clarify the role, if any, of airborne pollen. The results from exposure of greased slides, however, provided a clue to the role of air currents in pollen transport insofar as the conditions prevailing in our plots of *Z. furfuracea* are concerned. Four slides trapped an average of four pollen grains

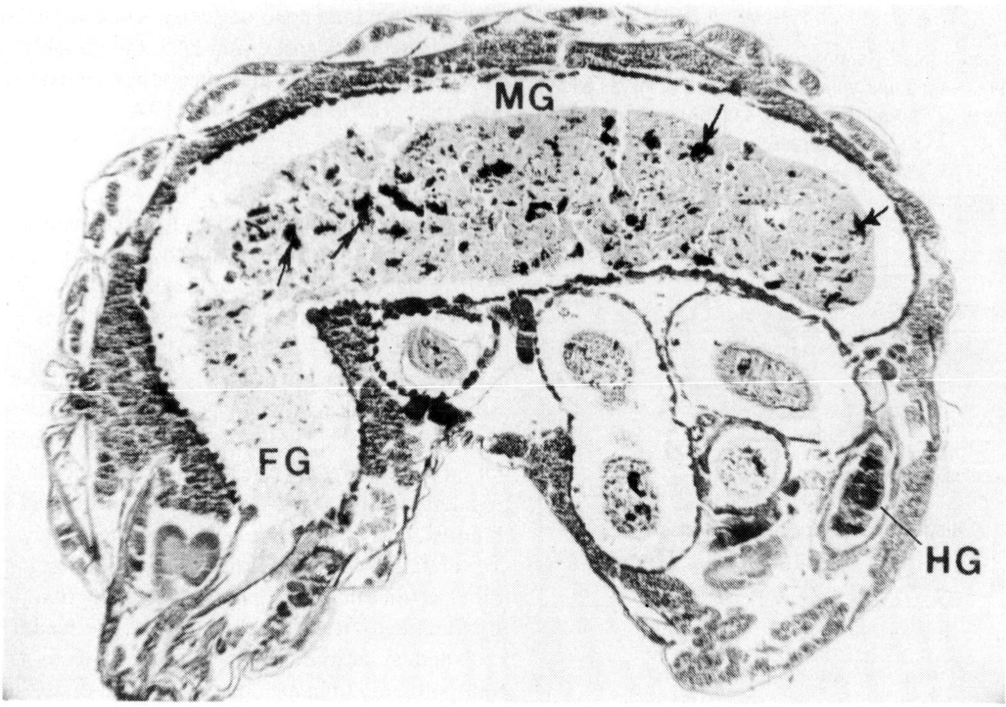

Figure 6: Larva of *Rhopalotria mollis* stained with NIN; mouth of larva is at lower left. Note few NIN+ particles in foregut (FG), progressively more in midgut (MG, arrows) and, especially, concentrated in hindgut (HG). This progressive concentration probably results from digestion of other ingested matter, leaving undigested idioblasts.

per cm - 3 - 5 cm being about the dimensions of the open area of a receptive female cone of *Z. furfuracea*. Given these conditions, in the highly unlikely circumstance that every pollen grain entering into a receptive female cone would impact upon a micropyle of an ovule, the fertility due to airborne pollen alone at best probably would not greatly exceed about 15%. Actually, the average fertility of four monitored cones in this planting on which slides were placed in 1988 was 82%, for slides 1 - 4, respectively, 93%, 93%, 67%, and 63%. There is no question, therefore, that in these examples fertility was due to insect pollination. Other experimental evidence showing weevils to be effective pollinators of *Z. furfuracea* are presented in Norstog et al. (1986).

Rewards in Pollination Biology

In pollination symbioses, a food reward is provided animal pollinators, often in the form of nectar and/or pollen, and its presence is signalled by odour or colour, or a combination of odour and colour (Faegri and van der Pijl, 1971; Kevan and Baker, 1984). since there is no evidence that either larvae or adults of *R. mollis* feed upon pollen of *Z. furfuracea*, we conclude that in this instance the reward, perhaps the sole reward, is starch. Because we have not observed weevils feeding within female cones we think their entry into female cones is a result of being lured by similarities of form, colour and, perhaps, odour between male and female cones. Possibly the presence of neurotoxin discourages feeding in female cones and its presence in the pupal cases protects the larvae against predation, as suggested by Norstog and Fawcett (1989), but evidence is lacking.

Dormancy of Weevils

Dormancy in insects is a mechanism whereby some kind of adversity is circumvented, whether of climate or of availability of food and/or oviposition sites (Burke, 1976; Denlinger, 1986). The dormant period, or diapause, may take place at any stage in the insect's life cycle, occurring at the larval stage in about 30% of examples. Its onset can be triggered by seasonal events such as drought, high rainfall, or changes in photo or thermoperiod. These factors often are less pronounced in the tropics but diapause occurs there about as frequently as in temperate zones.

Early in the breeding cycle of *R. mollis*, larvae have thin, transparent cuticles and form thin-walled pupal cases from which adults soon emerge; later on, about half the larvae become waxy and yellowish. These late-season larvae produce thick-walled pupal cases in which they remain dormant (Fig. 1f). In the first season of our observation of such dormant larvae, metamorphosis was not completed until April of the following year when adults began to emerge.

Rhopalotria mollis enters diapause toward the end of male-cone production. Within this brief time-frame there probably are few climatic clues, so diapause most likely is induced either by internal factors in the weevil itself or is induced by some change of form and function in the tissues of the cycad. Although factors responsible for the induction and breaking of diapause in *R. mollis* are conjectural, the fact that it occurs when it does is further evidence of the intimacy of this weevil/cycad symbiosis. Dormancy strategies in which an insect enters and remains in diapause throughout the non-reproductive phase of a specific host plant are well-documented (Denlinger, 1976).

We have not attempted to follow dispersal of adult weevils after they leave the host. They might feed on other plants, might even reproduce on other species, but we have no evidence of this. We think reproductive continuity of *R. mollis* from season to season hinges on year-to-year survival of larvae in diapause, probably in the remnants of pollen cones which are more-or-less persistent at the bases of male plants, rather than on survival of adults from year to year.

Evolution of Insect Pollination in *Zamia furfuracea*

The reproductive cycle of *R. mollis* appears almost uniquely adapted to exploitation of *Z. furfuracea* as a feeding and breeding site. Mating upon male cones alone, and only those about to mature, egg deposition in food-rich microsporophylls; excavation of a pupation site in the sporophyll stalk which precisely accommodates the larva, pupal and pre-emergent adult, all suggest an intimate and long-standing relationship involving the evolution of quite complex behavioural and structural adaptations on the part of the insect and the cycad.

Finally, we think the mode of female-cone opening is an adaptation for insect pollination which also tends to limit random visitation by other insects as well as entry of foreign pollen, mold spores, and the like. This feature also enhances protection of ovules from predation in a manner analogous to the closed carpels of flowering plants.

In summary, then, there appears to be a closely linked relationship between insect and cycad where toxin-free rewards such as shelter, food, and breeding sites are presented by male cones, but a paucity of food and diffusion of toxin in female cones ensure unpalatable grazing for weevils.

Acknowledgment

We dedicate this paper to Drs. Arthur Montgomery and John Popenoe for their long-time interest in and support of cycad studies at Fairchild Tropical Garden. Portions of this study were funded by a grant from CONACYT: Project P020CCOR-904133

References

***Burke, H.R.* 1976.** Bionomics of the anthomine weevils. Annu. Rev. Entomol. 21: 283 - 303.

***Comstock, J.H.* 1936.** An Introduction to Entomology, 8th ed. Comstock Publishing Co., Ithaca, N.Y.

***Denlinger, D.L.* 1986.** Dormancy in tropical insects. Annu. Rev. Entomol. 31: 239 - 264.

***Duncan, M.W., Kopin, I.J., Crowley, J.S., Jones, S.M., and Markey, S.P.* 1989.** Quantification of the putative neurotoxin α-amino-β-(methylamino) propanoic acid (BMAA) in Cycadales: analysis of the seeds of some members of the Family Cycadaceae. J. Analyt. Toxicol. 13: 169 - 176.

***Faegri, K., and van der Pijl, L.* 1971.** The Principles of Pollination Biology, 2nd ed. Pergamon Press, New York.

***Foster, A.S.* 1956.** Plant idioblasts: remarkable examples of cell specialisation. Protoplasma 46: 184 - 193.

***Greguss, P.* 1968.** Xylotomy of the Living Cycads. Akademiai Kiado, Budapest.

***Jensen, W.A.* 1962.** Botanical Histochemistry Principles and Practices. W.H. Freeman and Co., San Francisco.

***Kevan, P.G., and Baker, H.G.* 1984.** Insects on flowers. In Huffaker, C.B., and R.L. Rabb (eds.), Ecological Entomology, pp. 607 - 631. John Wiley and Sons, New York.

***Moretti, A., Sabato, S., and Siniscalco Gigliano, G.* 1983.** Taxonomic significance of methylazoxymethanol glycosides in the cycads. Phytochemistry 22: 115 - 117.

***Niklas, K.J., and Norstog, K.* 1984.** Aerodynamics and pollen grain depositional patters on cycad megastrobili: implications on the reproduction of three cycad genera (*Cycas, Dioon*, and *Zamia*). Bot. Gaz. 145: 205 - 222.

***Norstog, K.* 1987.** Cycads and the origin of insect pollination. Amer. Sci. 75: 270 - 279.

***Norstog, K., and Stevenson, D.W.* 1980.** Wind? Or insects? The pollination of cycads. Fairchild Trop. Gard. Bull. 35: 28 - 30.

***Norstog, K., Stevenson, D.W., and Niklas, K.J.* 1986.** The role of beetles in the pollination of *Zamia furfuracea*. L. fil. (Zamiaceae). Biotropica 18: 300 - 306.

***Norstog, K., and Fawcett, P.K.S.* 1989.** Insect-cycad symbiosis and its relation to the pollination of *Zamia furfuracea* (Zamiaceae) by *Rhopalotria mollis* (Curculionidae). Amer. J. Bot. 76: 1380 - 1349.

***Rothschild, M., Nash, R.J., and Bell, E.A.* 1986.** Cycasin in the endangered butterfly *Eumaeus atala florida*. Phytochemistry 25: 1853 - 1854.

***Spencer, P.S., Nunn, P.B., Hugon, J., Lundolph, A.C., Ross, S.M., Roy, D.N., and Robertson, R.C.* 1987.** Guam amyotrophic lateral sclerosis-Parkinsonism-dementia linked to a plant excitant neurotoxin. Science 237: 517 - 522.

***Vega, A.*, and *Bell, E.A.* 1967.** α-Amino-β-methylaminopropionic acid, a new amino acid from seeds of *Cycas circinalis*. Phytochemistry 6: 759 - 762.1.

***Zobel, A.M.* 1986.** Localisation of phenolic compounds in tannin-secreting cells of *Sambucus racemosa* L. shoots. Ann. Bot. 57: 801 - 810.

α-Amino-β-Methylaminopropionic Acid: A Neurotoxin From *Cycas* Species

Peter B. Nunn and Malarvizhi Ponnusamy
Department of Biochemistry
Division of Biomelecular Sciences
King's College London, Strand, London WC2R 2LS, UK.

Abstract

α-Amino-β-methylaminopropionic acid was first isolated from *Cycas circinalis* seed and was subsequently detected in all *Cycas* species that were examined. This amino acid is acutely neurotoxic and its presence in seed that is used for human consumption on Guam gave rise to speculation that the compound might be an aetiological agent in a chronic neurological disease that occurs on that island. Attempts to demonstrate chronic toxicity in rats were unsuccessful, but when macaques were fed the compound they developed a central-nervous-system motor dysfunction. This paper describes progress in exploring the mechanisms by which the neurotoxicity of this compound might be expressed.

Background

L-α-amino-β-methylaminopropionic acid (L-MeDAP, synonym β-methylamino-L-alanine, L-BMAA; Fig. 1) was first isolated from *Cycas circinalis* seed (Vega and Bell, 1967) during a purposeful search for neurotoxic amino acids in this species (see Bell and Nunn, 1988). All species of *Cycas* that were examined contained MeDAP, with an indication that a soluble, covalently-bound, form of the compound also existed (Polsky, Nunn and Bell, 1972; Dossaji and Bell, 1973). The amino acid was acutely neurotoxic to chicks, rats and mice (Bell, Vega and Nunn, 1967), an observation that stimulated speculation that it might be implicated in the aetiology of the amyotrophic lateral sclerosis-parkinsonism-dementia complex (ALS-PD), a chronic neurological disease that occurs on the island of Guam in the Western Pacific. Some aspects of this disorder resemble clinical features that are seen in patients with sporadic amyotrophic lateral sclerosis (motor neurone disease), Parkinson's disease or Alzheimer's dementia. The suggestion of Whiting (1963), that a high incidence of ALS-PD on the island might be related to the practice of native Chamorro people consuming flour derived from *C.circinalis* seed, implied that one or more components of the seed were aetiological factors.

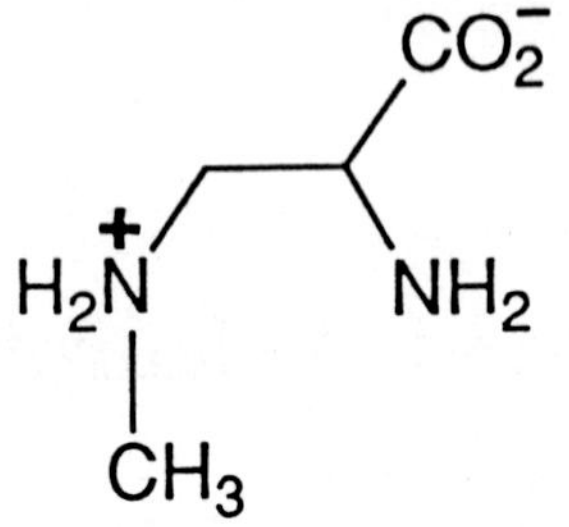

(I)

Figure 1: Structure of L-α-amino-β-methylaminopropionic acid (I, L-MeDAP; synonym β-methylamino-β-alanine, L-BMAA). Ionisation of the amino groups is complex (see text and Fig. 4 in which the ionisation of these groups is depicted). The structure shown is that which predominates at physiological pH values.

Attempts to induce chronic neurological signs by administering the amino acid to rats were not successful (Polsky et al., 1972), but subsequent experiments in which the compound was fed to macaques during a period of some weeks caused a motor dysfunction of the central nervous system, with some parkinsonian features (Spencer et al., 1987). Damage to upper (mainly) and to lower motor neurones was observed histologically. The compound was also neuronotoxic to cultured foetal mouse spinal-cord (Nunn et al., 1987). The acute toxicity of MeDAP to rats was associated subsequently with damage to specific cerebellar cell types (Seawright et al., 1990). In macaques no cerebellar pathology was observed (P.S. Spencer, personal communication).

At this juncture it is impossible to evaluate the extent to which MeDAP may be involved in the aetiology of Guamanian ALS-PD and current debate concerns whether sufficient of the compound is present in flour derived from *C. circinalis* seed to justify serious consideration of the substance as a food-borne neurotoxin (see Duncan et al., 1990). However, MeDAP is one of a number of environmental compounds that precipitate nerve-cell death and a knowledge of its mechanisms of action is relevant to the aetiology of chronic, sporadic, neurodegenerative diseases generally. A discussion of the mechanisms of action of the compound that have been discovered thus far is the subject of this paper.

Receptor-Mediated Mechanisms

The 'excitoxic theory of cell death' proposed by Olney (1978) considers that neurotoxicity elicited by acidic amino acids follows over-excitation of receptors on nerve-cell bodies and dendrites that is accompanied by sustained membrane depolarisation and the influx of Na^+, $C1^-$ and water into the cell. Pathological sequelae may result from a reduction in cellular [ATP], osmotic disruption, or both. An elaboration of this scheme proposes that the entry of Ca^{2+} into neurones is required to cause the irreversible damage that results in nerve-cell death (Choi, 1985).

Glutamate is recognised as the main excitatory amino acid in the central nervous system (see Watkins, Krogsgaard and Honore, 1990) and this activity appears to be mediated by three major sub-types of receptors, which have been named for the compounds that activate them most specifically (see Moynahan, Bridges and Cotman, 1989). The N-methyl-D-aspartate (NMDA) receptor regulates an ion channel that conducts Na^+, $C1^-$ and Ca^{2+}; the α-amino-β-hydroxy-5-methyl-4-isoxazolepropionate (AMPA) and kainate (K) receptors regulate ion channels that conduct only Na^+, although they probably also increase cellular Ca^{2+} content by indirect means. Activation of NMDA receptors may be distinguished from activation of non-NMDA receptors (AMPA and K) by the sensitivity of the former to antagonism by α-amino-7-phosphonoheptanoate (AP7, a competitive antagonist) and MK-801 (a non-competitive antagonist).

Like the non-protein amino acid from *Lathyrus sativus*, β-N-oxalyl-L-α,β-diaminopropionic acid (β-ODAP; synonym β-oxalylamino-L-alanine, BOAA), MeDAP causes typical excitotoxic damage to neurones in the mouse spinal-cord/dorsal root ganglion

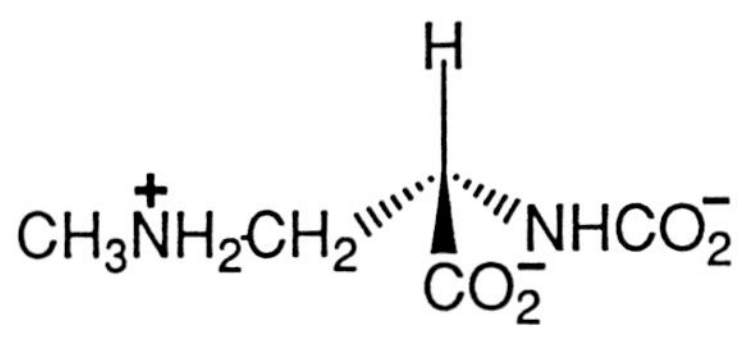

(II) (III)

Figure 2: The structure of α-carbamate of L-α-amino-β-methylaminopropionic acid (II) compared with that of N-methyl-D-aspartate (NMDA, III). The structures are depicted as they would exist at physiological values of pH. Compared with the structure given in fig. 1 the α-carbamate of L-MeDAP is depicted with the α-hydrogen atom pointing directly upward. The structure of NMDA has been presented similarly. The formation of the α-carbamate of L-MeDAP causes an apparent inversion of configuration about the α-carbon atom.

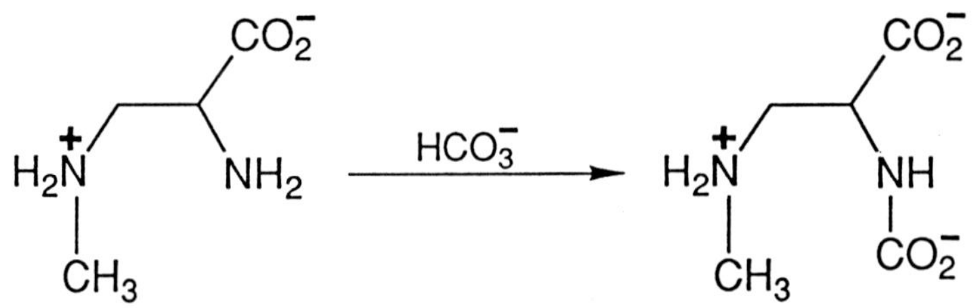

(I) (II)

Figure 3: Reaction of the α-amino group of L-MeDAP (I) to form the α-carbamate (II).

preparation. The effect was stereo-specific for both compounds, the naturally-occurring L-forms only being effective. However, the preparation responded differently to the two amino acids in that L-MeDAP acted more slowly and at higher concentrations than did L-β-ODAP (Nunn et al., 1987).

In mouse organotypic cerebral-cortex explants post-synaptic vacuolation was accompanied by the appearance of unidentified, dark, shrunken, cells and both responses were concentration-dependent (Ross, Seelig and Spencer, 1987). Post-synaptic vacuolation was antagonised by AP7 and by MK-801, suggesting that this feature was dependent upon activating NMDA receptors.

Although L-MeDAP appears to activate NMDA receptors in these experiments it lacks the monoaminodicarboxylate structure that characterises agonists to that receptor (see Watkins et al., 1990). The discovery that this activity was dependent upon the presence of bicarbonate (in excess of 6mM) offered an explanation of this observation (Weiss and Choi, 1988). Neuronotoxicity induced by NMDA, β-ODAP and quisqualate (which acts in part via AMPA receptors; see Monaghan et al., 1989) was not bicarbonate-dependent under the same experimental conditions.

The most likely explanation of this result is that bicarbonate reacts with the α-amino group of L-MeDAP to form a carbamate that is a structural ana-

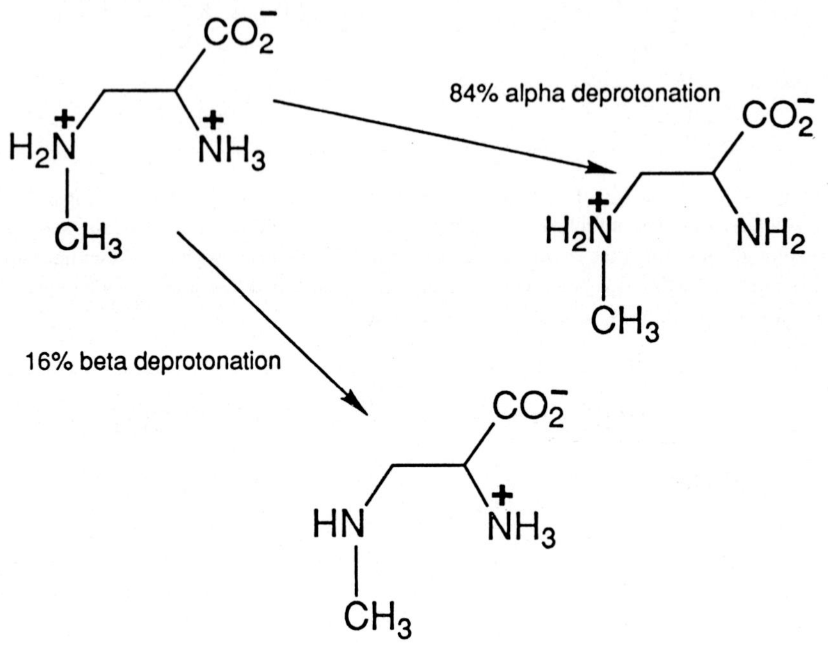

Figure 4: Complex ionisation of L-MeDAP. The predominance of the uncharged α-amino group at any value of pH favours the reaction of that group with bicarbonate.

logue of NMDA (Nunn and O'Brien, 1989; Fig. 2). This structure results from an apparent inversion of configuration that occurs on α-carbamate formation. The reaction of bicarbonate with the α-amino group of MeDAP (Fig. 3) is a consequence of the complex ionisation of the amino acid, which displays differential ionisation of its α-amino and β-methylamino groups (Fig. 4). The reaction does not proceed to completion; only about 26% of the amino acid will be in the α-carbamate form at physiological H^+ and bicarbonate concentrations. Thus, 3mM L-MeDAP would yield approximately 0.78mM α-carbamate (Nunn and O'Brien, 1989).

This explanation is convincing for acute experiments, but it is clear that L-MeDAP reacts with sodium bicarbonate in a complex fashion (Nunn and O'Brien, 1989) and that mixtures of the two in D_2O exhibit elaborate ^{1}H-nuclear magnetic resonance spectra when the reaction is permitted to occur for prolonged periods (O'Brien and Nunn, unpublished data). Since the α-carbamate is chemically unstable it is difficult to explore its pharmacological specificity and the apparent structural resemblance of the adduct to NMDA may not exclude it acting also as an agonist of non-NMDA receptors. It may be significant that, in longer-term (24h) experiments, the compound kills a group of neurones, characterised by their containing the enzyme NADPH-diaphorase, by a mechanism that appears to be mediated by non-NMDA receptors (Weiss, Koh and Choi, 1989). This effect was seen at lower concentrations than when the same tissue preparation was exposed to L-MeDAP for only 1h. Thus, NMDA receptor-mediated reactions may be more important under some experimental conditions, whereas activity against AMPA and K receptors may be involved in the expression of neurotoxicity of MeDAP under other conditions that may better reflect chronic neurotoxicity.

Chelation-Mediated Mechanisms and Metabolism

MeDAP is one of the most effective amino acid chelators of Zn^{2+} and Cu^{2+} (Nunn et al., 1989). The involvement of Zn^{2+} in deactivating NMDA receptors (Peters, Koh and Choi, 1987) suggests that compounds that chelate Zn^{2+} more effectively than does the receptor protein itself would remove the ion and indirectly activate the receptor. Whether MeDAP acts in this fashion has yet to be evaluated experimentally. However, Zn^{2+} has a number of other functions in the central nervous system (see Koh and Choi, 1988; Nunn et al., 1989), at least one of which is tissue-specific, and it is conceivable that MeDAP might modify the speciation of Zn^{2+} in such a way as to perturb the normal metabolism of these ions. It appears more likely that changes in zinc speciation brought about by the presence of MeDAP in the central nervous system would be a vehicle for the expression of chronic rather than acute toxicity. However, it is known that Cu^{2+} is abundant in the cerebellum, where its function is unknown (see Nunn et al., 1989). Thus, it is possible that some of the acute effects of MeDAP toxicity in rats are mediated by the chelation of Cu^{2+}, as well as by activating excitatory amino acid receptors. Chelation of Cu^{2+} might also contribute towards the chronic neurotoxicity of the compound.

Evidence for the metabolism of MeDAP is meagre, but is currently being studied in our laboratory. The compound is known to be α-acetylated in the rat, presumably via a reaction requiring acetyl-CoA (Reece and Nunn, 1989).

Acknowledgments

PBN thanks the Motor Neurone Disease Association (UK), the Wellcome Trust and Cycad '90 for financial support.

References

***Bell, E.A., Vega, A. and Nunn, P.B.* 1967.** A neurotoxic amino acid in seed of *Cycas circinalis. In* Toxicity of Cycads: Implications for Neurodegenerative Diseases and Cancer. Fifth Conference. pp. XI1 - XI7, ed. by M.G. Whiting, Third World Medical Research Foundation, New York.

***Bell, E.A. and Nunn, P.B.* 1988.** Neurological diseases in man: Are plants to blame? Biologist 35: 39 - 43.

***Choi, D.W.* 1985.** Glutamate toxicity in cortical cell culture is calcium dependent. Neurosci. Lett. 58: 293 - 297.

***Dossaji, S.F. and Bell, E.A.* 1973.** Distribution of α-Amino-β-(methylamino)-propanoic acid (BMAA) in cycad flour: an unlikely cause of amyotrophic lateral sclerosis and parkinsonism-dementia of Guam. Neurology 40: 767 - 772.

***Koh, J-y. and Choi, D.W.* 1988.** Zinc alters excitatory amino acid neurotoxicity in cortical neurones. J. Neurosci. 8: 2164 - 2171.

***Monaghan, D.T., Bridges, R.J. and Cotman, C.W.* 1989.** The excitatory amino acid receptors: their classes, pharmacology and distinct properties in the function of the central nervous system. Annu. Rev. Pharmocol. Toxicol. 29: 365 - 402.

***Nunn, P.G., Seelig, M., Zagoren, J.C. and Spencer, P.S.* 1987.** Stereospecific acute neuronotoxicity of 'uncommon' plant amino acids linked to human motor-system diseases. Brain Research 410: 375 - 379.

Nunn, P.G. and O'Brien, P. 1989. The interaction of L-β-methylaminoalanine with bicarbonate: an 1H n.m.r. study. FEBS Letters 251: 31 - 35.

Nunn, P.G., O'Brien, P., Pettit, L.D. and Pyburn 1989. Complexes of zinc, copper and nickel with the non-protein amino acid α-amino-β-methylamino-propionic acid: a naturally-occurring neurotoxin. J. Inorg. Biochem. 37: 175 - 183.

Olney, J.W. 1978. Neurotoxicity of excitatory amino acids. *In Kainic Acid as a Tool in Neurobiology*, pp. 95 - 121 ed. by E.G. McGeer, J.W. Olney and P.L. McGeer, Raven Press, New York.

Peters, S., Koh J-y and Choi, D.W. 1987. Zinc selectively blocks the action of N-methyl-D-aspartate on cortical neurones. Science 236: 589 - 593.

Polsky, F.I., Nunn, P.B. and Bell, E.A. 1972. Distribution and toxicity of α-amino-β-methylamino-propionic acid. Fed. Proc. 31: 1473 - 1475.

Reece, D.M. and Nunn, P.B. 1989. Synthesis of ^{14}C-labelled L-α-amino-β-methylaminopropionic acid and its metabolism in the rat. Biochem. Soc. Trans. 17: 203 - 204.

Ross, S.M., Seelig, M. and Spencer, P.S. 1987. Specific antagonism of excitotoxic action of 'uncommon' amino acids assayed in organotypic mouse cortical cultures. Brain Research 245: 120 - 127.

Seawright, A.A., Brown. A.W., Nolan, C.C. and Cavanagh, J.B. 1990. Selective degeneration of cerebellar cortical neurones caused by cycad neurotoxin, L-β-methylaminoalanine (L-BMAA), in rats. Neuropath. Appl. Neurobiol. 16: 153 - 169.

Spencer, P.S., Nunn, P.B., Hugon, J., Ludolph, A.C., Ross, S.M., Roy, D.N., and Robertson, R.C. 1987. Linkage of Guam amyotrophic lateral sclerosis-parkinsonism-dementia to a plant excitant neurotoxic. Science 237: 517 - 522.

Vega, A. and Bell, E.A. 1967. α-amino-β-methylaminopropionic acid, a new amino acid from seeds of *Cycas circinalis*. Phytochemistry 6: 759 - 762.

Vega, A., Bell, E.A. and Nunn, P.B. 1968. The preparation of L- and D-α-amino-β-methylamino-propionic acids and the identification of the compound isolated from *Cycas circinalis* as the L-isomer. Phytochemistry 7: 1885 - 1887.

Watkins, J.C., Krogsgaard, P. and Honore, T. 1990. Structure-activity relationships in the development of excitatory amino acid receptor agonists and competitive antagonists. Trends Pharmacolog. Sci. 11: 25 - 33.

Weiss, J.H. and Choi, D.W. 1988. β-N-methylamino-L-alanine neurotoxicity: requirement for bicarbonate as a cofactor. Science 241: 973 - 975.

Weiss, J.H., Koh, J-y. and Choi, D.W. 1989. Neurotoxicity of β-N-methyl-L-alanine (BMAA) and β-N-oxalylamino-L-alanine on cultured cortical neurones. Brain Research 497: 64 - 71.

Whiting, M.G. 1986. Toxicity of Cycads. Econ. Bot. 17: 271 - 302.

Potential Errors in the Determination of Methylazoxymethanol Glycosides in Plant Material

R. Osborne & J.J. Nair
Department of Chemistry, University of Natal
King George V Ave., Durban, 4001, South Africa

Abstract

The published results on the quantitative determination of methylazoxymethanol (MAM) glycosides in cycad tissue are generally based on one of three methodologies. One method is a spectrophotometric assay via formation of a coloured complex between liberated formaldehyde and chromotropic acid (CTA). The other methods are chromatographic; either by direct HPLC or by GLC or derivitised samples. Our results have shown that both the CTA and the HPLC methods are subject to significant interferences which lead to grossly high results for MAM-glycoside concentrations in cycad tissue. Analyses which report the presence of 'cycad toxins' in non-cycad plants, e.g. *Podocarpus* spp., are thus considered invalid.

Methods

Extraction of Plant Material

Approximately 0.5g of fresh plant material was macerated in 50% ethanol and the volume made up to 25ml and the resulting suspension allowed to sediment overnight to give the primary extract.

CTA Method (*Matsumoto & Strong, 1963*)

A 1.0ml aliquot of the primary extract was made up to 10.0ml with 50% ethanol to provide the secondary extract. An amount of 0.3ml of the secondary extract was withdrawn and 3.0ml CTA reagent added to it. The resulting solution was heated for 30 min in boiling water bath, cooled and the absorbance read at 570nm. A blank solution consisted of 0.3ml secondary extract with 3.0ml 12.5M H_2SO_4. A calibration curve was constructed using known amounts of formaldehyde.

HPLC Method (*Yagi et al., 1983, 1987*)

A 1.0ml aliquot of the primary extract was taken, to which 1.0ml 0.02% 1-naphthol was added and the volume made to 10.0ml with methanol. The resulting solution was filtered through a 0.45μm nylon-mesh unit. A sample of 10μl was injected onto a analytical C-18 reverse phase HPLC column and eluted with He-degassed 80% methanol. Peaks were detected by absorbance at 215μm and calibrated against known amounts of cycasin and macrozamin.

Results

The apparent results, using the two methods outlined above, for a selection of different plant materials from different taxa, are given in Table 1.

Discussion

Examination of the data in Table 1 - and the full set of results now shown here - reveals that (A) the toxin concentrations are generally much higher than those reported by other workers (De Luca et al., 1980; Moretti et al., 1981, 1983; Palekar and Dastur, 1965; Yagi et al., 1983, 1987), (b) there is a very poor correlation between the toxin concentrations determined by the two methods and (c) the presence of 'cycad toxins' in non-cycad material, e.g. *Podocarpus* samples, indicates that the results are to be regarded sceptically. Further analyses, not shown here, showed cycad toxins, as assessed by these two methods, are 'apparently present' in non-gymnospermous plant material.

Our conclusion is that these simple methods are unsatisfactory and that at least two sources of major interference are present. We speculate that (a) the plant material contains pentosans which generate furfural when heated in strongly acidic conditions and that the furfural is detected as formaldehyde by the CTA method, and that (b) the plant material contains other mono and di-glycosides which co-elute in simple HPLC column systems with cycasin and macrozamin and further that these co-eluted glycosides have an aglycone moeity which is also chromophoric in the UV range.

We thus express reservations about all published cycad toxin levels derived by the above CTA and HPLC methods. A more satisfactory assay is the GLC method, involving TMS-derivitisation, as used by Wells et al. (1968) and subsequently by De Luca et al. (1980) and Moretti et al. (1981).

References

***Dastur, D.K., and Palekar, R.S.* 1966.** Effect of boiling and storing on cycasin content of *Cycas circinalis* L. Nature No. 5039, pp. 841 - 843.

***De Luca, P., Moretti, A., Sabato, S., and G. Siniscalco Gigliano.* 1980.** The ubiquity of cycasin in cycads. Phytochemistry 19: 2230 - 2231.

***Matsumoto, H., and Strong, F.M.* 1963.** The occurrence of methylazoxymethanol in *Cycas circinalis* L. Arch. Biochem. Biophys. 101: 299 - 310.

***Moretti, A., Sabato, S., and G. Siniscalco Gigliano.* 1981.** Distribution of macrozamin in Australian cycads. Phytochemistry 20: 1415 - 1416.

***Moretti, A., Sabato, S., and G. Siniscalco Gigliano.* 1983.** Taxonomic significance of methylazoxymethanol glycosides in the cycads. Phytochemistry 22: 115 - 117.

***Palekar, R.S., and Dastur, D.K.* 1965.** Cycasin content of *Cycas circinalis*. Nature No. 4491, pp. 1363 - 1365.

***Wells, W.W., Yang, M.G., Bolzer, W., and Mickelsen, O.* 1968.** Gas-liquid chromatographic analysis of cycasin in cycad flour. Analyt. Biochem. 25: 325 - 329.

***Yagi, F.,* and *Tadera, K.* 1987.** Azoxyglycoside contents in seeds of several cycad species and various parts of Japanese cycad. Agric. Biol. Chem. 51: 1719 - 1721.

***Yagi, F., Tadera, K. and Kobayashi, A.* 1983.** Simultaneous determination of azoxyglycosides by high performance liquid chromatography and their contents in growing leaves of Japanese cycad. Agric. Biol. Chem. 47: 137 - 139.

Table 1

Apparent 'Cycad Toxin' Concentrations from Various Plant Sources

(All figures quoted are as percentage fresh weight the CTA results are expressed as cycasin)

(C. = Cycas, E. = Encephalartos, S. = Stangeria, P. = Podocarpus)

Material	Species	HPLC Method Results			CTA Results
		Cycasin	Macrozamin	Total	Total
Kernel	*C.thouarsii*	0.62	1.19	1.81	2.66
	C.revoluta	0.94	0.85	1.79	0.99
	E.inopinus	0.81	1.40	2.21	0.20
	E.natalensis	0.49	0.73	1.22	0.60
	S.eriopus	1.22	2.80	4.02	5.41
	P.henkelii	0.73	0.42	1.15	0.61
Sarcotesta	*C.thouarsii*	1.39	1.21	2.60	0.96
	C.revoluta	1.04	1.46	2.50	0.23
	E.inopinus	0.92	1.08	2.00	0.50
	E.natalensis	0.71	1.03	1.74	0.72
	S.eriopus	0.77	0.90	1.67	2.82
	P.henkelii	1.64	3.82	5.46	0.31
Female Cone Scale	*C.revoluta*	0.81	1.40	2.21	0.20
	E.inopinus	0.63	1.19	1.81	0.19
	E.natalensis	0.61	0.60	1.21	0.56
	S.eriopus	0.45	0.88	1.33	1.21
Male Cone Scale	*C.thouarsii*	1.25	0.45	1.70	0.45
	E.natalensis	0.35	0.68	1.03	0.38
	E.villosus	0.55	3.66	4.21	0.47
	S.eriopus	1.51	1.31	2.82	0.96
Pollen	*C.thouarsii*	0.73	1.01	1.74	1.28
	E.lebomboensis	1.88	1.06	2.94	0.65
	E.natalensis	0.52	0.78	1.30	0.56
	E.transvenosus	0.51	0.81	1.32	0.94
	E.villosus	0.17	0.20	0.37	0.37
	Z.furfuracea	2.31	2.15	4.46	0.83

Note: The following comparative figures are abstracted from the literature:

C.thouarsii cycasin 0.06, macrozamin 0.31 by GLC of TMS derivatives (Moretti et al. 1983).
C.revoluta cycasin 0.28, macrozamin 0.26 by GLC of TMS derivatives (Moretti et al. 1983).
C.revoluta cycasin 0.90-1.30 by HPLC on Shodex resin (Yagi and Tadera, 1987).
S.eriopus cycasin 0.02, macrozamin 4.70 by GLC of TMS derivatives (Moretti et al. 1983).

Material	Species	HPLC Method Results			CTA Results
		Cycasin	Macrozamin	Total	Total
Leaf	*C. thouarsii*	5.67	5.52	11.20	0.88
	C. revoluta	1.04	0.81	1.85	0.66
	E. nopinus	0.07	0.17	0.24	0.41
	E. natalensis	3.05	4.70	7.75	0.29
	E. transvenosus	2.69	2.67	5.36	0.31
	E. villosus	2.77	0.70	3.47	0.55
	S. eriopus	2.25	3.52	5.77	1.24
	Z. furfuracea	1.42	2.35	3.77	0.58
	P. henkelii	3.55	6.08	13.40	0.62
Coralloid Roots	*C. thouarsii*	0.17	0.47	0.64	0.65
	E. natalensis	0.76	0.50	1.26	0.76
	E. villosus	0.21	0.47	0.68	0.83
Insects	*Antliarhinus zamiae* (beetle on *E. natalensis*)	1.80	1.83	3.63	0.86
	Zeronopsis leopardina (larvae on *C.thouarsii*)	1.51	1.33	2.84	0.31
	(larvae on *E. villosus*)	2.62	1.03	3.65	0.30
	(adult moth)	4.24	5.62	9.86	2.04

Cycad Toxicity in Domestic Animals - What Agent is Responsible?

A. Seawright
Department of Veterinary Pathology
University of Queensland, Brisbane, Australia

A.W. Brown, C.C. Nolan and J.B. Cavanagh
MRC Toxicology Unit
MRC Laboratories, Carshalton, Surrey, England

Abstract

Cycad intoxication of domestic animals, mainly cattle, most commonly occurs in Australia although poisoning is known to occur in the West Indies, New Guinea and Japan. There are four cycad genera in Australia and each has been implicated in livestock toxicity. Two syndromes are recognised, namely a posterior ataxia with hindquarter weakness known as 'zamia staggers' due to consumption of leaves and an acute liver injury with gastrointestinal haemorrhage due mainly to eating the seeds. Methylazoxymethanol glycosides, mainly cycasin and macrozamin are believed to be the cause of the liver damage when it occurs but the cause of the nervous condition is unknown. The neurotoxic non-protein aminoacid, L-BMAA is found in *Cycas* spp. at up to about 130 mg/100g of leaf or seeds but it is not thought to be sufficiently toxic to cause the condition. In any case, L-BMAA when dosed intraperitoneally to young rats causes a selective degeneration of various GABAergic cerebellar molecular layer neurones. This is not the neurological lesion seen in cattle affected by eating cycad plants, and the compound has not been found for certain in other cycad genera which cause the disorder. The agent causing 'zamia staggers' is therefore unknown but it may be derived through decomposition or metabolism *in vivo* of the MAM glycosides.

Introduction

Although intoxication of grazing and foraging animals following ingestion of cycad plants has been reported in other tropical and sub-tropical to temperate areas of the earth (Anderson and Hall, 1964; Mason and Whiting, 1968; Kobayashi et al., 1984), poisonings have been by far of the greatest significance in Australia. In Australia 4 of the 11 genera of cycads are represented, namely *Cycas, Macrozamia, Lepi-*

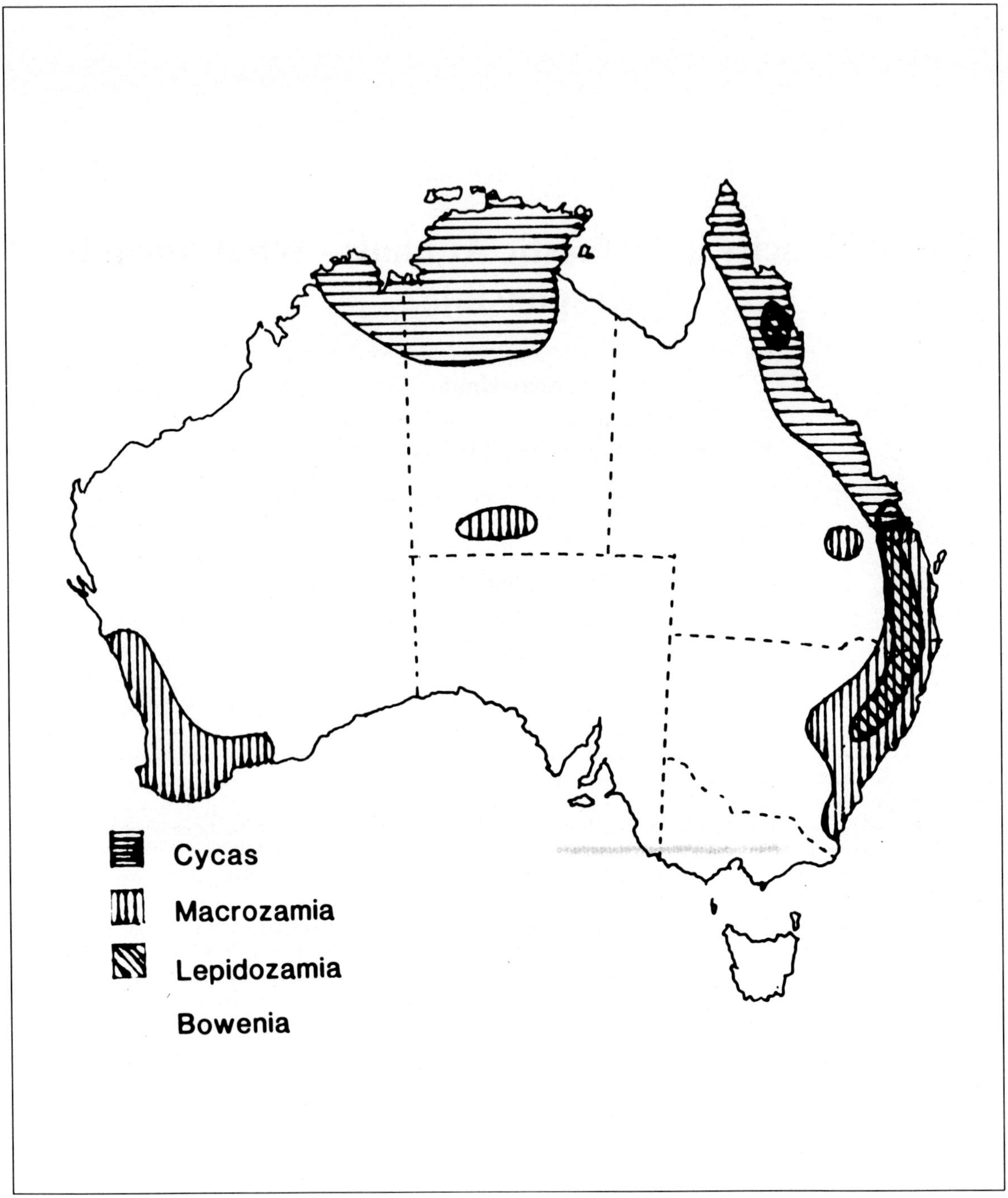

Figure 1: The distribution of the cycad genera in Australia. Toxicity in livestock is likely to occur in Australian wherever cycads grow.

dozamia and *Bowenia*. Their approximate distribution is shown in Figure 1 (Johnson, 1959). Wherever cycad species are established in Australia, intoxication of livestock has usually at some time been reported (Everist, 1981).

Field Experience in Australia

Although it is almost certainly true that cycad intoxication of grazing livestock has occurred from the time the animals were first introduced to Australia, it was not until the 1890's that reference was made to suspect cycad-related diseases in official animal heath reports. A brief summary of these early observations is given by Hurst, (1942). Two distinct syndrome types were recognised, namely a chronic condition characterised by hindquarter ataxia known as 'wobbles' or 'zamia staggers' in cattle due to consumption of the leaves, and a more acute condition in which there was liver damage, jaundice and alimentary tract lesions due mainly to consumption of the seeds. The ataxia was considered to be permanent although in animals that were cared for, partial recovery occurred and wholesome meat and milk could subsequently be derived from them (Hall, 1957). Sheep and pigs usually became intoxicated through consumption of the seeds and frequently succumbed to liver damage and jaundice. Following consumption of cycad seeds, pigs characteristically vomit (Seddon et al., 1931). It has been stated that sheep may develop a posterior ataxia similar to that seen in cattle following consumption of leaves (Gabbedy et al., 1975) but this is not normally reported as a feature of naturally occurring cycad poisoning in this species. Cattle with zamia staggers are occasionally found with some previous evidence of liver injury (Figs. 2, 3). In a single report of cycad intoxication in dogs caused by ingestion of *Zamia floridiana* seeds, the condition was characterised by vomiting, liver damage and a generalised haemorrhagic syndrome (Senior et al., 1985).

The most serious loss of livestock due to cycad poisoning so far recorded was the death of 2,200 of 6,000 sheep put to graze in *Philliga* scrub near Coonabarabran in New South Wales in May, 1929. The species considered responsible was *Macrozamia heteromera* and this was subsequently confirmed by a series of feeding trials in cattle, sheep and pigs using seeds collected from the *Philliga* scrub at the time of the outbreak (Seddon et al., 1931). Since that time numerous other *Macrozamia* spp. have been implicated in outbreaks of either or both the forms of cycad toxicity from central Queensland to the south western of Western Australia. In northern Australia *Cycas* spp. are the main cause of cycad intoxication of livestock (Hall, 1964; Hooper et al., 1974).

Experimental Intoxication

In many of the earlier experiments with feeding cycad foliage to cattle carried out over 90 years ago in Queensland and New South Wales (Hurst, 1942) clinical signs of toxicity were reported to occur after only a few days of feeding and it is not clear from many of these accounts as to whether the condition was the ataxia of 'zamia staggers' or the results of generalised weakness due to severe liver and gastrointestinal disease.

In a series of more recent studies, Hall (1957, 1964) confirmed the toxicity for cattle of *M. pauli-guilielmi*, *M. miquellii* (Stanley and Ross, 1989) and *M. lucida* which grow in South Queensland, and *Cycas media* and *Bowenia serrulata* which grow in northern and/or central Queensland. Again two distinct syndromes were produced. One was dominated by severe liver damage which was apparent after consumption of only 3 to 4 lb (1,359 to 1,702g) of fresh young leaves of *M. pauli-guilielmi* given over 6 days with death occurring some 3 weeks or so later without the appearances of ataxia. It was noted that if the leaves were allowed to dry, then consumption of the leaf at similar dose rates for up to about 3 weeks was without untoward effect. When fresh-green more-mature leaf was fed at the same dose rate, no effects were seen up to 10 days. However, when 154lb of the latter plant was fed over 100 days ataxia developed but there was no sign of liver damage. Feeding smaller amounts of this plant for the same period was without effect (Hall, 1957).

In the experiments with *M. lucida* (Hall, 1957) fresh young leaves fed at up to 1-oz (480g) per day for 34 days resulted in typical zamia staggers. When the

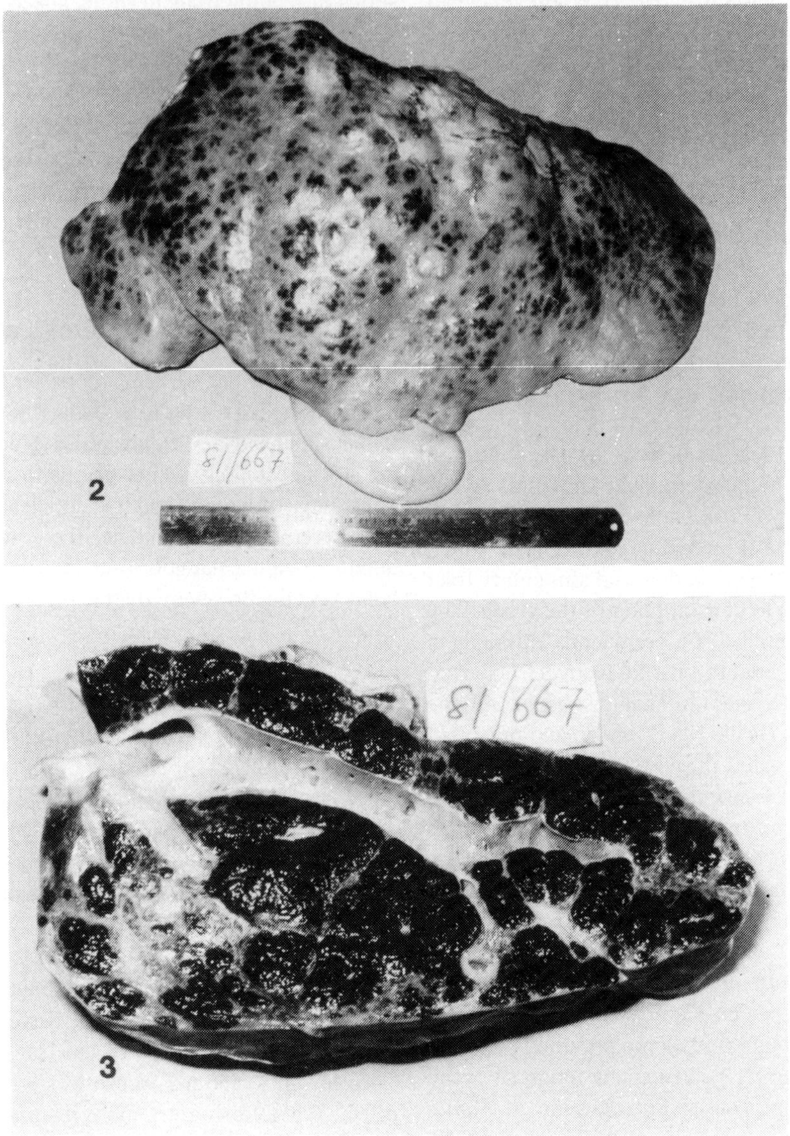

Figures 2,3: *Steer Liver Necrosis*

Figure 2: Extensive post necrotic scarring of the liver of a steer affected with zamia staggers. Liver injury due to ingestion of the foliage of the plants in the field is not common in animals that later develop ataxia.

Figure 3: A cut section of the liver shown in Figure 2. There is good regeneration of the parenchyma and there was no indication of liver failure.

leaves were allowed to dry no effects were observed up to 103 days in one beast while in another zamia staggers occurred in 87 days. Liver lesions were not observed in the trials with this plant. At this stage no lesions in nervous tissue had been described. In a further feeding trial using *M. lucida* a mean level of 8oz (240g) or less of dry leaf was fed for 53 days with signs of ataxia appearing at 46 days. Again no liver damage occurred (Hall and McGavin, 1968).

In an associated feeding trial in which 6oz (about 200g) of *B. serrulata* leaf were fed daily for 20 days, ataxia was present on day 18 and was well established by day 21 (Hall and McGavin, 1968). No liver lesions occurred. In this study there was well defined bilaterally symmetrical degeneration of nerve fibres in the fasiculis gracilis of the dorsal columns especially anterior to T7 and of the fibres of the dorsal spinocerebellar tracts extending into the posterior cerebellar peduncles and into the cerebellum. Bilaterally symmetrical degeneration of nerve fibres was demonstrated in all funiculi throughout the length of the spinal cord by the Marchi method (Fig. 4). No lesions were observed in the midbrain or reported to occur in higher centres.

Zamia staggers has been produced in cattle by feeding about 1lb (680g) fresh *Cycas media* leaf daily in chaff for 50 days (Hooper et al., 1974) and with similar dose rates of *C. revoluta* leaf for 36 and 82 days respectively (Kobayashi et al., 1985). No liver damage was reported except that there were transient elevations of δ-GTP recorded soon after the start of feeding in the latter animals. In these studies the neuropathological findings in the spinal cord recorded in affected cattle reported initially by Hall and McGavin (1968) were confirmed and extended.

In a recent unpublished study by the authors, a young Friesian bull calf weighing 120kg was dosed daily through a rumen fistula with 200g of fresh *B. serrulata* leaf in 6 out of 10 days. Death occurred due to acute liver damage (Fig. 5) and marked gastrointestinal haemorrhage without the appearance of any sign of zamia staggers. Pathological study revealed no degenerative lesions in the spinal cord.

The Toxic Principles - MAM Glycosides

The toxic principles in cycads responsible for the extraneural lesions are considered to be the MAM glycosides, mainly macrozamin and cycasin (Hirono, 1987). While both these glycosides are found in the seeds in all cycad genera, the levels of macrozamin in *Bowenia* spp. are by far the highest (up to 5.04% of the wet weight) although levels in some *Macrozamia* spp. are almost as high (3.88% of wet weight for *M. miquellii*) (Moretti et al., 1983). Levels of cycasin in these species are normally only a tenth of those for macrozamin. If it can be presumed that these macrozamin levels in seeds reflect the levels of the compound in the leaves then *Bowenia* spp. can be expected to be the most toxic of the cycads in relation to their capacity to produce animal disease through ingestion of foliage. This is supported by the recent observation in the Friesian calf dosed with *Bowenia* leaf described above and in the earlier study by Hall and McGavin (1968). The acutely hepatotoxic effects induced by feeding fresh green leaf of *M. pauli-guilielmi* by Hall (1957) also indicated that this plant is similarly effective. Why then did zamia staggers without liver damage occur within the short time of 18 days in the beast described by Hall and McGavin (1968) in which a similar daily dose rate of *B. serrulata* was fed?

It has been shown that MAM glycosides are hydrolysed in the alimentary tract by the action of β-glucosidoses present in the contents. The MAM released by such hydrolysis reaches the liver in which through further metabolism it causes acute hepatocellular injury. Although variable proportions of the MAM-glycoside intake may be absorbed from the gut unchanged, this is not in general believed to be toxic following absorption because tissues do not contain the β-glucosidases necessary to release the MAM. The absorbed glycosides are then eliminated harmlessly in the urine (Hirono, 1987). In respect of the earlier study with *B. serrulata* it is presumed that insufficient amounts of the MAM released through ruminal metabolism of the glucoside can have reached the liver to cause damage. This may have been due to loss of the glycosides in the plant during drying, by a more com-

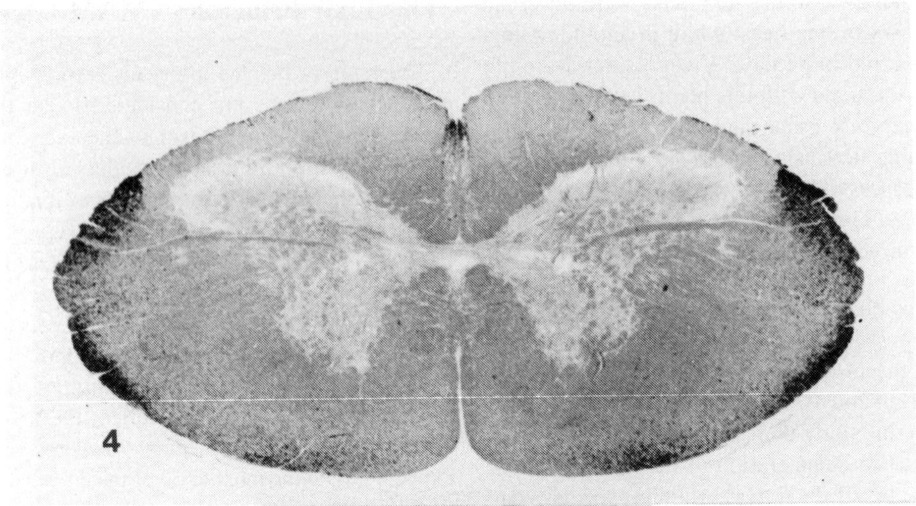

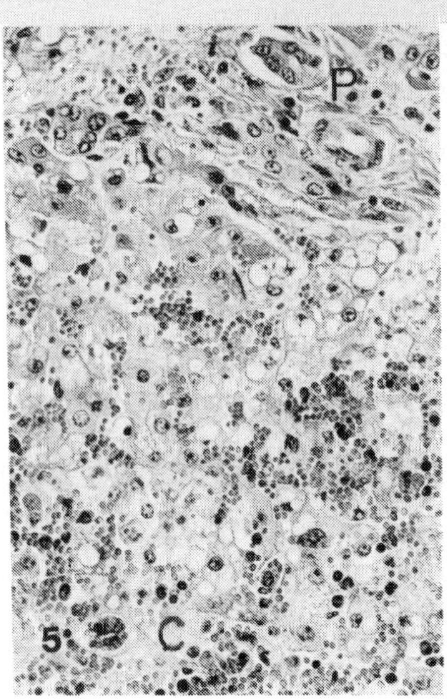

Figures 4,5: *Effects from Bowenia Serrulata*

Figure 4: A transverse section of the spinal cord at the junction of the cervical cord and the medulla oblongata of a steer fed about 200g *bowenia serrulata* leaf daily in the feed for 18 days, causing typical zamia staggers without liver damage. Symmetrical black stained areas are present in each of the columns but the effect is most marked in the fasiculis gracilis and the spinocerebellar tracts. Marchi stain, x10. Section kindly supplied by W.T.K. Hall.

Figure 5: A liver of a calf dosed through a rumen fistula with 200g fresh *Bowenia serrulata* leaf for 6 days in 10 showing extensive degeneration and necrosis with haemorrhage of the parenchyma. There was not ataxia nor spinal cord lesions. P = portal tract, C = central vein. H & E, x250.

plete metabolism of the released MAM by specific rumen microflora, or else the rate of ingestion of the glycoside was slow enough throughout the whole day's feeding to enable effective detoxication of it both in the gut and/or the liver to take place. In the only known study in which an MAM glycoside (cycasin) was given in multiple daily doses to a ruminant, (goat) both hepatic and spinal lesions qualitatively similar to those of zamia staggers of cattle were seen (Shimizu et al, 1986). The doses used were 4 and 6 mg/kg. Two goats given 4mg/kg/day for 24 days became anorexic and emaciated. Following a non-dosing period of 7 days, they were then given 6mg/kg/day until necropsy at 135 and 156 days respectively. Two other goats were given 6mg/kg/day for 89 and 103 days respectively. Liver lesions of varying degrees of severity were recorded in all animals. It was suggested that the spinal-cord lesions could have been a consequence of the liver damage. This can lead to hyperammonaemia and in ruminants this has been shown to lead to degenerative changes in the brain and cord (Hooper, 1975). The significance of the effects of oral cycasin in causing neurological damage in these goats therefore was not categorically determined.

If the macrozamin level in *B. serrulata* leaf is similar to that of the seed, i.e. 4.33% of fresh weight, then the individual doses of macrozamin for the calf above (contained in some 200g fresh leaf) could have been about 72mg/kg. It was found that when a goat was given three oral doses of 20mg/kg of cycasin at 7-day intervals, marked hepatic necrosis and death occurred (Shimizu et al., 1986). The oral LD_{50} of cycasin for the rat is reported to be 562 mg/kg, but species differences in the acute oral toxicity of the compound can be large (Hirono, 1972). For example those for the mouse, rabbit and guinea pig are 500, 30 and 20mg/kg respectively. The goat thus appears to be among the more highly susceptible species to cycasin (or macrozamin), and calves appear to be similar.

Studies in a variety of laboratory animal species dosed with cycasin or MAM acetate have indicated that while neuropathological changes have been produced, none of these resembles either ALS in man or zamia staggers in cattle (Jones et al., 1972; Rabe and Haddad, 1972; and, Sanger et al., 1972). Notwithstanding the findings of oral dosing of goats with cycasin (Shimizu, 1986), the different effects of dosing cattle with *B. serrulata* leaf, make it seem unlikely that the MAM glycosides and the MAM *per se* are directly responsible for the neuropathological changes seen in cattle. Methylazoxymethanol is however a substrate for alcohol dehydrogenase with the formation of methylazoxyformaldehyde (Grab and Zedeck, 1977). The potential for neurotoxic injury caused by this compound remains to be determined.

L-β-Methylaminoalanine (l-BMAA)

The non-protein amino acid L-BMAA which occurs in *Cycas* spp was initially shown to cause motor disturbances in rats, mice and chickens when given at high dose rates (Polsky et al., 1972), but until recently no neuropathological changes in any species had been described. The compound is present in these plants at relatively low concentrations (not more than 100mg/100g green plant) and was originally thought to be of no toxicological significance, either in respect of amyotrophic lateral selenosis (ALS) (Polsky et al., 1972) or zamia staggers (Seawright, 1989). However, recently it was shown in a brief report that when the compound was given orally at up to 380 mg/kg to monkeys, a motor dysfunction resembling human ALS occurred after several weeks' feeding and this was accompanied by degenerative changes in the Betz cells of the motor cortex but without cortico-spinal tract changes (Spencer et al., 1987).

Subsequently, we carried out a study in the rat in which intraperitoneal disease in the range originally used by Vega et al., (1968) resulted in disturbances of motor function as initially reported. Pathological studies revealed a selective necrosis of GABAergic neurons of the cerebellar cortex (Seawright et al., 1990). The affected cells were initially the stellate, basket and Golgi neurons later followed by Purkinje cells, with vacuolation of the neuropil of the cerebellar-roof nuclei. These changes were most severe in the anterior vermis but with higher doses, extended to include all the cerebellar lobes. No changes were observed in any other area of the brain, spinal cord or dorsal root ganglia. The compound was thought to act as a glutamate agonist possibly for NDMA receptors on the neurons since effects due to the compound *in vitro* on mouse cortical tissue could be prevented or ameliorated by treatment with an NMDA receptor antagonist (Ross et l., 1987; Weiss and Choi, 1988). Although clinical signs suggestive of cerebellar motor dysfunction also occurred in day-old chicks, in mice and in Mongolian gerbils dosed at similar levels with L-AA, no histological lesions have been found in these species (unpublished data).

The changes in the rat neurons were considered to be due to the excitotoxic action of the compound present in the brain in high concentration. Based on the published levels of L-BMAA in *Cycas* spp, however, it is estimated that the cattle which developed zamia staggers after several weeks of consumption of feed containing *Cycas media* would have consumed no more than 1mg/kg of L-BMAA daily (Seawright et al., 1990). This is substantially below that needed to cause cerebellar lesions in the rat when the compound was also given by the intraperitoneal route. Also since L-BMAA has not been confirmed as being present in other cycad genera which also cause zamia staggers it seems most unlikely that this compound is responsible fo the neurological disturbance in cattle. Since species susceptibility to the effects of L-BMAA may occur as for other toxins, it remains to be confirmed that a parenteral dose rate of 1mg/kg/day for some weeks of L-BMAA is actually without effect in cattle. It is worth noting further that chronic daily dosing of rats with L-BMAA for 87 days appeared not to result in any neurological signs (Polsky et al., 1972).

Other Possible Toxins

It was shown by Louw and Oelofsen (1975) that there was contained in the aqueous extract of *Encephalartos altensteinii* seeds an agent of molecular weight in excess of 1,000, which at a daily dose rate of about 1g/kg bodyweight for 15 to 30 days caused a 'chronic paralytic condition' in guinea pigs. The latter condition was believed to resemble zamia staggers in cattle. Similar high molecular weight preparations have been produced from leaves and seeds of *Cycas revoluta*. When fed to guinea pigs, the preparations from the leaves were without effect. Preparations from the seeds dosed daily over 5 days appeared toxic, producing some ataxia but the conditions of the experiment and the descriptions of the affected animals are not clear (Kobayashi et al., 1983). Further studies on the toxicity of such preparations from other cycad species are in progress and will be reported elsewhere. Should they prove to be neurotoxic in the guinea pig, as has been described for *E. altensteinii*, their relevance to zamia staggers will need to be evaluated.

The fact that zamia staggers occurs soonest after dosing with *Bowenia*, which is highest in MAM glycosides, and that spinal cord lesions occurred in goats after chronic daily oral dosing with cycasin suggests nevertheless that the presence of MAM glycosides in the rumen is somehow responsible. In the absence of any other credible neurotoxic agent in the plant, a compound which can be derived from these glycosides in the rumen and which is not hepatotoxic may be the neurotoxicant. Possibilities include products formed spontaneously from MAM in the more-or-less neutral aqueous environment of the rumen contents (Nagasawa et al., 1978) as well as compounds derived from the glycosides or MAM through microbial action in the rumen and/or metabolic activity in the liver.

References

***Anderson, J.L. and Hall, W.T.K.* 1964.** Neurotoxic effects from cycad leaves. Fed. Proc. 23: 1349.

***Everist, S.L.* 1981.** Poisonous Plants of Australia. ed. 2. pp. 226 - 248. Angus and Robertson, Sydney.

***Gabbedy, B.J., Meyer, E.P. and Dickson, J.* 1975.** *Zamia* palm (*Macrozamia riedlei*) poisoning of sheep. Aust. Vet. J. 51: 303 - 305.

***Grab, D.J. and Zedeck, M.S.* 1977.** Organ-specific effects of the carcinogen methylazoxymethanol related to metabolism by nicotinamide adenine nucleotide-dependent dehydrogenases. Cancer Res. 37: 4182 - 4189.

***Haddad, R.K., Rabe, A. and Dumas, R.* 1972.** Comparison of effects of methylazoxymethanol acetate on brain development in different species. Fed. Proc. 31: 1520 - 1523.

***Hall, W.T.K.* 1957.** Toxicity of the leaves of *Macrozamia* spp. for cattle. Qld. J. Agric. Sci. 14: 41 - 52.

***Hall, W.T.K.* 1964.** Plant toxicosis of tropical Australia. Aust. Vet. J. 40: 176 - 182.

***Hall, W.T.K. and McGavin, M.D.* 1968.** Clinical and pathological changes in cattle eating the leaves of *Macrozamia lucida* or *Bowenia serrulata* (Family Zamiaceae) Path. Vet. 5: 417 - 436.

***Hirono, I.* 1972.** Carcinogenicity and neurotoxicity of cycasin with special reference to species differences. Fed. Proc. 31: 1493 - 1497.

***Hirono, I.* 1987.** Cycasin. *In* Naturally Occurring Carcinogens of Plant Origin, Toxicology Pathology and Biochemistry, Ed. I. Hirono pp. 3 - 24. Elsevier, Tokyo.

***Hooper, P.T.* 1975.** Spongy degeneration in the central nervous system of domestic animals. Part 1. Morphology. Acta Neuropathol (Berl.). 31: 325 - 334.

***Hooper, P.T., Best, S.M. and Campbell, A.* 1974.** Axonal dystrophy in the spinal cords of cattle consuming the cycad palm, *Cycas media*. Aust. Vet. J. 50: 146 - 149.

***Hurst, E.* 1942.** Family Cycadaceae. *In* The Poisonous Plants of New South Wales, pp. 1 - 6. Published by the Poisonous Plants Committee of New South Wales, Department of Agriculture, Sydney.

***Johnson, L.A.S.* 1959.** The families of cycads and the Zamiaceae of Australia. Proc. Linn. Soc. N.S.W. 84: 64 - 117.

***Jones, M., Yang, M. and Mickelsen, O.* 1972.** Effects of methylazoxymethanol glucoside and methylazoxymethanol acetate on the cerebellum of the post natal Swiss albino mouse. Fed. Proc. 31: 1508 - 1511.

***Kobayashi, A., Tadera, K., Yagi, F., Kono, I., Sakamoto, T., and Yasuda, N.* 1984.** Studies on poisoning of grazing cattle due to ingestion of cycad, *Cycas revoluta* Thunb. Incidence of cycad-poisoning of cattle in Okinawa. Bull. Fac. Agric. Kagoshima Univ. 34: 19 - 129.

***Kobayashi, A., Tadera, K., Yagi, F., Kono, I., Sakamoto, T., and Yasuda, N.* 1983.** Cattle poisoning due to ingestion of cycad leaves. Neurotoxic effects causing paralysis in hindquarters. Toxin, Suppl. 3: 229 - 232.

***Louw, W.K.A., and Oelofsen, W.* 1975.** Carcinogenic and neurotoxic components in the cycad *Encephalartos altensteinii* Lehm. (Family Zamiaceae) Toxin 13: 447 - 452.

***Mason, M.M., and Whiting, M.G.* 1968.** Caudal motor weakness and ataxia in cattle in the Caribbean area following ingestion of cycads. Cornell Veterinarian 58: 541 - 5543.

***Moretti, A., Sabato, S. and Gigliano, G.S.* 1983.** Taxonomic significance of methylazoxymethanol glycosides in the cycads. Phytochemistry 22: 115 - 117.

***Nagasawa, H.T., Shirota, F.N. and Matsumoto, H.* 1972.** Decomposition of methylazoxymethanol, the aglycone of cycasin in D_2O. Nature 236: 234 - 235.

***Polsky, F.L., Nunn, P.B. and Bell, E.A.* 1972.** Distribution and toxicity of α-amino-β-methylaminopropionic acid. Fed. Proc. 31: 1473 - 1475.

***Rabe, A. and Haddad, R.K.* 1972.** Methylazoxymethanol-induced micrencephaly in rats: behavioural studies. Fed. Proc. 31: 1536 - 1539.

***Ross, S.M., Seelig, M., and Spencer, P.S.* 1987.** Specific antagonism of excitotoxic action of 'uncommon' amino acids assayed in organotypic mouse cortical cultures. Brain Research 425: 120 - 127.

***Sanger, V.L., Yang, M., and Mickelsen, O.* 1972.** Cycasin-induced central nervous system lesions in postnatal mice. Fed. Proc. 31: 1524 - 1528.

***Seawright, A.A.* 1989.** Cycads *In* Animal Health in Australia, Vol. 2, ed. 2: Chemical and Plant Poisons, pp. 46 - 48. Australian Government Printing Service, Canberra.

***Seawright, A.A., Brown, A.W., Nolan, C.C., and Cavanagh, J.B.* 1990.** Selective degeneration of cerebellar cortical neurones caused by cycad neurotoxin, L-β-methylaminoalanine (L-BMAA), in rats. Neuropath. Appl. Neurobiol. 16: 153 - 169.

***Seddon, H.R., Belschner, H.G. and King, R.O.C.* 1931.** Poisoning of sheep by the seeds of burrawang (*Macrozamia spiralis*). *In* Veterinary Research Report No. 6. pp. 70 - 80. Department of Agriculture, New South Wales, Sydney.

***Senior, D.F., Sundlof, S.F., Buergelt, C.D., Hines, S.A., O'Neil-foil, C.S. and Meyer, D.J.* 1985.** Cycad intoxication in the Dog. J. Amer. Anim. Hosp. Assn. 21: 103 - 109.

***Shimizu, T., Yasuda, N., Kono, I., Yagi, F., Tadera, K., and Kobayashi, A.* 1986.** Hepatic and spinal lesions in goats chronically intoxicated with cycasin. Jpn. J. Vet. Sci 48: 1291 - 1295.

***Spencer, P.S., Nunn, P.B., Ludolph, A.C., Ross, S.M.,* Roy, D.N. and *Robertson, R.C.* 1987**. Guam amyotrophic lateral sclerosis Parkinsonism-Dementia linked to a plant excitant neurotoxin. Science 237: 517 - 522.

***Stanley, T.D.,* and *Ross, E.M.* 1989.** Flora of South-eastern Queensland, Vol. III, pp. 448 - 453. Queensland Department of Primary Industries, Brisbane.

***Vega, E., Bell, E.A., and Nunn, B.P.* 1968.** The preparation of L-and D-α-amino-ß-methylaminopropionic acids and the identification of the compound isolated from *Cycas circinalis* as the L-isomer. Phytochemistry 7: 1885 - 1887.

***Weiss, J.H. and Choi, D.W.* 1988.** β-N-methylaminoalanine neurotoxicity: requirement of bicarbonate as a cofactor. Science 241: 973 - 975.

***Yasuda, N., Kono, I., Shimizu, T., Kobayashi, A., Tadera, K., and Yagi, F.* 1984.** Pathological studies on poisoning of grazing cattle due to ingestion of cycad, *Cycas revoluta* Thunb. The lesions and their distribution in the spinal cord. Bull. Fac. Agric. Kagoshima Univ. 34: 131 - 137.

***Yasuda, N., Kono, I., and Shimizu, T.* 1985.** Pathological studies on cycad poisoning of cattle experimentally caused by feeding with leaves of cycad, *Cycas revoluta* Thunb. Bull. Fac. Agric. Kagoshima Univ. 35: 171 - 178.

Neurotoxic Properties of Cycads - An Occupational Hazard for Botanists

Peter S. Spencer
Centre for Research on Occupational & Environmental Toxicology
and Department of Neurology, School of Medicine
Oregon Health Sciences University, Portland, Oregon, USA

The physical beauty of cycads - so often exploited in public settings - belies the toxic potential of virtually all components of these plants. The leaves are an established cause of a neuromuscular disorder in ruminants, and *Cycas* spp. seed has been linked to a devastating neurological disorder in humans. Experimental studies have demonstrated that cycads contain principles with neurotoxic, carcinogenic, mutagenic and teratogenic potential. Were these chemicals generated by industry and deployed in the environment, their properties would be sufficient to stimulate reams of environmental controls. By contrast, *Cycas* plants are widely appreciated by the public, sold in supermarkets and kept in the home (USA), and the poisonous seed are used for medicine (Japan) and food (Guam). Cycad botanists should be alert to the toxic properties of the subjects of their professional interest and use appropriate protective clothing to avoid dermal, inhalation or oral exposure to cycad chemicals.

The toxic potential of cycads has been reviewed recently (Spencer and Dastur, 1989). There is no concrete evidence that cycads have the potential to cause degeneration of the human nervous system, but it is well established that cattle and sheep grazing on leaves and other components of these plants developed a poorly studied neuromuscular disease (Whiting, 1963). Various genera and species are implicated. In Australia, *Cycas* as well as *Bowenia* and *Macrozamia* are held responsible; in New guinea, *Cycas circinalis*; in the Dominican Republic and Puerto Rico, *Zamia*, and in Mexico, *Dioon edule*. Experimental feeding of fresh *Cycas circinalis* leaves to a cow produced a motorsystem disorder (cycadism) after 85 days (Spencer and Dastur, 1989).

Acute effects of cycad poisoning in animals include hepatotoxicity, enterotoxicity and death. Rapid twitching of the eyelids, nostrils, lips, jaw muscles, with periodic tremors of the body, are reported in sheep, and muscle fasciculation has been seen in heifers (Whiting, 1963). Delayed effects of cycad consumption are characterised by a progressive and irreversible paralysis of the hindlimbs. Initially, there is a staggering weaving gait, with crossing of the legs, incoordination and ataxia.

More severe forms are characterised by posterior motor weakness, dragging of extended hindlimbs and, occasionally, a stringhalt-like action of the hocks (Spencer and Dastur, 1989; Whiting, 1963; Hooper, 1983). Neuropathological examination reveals degeneration of descending spiral tracts, readily observed in the lumbar region, with similar involvement of the fasciculus gracilis and dorsal spinocerebellar tracts, most prominent in the cervical region (Jubb et al.,

1985). Changes in the brain and peripheral nerves have not been described. A disorder similar to cycadism occurs in goats experimentally fed Cycasin (Shimizu et al., 1986), the principal cycad toxin.

Acute cycadism is similar in humans and animals. Cycad poisoning in humans is known from a single family which consumed food prepared from incompletely detoxified sago prepared from the stem of *Cycas revoluta* (Hironi et al., 1970). The description presumably reflects the toxic properties of remaining cycad chemicals rather than those associated with myocotoxins or other possible food contaminants. Ingestion was followed by a latent period of 12 - 40 hours before the appearance of symptoms. Headache, nausea and vomiting appeared suddenly. Convulsions and unconsciousness followed, and five of six affected family members died. Liver enlargement was found in all subjects, a phenomenon likely related to the hepatotoxic properties of Cycasin.

Whether cycad chemicals induce a neuromuscular disorder in humans, as in animals, has been the subject of much recent attention (Spencer et al.). Epidemiological evidence points to a relationship between the regional use of *Cycas* seed kernel as food and medicine in parts of Asia and Oceania and the associated high prevalence of motor neuron disease (amyotrophic lateral sclerosis, ALS) and Parkinsonism-dementia (P-D). O-D is characterised by slowed body movements, tremor, increased muscle tone and loss of memory and personality; ALS by progressive muscle weakness, with consequent loss of limb use and the ability to speak and swallow. Both disorders are present to differing degrees in three high-incidence disease foci of the western Pacific region (Garruto et al.). ALS/P-D has been particularly common among the Chamorro people of Guam and Rota in the Mariana Islands, Japanese residents of the Kii Peninsula of Honshu Island, and Auyu and Jaqai linguistic groups in southeastern Irian Jaya (west New Guinea). Decline in disease incidence has mirrored regional development and the concurrent disappearing use of *Cycas* seed for oral (Japan) and topical medicine (Guam, Irian Jaya). Chamorros have also used incompletely detoxified seed kernel (female gametophyte) for food (Spencer, 1987). The latter contains detectable quantities of two compounds with toxic potential Cycasin and β-*N*-methylamino-L-alanine (BMAA) (Duncan et al., 1990; Kisby et al., un-published). Fermented seed kernel has also been consumed by Japanese residents of the Ryukyu island chain: here, food samples lack detectable concentrations of the principal toxin (Cycasin) (Kobayashi, 1972), and ALS/P-D has not been recognised to occur in high incidence.

Direct demonstration of a relationship between cycad chemicals and the induction of human neurological disease is unavailable. There are no known cases in which an individual has been acutely intoxicated, survived, and developed motor neuron disease later in life. There is one case in which a female developed ALS after heavy exposure to the seed of *Macrozamia*, but a cause-effect relationship cannot be assumed from this incident. Forging a link between such events is extremely challenging because a considerable amount of time may separate exposure and disease onset. Epidemiological data from Guam indicate that individuals develop clinically apparent ALS/P-D years or decades after departing the island and apparently terminating exposure to local environmental factors (Garruto, et al., 1981), including cycad seed. In these circumstances, it is necessary to turn to laboratory animals as human surrogates to determine whether cycad chemicals induce motor neuron disease. An early study proved positive in this regard, a single macaque developing motor neuron degeneration and arm weakness after prolonged consumption of cycad flour prepared in Chamorro style (Dastur et al, 1964, 1990). However, a subsequent primate study examining the carcinogenic properties of cycad seed failed to mention changes suggestive of neurological disease (Sieber et al., 1980). More recent studies have employed laboratory animals and tissue-culture systems to examine and define the neurotoxic properties of Cycasin and BMAA.

BMAA is a stereospecific excitant neurotoxin present as the free amino acid in cycad seed kernel (Ross et al., 1987; Weiss et al., 1988, 1989). Small and variable concentrations of BMAA are detectable in Chamorro-style cycad colour (Duncan et al., 1990; Kisby et al., un-published). Animals treated with very large daily doses of BMAA develop neurological diseases that are species-specific. Rats display neuronal degeneration in the cerebellum (Seawright et al., 1990), while primates develop a motorsystem disorder with features reminiscent of ALS/P-D (Spencer et al., 1987). However, the primate disorder lacks certain key characteristics of the human disease, notably the progressive downhill clinical course in the absence of continued cycad exposure.

Cycasin is the β-glucoside of Methylazoxymethanol (MAM), a potent genotoxin that alkylates DNA and perturbs cellular function (Kobayashi et al., 1965; Laqueur, 1968). Cycad flour prepared by Chamorro families contains on average approximately ten times more Cycasin than BMAA (Kisby et al., un-published). While oral Cycasin is metabolised to MAM by gut flora, the intact molecule may enter the bloodstream and brain of rats (Dastur and Palekar, 1974). Recent studies have demonstrated that Cycasin is taken up into nerve tissue *in vitro* by the glucose-transport system. MAM is liberated in the process and the nerve cell dies (Kisby et al., un-published). Animal studies with Cycasin have been few, and systematic attempts to examine the neurotoxic properties of this compound are lacking. Most attention has focused on the experimental hepatocellular toxicity and carcinogenicity of MAM (Dastur and Palekar, 1974).

The presence of chemicals with toxic potential in virtually all parts of cycads dictates the need for care in handling these plants. Botanists working with cycads should recognise the hazard and minimise exposure. Chemical gloves should be worn when handling plant components, and care should be taken to avoid dermal contact with seed juice and gum. Gloves also provide mechanical protection when handling cycad leaves. A tight-fitting mask should be used to avoid inhalation exposure to pollen. Soiled clothing should be replaced at the earliest opportunity and exposed skin washed with soap and water. These simple precautions should minimise exposure to cycad chemicals.

References

***Dastur, D.K.* 1964.** Cycad toxicity in monkeys: Clinical, pathological and biochemical aspects. *Fed. Proc.*j 23: 1368.

***Dastur, D.K., and Palekar, R.S.* 1974.** The experimental pathology of cycad toxicity with special reference to oncogenic effects. *Ind. J. Cancer* 2: 33.

***Dastur, D.K., Palekar, R.S. and Manhani, D.K.* 1990.** Toxicity of various forms of *Cycas circinalis* in rhesus monkeys., Pathology of brain, spinal cord and liver. In Rose C., Norris F. *ALS: New Advances in Toxicology and Epidemiology*. London, Smith-Gordon.

***Duncan, M.W., Steele, J.C., Kopin, I.J. and Markey, S.P.* 1990.** α-Amino-β-(methylamino)-propionic acid (BMAA) in cycad flour: An unlikely cause of amyotrophic lateral sclerosis and parkinsonism-dementia of Guam. *Neurology* 40: 767.

***Garruto, R.M. and Yase, Y.* 1986.** Neurodegenerative disorders of the western Pacific: the search for mechanisms of pathogenesis. *Trends Neurosci.* 9: 368.

***Garruto, R.M., Gajdusek, D.C. and Chen K-M* 1981.** Amyotrophic lateral sclerosis among Chamorro migrants from Guam. *Ann Neurol* 10: 341.

***Hironi, I., Kachi, H., and Kato., T.* 1970.** A survey of acute toxicity of cycads and mortality rate form cancer in the Miyako islands. *Acta Pathol Jap.* 1970; 20: 327.

***Hooper, P.T.* 1983.** Cycas poisoning. In Keeler, R.F., Tu A.T., eds. *Handbook of Natural Toxins. I Plant and Fungal Toxins*. New York, Marcell Dekker, p. 463.

***Jubb, K.V.F., Kennedy, P.C. and Palmer, N.* 1985.** *Pathology of Domestic Animals*, 3rd ed. Orlando, Academic Press, 360 pp.

***Kisby, G.E., Ellison M. and Spencer, P.S.* (Submitted for publication).** Content of the neurotoxins Cycasin (methylazoxymethanol β-D-glucoside) and BMAA (β-*N*-methylamino-L-alanine) in cycad flour prepared by Guam chamorros.

***Kobayashi, A.* 1972.** Cycasin in cycad materials used in Japan. *Fed Proc.* 31: 1476.

***Kobayashi, A. and Matsumoto, H.* 1965.** Studies on methylazoxymethanol, the aglycone of Cycasin. Isolation, biological and chemical properties. *Arch. Biochem. Biophys.* 37: 217.

***Kisby, G.E., Ross, S.M., Gold, B.G. and Spencer, P.S.* (Submitted for publication).** Cycasin and BMAA: Candidate neurotoxins for western Pacific amyotrophic lateral sclerosis/Parkinsonism-dementia complex.

***Laqueur, G.L.* 1968.** Toxicology of Cycasin. *Cancer Research* 28: 2262.

***Ross, S.M., Seelig, M. and Spencer, P.S.* 1987.** Specific antagonism of excitotoxic action of 'uncommon' amino acids assayed by organotypic mouse cortical cultures. *Brain Res.* 425: 120.

***Ross., S.M. and Spencer, P.S.* 1987.** Specific antagonism of behavioural action of 'uncommon' amino acids linked to motor-system disease. *Synapse* 1: 248.

***Seawright, A.A., Brown, A.W., Nolan, C.C. and Cavanagh, J.B.* 1990.** Selective degeneration of cerebellar cortical neurons caused by cycad neurotoxin, L-β-methylaminoalanine (L-BMAA), in rats. *Neuropathol Appl. Neurobiol.* 16: 153.

***Shimizu, T., Yasuda, N., Kono, I., Yagi, F., Tadera K., and Kobayashi, A.* 1986.** Hepatic and spinal lesions in goats chronically intoxicated with Cycasin. *Japan J. Vet. Sci.* 8: 1291.

***Sieber, S.M., Correa, P., Dalgard, D.W., McIntire, K.R., and Adamson R.H.* 1980.** Carcinogenicity and hepatotoxicity of Cycasin and its aglycone methylazoxymethanol acetate in non-human primates: *J. Nat. Cancer Inst.* 65: 177.

Spencer, P.S., Ross, S.M., Kisby, G. and Roy, D.N. Western Pacific amyotrophic lateral sclerosis: Putative role of cycad toxins. In Hudson A.J., ed. *Amyotrophic Lateral Sclerosis: Concepts in Pathogenesis and Etiology*. Toronto, University of Toronto Press, p. 263.

***Spencer, P.S.* 1987.** Guam ALS/Parkinsonism-dementia: A long-latency neurotoxic disorder caused by 'slow toxin(s)' in food? *Can J. Neurol Sci.* 14: 347.

***Spencer, P.S. and Dastur, D.K.* 1989.** Neurolathyrism and neurocycadism. In Dastur D.K., Shahani M., Bharucha E.P., eds. *Neurological Sciences: An Overview of Current Problems.* New Delhi, Interprint, p. 309.

***Spencer, P.S., Nunn, P.B. Hugon, J., Ludolph, A.C., Ross, S.M., Roy, D.N. and Robertson, R.C.* 1987.** Guam amyotrophic lateral sclerosis-parkinsonism-dementia linked to a plant excitant neurotoxin. *Science* 237: 517.

***Weiss, J.H., and Choi, D.W.* 1988.** β-*N*-Methylamino-L-alanine neurotoxicity: Requirement for bicarbonate as a cofactor. *Science* 241: 973.

***Weiss, J.H., Koh, J.Y., Choi, D.W.* 1989.** Neurotoxicity of β-*N*-methylamino-L-alanine (BMAA) and β-*N*-oxalylamino-L-alanine (BOAA) on cultured cortical neurons. *Brain Res.* 496: 64.

***Whiting, M.G.* 1963.** Toxicity of cycads. *Econ Bot.* 17: 271.

Part Three

Population Ecology

Genetic Diversity in the Cycad *Macrozamia riedlei*

Byrne, M. and James, S.H.
Department of Botany, The University of Western Australia,
Nedlands, Western Australia

Macrozamia riedlei (Gaud.) C.A. Gardn. is native to Western Australia. A survey of 15 populations with 9 isozyme systems revealed 14 consistently interpretable loci, 13 polymorphic. The number of alleles per locus, percentage of polymorphic loci, and expected panmictic heterozygosity were very high when compared to values reported for 473 plant species representing a wide range of life history characteristics (Hamrick and Godt, 1990) and were higher than those reported for a cycad from eastern Australia, *M. communis* (Ellstrand, Ornduff and Clegg, 1990) (Table 1). Percentage of polymorphic loci and number of alleles per locus increased in a north-south direction through the population. Seventy of the 110 loci within populations exhibited an excess of heterozygotes compared to random mating expectations. The level and distribution of diversity within and between populations were similar to those for outcrossed, wind-pollinated plants, but quite different to those for *M. communis* (Table 2). Estimate of gene flow gave Nm=2.47, a value which indicates substantial gene migration between populations. *M. riedlei* is a highly polymorphic species that maintains most of its genetic diversity within populations and undergoes selection for heterozygous individuals in the adult plants.

REFERENCES

***Ellstrand, N.C., R. Ornduff and J.M. Clegg* 1990.** Genetic structure of the Australian cycad *Macrozamia communis* (Zamiaceae). Amer. J.Bot. 77:677-681.

***M.T. Clegg, A.L. Kahler and B.S. Weir* (eds).** Plant population genetics, breeding and genetic resources, Sinnaur Associates Inc., Massachusetts, pp43-63.

Table 1

Mean parameters of diversity for *M. riedlei, M. communis* and other plant species

A, mean number of alleles
P, mean percentage of polymorphic loci
H_o, observed heterozygosity
H_e, expected panmictic heterozygosity
F, fixation index

	A	P	H_o	H_e	F
M.riedlei	2.4	93.0	0.263	0.274	-0.08
Avg. of 473 plant species	1.96	50.5		0.149	
M.communis	1.61	50.0		0.080	

Table 2

Total genetic diversity and the distribution of diversity within and between populations of *M. riedlei, M. communis* and other wind pollinated species

H_T, total genetic diversity
H_S, mean genetic diversity within populations
D_{ST}, mean genetic diversity between populations
G_{ST}, mean proportion of diversity between populations

	H_T	H_S	D_{ST}	G_{ST}(%)
M.riedlei	0.295	0.268	0.027	9.2
Avg. of wind - pollinated species.	0.293	0.259	0.034	9.9
M.communis	0.190	0.090	0.100	27.0

The Roles of *Tranes lyterioides* and *T. sparsus* Boh (col., Curculiodidae) in the Pollination of *Macrozamia communis* (Zamiaceae)

C.E. Chadwick
Australian Museum, 68 College Street
Sydney, New South Wales, Australia

Abstract

The Australian cycad *Macrozamia communis* is confined to New South Wales, where it is restricted to an area between the Tasman Sea and the Great Dividing Range. In male plants, pollen shedding can commence as early as mid-August and cones remain persistently open thereafter. In mature female plants, the megasporophylls separate vertically, leaving an entrance to the interior which remains open only for a few days, considered to be the receptive period. The sporophylls then come together tightly, apparently due to the enlargement of the seeds. Numbers of invertebrates frequent the plant. The weevil *Tranes lyterioides* occurs throughout and beyond the range of the plant. The larva develops inside the male, but not the female, cone: it consumes fleshy tissue; the adult feeds on pollen. The weevil has been found deep in the caudex, at the base of the fronds, throughout the year and in male and female cones for much more limited periods. *Tranes sparsus* is less widely distributed: the adult feeds on pollen, has been found only in male cones, and overwinters under bark of *Eucalyptus maculata* to a limited extent. The life cycle is unknown, and it has not been associated with pollination, although it carries pollen on its body. Adults of *T. lyterioides* were observed inside receptive female cones at night, but pollination is believed to occur only over a very limited period by large numbers acting in concert. As the weevil completes its life cycle in the male cone, the reason for its invasion and the pollination of the female cone is not fully known. The theory of wind pollination is examined and considered inadequate. Associations between other cycads and other beetles are mentioned.

Introduction

Little has been published on pollination in Australian cycads. Chamberlain (1913) did not mention pollination in *Macrozamia moorei*. Lawson (1926), dealing with *Bowenia serrulata* Andre, observed nothing that would suggest any other agency than wind for pollination. Baird (1939) believed that pollen was transferred by wind from male to female cone in

M. riedlei. She was convinced that insects played an important part in transferring pollen from the cracks between the sporophylls to the micropyle. However, she was unable to supply experimental proof of her convictions.

Brough and Taylor (1940) observed insects among the younger vegetative leaves and among sporophylls of both sexes, but concluded that their occurrence was 'fortuitous and sporadic' and that pollination was anemophilous in *M. communis*.

Studies on *Macrozamia communis*

The author has included observations on pollination in a study of *Macrozamia communis* and its fauna. This cycad occurs from the Macleay River (Big Hill, Lat. 30º 42' S) in the north to the northern bank of the Bega River (Mogareka Inlet, Lat. 36º 42' S), also to the west (Ulan, Long. 149º 45' E), Munghorn Gap, NE of Capertee, Badgerys Crossing, east of Braidwood, and along the coast in suitable areas) (Fig. 1). The main study areas were at Noraville and Seven Mile Beach, but plants in the Sydney suburbs Avalon (Careel Bay), Balgowlah, Chatswood and Ryde and in country areas were also visited.

Numerous species of invertebrates, mostly insects, were found in the caudex, cones and fronds; some are undescribed species. Insects obviously capable of flight found in male cones were considered the most likely to be involved in pollen transfer. These included the curculionids *Tranes lyterioides* and *T. sparsus*; the braconids *Apanteles* sp., the nitidulid *Brachypeplus barronensis*. Another braconid, *Chelonus* sp., was seen enclosed in a mass of pollen, and an undescribed species of thrip often almost invisible in pollen. Other invertebrates, some wingless, were found in much smaller numbers and were much less likely to be involved in pollination.

Male Cone

The mature male cone has two zones, a small basal sterile area and the rest of the cone bearing microsporangia below the microsporophylls; the terminal microsporophylls may be sterile. At maturity, spaces develop between the microsporophylls beginning at the apex of the cone, due to the elongation of the rachis.

Mature cones progress from green to yellowish-green, to purplish or blackish and, when dried out, to straw colour. Pollen is shed during the colour change, before the cone dries out. Dried out sporangia open longitudinally, but if moistened will close within minutes. This would ensure that pollen shedding is reduced in wet weather, or perhaps even in conditions of high humidity.

Many male cones were observed to shed pollen before the female cones were receptive. In one case, a male cone was shedding pollen at Noraville on 10 August, 1980; and, at the other extreme, one small stunted cone was still unopen on 3 January 1979 at Mt Sugarloaf, near Wallsend. Very frequently, deposits of pollen were found on the ground, and/or peduncles, pinnules, cones, weeks and/or other vegetation, below the male cone when pollen was being shed.

Male cones remain permanently open after the pollen is shed. Pollen is protected from rain by the closing of the sporangia. Occasional male cones were observed with the apex black. This spread slowly downwards; the sporophylls did not open normally. Pollen grains are illustrated in Fig. 1. A distinct surface patten is visible. Grains rapidly assume a spherical shape when moistened with water.

The weevil *Tranes lyterioides* Pasc. (1875) is the most common insect associated with male cones. It appears to be present throughout the geographical range of the plant and beyond and is not restricted to *M. communis*, having been found outside the range of the species in the Warrumbungle Mountains, where *M. diplomera* and *M. heteromera* occur. The type came from Gayndah, Queensland (Pascoe 1875). Ornduff (1989) found large numbers of the species in the male cones of *Lepidozamia peroffskyana* of Coffs Harbour in January, 1987. Large numbers of the beetle were found on cones of this cycad in bright sunshine at Grassy Head (NNE of Kempsey) and Mooball (SE of Murwillumbah) in December, 1990 (Kennedy, pers. commun.). Microsporophylls were riddled by larvae, accompanied by an adult, at Bellingen in January, 1991 (Trnski, pers. commun.). It is predominantly

Figures 1-6: Pollen of *Macrozamia communis* in relation to *Trans lyterioides*. Scale bars indicate true dimensions. **1**: Pollen grain of *M. communis*. Figures. **2-6**: *T. lyterioides*. **2**: Pollen grains and scales on right side of head and protnotum. **3**: Mouth parts, ventral surface. **4**: Faecal pellet, when pollen available. **5**: Faecal pellet, high magnification. **6**: Faecal pellet lacking pollen grains.

nocturnal, having been seen moving in the field after dark and on numerous occasions at varying times throughout the night under laboratory conditions over a period of weeks. It has well developed wings, flies to light, but does not fly readily during the day.

The adult *T. lyterioides* congregates in the caudex between the bases of the fronds and in the axils of the pinnules inside the caudex throughout the year. In the colder weather, the weevils retreat to the bases of the fronds, where individual fronds meet, leaving a narrow space having the protection of the sharp cataphylls. They become comatose and only move when handled and/or exposed to light. The beetle has only been found in well developed plants, and is rare in solitary plants. Younger plants do not have enough shelter (fronds and pinnules) in the caudex to harbour insects even of moderate size, such as *T. lyterioides* (e.g. a potted plant which was a seedling with one frond in February 1982, still lacks adequate shelter).

The weevil has a one year life cycle. Although oviposition has not actually been observed, it seems likely that it occurs in the spring in the peduncles, microsporophylls and central rachis of the cone, resulting in exudations of frass and gum, particularly noticeable in the peduncle. Subsequent larval growth would be rapid.

The mature larva constructs an elongated oval cocoon, sometimes as a solitary pupal chamber, sometimes as part of an aggregation of several cells, and pupates. The interior of the peduncle and the rachis often become so filled with frass from feeding by the larvae and so weakened, that they snap when very little pressure is applied.

Of larvae collected at Seven Mile Beach on 24 December, 1978, three took respectively: 6 days (7/1 - 13/1/79), 7 days (7/1 - 14/1/79) and 8 days (7/1 - 15/1/79) from pupation to emergence. Of ten pupae from the same area, the following emerged as beetled: 1 on 9th, 2 on 11th, 3 on 13th, 1 on 14th and 1 on 15th January, the latter being the last date of emergence. Insects collected in the same area on 18 December, 1990 consisted of 14 larvae and 6 pupae. By mid-January, male cones are normally dried out and not frequented by weevils, which would be expected to enter caudices, overwinter and remain there until warmer weather returned.

At monthly intervals, between July, 1979 and July, 1980 inclusive, insects associated with the bases of 100 fronds were taken at Seven Mile Beach. There was an obvious decrease in numbers of this weevil in November, 1979 and a slighter decrease in March, 1980. A similar study at Noraville between October, 1979 and April, 1980 showed a conspicuous decrease in October and November. These results could indicate a movement of many, but not all, *T. lyterioides* to open male cones during the period when pollination is occurring.

Considerable numbers of this species may occur in male cones, e.g. 185 specimens (90 males, 95 females) were found in one cone at Bermagui on 27 October, 1978. At Noraville on 19 November, 1980, 65 weevils (16 males, 49 females) were found in two cones; on 9 October, 1988, 441 weevils (188 males, 153 females) in three cones; on 13 November, 1988, 145 weevil (96 males, 49 females) in three cones. On the latter date, only five male cones were found not to be dried out in the general area studied.

At times, the weevil has been seen crawling in the open in daylight, e.g. 33 adults (males and females) were found during the day on a small, unopened male cone at Mt Sugarloaf on 3 January, 1979. All other male cones in the area had opened and shed their pollen at this time.

Pollen grains will adhere to practically any part of the body, e.g. body hairs, elytra, tarsi, the body surface (Fig. 2) and may be visible to the naked eye as a whitish coating.

Tranes lyterioides adults feed on pollen grains as was demonstrated by scanning electron micrographs of faecal pellets (Figs. 4 and 5). As the mandibles lack an effective molar area (Fig. 3) and the pollen grains are passed with the exine entire, it appears that the pollen grains are not masticated, but are swallowed whole. Conceivably, the gastric juices either dissolve or pass through the intine to extract the cell contents. It has also been observed that faecal pellets voided

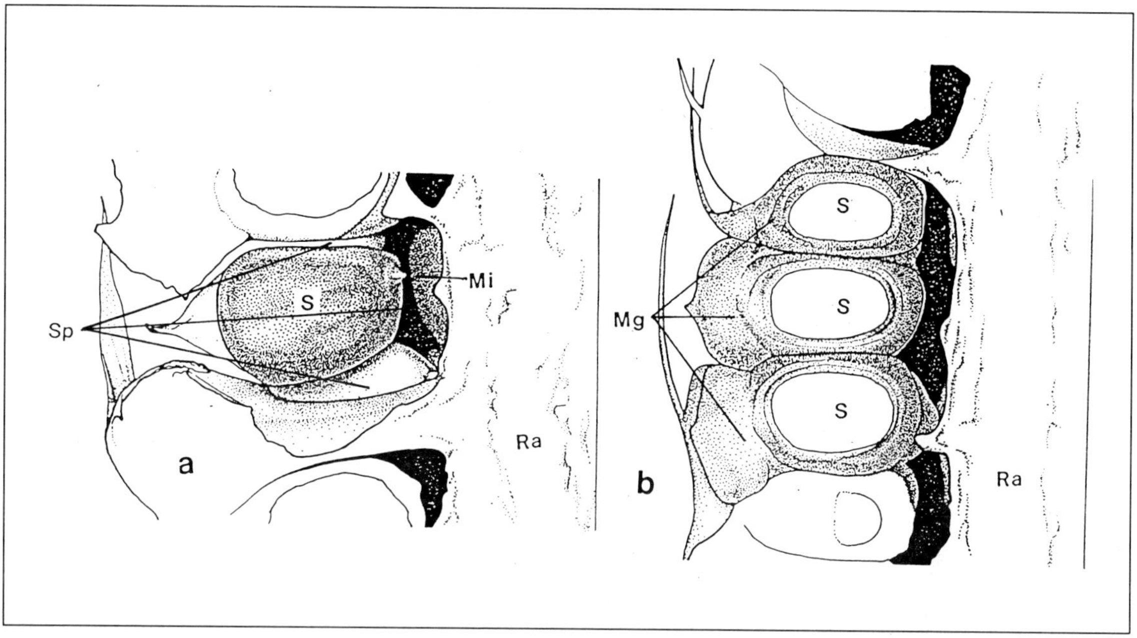

Figure 7: Sections of female cone of *Macrozamia communis*, S = seed; Mg = megasporophyll; Mi = micropyle; Ra = cone rachis; Sp = sporophyll. X1. A, surface view of seed and micropyle; B, sectional view of three adjacent seeds.

outside the pollination period or when all the pollen has been ingested are amorphous (Fig. 6).

E.C. Zimmerman (pers. commun.) has found that the species formerly known as *Tranes sparsus* Boh. is the female of *T. lyterioides* (Pasc.) and that the name *Tranes sparsus* should be applied to another, larger species also associated with *M. communis*. *Tranes sparsus* has well developed wings, is nocturnal and will fly to light. It has been found in large numbers in male cones at Noraville, mainly in October and November, e.g. 215 were found in two cones on 19 November, 1990; 102 specimens in three cones on 9 October, 1988, but only five in three cones on 13 November, 1988. It also can transport pollen (Fig. 8), even though I have not found the weevil in female cones.

Small numbers (1 to 6) of *T. sparsus* have been observed under the bark of *Eucalyptus maculata* in the cooler months, but hardly in numbers commensurate with numbers found in mature cones during the warmer weather. SEMs of faecal pellets show microspores, as in *T. lyterioides*. There has been no indication of where it breeds.

Three light green male cones collected at Noraville on 9 October, 1988 contained 441 *T.l yterioides* and 103 *T. sparsus*, a total of 554 specimens. On 13 November, 1988, only three greenish yellow male cones remained in the same area, the other cones being brown and dried out. These cones contained 145 *T. lyterioides* and five *T. sparsus*, a total of 150 specimens. The beetles were noticeably abraded, and lacked scales compared with those taken a month earlier. No pollen was seen in any cone. It has frequently been observed that dried out cones lack a weevil population. No doubt the same thing could have happened in this area. On 22 October, 1989, two cones taken at Noraville contained respectively 17 (6 male, 11 female) *T. lyterioides* and 16 *T. sparsus* and 75 (29 male, 46 female) *T. lyterioides* and 119 *T. sparsus*. In sampling at the bases of fronds one specimen of *T. sparsus* was found at Noraville in February, 1980. The species was not seen at Seven Mile Beach, but specimens were found at Avalon (Careel Bay) December, 1985 and November, 1990.

I have bred *Tranes internatus* from caudices of *M. communis* found in suburban sites at Kurnell and Cromer, but have not found it in either male or female cones. However, Ornduff (1990) states that he found it in female cones at Seven Mile Beach in January, 1987.

The female cone and the pollination process

In the case of maturing female cones, Brough and Taylor (1940) stated that sporophylls became "slightly separated by the elongation of the cone axis". Baird (1939) made a similar observation on *M. riedlei*. Pollination would no doubt occur at this time. I have observed that, after a time, in *M. communis*, the spaces between the megasporophylls close up, probably due to the growth of the developing seeds (Fig. 7). Thus, insects and other potential pollinators could only enter the female cone readily while space is available between the megaporophylls. Access before and after would be very difficult, and insects, except the very small, e.g. *Collembola*, *Sericoderus* (Corylophidae), minor worker ants, nymphs of *Aspidiotus hederae*, early instars of *Pseudococcus zamiae* not departed before closure would be imprisoned for weeks, due to growth of the cone.

Some weeks later, by elongation of the cone axis, open spaces develop between the megasporophylls, beginning at the apex of the cone and the sporophylls become loose at their point of attachment to the rachis. The megasporophylls eventually fall to the ground or come to rest on underlying vegetation and the seeds separate from the rest of the megasporophyll.

For present purposes, the development of the female cone may be said to have four phases:

1. growth - from the initiation of growth;

2. the receptive stage, when the rachis elongates producing vertical spaces between the megasporophylls;

3. the closed stage, when the external spaces between the megasporophylls are eliminated; and

4. the dispersal stage, when the megasporophylls fall from the rachis and the seeds are detached.

Unfertilised female cones, e.g. on isolated plants, have been observed not to disintegrate as do normal fertilised cones, but to retain their sporophylls, which turn purplish and shrink, leaving visible spaces between the sporophylls, ultimately turning brown, drying out and disintegrating over a period of months.

At Wamberal, an isolated, i.e. unpollinated, roadside plant produced cones as follows: 1980 - 2 cones; 1081 - 0; 1982 - 2; 1983 - 2; 1984 - 2; 1985 - 0; 1986 - 3; 1987 - 2; 1988 - 0; 1990 - 2.

In phase 1 there is considerable space inside well-developed female cones. In mid-November, 1979 a typical unopened/unreceptive cone was cut transversely about one-third of the way below the apex; the distance between the rachis and the tip of the micropyle was 5mm; the distance between stalks of the megasporophylls at their distal ends could exceed 15mm. There would thus be abundant room for both vertical and horizontal movement for small insects in a cone (Fig. 7).

In phase 2, the vertical distance in millimetres between ten pairs of megasporophylls of two plants at Noraville was as follows:

1: 4.3, 3.0, 3.3, 3.5, 3.0, 5.2, 3.5, 3.0, 4.6, 3.2

Average 3.66, maximum 5.2, minimum 3.0, mean 4.1;

2: 5.0, 4.2, 3.0, 4.5, 3.5, 3.0, 3.5, 4.0, 4.5, 4.0

Average 3.92, maximum 5.0, minimum 3.0, mean 4.0.

Other measurements ranged up to 5.5mm.

The body depth of five males and five females of *T. lyterioides* collected at Balgowlah on 3 December, 1979 was as follows:

Male: 1.5, 1.9, 1.5, 1.7, 1.5

Average 1.62, maximum 1.9, minimum 1.5, mean 1.4;

Female: 1.8, 1.8, 1.6, 1.6, 1.5

Average 1.66, maximum, 1.8, mean 1.65.

Thus, access to female cones presents no difficulties at this stage.

The writer cut a vertical section through the receptive stage of a cone of *M. communis* on 7 November, 1989, photographed the section and had it drawn (Fig. 7). There is a conspicuous space above, below and behind each seed. The micropyle is conspicuous as the most distal part of the seed and projects into a well-marked space. A similar section taken about six weeks later showed an absence of space between, but not behind, the seeds (Fig. 7).

Daily observations were not practicably to determine the length of time individual female cones remained open, but at Noraville a cone which was open on 21 September 1980 had closed on 5 October 1980. At Seven Mile Beach, 18 out of 48 cones were slightly open on 29 November 1981. Brough and Taylor (1940) at Narooma noted that in one cone the sporophylls had separated on 12 November 1939 and in another case the sporophylls were more widely separated on 10 November 1939. These data would indicate that the species as a whole is receptive to pollination during a period of several weeks over its geographical range.

Night observations at Noraville failed to find any weevil associated with *Macrozamia* cones in the spring of 1981.

Observations were then made at Seven Mile Beach where cones were also numerous. During the moonlit evening of 9 December, 1981, more than 50 female cones were examined. One cone had three *T. lyterioides* crawling over it; another had one weevil; all other cones were devoid of weevil.

On the following night, 39 female cones were examined at night, but no *Tranes* were found. During the night of 11 December, beetles were observed on a female plant between 8:45 and 10:05 pm (Eastern Australian Standard time). They were crawling on the cones, peduncles, petioles, pinnules and even resting on the tips of pinnules, also on the ground debris (mostly dried leaves). They seemed to be crawling upwards in most cases. Some beetles had a very obvious coating of pollen. At this time, the seeds in some cones had expanded and no visible spaces occurred between the megasporophylls, so that beetles could not enter. After 10:30 pm, 12 female cones were examined. Three *T. lyterioides* were seen on two cones, but no beetles were found on the other ten cones. Another female cone was placed in a sealed plastic bag overnight and, when examined the next day, contained 146 (63 male and 83 female) *T. lyterioides*, indicating a cone in the receptive phase and adding to the author's earlier conviction of *T. lyterioides*' involvement in pollination.

As adults of *T. lyterioides* move quite actively at night and may be present in large numbers in the vicinity of a female cone, over a hundred having been counted on a number of occasions, they would have no difficulty in pollinating a cone in a short time under suitable conditions. As weevils have been seen only in very small numbers i n female cones collected in daylight (Baird 1939; Brough and Taylor 1940, and the author) it would be more than possible that they leave the cones before light returns to take shelter in male cones or at the bases of fronds in the caudex.

The writer believes that weevils are attracted to the female cone, which they reach either by walking or flying, and deposit pollen grains from their bodies or mouthparts coming in contact with the micropylar drop. Considerable numbers of pollen grains may be deposited and reach the pollen chamber (Brough and Taylor 1940, see their Fig. 38).

It has been demonstrated that *T. lyterioides* can develop from larva to adult in the male cone: the larva consumes the fleshy tissue and the adult feeds on pollen. Thus, the beetle should not need to enter the female cone at all. However, there is no doubt that it does so in considerable numbers, in spite of the literature. Why both sexes enter the female cone is not obvious. Perhaps this cone supplies ingredients necessary for the reproductive processes. What stimulates the beetle to enter the female cone?

One might speculate that the beetle responds to a stimulus emitted by the plant when the female cone is

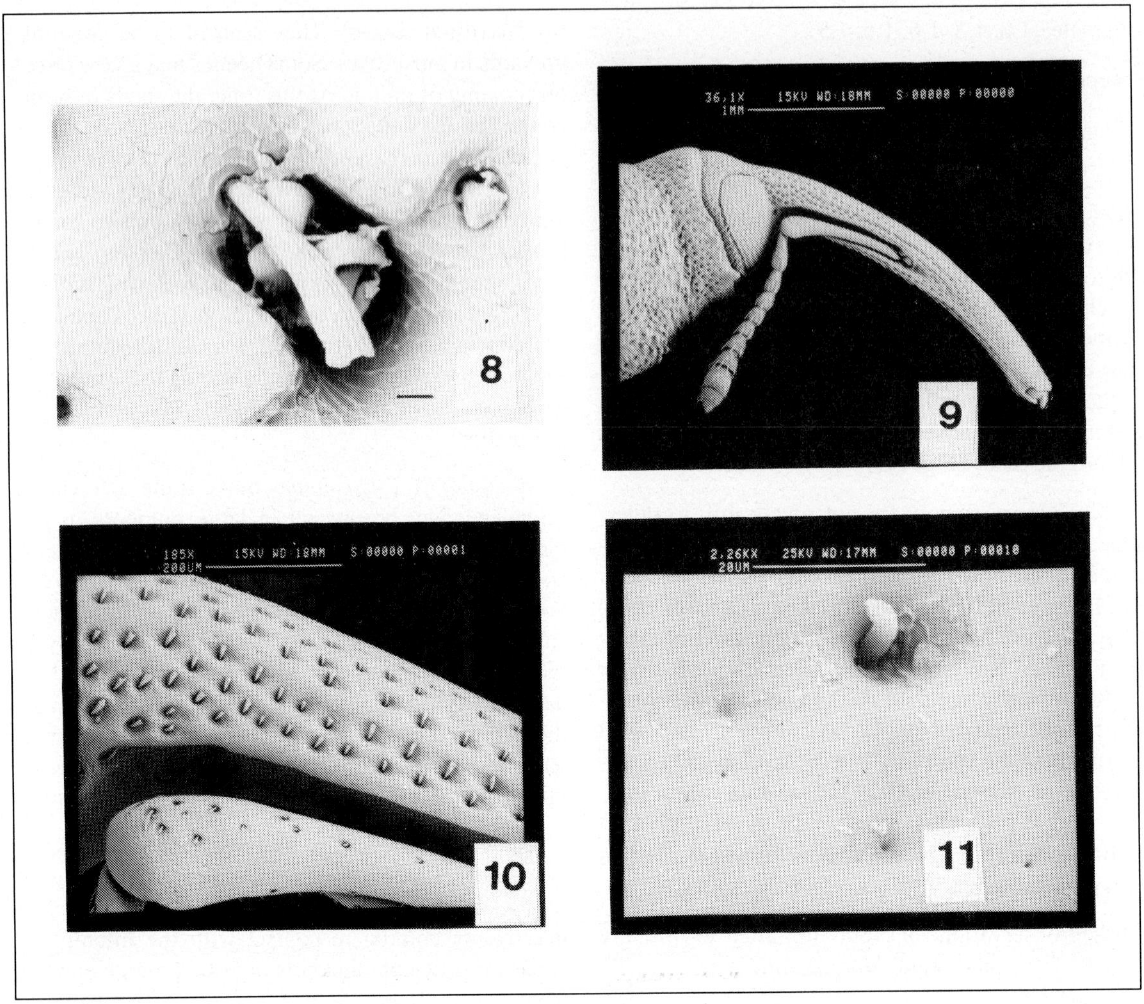

Figures 8-11: **8**: *Tranes sparsus*, portion of elytron. Pollen grains held by abraded scale. Scale bar = 10μm (photo by Cathy Miller). Figures **9-11** *T. lyterioides*, scale bars indicate true dimensions. **9** Right side, head and protorax. **10** Portion of rostrum and basal antennal segment (scrape), showing abraded scales and minute pores. **11** Three sizes of pores at distal end of rostrum.

ready for pollination, e.g. from the micropylar drop. Barth (1985) has summarised some of the work done on the senses of insects which indicates that some have extreme sensitivity to stimuli not detectable by human senses.

On the evening of 27 November, 1990, ten *T. lyterioides* were released into a plastic box containing 12 megasporophylls, each previously seen to contain at least one micropylar drop. No response was observed from the beetles. However, SEMs show that structures believed to be sensory in function are present on the head, rostrum and antennae of the species. The exact function of these is yet to be determined (Figs. 9 and 11), although studies quoted by Schneider and Steinbrecht (1968) and by Barth (1985) indicate olfactory possibilities.

Discussion

There has long been a difference of opinion among botanists as to whether anemophily or entomophily occurs in Gnetales and Cycadales. J.D. Hooker (1864) believed that pollination in *Welwitschia mirabilis* was due to insects. Strasburger (1872) quoted by Pearson (1906) was convinced that it was anemophilous, but that its ancestors may have been entomophilous.

Pearson (1906a), in relation to *Welwitschia mirabilis*, believed that the sticky pollen masses were unlikely to be removed by wind and that the plant was partially, if not entirely, insect-pollinated. Later (1907, 1909), he observed pollen on the abdomen and appendages of the pyrrhocorid bug *Odontopus sexpunctulatus*, which he considered an important, but not the only, pollen carrier.

Karsten (1892, 1893) quoted by Pearson (1906a) concluded that in *Gnetum gnemon*, flowers once anemophilous were in part adapted to insect pollination. Micropylar drops had a sweet taste and were visited by ants.

Jaccardi (1894) quoted by Pearson (1906a) stated that in *Ephedra helvetica*, wind was the principal pollinating agent, and that the part played by insects was purely fortuitous. Lotsij (1899) quoted by Pearson (1906a) stated that in *Gnetum gnemon*, pollination was mainly due to wind, but that insects might play a part in it.

Webber (1901) concluded that pollination in *Zamia floridana* and *Z.pulima* occurred mainly through the agency of wind.

J.D. Hooker (1904, in Oliver and Scott, 1904), referring to the fossil *Lagenostoma lomaxi*, expressed the opinion that cycads were entomophilous.

Pearson (1906b) believed entomophily possible in *Encephalartos villosus* and commented on pollen being carried by the weevil *Phloeophagus hispidus*. He regarded wind as an inefficient pollinating agent in *Stangeria* sp. (*S. katzeri* = *S. eriopus*). Rattray (1913) considered *E. villosus* entomophilous: in *E. altensteinii* both anemophily and entomophily were possible; in *Stangeria katzeri*, anemophily held good. Chamberlain (1913) firmly favoured anemophily in all cycads.

Marloth (1913) stated that the weevil found by Rattray on *E. altensteinii* and on *E. villosus* and called *Phloeophagus* was really *Antiliarhinus zamiae*, which bores into the seeds of both species and also pollinates the seeds. Another weevil, *Derelomus languidus*, was recorded from both male and female cones. Dyar (1965) conceded that the possibility of insect pollination could not be refuted, whether it occurred by design or by chance. However, very little had been added to knowledge of the subject since the publication by Marloth.

Norstog and Stevenson (1980), also Norstog (1987) pointed out that in Florida male and female plants of *Encephalartos* (African) and *Macrozamia* (Australian) in close proximity did not produce fertile seeds unless pollinated artificially. The native *Zamia pumila* and Mexican *Z. furfuracea* produced fertile seeds. Each of these species is attended by small weevils, which breed in microsporophylls and enter female cones. The authors drew attention to the work of Rattray (1913).

Norstog, Stevenson and Ninlas (1986), also Norstog and Fawcett (1989) demonstrated that the weevil *Rhopalotria mollis* is capable of pollinating *Z. furfuracea*. In a study of *Z. pumila*, Tang (1987a) concluded that the languriid beetle *Pharaxonotha zamiae* and the weevil *Rhopalotria slossoni* were probable pollinators of *Zamia pumila*. In their absence, viable seeds were not produced.

Baird (1940) was convinced that insects played an important part in the pollination of *M. riedlei* and mentioned a mealy bug *Dactylopius macrozamiae* and a small brown beetle. Crowson described a new species of boganiid beetle from Queensland and speculated on the possible role of Boganiidae in the pollination of *M. riedlei* (1990). I believe that an adequate study of the life history and habits of *Tranes roei*, a frequently recorded visitor to this cycad, would be well worthwhile making in this connection.

Neither Baird nor Brough and Taylor recorded any retention of insect material and the author has not located any relevant specimens.

Incidentally, L.A. Mound (pers. commun.) has speculated on a possible role for the undescribed thrips in the pollination of *M. communis*. I have not investigated this hypothesis.

In the literature, Ornduff (1989) recorded *T. lyterioides* from *Lepidozamia peroffskyana* near Coffs Harbour, and Covassi (1974) described damage by the weevil *Demysus meleoides* to *Macrozamia macdonnelli* and to *Encephalartos villosus* in Florence, Italy.

The complexity of wind movement in relation to pollination has been examined by Niklas (1985), and Niklas and Norstog (1984) used wind tunnels to study the deposition of pollen on cones. They concluded that the probability of direct pollen-to-micropyle impaction in *Cycas* and *Dioon* was low. In *Zamia furfuracea*, occasionally pollen could be deposited on ovules with inwardly facing micropyles. They stated (1984) that three species of *Encephalartos*, one *Microcycas*, two of *Dioon*, all exotic species, produced fertile seeds only by artificial pollination.

Norstog (1987) pointed out that there is little experimental evidence of the role of air currents in pollination and that only in the genus *Cycas* alone are ovules directly accessible to air-borne pollen. In all other cycads, the ovules are screened by megasporophylls.

For the theory of anemophily to obtain credence in *Macrozamia communis*, it would be necessary to demonstrate that in dense thickets, such as those at Seven Mile Beach, Nelligen (Clyde River), Nelson Beach, and Moon Beach, wind circulates at sufficient speed to carry pollen between individual plants. In addition, proof would be needed that by some means numbers of pollen grains are transported several centimetres from the exterior of the cone to the micropyle.

Bough and Taylor (1940) dismissed the possibility of entomophily and readily accepted anemophily without giving reasons for either decision. Like others, they gave no evidence of seriously investigating the possibilities sufficiently before making their pronouncements.

The transfer of pollen in *M. communis* by wind would have certain necessitates: mature pollen, fine weather, adequate wind velocity in the right direction, a female cone in the receptive stage, a wind current entering the cone and penetrating to the interior. Wind transfer would mean that pollen grains are transferred as individuals, not *en masse*.

Insect pollination would require mature pollen, dry weather and some form of sensory perception to attract insects to the interior of the receptive female cone; wind would not be needed. A beetle carrying pollen would transport many grains on a firm substrate, so that numerous grains would come into contact with the micropylar drop at the same time if body contact is made. The drawing by Brough and Taylor (1940, their Fig. 38) shows a vertical section with 11 pollen grains, probably only a fraction of the grains in a pollen chamber, which they stated were 'numerous'. In my view, this is more in keeping with insect transfer.

The micropyle in this and most other cycads is at the greatest possible distance from the exterior of the cone and could not be more unfavourably placed for wind pollination (Fig. 2a). This points to pollination from the inside of the cone, not from outside, which would necessitate a relatively long and tortuous, but not impossible, journey on the part of the pollen. It would also require an inward current of air, which presupposes an exit.

Most authors, including the writer, appear to have started with the idea that wind transfer was the accepted explanation for cycad pollination and therefore made no attempt to question the theory by making field observations under natural conditions, starting from first principles. It appears to have been assumed that pollination had to occur during the day. In Australia, authors, with one exception, have dismissed out of hand or with scant notice, insects as possible pollinators. In the present study, observed nocturnal activity and diurnal quiescence in the weevils aroused curiosity in relation to pollination and stimulated nocturnal observations. The writer has not found any record of this line of action in previous publications. In fact, except for the paper by Tang (1987b), very little, if anything, seems to have been done at night.

Norstog (1987) drew attention to the unreliability of techniques for excluding beetles from female cones and the ability of beetles to pass through seemingly impenetrable barriers. The mechanical structure of the weevil body in the shape of an elongated cone would give it leverage in penetrating flexible barriers. *Tranes lyterioides* and *T. sparsus*, although often passive during the day, are active at night. To be really satisfactory, an excluder would need to be constructed of inflexible material, and to be monitored for efficiency, 24 hours a day, at least initially. In any case, the use of excluders can hardly be accepted as a substitute for direct observations on what insects actually do. Their use is really an unintentional and unconscious recognition of nocturnal habits in the insect.

The writer believes that in Australia more attention should be given to the possible involvement of insects in cycad pollination by observing and collecting likely pollinators from the plants and retaining these permanently, fully documented, in curated insect collections for possible future studies related to pollination.

The Australian Museum insect collection includes beetles collected in association with cycads outside the distribution of *M. communis*, e.g. J.R. Machonochie found *Hapalips* n.sp (Languriidae) in crowns of *Cycas cairnsiana* SW of Cairns: P. Hind collected the weevil *Cossonidius* sp. on *C. media* east of Coen and on male cones of *Cycas* n.sp. discharging pollen east of Biloela. J. Lawrence (pers. commun.) states that the collection of the CSIRO, Canberra, contains numbers of beetles from cycads growing in northern Queensland. Numbers of specimens of *Tranes roei* in the collection of the Australian Museum came from *M. riedlei* in Western Australia.

Acknowledgements

The writer is indebted to R.H. Mulder for originally drawing his attention to the fact that *T. lyterioides* frequented the male cones of *M. communis*. P. Hind and J.R. Machonochie collected beetle specimens from *Cycas* spp. The following have supplied determinations: R.T. Thompson and E.C. Zimmerman, Curculionidae; G.A. Holloway, Hymenoptera; I.F.B. Common, Lepidoptera; L.A. Mound, Thysanoptera.

The Royal Botanic Gardens, Sydney, has made available additional data on the distribution of *Macrozamia communis*.

Australian Museum authorities have provided facilities enabling laboratory work to be done.

Scanning electron micrographs have been taken by G. Avern, who has also contributed useful discussions and other assistance. However, Cathy Miller was responsible for Fig. 7. G.A. Holloway and D.K. McAlpine have made it possible to obtain the SEMs. The drawings are the work of Lisa Parkinson.

References

***Baird, A.M.* 1939.** A contribution to the life history of *Macrozamia riedlei*. J. Roy Soc. W. Aust. 25:153-175.

***Barth, F.G.* 1985.** Insects and Flowers. The Biology of a Partnership. Princeton University Press, Princeton, New Jersey.

***Brough, P. and Taylor, M.H.* 1940.** An investigation of the life cycle of *Macrozamia spiralis* MIQ. Proc. Linn. Soc. N.S.W., 65(1): 494-524.

***Chamberlain, C.J.* 1913.** *Macrozamia moorei*, a connecting link between the living and fossil cycads. Bot. Gax. 55: 141-154.

***Covassi, M.* 1974.** Il *Femyrsus meleoides* Pascoe: un potenziale nemico Cicadee ornamentali introdotta in Italia (Cleopatra, Curculionidae). Redia, 60: 211-218.

***Crowson, R.A.* 1990.** A new genus of Boganiidae (Coleoptera) from Australia, with observations on glandular openings, cycad associations and geographical distribution in the family. Aust. Ent. Soc. 29(2): 91-99.

***Dyar, R.A.* 1965.** The cycads of Southern Africa. Bothalia, 8(4)(1): 405-429.

***Hooker, J.D.* 1864.** On *Welwitschia,* a new genus of Gnetaceae. Trans. Linn. Soc. Lond., 24: 1-48, p1. 1-16. (1904) in Oliver & Scott.

***Jaccardi, P.* 1894.** Rechersches Embryologiques sur l'*Ephedra helvetica*. Lausanne.

***Karsten, G.* 1892.** Beitrag zur Entwickelungsgeschichte einiger Gnetum-Aretn. Bot. Zeit., p.205. (1893) Zur Entwickelungs- Geschichte der Gatting Gnetum. Beitrage zur Biologie der Pflanzen (F. Cohn). 6, 337.

***Lawson, A.A.* 1926.** A contribution to the life history of *Bowenia*. Proc. Roy. Soc. Edinburgh, liv, 357-94.

***Lotsij, J.* 1899.** Contributions to the life history of the Genus *Gnetum*. 1. 1. The Grosser Morphology of the Production of *Gnetum gnemon* L. Ann. Jard. Buitenzorg, (Ser. 2), 1, 46.

***Marloth, R.* 1913.** Flora of South Africa. Vol. 1. Darter Bros, & Co., Cape Town.

***Niklas, K.J.* 1985.** Wind Pollination - a study in controlled chaos. Am. Sci., 73: 462-470.

***Niklas, K.J. and Norstog, L.* 1984.** Aerodynamics and pollen grain depositional patterns on cycad megastrobili: implications on the reproduction of three cycad genera (*Cycas, Dioon* and *Zamia*). Bot. Gaz., 145(1): 92-104.

***Norstog, K.* 1987.** Cycads and the origin of insect pollination. Am. Sci., 75, 270-279.

***Norstog, K. and Fawcett, P.S.K.* 1989.** Insect-cycad symbiosis and its relation to the pollination of *Zamia furfuracea* (Zamiaceae) by *Rhopalotria mollis* (Curculionidae). Amer. J. Bot. 76 (9): 1380-1394.

***Norstog, K. and Stevenson, D.W.* 1980.** Wind? or Insects? The pollination of cycads. Fairchild Tropical Garden Bull. Jan. 35: 28-30.

***Norstog, K., Stevenson, D.W. and Niklas, K.J.* 1986.** The role of beetles in the pollination of *Zamia furfuracea* L. fil. Biotropica 18(4): 300-306.

***Oliver, F.W. and Scott, D.H.* 1904.** On the structure of the Palaeozoic seed *Lagenostoma lomaxi* etc. Phil. Trans. (Roy. Soc. Lond.), B, 197: 193-247.

***Ornduff, R.* 1989.** Size, distribution and coning behaviour of the Australian cycad *Lepidozamia peroffskyana*. Aust. J. Ecol. 14: 241-245.

***Ornduff, R.* 1990.** Geographic variation in reproductive behaviour and size structure of the Australian cycad *Macrozamia communis* (Zamiaceae). Amer. J. Bot. 77(1): 92-99.

***Pascoe, F.P.* 1875.** Additions to the Australian Curculionidae. Part VIII. Ann. Mag. Nat. Hist., (4)xvi, 55-56.

***Pearson, H.H.W.* 1906a.** Some Observations on *Welwitschia mirabilis* Hooker, f. Phil. Trans. Roy. Soc. Lond., B, 198 (1905): 265-304. (1906b) Notes on South African Cycads. 1. Trans. S. Af. Phil. Soc. xvi, 341-354, p1, 6-8.

***Pearson, H.H.W.* 1907.** The Living *Welwitschia*. Nature, 75: 536.

***Pearson, H.H.W.* 1909.** Further Observations on *Welwitschia*. Phil. Trans. Roy. Soc. Lond., B, 200: 331-402. Pls. 22-30.

***Rattray, G.* 1913.** Notes on Pollination of some South African Cycads. Trans. Roy. Soc. S. Af., III: 259-270.

***Schneider, D. and Steinbricht, R.A.* 1968.** Checklist of Insect Olfactory Sensilla. Symp. Zool. Soc. Lond., No. 23: 279-297.

***Strasburger, E.* 1872.** Die Coniferen und die Gnetaceen. Jena.

***Tang, W.* 1987a.** Insect Pollination in the cycad *Zamia pumila* (Zamiaceae). Am. J. Bot., 74(1): 90-99.

***Tang, W.* 1987b.** Heat Production in Cycad Cones. Bot. Gaz. 148: 165-174.

***Webber, H.T.* 1901.** Spermatogenesis and Fecundation of *Zamia*. U.S. Dept. Agric. Bur. Pl. Ind. 2, 100.

* Not seen by the author.

Comparative Population Biology of the Australian Cycads *Macrozamia riedlei* and *M. communis*

S.W. Connell
Department of and Palaeoanthropology
Baron Hay Court, South Perth, 6151

P.G. Ladd
School of Biological and Environmental Sciences
Murdoch University, Murdoch 6150 Australia

Introduction

Macrozamia riedlei is the sole cycad found in south western Australia while *M. communis* is the most widespread of the eastern Australian cycads. The species are closely related (Johnson, 1958) and are morphologically and ecologically similar. They occur across similar latitudinal ranges though under very different climatic regimes; *M. riedlei* under Mediterranean conditions and *M. communis* in sub-tropical to temperate climates.

Methods

In Western Australia the study was undertaken at Hill River Nature Reserve (Lat. 30º 18', Long 115º11'). The vegetation was heathy woodland dominated by *Eucalyptus todtiana* and *Banksia* spp.). In eastern Australia, the study was conducted in Yalwal State Forest (Lat. 34º 52', Long. 150º 30'). Vegetation was dry to wet sclerophyll forest dominated by *Eucalyptus* spp.

Studies were undertaken in large permanent plots (to 1 ha) in which all individuals were counted and tagged. Plant measurements included some or all of the following, depending upon accessibility of the individual plant (i.e. not buried in the case of trunk measurements); number of leaves, length of the longest leaf, maximum trunk diameter, and trunk height. Observations of the reproductive attributes of mature plants included sex, number of cones and their lengths, number of ovule and seeds (females). Individual plants were excavated outside the permanent plot to obtain morphological measurements. Population sizes and stage structures were determined from ovule through to adult. Demographic studies of *M. riedlei* are based on monitoring over 4 years while *M. communis* information was obtained over a period of 6 months.

Observations

In *M. riedlei* (and undoubtedly in *M. communis*) the trunk contributes the major part of the plant biomass (Halliday and Pate, 1976). Size classes based on trunk length or girth should be used to structure populations. Trunk diameter is an appropriate size index in all but the most mature plants. In the latter case, the apical

Table 1
Survivorship between life history stages in *M. riedlei and M.communis*
(see text for definitions of size classes)
Percentages in parentheses are cumulative survivorship

Life History Stage	*M.riedlei*	*M.communis*
Ovule→ Seed	46%	76%
Seed→Seedling	17% (8%)	15% (11%)
Seedling→Juvenile	31% (2.4%)	38% (4.3%)
Juvenile→Adult	65% (1.6%)	68% (2.9%)

Table 3

Reproductive attributes of *M. riedlei* and *M.communis*

Attribute	*M.riedlei*	*M.communis*
Cones per male	4.2 ± 2.0	1.9 ± 0.8
Cones per female	2.4 ± 1.0	1.8 ± 0.5)
Ovules/cone	202 ± 58	160 ± 17
Seed weight (g)	27.6 ± 2.4	11.4 ± 1/5
Seed set (%)	46	76

meristem will have reached a presumably genetically determined maximum size (Stevenson, 1980b). At this stage the only increase in plant size is in trunk length. Trunk length is a deceptive measurement of size for most plants, as the contractile behaviour of the tap root and stem means that the shoot apex is continually pulled down (Stevenson, 1980a). Variability in the resistance of the substrate to penetration will affect the degree to which the apex of the plant can be drawn down. Size indices based on trunk girth initially and, perhaps, trunk height later (once maximum apical meristem size has been reached) are required to reliably examine population structure in these two species. As the majority of plants possess an underground trunk in many situations, an algometric relationship

Table 2

Temporal Population Dynamics of *M.riedlei*

Year	Ovules	Seeds	Germinants	Seedlings	Juveniles	Adults		
						?	Male	Female
1984	?	?	?	>650	>200	60	39	31
1985	0	?	?					
1986	0	0	?					
1987 R	8443	0	0	650	200	46	32*	20*
1988	0	3926	0					
1989	0	0	666					
1990	0	0	0	447 + <650	<200	46	46	38

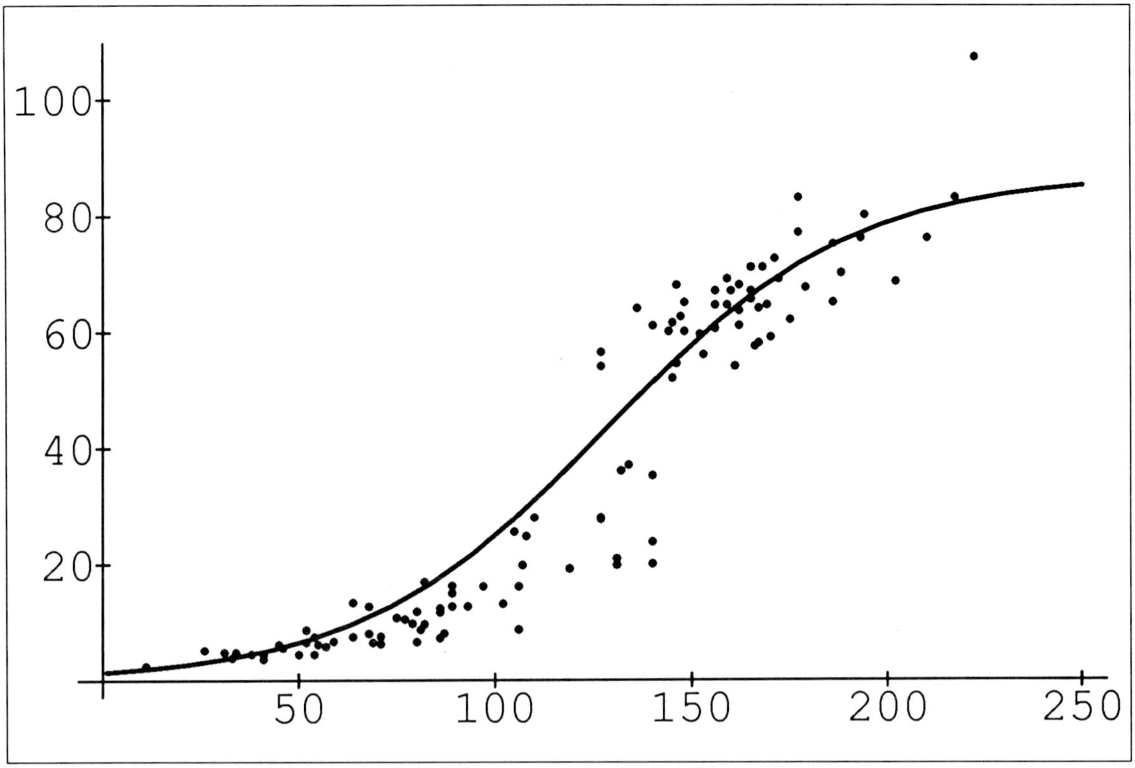

Figure 1: Leaf Length as an index of trunk diameter in *M. riedlei*
x axis label = Leaf length (cm)
y axis label = Trunk diameter (cm)

must be found that is an accurate key to trunk girth. The population of *M. riedlei* is most accurately size-structured using a logistic regression model with leaf length as a predictor of trunk diameter (Fig. 1). In contrast to previous results of Clark and Clark (1988), Ornduff (1985, 1990) and Newell (1983), the number of leaves possessed by a plant was a poor indicator of its size (i.e., its trunk size). This was particularly the case for small plants where only one to three leaves were possessed by plants varying in size from <1 to 20cm diameter.

Shifting patterns of resource allocation may be suggested from Figure 1. Initially leaf length (an indication of leaf area) increases more rapidly than does trunk girth (though perhaps not stem and/or root length). In the second phase, increments in trunk diameter are similar to increments in leaf length (though perhaps not in leaf number). In the third phase, the asymptotic reproductive phase, trunk diameter increment rates diminish. At this time energies formerly allocated to trunk biomass are possibly diverted to reproduction. Trunk height increase, which is essentially a function of leaf production, continues. These observations, though crude, indicate that a proper understanding of size, growth and reproduction, in *Macrozamia* at least, will require measuring the changes in leaf number and size, trunk diameter and height and the reproductive status of individual plants. A multiple attribute system would then be produced which would enable very accurate size structuring of populations and deeper understanding of the processes involved in the population and reproductive ecology of these species.

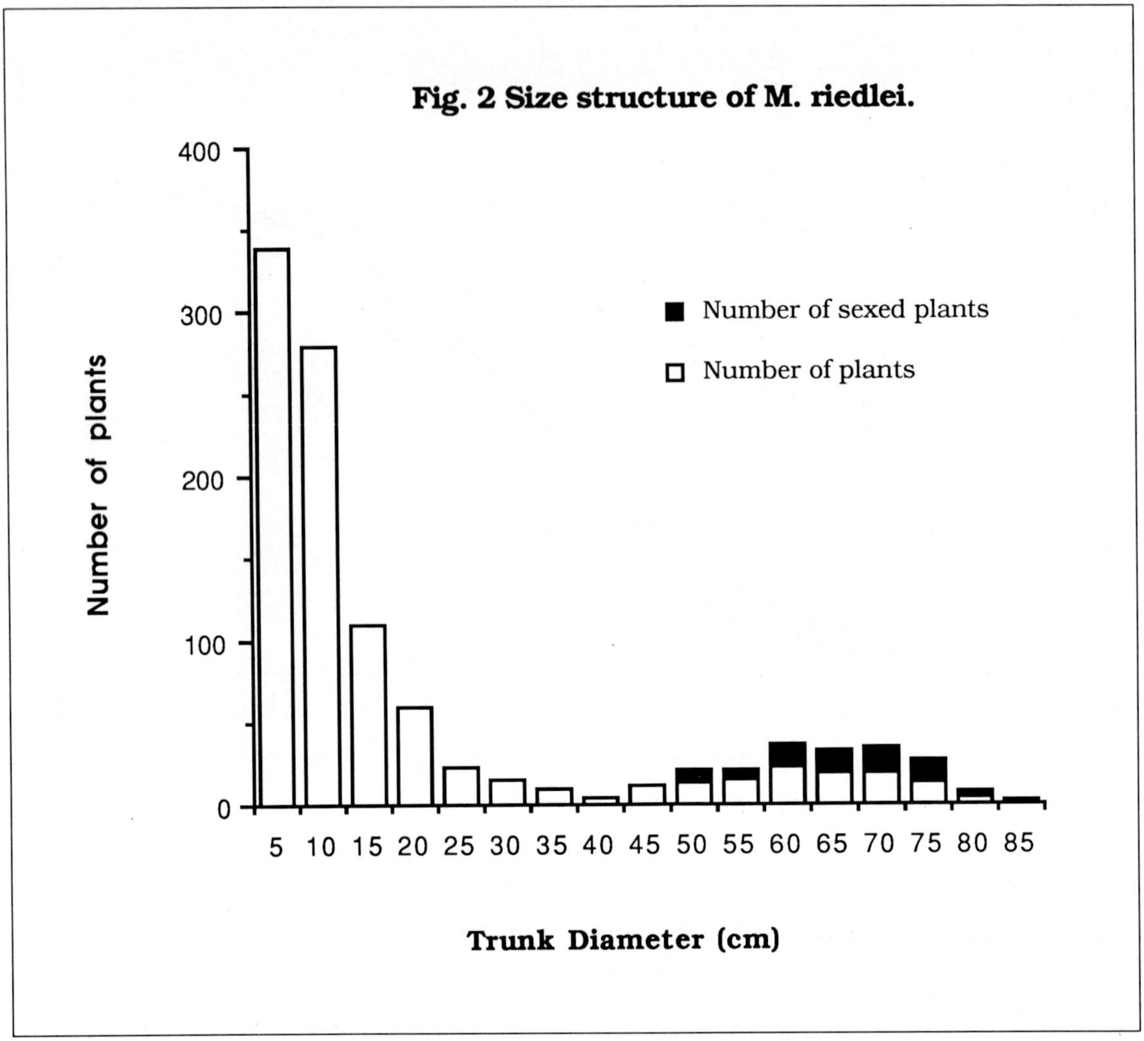

Figure 2: Size structure of *M. riedlei*

Both species experience irregular recruitment related to the incidence of fire (e.g. *M. riedlei*, Fig. 2). Seeds are dispersed by birds and small mammals which remove the coloured fleshy sarcotesta either within the digestive system (mostly in large birds) or by gnawing the seed. Sarcotesta removal is not necessary for seed germination. In contrast to the studies of Tang (1989) with *Zamia* seed dispersal patterns show that, even after animal processing, the majority of seeds remain within or just outside the canopy of the parent plant (Fig. 3). Studies of *M. riedlei* by Burbidge and Whelan (1982) and of *M. communis* by Ballardie and Whelan (1986) detected similar patterns of seed dispersal from parent plants. Density patterns of seedlings (trunk girth cm) and juvenile plants (trunk girth <5 - 20cm) suggest that survivorship of recruits is greater outside the female canopy. Competition for water, nutrients and light between siblings and with the adult plant is probably involved in the spatial patterning of mortality. Both species exhibited nega-

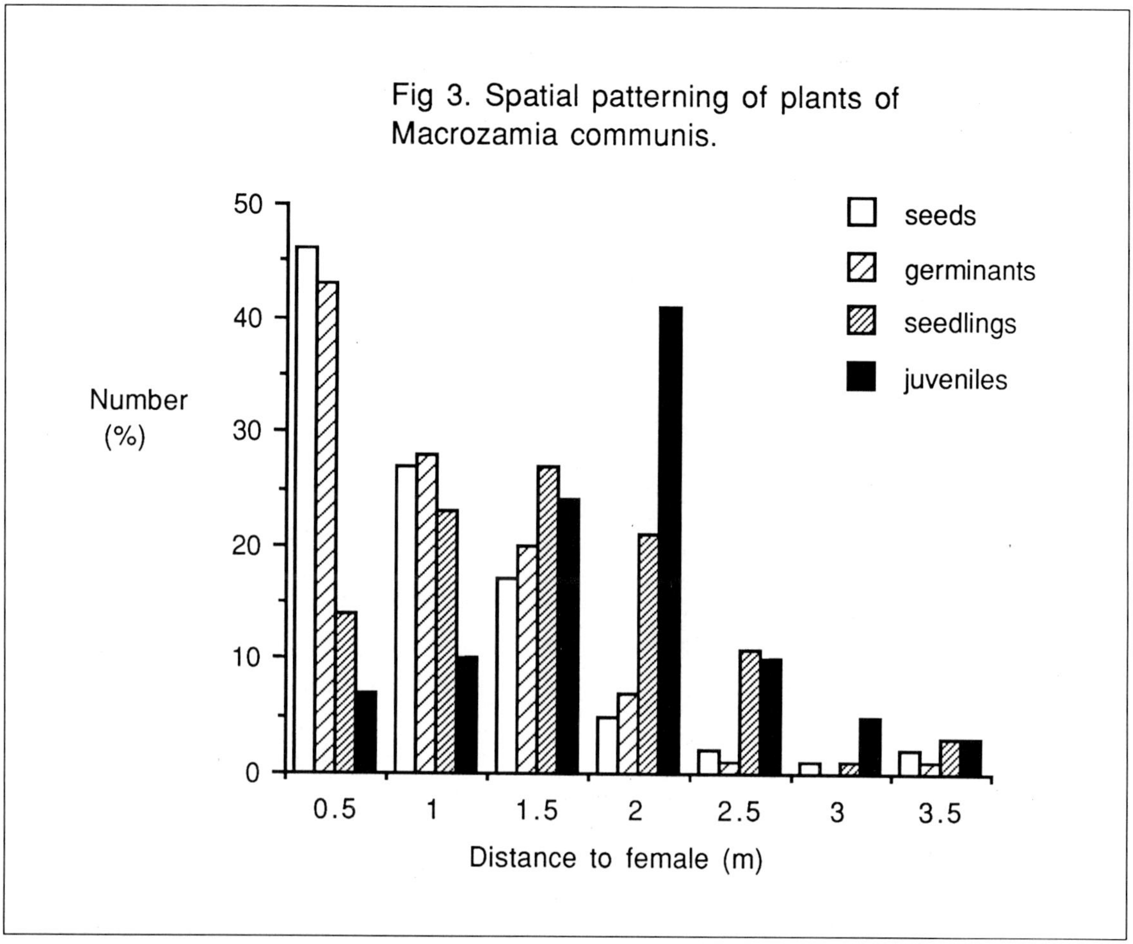

Figure 3: Spatial patterning of plants of *M. communis*

tive exponential mortality rates such that less than one percent of ovules survived to become adults in *M. riedlei* (Fig. 2 and Table 1). Survivorship was greater in *M. communis* but this may be an over-estimate due to the short time of the study (Table 1). Overall patterns of survivorship were similar with seed germination being the time of greatest mortality.

Phenotypic sex ratios were male biased but it is suggested that the genotypic sex ratios were one in both species (Table 2). The male bias of individual reproductive events may relate to the different cost of reproduction to males and females and to environmental variation. Studies with a range of cycad species have reached a similar conclusion (e.g. Ornduff 1990). Spatial patterning of the sexes and reproduction in dioecious species has been documented by (Clark and Clark, 1987, 1988; Meagher, 1981) while temporal patterns of male and female reproduction are poorly understood (Oyama, 1990; Clark and Clark, 1988). Given the long reproductive cycle of cycads, greater than one year from strobilus initiation to disintegration, and the intermittent nature of coning, annual sex ratios may vary considerably and the only way to determine the true sex ration of the adult plants is by long term study. This will not answer questions concerning differential spatial and temporal mortality patterns of the sexes. These issues may be resolved by genetic studies, though as yet sex chromosomes have not been illustrated for any cycad species.

Reproduction, and consequent seedling recruitment, were synchronised (probably by fire) and individual plants had a reproductive periodicity of approximately four years (Table 2). Similar periodicities have been detected in *Zamia skinneri* by Clark and Clark (1987) and have been informally reported by Giddy (1974) for *Encephalartos*. The rhythm of reproduction in cycads is in part due to the long time between cone initiation and seed release by females; approximately 15 months in *M. riedlei* (Baird, 1938). Because the one reproductive event takes more than one year to complete in females, individual plants may not be able to reproduce in consecutive years. If co-occurring males were to reproduce every year then at least every second event would be energetically profligate. In *Macrozamia* species and other gymnosperms from fire-prone environments, reproduction is probably synchronised by fire. Once synchronised, individual plants may remain in phase until the next fire. Such synchronisation would ensure that males and females in close proximity would reproduce together; potentially increasing the probability of a plant's genes being transferred to the next generation. As the production of individual cones is probably more expensive in females this may account for the observation that males produce more cones than females (Table 3, see also Clark and Clark 1987; Ornduff, 1985). Simple mathematical models testing the number of potential mates under differing scenarios (i.e. random temporal reproduction versus intermittently synchronised reproduction, etc.) confirm the increased probability of mating in the above situation (Connell and Ladd, unpub.).

Average number of cones, ovule production and seed weight were greater in *M. riedlei* than in *M. communis*. Percentage seed set, while varying considerably between individuals in both species, was on average similar (Table 3). Though the periodicity of reproduction is still unclear due to the short term of the study of *M. communis*; it would appear that plants of *M. riedlei* devote more energies to reproduction than do *M. communis* plants. This may be related to the differing climatic situations of the two species. The mediterranean climate of south western Australia produces an annual summer drought. Large seed size may confer a greater probability of establishment to individual seedlings.

References

***Baird, A.M.* 1938.** A contribution to the life history of *Macrozamia riedlei*. J. Roy. Soc. West. Aust. 25: 153 - 169.

***Ballardie, R.T., and Whelan, R.J.* 1986.** Masting, seed dispersal and seed predation in the cycad *Macrozamia communis*. Oecologia 70: 100 - 105.

***Brough, P. and Taylor, M.H.* 1940.** An investigation of the life cycle of *Macrozamia spiralis* Miq. Proc. Linn. Soc. N.S.W. 65: 494 - 525.

***Burbidge, A.H. and Whelan, R.J.* 1982.** Seed dispersal in a cycad, *Macrozamia riedlei*. Aust. J. Ecol. 7: 63 - 67.

***Clark, D.A. and Clark, D.B.* 1987.** Temporal and environmental patterns of reproduction in *Zamia skinneri*, a tropical rain forest cycad. J. Ecol. 75: 135 - 149.

***Clark, D.A. and Clark, D.B.* 1988.** Leaf production and the cost of reproduction in the neotropical rain forest cycad. *Zamia skinneri*. J. Ecol. 76: 1153 - 1163.

***Giddy, C.* 1974.** Cycads of South Africa. Purnell, Capetown.

***Halliday, J. and Pate, J.S.* 1976.** Symbiotic nitrogen fixation by coralloid roots of the cycad *Macrozamia reidlei*: physiological characteristics and ecological significance. Aust. J. Plt. Physiol. 3: 349 - 358.

***Johnson, L.A.S.* 1959.** The families of cycads and the Zamiaceae of Australia. Proc. Linn. Soc. N.S.W. 84: 64 - 117.

***Meagher, T.R.* 1981.** Population biology of *Chamaelirium luteum*, a dioecious lily. II. Mechanisms governing sex ratios. Evolution 35: 557 - 567.

***Newell, S.J.* 1983.** Reproduction in a natural population of cycads (*Zamia pumila* L.) in Puerto Rico. Bull. Torrey Bot. Club 110: 464 - 473.

***Ornduff, R.* 1985.** Male-biased sex ratios in the cycad *Macrozamia riedlei* (Zamiaceae). Bull. Torrey Bot. Club 112: 393 - 397.

***Ornduff, R.* 1989.** Size distribution and coning behaviour of the Australian cycad *Lepidozamia peroffskyana*. Aust. J. Ecol. 14: 241 - 245.

***Ornduff, R.* 1990.** Geographic variation in reproductive behaviour and size structure of the Australian cycad *Macrozamia communis* (Zamiaceae). Amer. J. Bot. 77: 92 - 99.

***Oyama, K.* 1990.** Variation in growth and reproduction in the neotropical dioecious palm *Chamaedorea tepejilote*. J. Ecol. 78: 648 - 663.

***Stevenson, D.W.* 1980a.** Observations on root and stem contraction in cycads (Cycadales) with special reference to *Zamia pumila* L. Bot. J. Linn. Soc. 81: 275 - 281.

***Stevenson, D.W.* 1980b.** Radial growth in the cycadales. Amer. J. Bot. 67: 465 - 475.

***Tang, W.* 1989.** Seed dispersal in the cycad *Zamia pumila* in Florida. Can. J. Bot. 2066 - 2070.

Pollination Biology of *Macrozamia riedlei* - The Role of Insects

S.W. Connell
Weed Sciences
Department of Agriculture
Baron Hay Court, South Perth, 6151 Australia

P.G. Ladd
School of Biological and Environmental Sciences
Murdoch University, Murdoch 6150 Australia

Abstract

Recent studies have explored the role of beetles (Coleoptera) as pollen vectors in a range of cycad species. An experimental study of the pollination biology of *Macrozamia riedlei* is reported. The experiments were undertaken in natural populations and involved insect and/or wind exclusion bags placed over megastrobili. Seed set was visually assessed just before megastrobilus disarticulation. Treatment effects were not statistically significant and it is concluded that pollen was moved by both insects and wind, either independently or sequentially. Seed set as a function of average distance to the nearest male plants (pollen sources) substantiated this conclusion.

Introduction

The role of insects in cycad pollination biology has been historically ambiguous. Early studies reported observations of insects being associated with strobili of both sexes in a range of genera (Baird, 1938; Rattray, 1913); though the conventional wisdom has been that pollen is transported by wind as in most other gymnosperms (Chamberlain, 1935). Scientific investigation of the relative importance of wind and insects as pollen vectors in cycads did not commence until more recent times. Comparative wind-tunnel experiments indicated that megastrobilus morphology influenced pollen deposition patterns (Niklas and Norstog, 1984). The smooth-scaled megastrobili of taxa such as *Dioon* and *Zamia* had different depositional patterns from the rough-scaled megastrobili of *Cycas*. Studies of heat and scent production by cycad strobili revealed differences between genera, species within genera, and sexes and populations within species

(Tang, 1987). Experimental studies with species of *Zamia* found that beetles were the main pollen vectors, with wind transport being of secondary, almost negligible, importance (Norstog and Fawcett, 1989; Tang, 1987; Norstog, Stevenson and Niklas, 1986).

The morphological and ecological diversity of cycads suggest that discoveries based on one genus (or species) may have limited correlations with the biologies of other taxa (e.g. seed dispersal syndromes; Dehgan and Yeo, 1983, Bauman and Yokoyama, 1976). This paper reports on the pollination biology of *Macrozamia riedlei*; an Australian species which differs considerably in morphology and ecology from species of *Zamia* (Ornduff, 1985; Burbidge and Whelan, 1982; Johnson, 1958; Baird, 1938).

Species Description

Macrozamia riedlei (Fischer ex. Gaudich) C. Gardner is the sole cycad found in south western Australian (Johnson, 1958). It is widely distributed and locally common in a range of vegetation types. Large plants may reach a trunk height of 3 - 4 metres; though often the trunk remains underground. The non-reproductive period is prolonged (in terms of decades) by comparison with many other cycads. Both male and female plants may produce numerous large strobili at infrequent intervals (4 - 5 years); with coning being encouraged by burning of the plant (Connell and Ladd, unpublished data). Developing strobili become visible on plants in late autumn and are receptive in spring. The pollen is dry and sheds freely from male strobili, while female strobili at the receptive stage expand so that the broad sporophyll apices separate by 2 - 3mm. Sporophyll separation is even over the strobilus; unlike the situation described by Norstog and Fawcett (1989) in *Zamia furfuracea* where only a single line of sporophyll separation occurs to allow beetle access. After several days the gaps between the sporophylls close. During this time, the strobili become warmer than ambient temperature, at least at night; although this feature has yet to be quantified. Both male and female strobili have a heavy, 'fermenting', slightly sour odour.

Study Sites

Three study areas were used during the course of these experiments. In the first year (1987), studies were undertaken at Hill River Nature Reserve (Lat. 30º 18', Long. 115º11') and at Banganup (Lat 32º 10', Long. 115º 49'). Replication of the first year's study was required as results were inconclusive. In the following year no plants were reproducing at either of the above sites, so a third area (Kardinya; Lat. 32º 4', Long. 115º 48') was selected. The vegetation at all study sites consisted of heath woodland dominated by *Banksia* and *Eucalyptus* species. The climate is dry Mediterranean and all sites occurred on white sand substrates.

Methods

The experimental strategy followed Tang (1987) and Norstog and Niklas (1986). Pollen-exclusion bags were of three types; two insect exclusion and one wind exclusion. Initial insect exclusion bags were of 1mm mesh (fine mesh type). A second type using 2mm mesh (coarse mesh type) was introduced during the second year of the study as it was considered that the finer mesh may have interfered with wind transport of pollen around the sporophylls. Wind-exclusion bags were made of calico and completely covered the strobilus; being open at the bottom to permit insect entry. As many female plants produced more than one strobilus, it was often possible to obtain treatment and control strobili from the one individual. Distances to the nearest four reproducing male plants (and the number of strobili possessed by each male) were measured for each experimental female.

Seed set was determined by strobilus dissection prior to disintegration when strobilus length was also measured. Total ovule counts were made and the number of mature seeds determined by visual assessment of seed size and sarcotesta development.

Insects associated with male and female strobili were collected during the course of the study and identified by the Western Australian Museum. Data from the three study sites and over the two years of the investigation have been pooled to yield greater statis-

Table 1

Percentage Seed set in *M.riedlei* following various pollen-exclusion treatments. Values identified by the same letter are not significantly different (Pr.). Results are Mean + S.D. (sample size)

Site	Control	Wind	Insect (Fine)	Insect (Coarse)
Hill River 1987	55.4 ± 22.5 (12)	11.1 ± 22.5 (9)	14.5 ±18.7 (10)	
Banganup 1987	81.0 ± 22.1 (7)	82.5 ± 8.2 (4)	35.1 ± 34.4 (3)	
Kardinya 1988	50.0 ± 34.7 (58)	37.8 ± 8.2 (4)	33.2 ± 36.1 (20)	45.8 ± 33.6 (21)
TOTALS:	A 53.7 ± 34.2 (77)	B 36.3 ± 35.3 (37)	B 27.6 ± 31.9 (33)	AB 45.8 ± 33.6 (21)

tical confidence. Statistical analyses were undertaken using the SAS numerical package (SAS Institute Inc. 1982). Percentage data were Arc-sin transformed where necessary.

Results and Discussions

Table 1 presents percentage seed set versus experimental treatment using pooled information from all study sites. Average percentage seed set in control strobili ranged from 0 to 100% (mean 53.7%); comparable to levels of seed set documented in natural populations of *M. riedlei* by Burbidge and Whelan (1982). All experimental treatments except the coarse insect exclusion treatment significantly reduced seed set (Duncan's Multiple Range Test, *Pr*). Between-treatment comparisons were not significant. This suggests that insects may be less important in pollen transport than wind; assuming that the fine insect exclusion treatment interfered also with wind transmission. However as seed was set in all treatments it would appear that pollen transport was not confined to one avenue (insects or wind). Initial transport of pollen by wind to the megastrobilus followed by insect-facilitated movements of pollen to the ovules may account for the observations made. These conclusions are in contrast to those of Norstog and Fawcett (1989), Tang (1987) and Norstog and Niklas (1986) working with *Zamia* species where seed set was negligible in insect exclusion treatments.

The variation in percentage seed set evident in the large standard deviation values of Table 1 could not be attributed to any one cause. Percentage seed set was not related to strobilus size (Fig. 1) nor to the number of strobili produced by the female plant (Table 2). The chaotic relationship between percentage seed set and distance to the average nearest pollen source (Fig. 2) clearly indicates the unpredictability of pollen transmission in *M. riedlei*. Numerous data transformations and dispersal models were applied to these data; e.g., a negative exponential decay model based on distance and size (number of microstrobili) of the four nearest reproducing males. These results are not present as no transformation or model produced a statistically significant result.

Adult specimens of three species of beetles were collected from both micro- and megastrobili. *Tranes vigorsii* Bohem. and *T. roei* (Bohem.) Pascoe (Coleoptera, Curculionidae) are large species (adults to 1.5cm

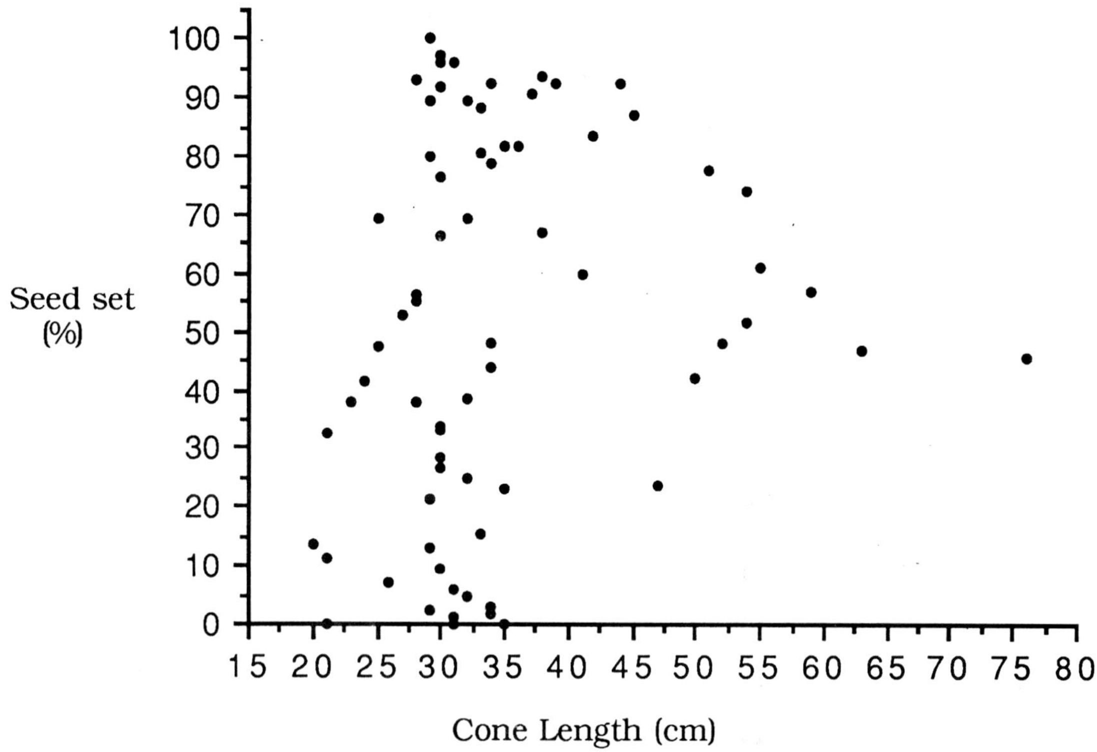

Figure 1: Seed set in relation to strobilus length

in length). Records of *Tranes* species associated with cycads have been published by Ornduff (1989, 1990). The third beetle species, *Paracucujus rostratus* Sen Gupta and Crowson (Coleoptera, Boganiidae) was the smallest in size (adults to 5mm). This species is closely related to a south African genus associated with microstrobili of *Encephalartos* (Endrody-Younga and Crowson 1986). The systematic relationships of these beetles are distinct from the beetles associated with *Zamia* strobili (Norstog and Fawcett,1989; Tang, 1987; Norstog and Niklas, 1986. *Tranes* and *Paracucujus* larvae were observed tunnelling within the rachises and sporophylls of both micro- and megastrobili and within the sarcotesta of some seeds. There was no evidence of insect damage to seed kernel.

In *Zamia* it appears that pollination is entirely dependent upon beetle transmission of pollen (Norstog and Fawcett, 1989; Tang, 1987). The species investigated by these researchers produce small, smooth-scaled strobili. The plants themselves tend to be small forest ground storey plants which produce strobili within 30cm or so of the soil surface. Plants of *M. riedlei*, in contrast, produce large rough-scaled strobili, well above the ground. From aerodynamic studies of the rough-scaled strobili of *Cycas* it was concluded that air currents could potentially deposit

Table 2

Seed Set (Mean + S.D.) and Strobilus Production in *M. Riedlei*

No. of Strobili	**1**	**2**	**3**	**4**	**5**	**8**
Sample Size	5	33	7	6	5	3
%Seed Set	68 ± 22	64 ± 26	54 ± 21	54 ± 16	49 ± 15	67 ± 8

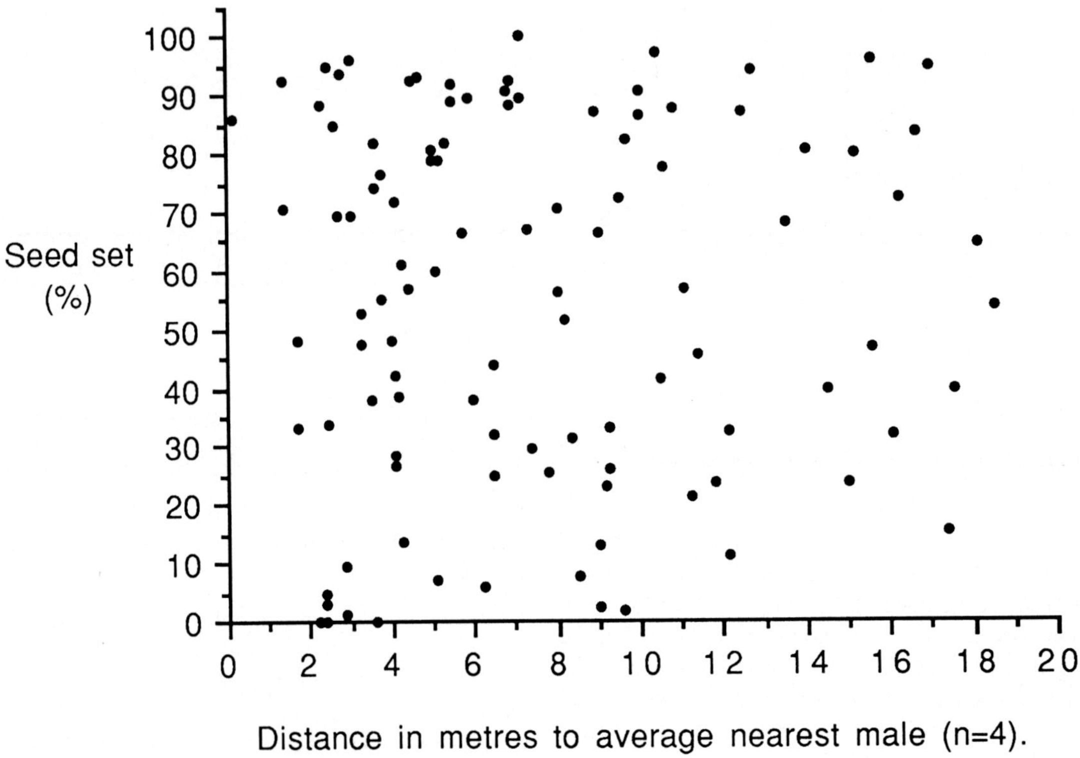

Figure 2: Relationship between seed set and distance to average nearest male in *M. riedlei*.

pollen on the sporophylls where further movement, either by sifting, water flow or insect activity, could deliver pollen to the micropyles (Norstog and Niklas, 1986).

Seed set in *M. riedlei* is erratic. Resource limitation does not appear to be an important determinant of fertility as strobilus size (Fig. 1) and the number of strobili produced (Table 2) were not related to seed set. Spatial and temporal variation in population densities of the beetle species and their different body sizes may account for some of the variation in seed set data encountered. A small species such as *P. rostratus* may be important in the direct transfer of pollen from male plants in close proximity to females, while the larger, more robust, *Tranes* beetles may transfer pollen longer distances. This is in addition to movement of pollen initially transported by wind to the megasporophylls. Spatial variations in pollen transfer within a guild of beetles associated with a single plant species have been documented (Young, 1988). The nature of the involvement of the beetle species in pollen transfer of *M. riedlei* remains to be determined.

References

***Baird, A.M.* 1938.** A contribution to the life history of *Macrozamia riedlei*. J. Roy. Soc. West. Aust. 25: 153 - 169.

***Bauman, A.J. and Yokoyama, H.* 1976.** Seed coat carotenoids of the cycad genera *Dioon, Encephalartos, Macrozamia* and *Zamia*: evolutionary implications. Biochem. Syst. Ecol. 4: 73 - 74.

***Burbidge, A.H. and Whelan, R.J.* 1982.** Seed dispersal in a cycad, *Macrozamia riedlei*. Aust. J. Ecol. 7: 63 - 67.

***Chamberlain, C.J.* 1935.** Gymnosperms: structure and evolution. University of Chicago Press, Chicago.

***Dehgan, B. and Yeo, C.K.K.H.* 1983.** Seed morphology in relation to dispersal, evolution and propagation of *Cycas*. Bot. Gazz. 144: 412 - 418.

***Endrody-Younga, S. and Crowson, R.A.* 1986.** Boganidae, a new beetle family for the African fauna (Coleoptera: Cucujoidea). Ann. Transvaal Mus. 34: 253 - 273.

***Johnson, L.A.S.* 1959.** The families of cycads and the Zamiaceae of Australia. Proc. Linn. Soc. N.S.W. 84: 64 - 117.

***Niklas, K.J. and Norstog, K.* 1984.** Aerodynamics and pollen grain depositional patterns on cycad megastrobili: Implications on the reproduction of three cycad genera (*Cycas, Dioon* and *Zamia*). Bot. Gaz. 145: 92 - 104.

***Norstog, K.* 1987.** Cycads and the origin of insect pollination. Amer. Sci. 75: 270 - 279.

***Norstog, K and Fawcett, P.K.S.* 1989.** Insect-cycad symbiosis and its relation to the pollination of *Zamia furfuracea* (Zamiaceae) by *Rhopalotra mollis* (Curculionidae). Amer. J. Bot. 76: 1380 - 1394.

***Norstog, K.J., Stevenson, D.W. and Niklas, K.J.* 1986.** The role of beetles in the pollination of *Zamia furfuracea* L. fil. (Zamiaceae). Biotropica 18: 300 - 306.

***Ornduff, R.* 1985.** Male-biased sex ratios in the cycad *Macrozamia riedlei* (Zamiaceae). Bull. Torrey Bot. Club 112: 393 - 397.

***Ornduff, R.* 1987.** Sex ratios and coning frequency of the cycad *Zamia pumila* L. (Zamiaceae) in the Dominican Republic. Biotropica 19: 361 - 364.

***Ornduff, R.* 1989.** Size distribution and coning behaviour of the Australian cycad *Lepidozamia peroffskyana*. Aust. J. Ecol. 14: 241 - 245.

***Ornduff, R.* 1990.** Geographic variation to reproductive behaviour and size structure of the Australian cycad *Macrozamia communis* (Zamiaceae). Amer. J. Bot. 77: 92 - 99.

***Rattray, G.* 1913.** Notes on the pollination of some South African cycads. Trans. Roy. Soc. S. Afr. 3: 259 - 270.

***SAS Institute Inc.* 1982.** SAS user's guide: statistics. SAS Institute Inc., Cary, N.C.

***Sen Gupta, T. and Crowson, R.A.* 1966.** A new family of cucujoid beetles based on six Australian and one New Zealand genera. Ann. Mag. Nat. Hist. Sr. 13 Vol. 9: 61 - 85.

***Tang, W.* 1987.** Heat production in cycad cones. Bot. Gaz. 148: 165 - 174.

***Tang, W.* 1987.** Insect pollination in the cycad *Zamia pumila* (Zamiaceae). Amer. J. Bot. 74: 90 - 99.

***Young, H.J.* 1988.** Differential importance of beetle species pollinating *Dieffenbachia longispatha* (Araceae). Ecology 69: 832 - 844.

Insect Predation of Ovules in the South African Species of *Encephalartos* (Cycadales: Zamiaceae)

John S. Donaldson
National Botanical Institute
Kirstenbosch, P/Bag X7, Claremont, South Africa

Abstract

South African species of *Encephalartos* reputedly suffer considerable ovule mortality as a result of insect predation. In some species, such as *E. lehmannii*, 100% of the ovules in a single cone may be preyed upon (Giddy, 1974). The resultant reduction in potential seed set could have serious implications for the conservation of cycads, particularly for species which are already threatened. However, the extent of ovule predation in different species of *Encephalartos* and the potential for ovule predators to colonise new cycad hosts are unknown. In this study, ovule and seed predation were examined in 19 species of *Encephalartos*. Six ovule predators were identified. Four of these insect species preyed facultatively on the megagametophyte of their host species and seldom caused substantial mortality. In contrast, two specialist ovule predators, *Antliarhinus zamiae* and *A. signatus* together destroyed between 2 and 96% (mean values) of their host's ovules and may therefore have a considerable influence on seed production. Both species were absent from the Transvaal region of South Africa but may yet colonise species of *Encephalartos* that grow here. However, successful colonisation may depend on the attributes of *E. cycadifolius* and *E. friderici-guilielmi* were never preyed upon by *A. zamiae* or *A. signatus* even when they occurred within the distributional range of these beetles.

Introduction

South African species of the cycad genus *Encephalartos* (Zamiaceae) are perhaps unique among cycads worldwide because their seeds and ovules are apparently heavily preyed upon by insects (Rattray, 1913; Giddy, 1974; Goode, 1989). In some instances, 100% predation of seeds collected in the field has been reported (Giddy, 1974). Such seemingly extraordinary levels of predation are of concern for the conservation of these cycads since many populations of *Encephalartos* are already vulnerable (Giddy, 1974; Goode, 1989). More specifically, insect predation is perceived to pose two potential threats to cycad survival. (1) That seed and ovule predators may colonise new hosts thereby placing additional pressures on species of *Encephalartos* that may already have poor recruitment. (2) That predation could have an increased impact on recruitment in populations that are declining due to other factors such as habitat destruction and the removal of mature plants by collectors. In other words, as population numbers decline, insect predators may destroy more significant proportions of the potential seed crop. This may be particularly true of specialist seed predators that are restricted to cycad species.

The aims of this study were to determine the following; (1) the identify of insect ovule and seed predators and their distribution on South African species of *Encephalartos*; (2) to establish the potential for colonisation of new hosts by the most damaging ovule predators; and, (3) to determine whether ovule and seed predation was higher in small populations.

Methods

To determine the extent of ovule and seed predation in species of *Encephalartos*, megasporangiate cones from 19 species (see Table 1) were collected between November 1988 and November 1990. A total of 232 cones from 71 localities were examined. Each cone was dissected and all seeds and ovules were placed in a basin of water. 'Floater' seeds and ovules were dissected to confirm insect predation and to identify the insect predator. A sample of 'sinker' seeds and ovules was opened to confirm the absence of seed predators.

Populations of *E. altensteinii* and *E. villosus* were most extensively sampled to determine the effects of population size on ovule and seed predation. Cones were collected from small populations (less than approximately 200 mature plants) and larger populations. In total, 47 megasporangiate cones of *E. altensteinii* and 36 cones of *E. villosus* were examined.

To establish host specificity in the most damaging ovule predators, *Antliarhinus zamiae* (Thunberg) and *A. signatus* Gyllenhal (Coleoptera: Curculionoidea), 50 adults of each sex were confined in mesh bags on the cones of non host species in Kirstenbosch Botanic Garden, Cape Town, or on to isolated cones collected from natural populations. *Antliarhinus zamiae* adults were released on to *E. cycadifolius, E. friderici-guilielmi, E. laevifolius, E. lanatus* and *E. transvenosus*. *Antliarhinus signatus* adults were released on to the same plants but with the addition of *E. caffer, E. princeps* and *E. villosus*, species from which they were not collected in nature. In addition both species were released on to cones of *E. altensteinii* as a control. Only complete development of the insect predators on the host plant was interpreted as successful colonisation.

Results and Discussion

Extent of Insect Predation of Ovules

Six species of insects were found that killed ovules by feeding on them (see Table 1). In all cases, penetration of the integument and at least partial consumption of the megagametophyte appeared to be necessary to kill the ovule. Four of the insects, *Zerenopsis leopardina* Felder (Lepidoptera: Geometridae), *Platymerus eckloni* Gyllenhal (Coleoptera: Curculionidae) fed facultatively, and only occasionally, on the megagametophyte. They fed more commonly on the megasporophylls (*Z.leopardina* and *Amorphocerus* spp.) or the fleshy integument of the ovule (*Amorphocerus* spp. and *P. eckloni*). As a result, these facultative ovule predators accounted for relatively little direct mortality of ovules (Table 1). In some instances, feeding of these insects appeared to facilitate pathogenic infection of the ovules so that they may be indirectly responsible for far greater mortality of ovules. However, without detailed studies of the sequence of infection, it is groundless to speculate on the role of predation in the onset of disease.

The most devastating ovule predators were *Antliarhinus zamiae* and *A. signatus* (Table 1), two weeviloid species that probably have an ancient association with species of *Encephalartos* (Crowson, 1981). *Antliarhinus zamiae*, alone, accounted for more than 60% of ovule mortality in eight host species. In combination with *A. signatus*, mean ovule mortality was often around 80% (Table 1) and it was not uncommon for all the ovules in a single cone to be predated. The exceptionally high mean mortality for *E. horridus* ovules (96%) is testament to the potential

Table 1
Percentage of Ovules or Seeds destroyed by Insect Predators in 19 Species of Encephalartos. For each cycad species, the mean percentage of ovules preyed upon by Zerenopsis leopardina (Lepidoptera), Amorphocerus sp. 1 and sp. 2, Platymerus eckloni, Antliarhinus zamiae and Antliarhinus signatus (all Coleoptera) is given. In addition, the total percentage of ovules preyed upon and the number of megasporangiate cones sampled are provided. Percentages were rounded off to the nearest integer except where the value was less than 1.

Species of Encephalartos	Zerenopsis leopardina	Amorphocerus		Platymerus eckloni	Antliarhinus		Total predation (%)	No. of Cones
		sp.1	sp. 2		zamiae	signatus		
Altensteinii	0	0.02	0	0	63	12	75.02	47
Arenarius	0	0	0	0	74	8	82	5
Caffer	0	0	0	0	2	0	2	10
Cycadifolius	0	0	0	0	0	0	0	8
Dolomiticus	0	0	0	0	0	0	0	3
Ferox	0	0	0	0	62	0	62	2
Friderici-guilielmi	0	0	5	7	0	0	12	30
Horridus	0	0	0	0	87	9	96	9
Laevifolius	0.2	0	0	0	0	0	0.2	16
Lanatus	0	0	0	0	0	0	0	6
Lebomboensis	0	0	0	0	51	0	51	2
Lehmannii	0	0.4	0	0	41	13	54.4	16
Longifolius	0	0.3	0	0	4	13	17.3	14
Natalensis	0	0	0	0	61	13	74	6
Princeps	0	6	0	0	82	0	88	5
Transvenosus	0	0	0	0	0	0	0	3
Trispinosus	0	0	0	0	53	10	63	12
Umbeluziensis	0	0	0	0	81	0	81	2
Villosus	0.6	0	0	0	87	0	87.6	36

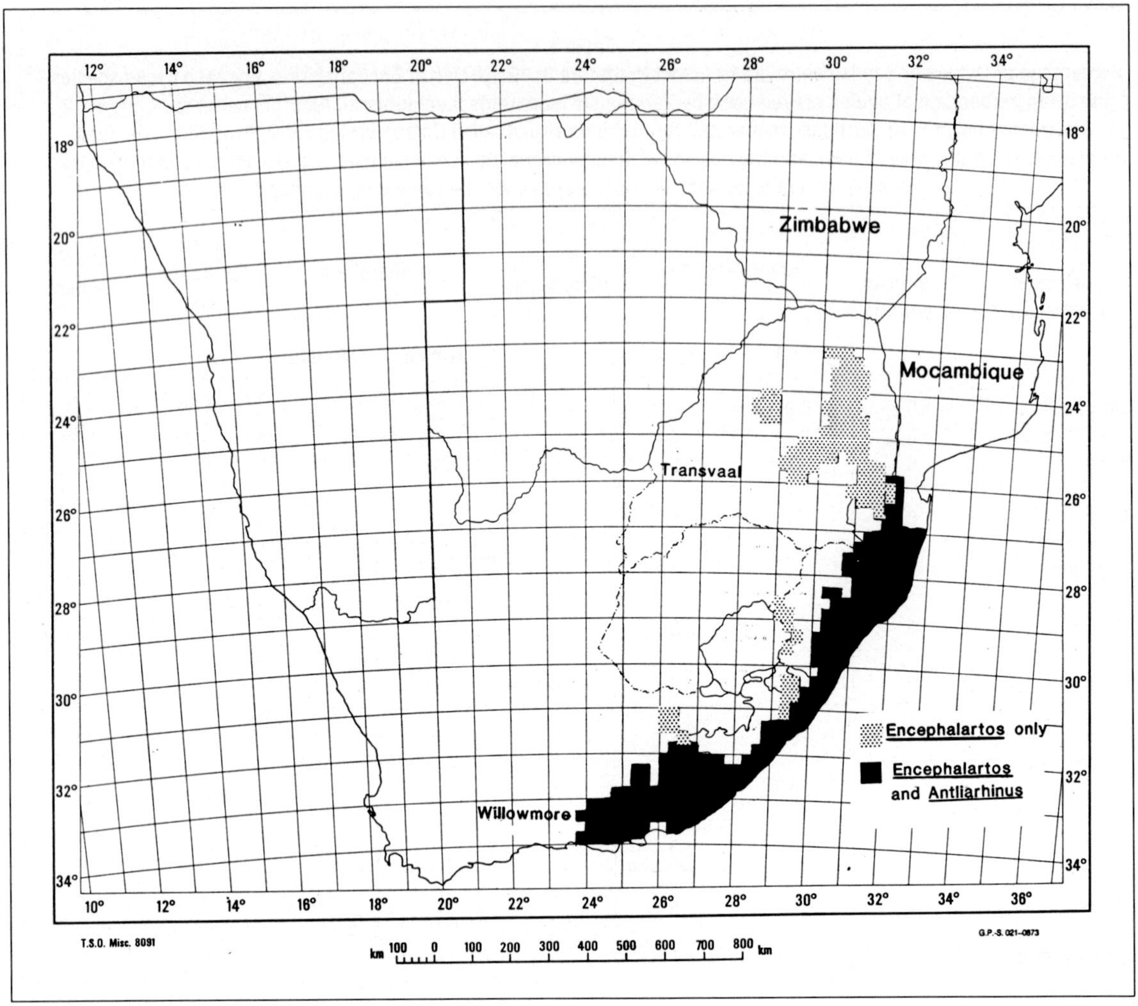

Figure 1: Map of South Africa showing the general distributions of the cycad genus *Encephalartos* and the cycad weevils *Antliarhinus zamiae* and *Antliarhinus signatus*.

effects of predation by *A. zamiae* and *A. signatus* on seed production.

Thus, in general, these data support previous claims that *A. zamiae* (Giddy, 1974; Annecke and Moran, 1982; Goode, 1989) and to a lesser extend *A. signatus* (Goode, 1989) are highly destructive of *Encephalartos* ovules.

The actual effects of high ovule mortality on recruitment are unknown. If significant, density dependent, mortality of seeds occurred during or after dispersal, then it is possible that predispersal mortality of ovules would have a negligible effect on reproduction (Harper, 1977). However, in the absence of such data it may be prudent to assume that very high ovule mortality has some effect on reproduction.

Distribution of *A. zamiae* and *A. signatus*

Although *A. signatus* and, more particularly, *A. zamiae* have relatively broad host ranges, neither species was recorded on all species of *Encephalartos* (Table 1). They were notably absent from almost all cycad species in the Transvaal province of South Africa (Fig. 1) despite close taxonomic affinities between some Transvaal species of *Encephalartos* and species occurring elsewhere that were preyed upon by *A. zamiae* or *A. signatus* (e.g. *E. dolomiticus* from the Transvaal and *E. lehmannii* from the eastern Cape Province). It is possible that the beetles have not colonised cycads in these areas simply due to an unsuitable climate or due to geographical barriers. However, unconfirmed reports from collectors in the Transvaal claim that *A. zamiae* and *A. signatus* have become naturalised in gardens in the region presumably through the importation of infested ovules into the area. Although garden situations cannot be equated with natural habitats, it remains a possibility that these predators could spread to Transvaal cycads unless they are prevented from doing so by attributes of the potential host plant.

Antliarhinus zamiae and *A. signatus* were also never recovered from species groups that included taxa with particularly woolly cones, i.e., *E. cycadifolius, E. friderici-guilielmi, E. ghellinckii, E. laevifolius*, and *E. lanatus*. Cones of the related species, *E. humilis*, were not examined but I have not come across any records of ovule predation in this species. At least some of these cycads, e.g., *E. cycadifolious* and *E. friderici-guilielmi*, are distributed within the general range of *A. zamiae* and *A. signatus* so that they are not excluded as hosts by geographical or climatic barriers. The most plausible explanation for their omission from the host range of *A. zamiae* and *A. signatus* appears to be incompatibility between the insects and these cycads. Similarly, the absence of *A. signatus* from *E. caffer* and *E. villosus* (Table 1) may be caused by inherent qualities of these plants. This conclusion was confirmed by the host specificity tests carried out in the Kirstenbosch Botanic Garden.

Host specificity in *A. zamiae* and *A. signatus*

Antliarhinus zamiae and *A. signatus* developed normally on at least one species of *Encephalartos* from the Transvaal when confined to plants in Kirstenbosch Botanic Garden. Their release on to cones of *E. transvenosus* resulted in more than 60% predation of ovules by *A. zamiae* and about 15% mortality as a result of predation by *A. signatus*. The number of ovules preyed upon was not significantly different from predation in *E. altensteinii* ($P > 0.05$). This result confirms the prediction that *A. zamiae* and *A. signatus* have the potential to colonise some Transvaal species. Cones from *E. dolomiticus* or its near relatives were not available for testing, but a similar result would be expected.

Antliarhinus signatus also developed successfully on *E. princeps*. Since *E. princeps* is distributed within the range of *A. signatus*, the absence of the weevil from *E. princeps* may reflect a temporary absence from the populations sampled in this study. In this case *A. signatus* may be recovered from *E. princeps* populations in future.

Neither *A. zamiae* or *A. signatus* developed successfully on *E. cycadifolius, E. friderici-guilielmi, E. laevifolius*, or *E. lanatus*. Dissection of these cones showed that adults had made no attempt to oviposit into the ovules and that they did not recognise these cycad species as host plants. Oberprieler (1989) has attributed the absence of *A. zamiae* and *A. signatus* from ovules of *E. friderici-guilielmi* to the woolly covering present on the megasporophylls. However, the tomentum has little effect on the movement of beetles on the cone (Donaldson, unpublished data) and is therefore unlikely to hinder oviposition. It is more likely that major differences in coning phenology between these cycads and other species of *Encephalartos* have resulted in the absence of predation by species of *Antliarhinus* (Donaldson, unpublished data).

Antliarhinus signatus also failed to colonise *E. caffer* and *E. villosus* in the botanic garden thus confirming the field results in which this seed predator was always absent from these cycads. The most plausible explanation for this phenomena is that *A. signatus* females are not able to oviposit into the ovules of either *E. caffer* or *E. villosus*. *Antliarhinus signatus* females have a relatively short rostrum (ca. 3mm) which they use first to penetrate the ovule integument near to the attachment to the megasporophyll and then to excavate a hole in the gametophyte into which they lay their eggs (Donaldson, in press). Due to the limitations of snout length, female *A. signatus* are only able to penetrate integuments that are approximately 3mm thick in the regions near the attachment to the megasporophyll. As a result they could be excluded from cycads such as *E. caffer* and *E. villosus* which often have integuments greatly in excess of 3mm thick in this region (Donaldson, unpublished data).

In general, these host specificity data show that *E. friderici-guilielmi* and related plants are unlikely to be colonised by *A. zamiae* or *A. signatus* regardless of where they occur and the same could be said for colonisation of *E. caffer* and *E. villosus* by *A. signatus*. However, other species of *Encephalartos* that are currently not predated by *A. zamiae* or *A. signatus* should be considered as potential host plants, particularly *E. transvenosus* and probably *E. eugene-maraisii, E. dolomiticus*, and *E. dyerianus*.

Population Size and Ovule Predation

Significantly greater predation of ovules was recorded in small populations of *E. altensteinii* (mean predation = 75%) and *E. villosus* (mean predation = 93%) than in large populations (means of 61% and 84% for *E. altensteinii* and *E. villosus* respectively; X^2 on raw data, $P < 0.01$). The reasons for this difference were not investigated but could be due to predator satiation in large populations or to density dependent mortality of adult beetles in large populations. Adult *A. zamiae* and *A. signatus* seek shelter under bark between coning cycles (Rattray, 1913) and probably suffer their greatest mortality at this stage (Donaldson, unpublished data). If shelter sites are limited, then density dependent mortality of adult beetles would be expected. Since large populations of cycads are likely to host greater numbers of beetles, mortality of adult beetles may be higher in these populations. As a result, predation in the following season would be lower than in a smaller cycad population in which the adult beetles may have suffered relatively less mortality due to their lower initial density.

These results indicate that predispersal predation of ovules may increase as populations decrease in size. However, it should be stressed that this conclusion is based on a relatively small sample. Collecting over a longer period and from more populations is needed to confirm this result. In addition, the effects of predispersal ovule mortality on recruitment needs to be studied to determine whether increased predation in small populations has any real effect on reproduction.

Acknowledgments

I would like to thank Profession V.C. Moran of the University of Cape Town, the management of the National Botanical Institute and the South African organisers of Cycad '90 for providing financial assistance to present this work.

References

***Annecke, D.P. and Moran, V.C.* 1982.** Insects and Mites on Cultivated Plants in South Africa. Butterworths, Cape Town.

***Crowson, R.A.* 1981.** Biology of the Coleoptera. Academic Press, London.

***Donaldson, J.A.* (in press).** Adaptation for oviposition into concealed cycad ovules in the cycad weevils, *Antliarhinus zamiae* and *A. signatus* (Coleoptera: Curculionoidea). Biol. J. Linn. Soc. London.

***Giddy, C.* 1974.** Cycads of South Africa. Purnell, Cape town.

***Goode, D.* 1989.** Cycads of Africa. Struik Winchester, Cape Town.

***Harper, J.S.* 1977.** Population Biology of Plants. Academic Press, London.

***Oberprieler, R.G.* 1989.** *Platymerus*, the forgotten cycad weevil. Pelea 8: 50 - 54.

***Rattray, G.* 1913.** Notes on the pollination of some South African species of *Encephalartos*. Trans. R. Soc. South Africa 3: 259 - 270.

Zamia pumila in South Florida: A Preliminary Report on its Pollinators *R. slossoni*, a Snout Weevil and *P. zamiae*, a Clavicorn Beetle

Priscilla K.S. Fawcett
Fairchild Tropical Garden, Miami FL, 33156, USA

Knut J. Norstog
Missouri Botanical Garden, St Louis, MO, 63166, USA

Abstract

The native Florida cycad, *Zamia pumila* L. is effectively pollinated by two beetles. *Rhopalotria slossoni*, a snout weevil, swarms on male cones, lays eggs in cone tissue and completes its life cycle within ten days. It does not consume pollen but does shelter in male cones, becomes dusted with pollen, visits female cones and pollinates the ovules. The clavicorn beetle, *Pharaxonotha zamiae*, is a destructive pollinator, laying eggs among the pollen sacs of male cones. Both adults and larvae eat pollen but larvae also excavate male cone stalks, passing through about eight instars in male cones, then drop out of the cone stalk, burrow into soil, pupate and complete metamorphosis.

Introduction

In south Florida, the cones of *Zamia pumila* often harbour two species of weevils which have been shown to effectively pollinate female cones of the species (Tang, 1987). Male cones provide an egg-laying site, as well as food for both the adults and the larvae of *Rhopalotria slossoni* Schaeffer, a weevil (Fig. 1 A-F). Both beetles are able to live only in the male cones of the native Florida *Zamia pumila* or closely related Caribbean forms of this species.

Life Cycle of *Rhopalotria slossoni*

The life cycle of *R. slossoni* is similar to that we have described for *R. mollis*, a snout weevil which is the pollinator of *Zamia furfuracea* (Norstog and Fawcett, 1989). In that cycad, the weevils visit the male cones about one day before the pollen cone opens and the pollen sacs dehisce. At about this time the cones become aromatic and warm, and a few of the scales at the base begin to loosen and separate. The process, which occurs by starch- and lipid-fuelled metabolism (thermogenesis), is a general one among cycads and is considered as functioning both to open male and female cones and also produce fragrances (Tang, Sternberg and Price, 1987). During its congregational invasion of male cones of *Z. pumila*, the weevils mate, feed on the starchy sporophyll tissue, and lay their eggs in holes which they make when they feed. Both sexes exhibit a size graduation (Fig. 1 A-D). The front legs of most of the males have enlarged and muscular femurs with which they are able to fight off other males when mating (Fig. 1 A). The smaller males lack these powerful front legs (Fig. 1 C), but manage to mate by moving in quickly, often when two larger males are distracted by fighting with each other!

Feeding behaviour

After feeding and mating on the outside of the cone, both sexes go inside, entering between the loose scales at the base, and continue feeding on the bases of the sporophylls while waiting until another male cone is ready (Fig. 2 A-C). Meanwhile, the eggs hatch and the larvae feed on the sporophyll tissue, making their way towards the stalk (Fig. 2 D). When the larvae meet, one of them eats the other. In *R. slossoni* up to four eggs can be laid in a sporophyll but only one larva survives to build a pupa case from fecal material (Fig. 2 E). The pupae cases contain a high proportion of undigested storage cells or idioblasts (Norstog *et al*, in review). Among the idioblast contents may be a neurotoxin (BMAA) (Spencer *et al*, 1987), and cycasin, a demonstrated carcinogen (Rothschild, Nash and Bell, 1986). The pupa case is built from the inside, starting at the bottom, the larva then becomes a pupa which undergoes further metamorphosis and then leaves the pupa case by chewing its way out, emerging through the tip of the sporophyll (Fig. 2 F-G).

Pollination

Some of the weevils go to the female cones (Fig. 2 H), which also become warm and aromatic when they are ready for pollination. The tissue of the female cone has almost no starch, and contains idioblast cells that have released their contents (toxins?) into the cone tissue. We find no evidence of feeding by weevils, although there are nibble marks where they have tried to feed. We think this may be the result of diffusion of poison throughout the female cone tissues making them unpalatable and even poisonous to predatory inspects, including the two pollinating species. A weevil that has been waiting inside a male cone picks up a lot of pollen, particularly if *Pharaxonotha* is also present, as we will relate later on in this account. The female cone of *Z. pumila* opens along irregular cracks, so that once inside a weevil has to find its way out again in order to escape. It may be pointed out here that female cone opening in *Z. furfuracea* seems more highly specialised for insect pollination than does that of *Z. pumila*. In *Z. furfuracea* the cone opening is generally a single vertical crack between adjacent rows of cone scales (Norstog and Fawcett, 1989) and it is possible that this species is more highly evolved in the direction of insect pollination than is *Z. pumila*.

Dormancy

At the end of the cycad's reproductive season there is a change in both the behaviour of the weevils and in the nature of the sporophyll tissue of the male cone. The tissue at the ends of the sporophylls becomes greenish and acquires a preponderance of idioblast cells, as well as far less starch (Fig. 3 C). At the same time, there is a large increase in the weevil population as well as fewer maturing male cones, and since the green tissue is to some extent avoided by feeding and egg-laying weevils, only very few eggs are laid at the ends of the sporophylls then, for want of an egg-laying site, weevils begin to lay in the sides and base of the sporophylls. The larvae eat their way through the sporophyll tissue as before (Fig. 3 D), but some of the larvae become fat and waxy and build very thick pupae cases in which they go into diapause until the next season (Fig. 3 e-g). Others go rapidly through metamorphosis and emerge as adults within a few days (Fig. 3 E-G). Diapause, which is a type of dormancy consisting in an arrested state of development, can last for two or more years.

Rhopalotria slossoni differs from *R.mollis* (the pollinator of *Z. furfuracea*) in that it is less active during the day. In *Z. pumila* the male cone sporophylls are a little smaller than in *Z. furfuracea*, so that fewer eggs can be laid in the ends. While *R. slossoni* generally

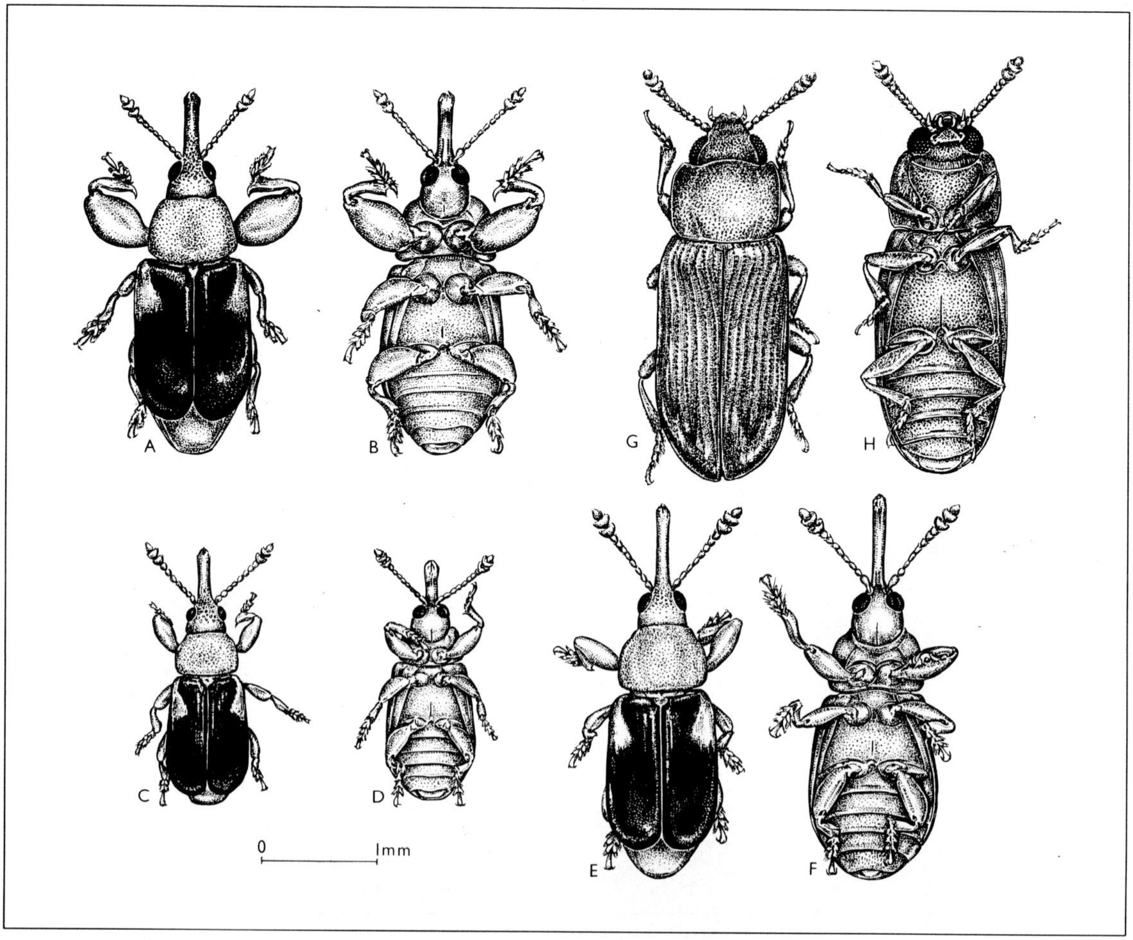

Figure 1: *Rhopalotria slossoni* and *Pharaxonotha zamiae*.

A: *R. slossoni* (dorsal view) large male with enlarged femurs used in fighting.

B: Ventral view of same male.

C: *R. slossoni*, small male (dorsal view), lacking enlarged femurs.

D: Ventral view of same male.

E: *R. slossoni*, large female (dorsal view).

F: Same female (ventral view).

G: *P. zamiae* (dorsal view).

H: *P. zamiae* (ventral view).

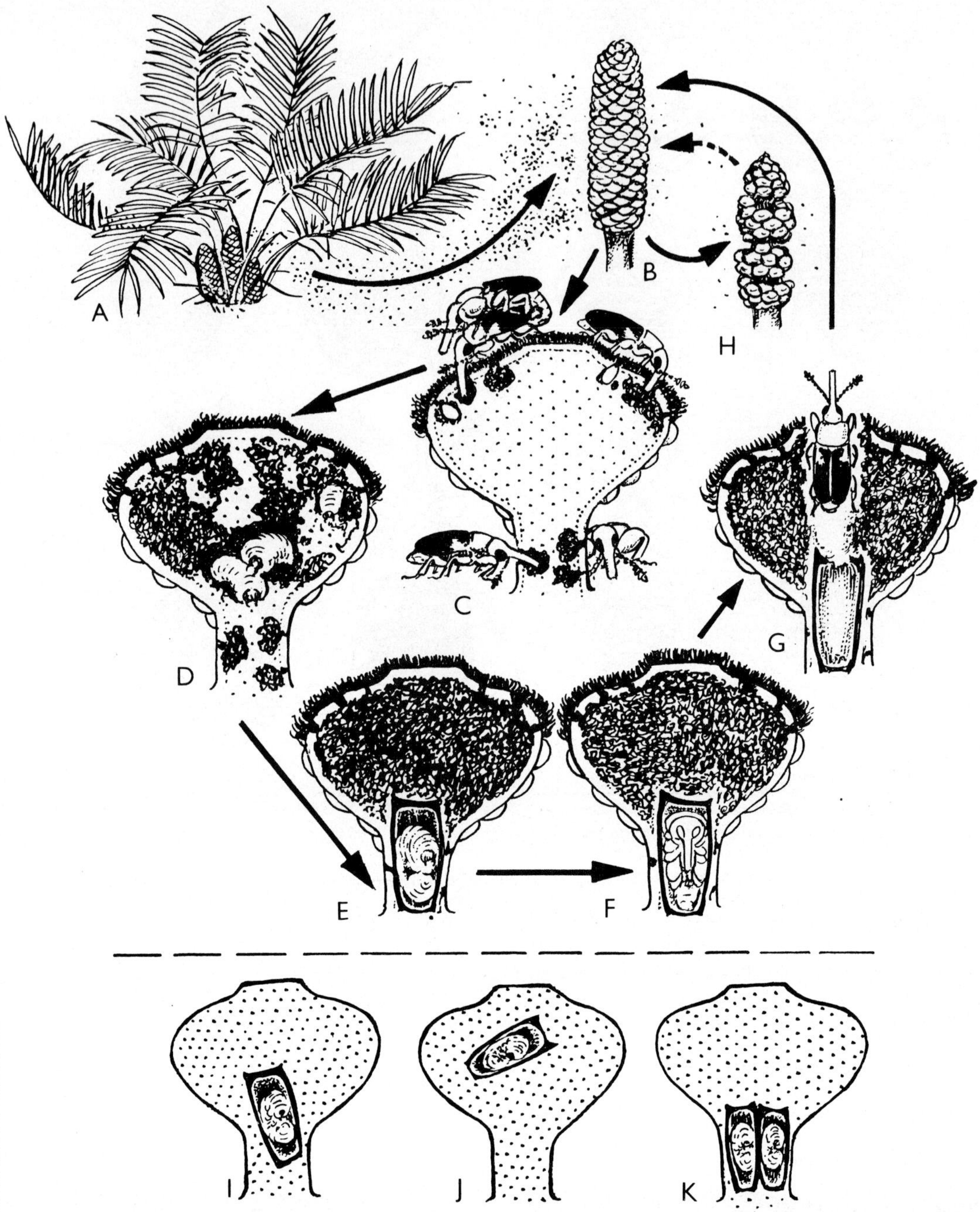
A
B
H
C
D
G
E
F
I
J
K

pupates at the base of the sporophylls like *R. mollis*, this is not always the case. Many pupate crosswise in the ends of the sporophylls (Fig. 2 I-J), and occasionally one can find two pupae cases side by side in the base of the sporophyll (Fig. 2 K). This behaviour has not been observed in the pupation of *R. mollis* in sporophylls of *Z. furfuracea* (Norstog and Fawcett, 1989), and even in *Z. pumila* it is quite rare.

At the end of the reproductive season of *Z. pumila*, its male cones containing larvae in diapause become fragmented, mostly because the sporophylls and cone axes have been completely eaten away by larvae of both *R. slossoni* and *P. zamiae*. Eventually, the sporophylls rot away completely, so that only the pupae cases containing larvae in diapause remain in the soil, looking like little, nearly black, cylindrical boxes (Fig. 3 e-g). We think that the year to year continuity of successful pollination of *Z. pumila* depends largely on the well being of these dormant pupae. Many larvae in diapause become infected with fungus, and although they generally metamorphose, they frequently die in the pupa case, surrounded by fungal fruiting bodies. Those which are able to fly to a cone are unable to breed because their abdomens contain only these fungal fruiting bodies, which they scatter through the cone (Fig. 5 C). Because of this high mortality, a much larger number of larvae must go into diapause than would be needed to repopulate the male *Zamia* cones in the following season.

Life Cycle of *Pharaxonotha zamiae*

Pharaxonotha zamiae (Fig. 1 G-H) goes to the male cones of *Zamia pumila* when the sporangia are about to dehisce, to feed and lay eggs between the pollen sacs, and sometimes inside a dehisced pollen sac (Fig. 4 A-C). The eggs are very large in proportion to the female, which carries up to six eggs at a time at various stages of maturity. The adults feed only on pollen, and when *Pharaxonotha* is present, there is pollen everywhere, so that all of the insects, including *R. slossoni*, which happen to be in the cone become covered with it. Otherwise, unless the sporangia of

←

Figure 2: Basic life cycle of *Rhopalotria slossoni*.

A: Adult weevil leaves the cones in which they were sheltering, and visit an attractive male cone (B).

C: There they feed on the tissue at the ends of the sporophylls, mate and lay their eggs, then go inside the cones to continue feeding at the base of the sporophylls until another male cone is ready for visitation.

D: The eggs hatch and the larvae eat their way towards the base of the sporophyll and, when they meet, one of them eats the other.

E: Only one larva generally survives to build a pupa case from fecal matter.

F: After 2-3 days, the larva metamorphoses into a pupa.

G: Next day, the mature *Rhopalotria* chews its way out of the pupa case and through the end of the sporophyll, then flies to an attractive male cone (B) to feed and breed.

H: Pollination takes place when some individuals on leaving a male cone in search of a new feeding breed site, enter a receptive female cone. There they try to feed, but find no food or egg-laying site. Some become trapped inside the cone, but probably many escape.

I-J: Some larvae do not pupate at the base of the sporophylls.

K: Occasionally two larvae pupate at the base of the same sporophyll, but this is rare.

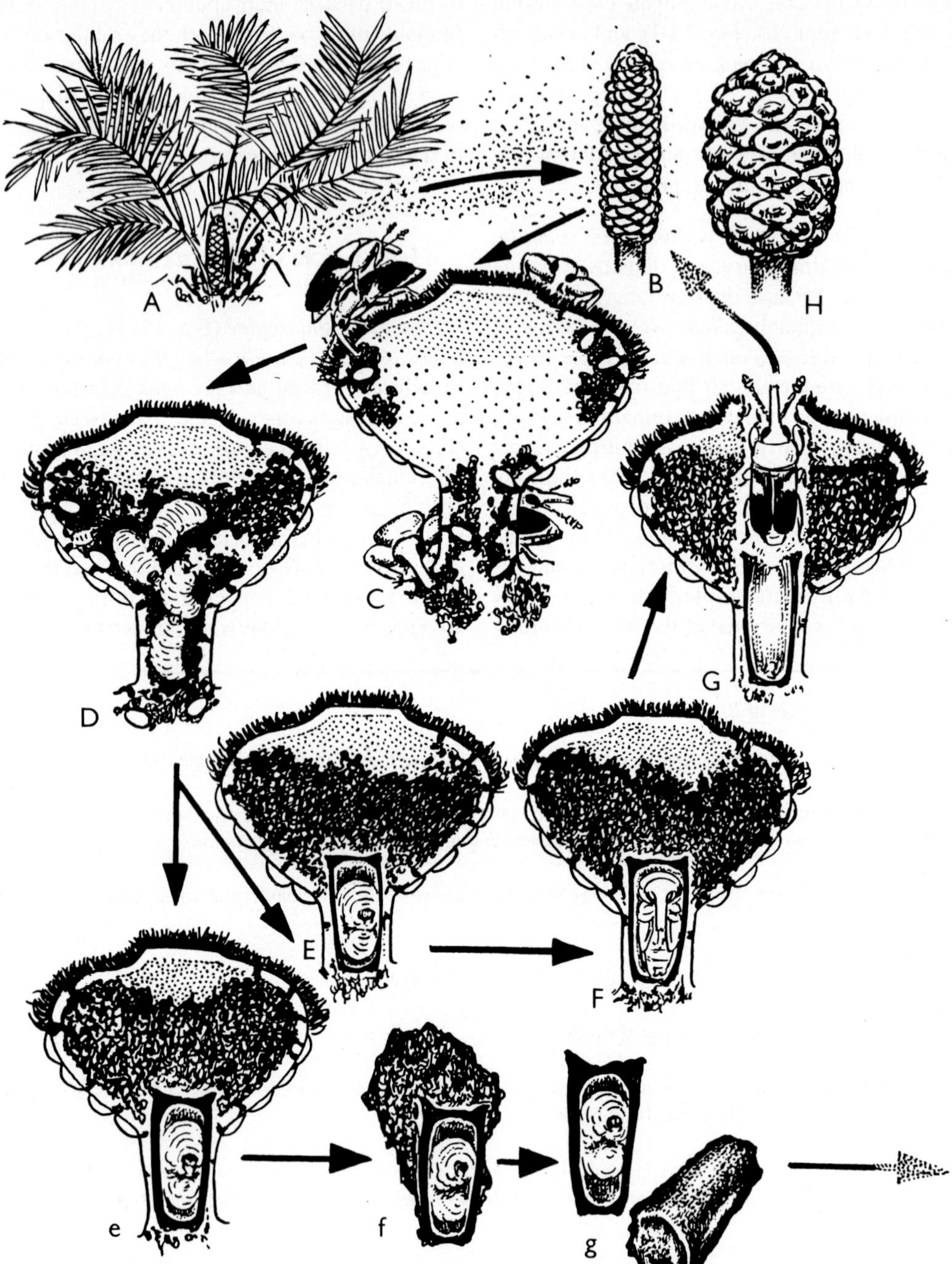
A
B
H
C
D
G
E
F
e
f
g

Zamia pumila are disturbed, the pollen tends to remain in the pollen sacs in little sticky clumps.

Larval stages

The oval eggs are greyish white and translucent, and about twice as long as they are wide. The red compound eyes of the larvae can soon be seen through the membrane of the egg. These eyes have five ommatidia. The larvae eat their way out of the egg and begin feeding on pollen (Fig. 4 D). Newly hatched larvae are almost white.

All of the larval instars look very similar (Fig. 4-H). Each segment has a pair of light brown patches on the dorsal side, so that the larvae seem to be striped. They have long stiff hairs all over their bodies, and all larval instars have simple compound eyes consisting of five ommatidia, and three pairs of soft legs, which are quite well developed and functional.

Feeding behaviour

It is only the first instars that eat pollen, and in fact there is very little pollen left in the cone after the first two instars. A larva that is ready to moult eats a hole in the back of one of the sporophylls and crawls inside, later emerging in the next instar. The later instars eat only tissue. They enter the main axis through a sporophyll, and eat their way down through the cone into the stem, which eventually becomes completely hollow (Fig. 4 G). They do not tunnel beyond the base of the stem. As both *Rhopalotria slossoni* in all feeding instars, and the later larval instars of *Pharaxonotha zamiae* eat sporophyll and stem tissue, *Pharaxonotha* must also have a way of avoiding the toxins. At present, we have no information on the possible toxicity of pollen. When they have finished feeding and are ready to leave the cone, they eat a little hole in the stem and drop down to the ground where they burrow

Figure 3: End of season life cycle of *R. slossoni*.

A: Adults leave in search of an attractive male cone in which to feed and breed.

B: Male cone.

C: Late cone has a green area at tip of sporophyll, which is low in starch and high in idioblast cells. Weevils tend to avoid the green tissue when feeding and egg laying, and the weevil population is larger at the end of the season, so there is a shortage of suitable space for feeding and breeding. At this time after feeding, mating and egg laying in non-green parts of sporophylls, weevils go inside cones and continue egg laying and feeding.

D: Eggs hatch and larvae converge towards the usual area for pupation in the stalk, eating each other when they meet on the way. Meanwhile, for want of more attractive male cones, eggs continue to be laid. At this point, the larvae can either follow the basic route (E, F, G) or the diapause route (e, f, g).

Basic route:

E: Surviving larva builds a pupa case.

F: 2-3 days later larva becomes a pupa.

G: Mature adult chews its way out and leaves the old cone in search of an attractive male cone, which it may or may not find.

H: At this time, female cones are bypassed as they have already been pollinated and have begun to expand and mature.

Diapause route:

e: Surviving larva which is yellowish and waxy looking, builds a very thick pupa case.

f: Cone fragments and sporophylls begin to rot in the ground.

g: Eventually only pupae cases remain, still housing unmetamorphosed larvae, which will not complete their development until one or more years later.

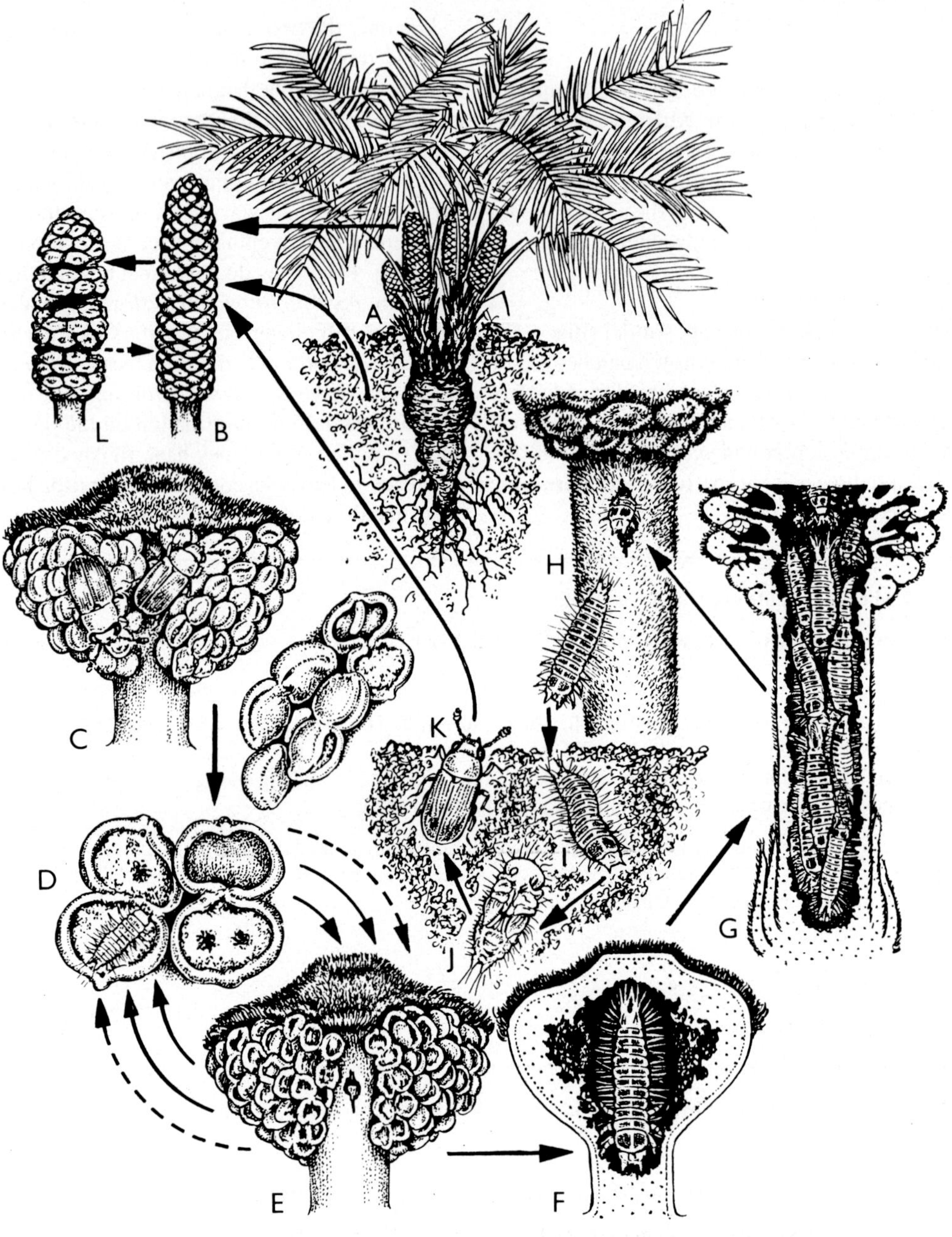
A
B
L
C
D
E
F
G
H
I
J
K

into the soil (Fig. 4 H). We have not watched *Pharaxonotha* larvae in the act of leaving a cone.

Pupation

After several days, *Pharaxonotha* larvae become pupae and remain in this stage for about four to seven days (Fig. 4 I-J). We made these observations late in the season, when there were scarcely any new cones to go to. At this time, the pupal metamorphose to form young adults which, instead of flying away to repeat the breeding cycle, burrow down into the soil (Fig. 4 J-K). When they are dug up, they seem to have no inclination to leave, but only to hide. It is also possible that they need a few days to harden their exoskeletons before emerging from the soil, and that what we saw was normal behaviour for a new adult. In order to find out whether they await the next season as larvae, pupae or mature beetles, we need to examine soil from the base of a male plant from time to time between seasons.

Mating

We are not sure where *P. zamiae* mates. It is possible that they mate underground, but we assume that they mate in the male cones when they go there to feed and lay their eggs.

Pollination

We have found *Pharaxonotha* in female cones of *Zamia pumila* in south Florida, and we assume they blunder into them in error as appears to be the case with *Rhopalotria*. In north Florida, the other species of *Pharaxonotha* appears to be the only pollinator of *Zamia*. *Rhopalotria* may have been there originally, but could have become extinct as a result of the 'coontie' industry, coontie being the Indian name for the Florida *Zamia*. At the turn of the century, white

---

←

Figure 4: Life cycle of *Pharaxonotha zamiae*.

A: Young adults leave the soil, and older adults leave spent male cones of *Zamia pumila* in search of a male cone in which the sporangia are beginning to dehisce.

B: Male cone.

C: In the cone the beetles feed on the pollen and lay their eggs between the sporangia and sometimes in sporangia from which the pollen has been eaten.

D: The eggs hatch, and the larvae feed on pollen. Often little holes in the clumps of pollen can be seen where a larva is feeding.

E: When the larvae are ready to moult, they eat a little hole in the sporophyll into which they go to moult. When they emerge in the next instar, they go back to feeding on pollen. Only the early instars feed on pollen and, by the third instar, there is not much pollen left.

F: The later larval instars eat only tissue. They enter the main axis through a sporophyll and eat their way towards the cone stem.

G: They continue feeding until the cone stem is hollow, but they do not eat into the stem tuber.

H: When the larvae are ready to metamorphose, they eat a hole in the stem, and drop to the ground.

I: The larvae burrow into the soil.

J: They pupate.

K: After 4-7 days the mature beetles leave the soil in search of a cone to feed and breed in.

L: Pollination takes place when some individuals enter a female cone, instead of a male cone, as is the case with *Rhopalotria*.

pioneers harvested coontie stem-tubers, from which they extracted the starch, exporting it to the north under the name of 'arrowroot'. We have seen references to a blue and orange *Rhopalotria* which may be extinct, but this could also be a *R. slossoni* preserved in some unorthodox substance.

Interactions of *Rhopalotria* and *Pharaxonotha*

Both *P. zamiae* and *R. slossoni* are often found together in the same male cone. However, there is no conflict because *Rhopalotria* prefers the sporophylls at the base of the cone while *Pharaxonotha* larvae generally prefer the sporophylls at the tip of the cone. The larvae of *Pharaxonotha* are not in any danger of being cannibalised by those of *Rhopalotria* because of their stout bristles and their three pairs of active legs. Furthermore, the eggs of *Pharaxonotha* are laid about a day after those of *Rhopalotria*, so the larvae are already starting to pupate by the time that the *Pharaxonotha* larvae are ready to moult and begin the second instar.

Other Pollinators of American Cycads

Bart Schutzmann (personal communication) has found different fragrances in receptive cones in different populations in north Florida, and these fragrances draw the appropriate *Pharaxonotha* species in baited McPhail traps, which would indicate that there could be either different ecotypes, or different species of *Zamia* in Florida.

A snout weevil, *Rhopalotria bicolor*, is associated with a similarly bicolored *Pharaxonotha* species in male cones of *Dioon califanoi* (Vovides, 1991), and a *Pharaxonotha* species is associated with *Rhopalotria mollis* in male cones of *Zamia furfuracea*, as well as a *Pharaxonotha*-like beetle in male cones of a *Ceratozamia* species (Vovides, personal communication). Deborah and David Clark, who are working on pollination in *Zamia skinneri* at La Selva in Costa Rica, have also found the probable pollinator to be a *Pharaxonotha* species (personal communication).

Other Insects in *Zamia* Cones

In addition to the two pollinating species, a number of other insects can be found in the cones of *Zamia pumila*. These include the eggs, larvae and pupae of two moths, which live on debris in the old cones, and the eggs of the *Eumaeus* butterfly, the larvae of which devastate the leaves (Fig. 5). Scale and other small insects are often also present.

Implications of Cycad Conservation

At Fairchild Tropical Garden, cultivated cycads have been sprayed with insecticides to control predatory inspects, and a wild population of *Z. pumila* has been reduced by collecting, fire and habitat conversion to other purposes. As a result, numbers of pollinators have also declined.

When a large *Zamia* is dug up, frequently small pieces of stem tuber remain to regenerate, and the extent of this regeneration can be remarkable. However, it takes several years for a regenerated plant to become big enough to cone. Without a series of cones to breed in, a weevil population cannot build up enough to provide an adequate number of larvae in diapause to provide for next year's population of pollinators. Even under normal conditions, many larvae in diapause die and, if the pollinators become extinct, the regenerating remnants of these cycads will cease to be fertile. We have seen evidence of much the same situation at Everglades National Park, in areas of pine-palmetto habitat where, in some areas, a practice of controlled burning every three years has been in effect. Here, populations of *Z. pumila* consist of small and weak plants apparently regenerated from fire-damaged specimens. Such fires not only destroy the tops of the stem tubers, and male cones where both pollinators breed, but also the top layers of humus in which the dormant insects remain from one season to the next.

In the burned areas in the Everglades, we saw small male plants with one, or occasionally two cones, containing a few of one or both pollinating beetle species, but we were only able to find a very few female cones. On the other hand, at the edge of one of the plots, next to the verge at the roadside, we saw a single beautiful many-crowned male *Zamia* which was coning profusely and appeared to have escaped burning. It appears to us that more than three years are required to accumulate the necessary resources for a fire-damaged female cycad to produce a cone. Probably healthy undisturbed zamias would survive a rare controlled burn after a wet period with no danger at all. Willy Tang has worked on the Everglades experimental plots, and has told us that the *Z. pumila* population was more vigorous before the once endangered *Eumaeus* butterfly was reintroduced. At Fairchild Tropical Garden and in the Everglades National Park, we have noticed that this butterfly often leaves plants undisturbed when they are growing among other vegetation, but devastates solitary plants. It appears to find the plants by sight. After a burn, there is a flush of tender, very green cycad leaves against a background of charred rock, which makes a highly visible target for this predatory butterfly!

All cycads have recently become popular all over the world, including the far north, where the smaller species can be grown in glasshouses. We have seen evidence of indiscriminate collection in several populations, and this, combined with the practice of 'slash and burn' agriculture in both the sophisticated 'first world' version and the unsophisticated 'third world' version, could very well result in the slow elimination of many cycad species because of the demise of pollinating beetles. Later, if these species were to be re-established in the wild from cultivated plants, and if they were to survive and reproduce, the appropriate pollinators would have to be re-established in the wild as well. There can be no permanent survival without reproduction!

References

***Norstog, K.J. and Fawcett, P.K.S.* 1989.** Insect-cycad symbiosis and its relation to the pollination of *Zamia furfuracea* (Zamiaceae) by *Rhopalotria mollis* (Curculionidae). Amer. J. Bot. 76: 1380-1394.

***Norstog, K.J., Vovides, A.P. Duncan, M.W. and Fawcett, P.K.S.* (in review).** Idioblast toxicity during cone maturation and pollination in *Zamia furfuracea*.

***Norstog, K.J., Fawcett, P.K.S., Vovides, A.P. and Duncan, M.W.* (in press).** Toxicity and insect-cycad symbiosis in *Zamia furfuracea*. Proceedings of Second International Conference on Cycad Biology, Cycad 90, Palm and Cycad Soc. of Australia, Milton, Queensland, Australia.

***Rothschild, M., Nash, R.J. and Bell, E.A.* 1986.** Cycasin in the endangered butterfly *Eumaeus atala florida*.

Phytochemistry 25: 1853-1854.

***Spencer, P.S., Nunn, P.B., Hugon, A.C., Ludolph, A.C., Ross S.M., Roy, D.N. and Robertson, R.C.* 1987.** Guam amyotrophic lateral sclerosis-Parkinsonism-dementia linked to a plant excitant neurotoxin. Science 237: 517-522.

***Tang, W.* 1987.** Insect pollination in the cycad *Zamia pumila* (Zamiaceae). Amer. J. Bot. 74: 90-99.

***Tang, W., Sternberg, L. and Price, D.* 1987.** Metabolic aspects of thermogenesis in male cones of five cycad species. Amer. J. Bot. 74: 1555-1559.

***Vovides, A.P.* 1991.** Insect pollination in some Mexican cycads in their natural habitat. Biotropica (in press).

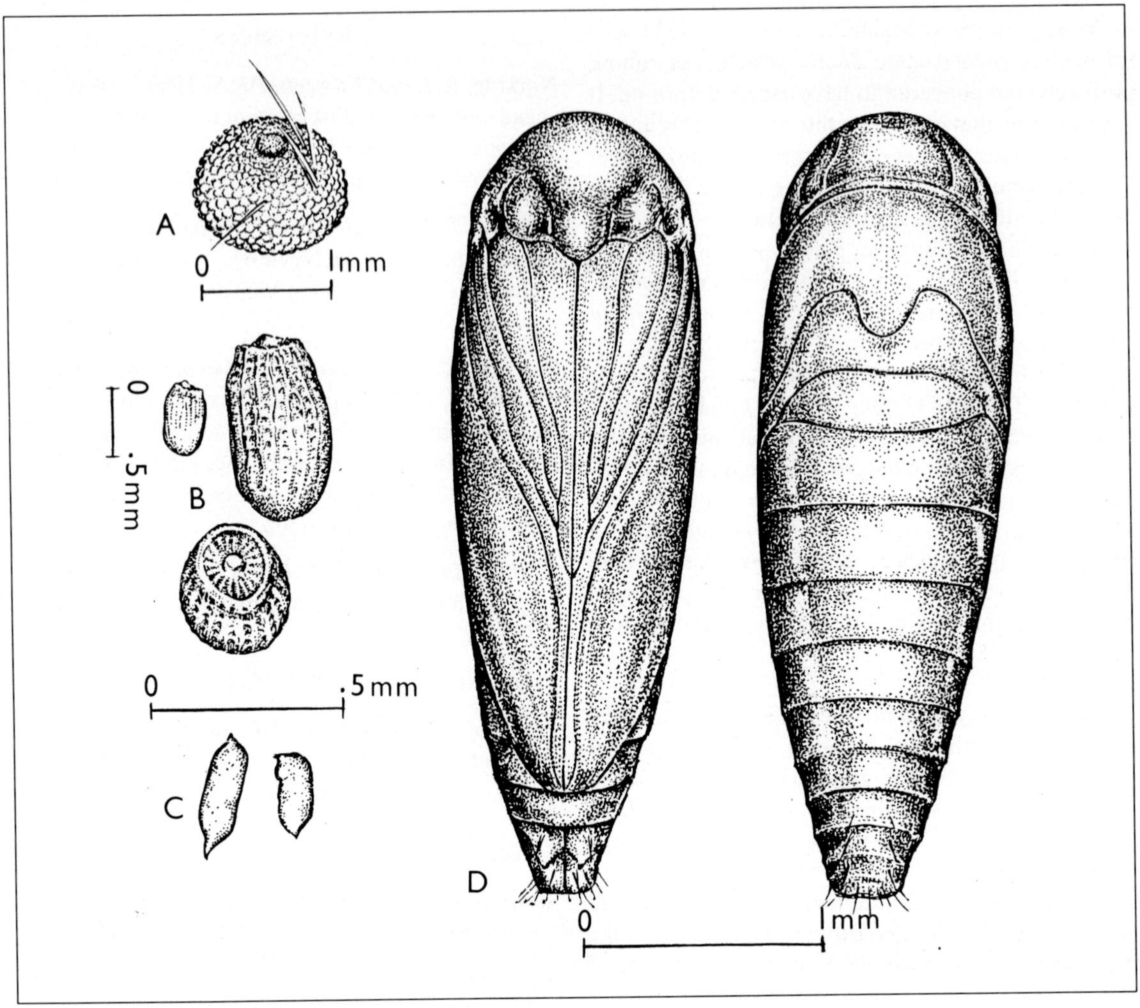

Figure 5: Organisms also found in the male cones of *Zamia pumila*.

A: Egg of the *Eumaeus* butterfly.

B: Egg of moth species.

C: Fungal fruiting bodies.

D: Pupa of moth species, ventral and dorsal views.

Features of the Reproductive Biology of *Macrozamia* Species and *Cycas media* in Australia

Robert Ornduff
Department of Integrative Biology
University of California, Berkeley, CA 94720 U.S.A.

Abstract

Although *Macrozamia communis* shows clinical variation in mean leaf numbers of reproductive plants, leaf numbers of males and females do not differ and generally cone numbers produced by plants of each sex do not differ. The sex ratio of plants in cone is slightly male-biased, but substantial numbers of putatively mature plants do not participate in a given reproductive episode. *Macrozamia riedlei* likewise shows geographic variation in mean leaf number, with males and females mostly producing equal numbers of leaves and cones, and a male-biased sex ratio. Fire appears to stimulate cone production in both species and, in its absence, cone production in *M. riedlei* appears to decrease markedly as a result of competition. Some populations of several *Macrozamia* species in New South Wales, including *M. communis*, entirely lacked cones during the period of study. Males and females of *Cycas media* did not differ in leaf number or trunk height, and not all putatively mature plants participated in a given reproductive episode. Weevils, other beetles, and the bee *Trigona carbonaria* were collected from cones of these cycads.

Introduction

This paper discusses the results of my field studies on *Macrozamia communis* L. Johnson and several other *Macrozamia* species in New South Wales, *M. riedlei* (Fisch. ex Gaud.) C.A. Gardner in Western Australia, and *Cycas media* R. Brown in Queensland. These studies have been concerned with an examination of differences between male and female plants in resource allocation to reproduction and possible mechanisms that may compensate for these differences; estimates of the frequency of reproduction of sexually mature plants in natural populations and a discussion of the factors that influence frequency; and a brief discussion of the insects encountered in the cones of these cycads and their possible role in pollination. Some of the work discussed here has been published previously, but many new observations are presented as well. Because the techniques utilised in the field studies of these three cycads differed somewhat, each species will be discussed separately.

Macrozamia communis

Populations of *Macrozamia communis* were studied in January, 1987 (Ornduff, 1990; Ellstrand, Ornduff and Clegg, 1990). Because of difficulties in determining the ages of plants in natural populations, and the suggestion that size rather than age *per se* may be an accurate predictor of sexual maturity, leaf numbers were counted for reproductive and non-reproductive individuals in ten populations of *M. communis* representing most of the range of this species in New South Wales. Mean leaf numbers of reproductive individuals varied clinically from north to south, with plants in southern populations having mean leaf numbers about 12 times greater than those near the northern end of the range of this species. It appears that size of female cones and mature ovule weights also increase along this gradient. Based on the assumption that plants having leaf numbers equal to or exceeding those of plants bearing cones are sexually mature, rough estimates are that from 37% to 100% of the individuals in the study populations were mature. The maximum number of presumed mature individuals participating in the reproductive episode that spanned January, 1987, was 68%. In one southern population sampled and in two other southern populations visited but not sampled, no male or female plants were observed to be reproductive. Study of climatic data for several years prior to January, 1987 offered no clue as to the cause of this phenomenon, and other workers have reported that reproduction in these three populations occurred prior to and subsequent to 1987. In nine of the ten study populations, more males than females were participating in the 1987 coning episode, resulting in an overall slight male-biased sex ratio (61%) for the species as a whole. The male bias incorporates only those plants with cones, and since a third or more of the putatively mature plants sampled lack cones, this male bias may disappear if the sexes of non-coning individuals were known. Long-term studies of marked plants would allow determination of the actual sex ratio within populations of this species.

Although weights of male and female cones were not determined, it is obvious that resource allocation to production of a mature female cone is considerably greater than that required to produce a mature male cone. Cone numbers produced by males and females were determined for five populations. In three of these, male and females produced equal cone numbers; in two populations, males produced more cones than females. Thus, despite the differences between males and females in resource allocation to cone production, these differences are not uniformly accommodated by females producing fewer cones per plant than males. In all populations sampled, leaf numbers of males and females were equal, suggesting that in this trait reproductive males and females are the same size.

Despite the interpopulation variation in numbers of mature plants participating in the 1987 coning episode, few plants with 1987 cones showed evidence of recent prior cone production. This suggests that both males and females 'rest' between successive coning episodes; the male bias in the sex ratios of coning plants in various populations may reflect a longer resting period for females than for males. Some observers have suggested that the timing of set fires used to suppress subsequent 'natural' fires may reduce the frequency of cone production in *Macrozamia communis* and, possibly, in some areas, prevent it entirely.

The shape of size curves for some populations of this species suggests poor recruitment. This may result from repeated removal of the entire seed crop from populations near urban centres, or may also result from the very poor seed dispersal that appears to characterise this species throughout its range.

The weevil *Tranes internatus* (Pascoe) was observed occupying the interior of the female cones of *M. communis*. Elsewhere at this conference, Chadwick has observed *T. internatus* (noted on *Lepidozamia peroffskyana* Regal in northern New South Wales (Ornduff, 1989)) associated with *M. communis* and explores the possibility that this species may be involved in the pollination of *M. communis*.

Macrozamia riedlei

Field studies of *Macrozamia riedlei* were carried out in 1983 and 1988 in southern Western Australia (Ornduff, 1985; in press). Eight populations were sampled in 1983; six of these populations contained plants producing cones during that year. Two populations contained no individuals with 1983 cones but contained plants showing evidence of prior cone production. The sex ratio of plants with current or prior season's cones was male-biased (66% to 79% of the coning individuals were male) in all but two samples. In these, the sex ratio was 1:1. The proportion of putatively mature plants participating in the 1983 coning episode ranged from 0 in two populations to ca. 45% in two populations.

Thus, this species resembles *M. communis* in the male bias of the sex ratio of coning plants and in the occurrence of some populations lacking any reproductive individuals during a coning episode. On the basis of indirect evidence, I have suggested that both male and female plants of *M. riedlei* 'rest' between successive coning episodes, and that males produce more frequently than females.

In November, 1988 I visited the populations of *Macrozamia riedlei* sampled in 1983 as well as two additional populations, to determine size classes, cone numbers, and to collect insects associated with mature male cones. Two samples made in 1983 consisted of individuals in two areas of a very large population in King's Park, Perth; since these two areas could not be precisely identified in 1988, the King's Park population was considered to be a single sample in 1988.

Of the five populations of *Macrozamia riedlei* having males and females with 1983 cones, three had no individuals of either sex with 1988 cones, one had males but no females with 1988 cones, and only one had both male and females with 1988 cones. This last population consists of plants growing in a municipally maintained lawn area, although the cycads there doubtless predate the time when other native plants were eradicated. Two additional populations lacking cones in 1983 also produced no 1988 cones, but both populations showed evidence of low levels of cone production between the two dates of observation.

One population with a male-biased sex ratio (based on plants with cones) in 1983 had a 1:1 male:female sex ratio in 1988. Two populations not sampled in 1983 had a male bias and a 1:1 male:female sex ratio, respectively, in 1988. Considering 1983 and 1988 collectively, a male bias prevailed, but sex ratios of coning plants in a population may change from one reproductive episode to another.

As in *Macrozamia communis*, populations of *M. riedlei* show substantial interpopulation differences in average leaf numbers of plant with cones, although these differences were not analysed on a geographical basis. In all but one population, males and females had equal leaf numbers; in one population, females had twice as many leaves as males. Cone numbers per plant produced by males equalled those of females except in two populations, where males produced more cones per plant than females. There was no correlation between leaf number and cone numbers for males, i.e. greater cone production was not associated with greater leaf numbers. This was also true for females in all but one population, where larger cone numbers were associated with larger leaf numbers.

These studies of *Macrozamia riedlei* at two periods separated by an interval of five years confirms the suggestion that individuals of both sexes commonly rest between successive reproductive episodes. They also indicate that, as in *M. communis*, some populations may lack any reproductive individuals during a period when cones are evident in other populations. The basis for the absence of reproduction in some populations of *M. communis* is unknown, but may, in part, be the result of the timing of artificial fires employed as a fire control measure. The lack of 1988 cones in some populations of *M. riedlei* is almost certainly due to shading and other competition effects on sites that have not been burned recently, although lack of cone production in other populations may be associated with the marginal environmental conditions under which these populations grow. Providing that fires sweep through populations when cone production is not taking place, such fires are believed to stimulate cone production by reducing competition from shrubs and trees and by increasing levels of nitrogen fixation by the Cyanobacteria that inhabit cycad roots.

In two populations of *Macrozamia riedlei*, male cones were shedding pollen, and insects present in these cones were collected and identified. Large weevils (*Tranes vigorsi* Boheman); small, unidentified weevils of the Cossoninae, *Paracucujus rostratus* Sen Gupta and

Crowson (Boganiidae); and one individual of *Xanocryptus tenebroides* Arrow (Languriidae) were present. These insects have not been observed to pollinate *M. riedlei*; further studies are needed to determine their role, if any, in pollination. The work of Norstog et al (1986), Tang (1987) and Norstog and Fawcett (1989), Chadwick (in this conference), and my own preliminary observations lead me to predict that insects are the preliminary pollen vector in *Macrozamia*.

Other Macrozamia species

In January, 1987, I visited several populations of various *Macrozamia* species in New South Wales, but did not make detailed studies of them. I visited two populations of *M. secunda* C. Moore in New South Wales; In a sample of 461 plants, no male only one female plant with a current season's cone was found. One population of *M. stenomera* L. Johnson was visited; of 228 individuals sampled, no current season's cones of either sex were present. One population of *M. fawcettii* C. Moore was visited; of 98 plants sampled, four male plants and one female plant with current season's cones were present. Five populations of *M. heteromera* C. Moore were visited; of 556 plants sampled, only two females with current season's cones were observed in one population and one male plant with a current season's cone was found in a second population. Three populations of *M. diplomera* (F. Muell.) L. Johnson were visited; of the 270 plants sampled, no individuals of either sex were found with current season's cones. One population of *M. pauli-guilielmi* W. Hill & F. Muell subsp. *flexuosa* (C. Moore) L. Johnson was visited; of the 483 plants sampled, 23 males and 14 females were found with current season's cones. Thus, with the exception of the last taxon, these species exhibited either no reproduction or very low levels of reproduction during 1987.

Cycas media

Field studies were conducted for populations of *Cycas media* in the vicinity of Cairns, Queensland, in November, 1988. Females and males did not differ in leaf number or stem height. Proportions of putatively mature individuals reproducing in a given population were 38% or lower. Leaf life for this species is very short, and many individuals defoliate naturally in the absence of fire. Mature male cones produce a fruity odour, and are visited by the beetles *Aethina unguiculata* Kirejtshuk (Nitidulidae) and *Ulomoides australis* Carter (Tenebrionidae). Numerous individuals of the bee *Trigona carbonaria* Smith were observed collecting pollen. Since *Trigona* is the oldest known fossil bee (Michener and Grimaldi, 1988), predating the origin of angiosperms, it is possible that this bee might have depended on various gymnosperms, including cycads, prior to the evolution of flowering plants.

References

***Ellstrand, N.C., Ornduff, R., and Clegg, J.J.* 1990.** Genetic structure of the Australian cycad, *Macrozamia communis* (Zamiaceae). Amer. J. Bot. 77: 677-681.

***Michener, C.D. and Grimaldi, D.A.* 1988.** The oldest fossil bees: Apoid history, evolutionary stasis, and antiquity of social behaviour. Proc. Nat. Acad. Sci. USA 85: 6424-6426.

***Norstog, K.J., Stevenson, D.W. and Niklas, K.J.* 1986.** The role of beetles in the pollination of *Zamia furfuracea* L. fil. (Zamiaceae). Biotropica 18: 300-306.

***Norstog, K.J. and Fawcett, P.K.S.* 1989.** Insect-cycad symbiosis and its relation to the pollination of *Zamia furfuracea* (Zamiaceae) by *Rhopalotria mollis* (Curculionidae). Amer. J. Bot. 76: 1380-1394.

***Ornduff, R.* 1985.** Male-biased sex ratios in the cycad *Macrozamia riedlei* (Zamiaceae) Bull. Torrey Bot. Club 112: 393-397.

***Ornduff, R.* 1989.** Size distribution and coning behaviour of the Australian cycad *Lepidozamia peroffskyana*. Aust. J. Bot. 14: 241-245.

***Ornduff, R.* 1990.** Geographic variation in reproductive behaviour and size structure of the Australian cycad *Macrozamia communis* (Zamiaceae). Amer. J. Bot. 77: 92-99.

***Ornduff, R.* (in press).** Size classes, reproductive behaviour and insect associates of *Cycas media* (Cycadaceae) in Australia. Bot. Gaz., in press.

***Tang, W.* 1987.** Insect pollination in the cycad *Zamia pumila* (Zamiaceae). Amer. J. Bot. 74: 90-99.

Biology of the S.W. Australian Cycad *Macrozamia riedlei* (fisch. Ex Gaudich) C.A. Gardn.

John S. Pate
Botany Department
University of Western Australia, Nedlands W. Australia 6009

Abstract

Age of large-trunked (1.5 - 3.5m) plants of *Macrozamia riedlei* was assessed by relating current rates of leaf production over a 13-year period to counts of dead leaf bases on the trunks. The study suggested that above ground trunk ages ranged from 670 - 1,530 years. Mean rate of leaf production of males (9.6 yr^{-1}) was similar to females (10.6) and plants produced massive flushes of leaves immediately after fire. Coning was strongly stimulated by fire and the ratio of reproducing males to females was close to 1:1 in plants experiencing a fire-free interval exceeding 10 years, prior to the fire inducing their reproduction. Conversely, plants of habitats burnt three times in an interval of 12 - 14 years showed coning ratio of males to females 9-12:1, suggesting a protective mechanism in females preventing over exhaustion of resources through too frequent reproduction. Observations on coralloid root development and assessments of plant dependence on symbiotic N_2 fixation using the ^{15}N natural abundance technique indicated that seedling and young adult plants of the species relied heavily (19 - 72%) and continuously on atmospheric N_2 under natural habitat conditions.

Introduction

Macrozamia riedlei, the only cycad known for the SW corner of Australia, is a very widely distributed species, with populations extending from Cape Arid on the SE right round the coast to Geraldton in the north of the SW land division (See Byrne and James (in press)). Populations of plants of notably different growth form, size and habitat preference are found across this range, suggesting the existence of a number of distinct genotypes warranting recognition to at least the sub species level. Greatest contrast in habit occurs between the large arboreal form, with trunks up to 1m in diameter and 4m in height (Fig. 1), encountered in open sandplain habitats of the NW and SE limits of the range, and the much smaller form (Fig. 2), normally with completely buried stem, typical of the understoreys of *Banksia* woodland and jarrah (*Eucalyptus marginata*) forest. As shown in previous studies the species regularly engages in symbiosis with cyanobacteria (Pate, 1976) and the resulting N_2 fixation can effect contributions of up to 3 - 20kg fixed $N.ha^{-1}$

in natural habitat (Halliday and Pate, 1976). The general morphology and anatomy (Pate, 1976; Pate, Lindblad and Atkins, 1988) of the NN_2-fixing coralloid roots have been described, and the biochemical pathway of NN_2 fixation utilised by the species, compared with that of other cycads (Pate et al., 1988).

This paper contributes further to the general biology of the species by presenting data determining the age of mature plants, evaluating the reproductive responses of male and female plants to fire and assessing the significance of N_2 fixation to the species using the ^{15}N natural abundance technique (see Ofori, Pate and Stern, 1987).

Age Determination of Mature Specimens of *M. riedlei*

This study involved assessments of annual leaf production by 50 male and 50 female members of the population of large (1.5 - 3.5m above ground trunk height) plants (Fig. 1) in the Badgingarra/Watheroo region of the Northern Sandplains. Using plastic coated wire rings (Fig. 3) to tag all current emerged leaves at the commencement of study (1976), counts were made of yearly leaf production through to 1989 by transferring rings after each annual count to the then innermost youngest leaves. Counts over the fire-free period 1976 - 1983 recorded a mean of 7.1 ± 2.3 (SE) leaves per year of the male plants and 7.9 ± 3.5 on the females. No plants died and only seven plants (5 male and 2 female) produced cones during this period.

A severe local bushfire at the beginning of 1984 killed 3 of the 100 plants and severely scorched the leaf crowns of all others. Virtually all remaining plants produced large flushes of leaves (mean of 57 ± 23 per plant in the females, 42 ± 17 for the males) in the same year (1984), followed by prolific coning of 92 of the study plants, including two of the males which had coned earlier (1976 or 1977). Leaf production was very low (2.9 in the males, 1.1 in the females) over the next year (1985), but thereafter (1986 - 9) returned to values (6.0 ± 2.1 in the males, 4.7 ± 3.5) close to those of the previous interfire period. Thus, for the whole 13 years of study, mean leaf production of males was 9.6, of females 10.6.

Selecting 20 plants (10 male and 10 female) representative of the full range of trunk heights encountered amongst the 100 members studied for current leaf production, estimates were then made of the total numbers of dead leaf bases on each trunk. This was accomplished by considering the trunk as a cylinder of known diameter and counting leaf base numbers per unit area of cylinder at randomly selected points up and around the trunk, using the grid shown in Fig. 4. The values obtained ranged from 6,760 leaf scars for the smallest (1.4m height) to 15,440 for the largest (3.3m) plant sampled. Then, assuming that leaf production had remained constant for 10.1 leaves per year (mean value for males and females, see above), the age range of the above-ground trunks fo the plants would be 670 to 1,530 years. Note that the considerable-below-ground part of the shoot could not be aged non destructively in this manner. However, inspection of the below-ground boles of three dead large individuals showed leaf base numbers of 1,370, 951 and 1,403, suggesting that at least 100 years might have elapsed before the plants commenced developing their above-ground trunk.

Reproduction of Male and Female Plants in relation to Fire Frequency

As demonstrated earlier by Orndorf (1985), induction of coning in *M. riedlei* is strongly induced by fire, and as shown above for the population at Badgingarra, male and female plants responded with equally high intensity when a fire occurred after a relatively long (exceeding 10 years) fire-free interval. Resource allocation to cones, in these particularly large plants was equivalent to only 1 - 3% of total shoot dry wt. in the case of males, and for females during the first season of cone growth 2 - 7%. Seed filling of the female cone in its second season with its attendant drain on possibly scarce nutrients such as phosphorus, then incurred an additional cost above that of a smaller ephemeral male cone. Assuming that such differences in reproductive effort are of physiological importance, one would expect to see differences of one kind or another in behaviour between plants of different sex. This was not evident, however, at Badgingarra possibly due to the large size of the plants concerned and the long recovery period from the last reproductive episode.

In marked contrast, populations in *Banksia* woodland at Yanchep and at two sites in jarrah (*Eucalyptus marginata*) forest near Dwellingup showed strong evidence of sex-related differences in reproduction. All these plants were relatively small (10 - 40kg fresh wt.), and female cone production estimated to be equivalent to 10 - 24% of total shoot dry wt. versus 7 - 11% in the case of male plants. The Yanchep site had received 3 fires in the past 12 years and in response to the last of these fires (1983) showed a coning ratio of male plants to female plants of 12:1. This was not due to an imbalance in sex ratio, since sexing the non reproducing adult plants, by looking for remains of an earlier generation of cones (male or female), or for cohorts of seedlings under previously coning female plants (fig. 5), indicated the population to be composed of closely similar numbers of male and female plants.

One of the sites at Dwellingup had experienced a fire-free interval of at least 10 years prior to the fire (1988) including coning, while the other had been burnt three times in the space of only 14 years., At the former site virtually all mature plants coned, 47% of female 53% male. At the second, heavily burnt site, the vast majority (91%) of females and a significant proportion of males (16%) failed to cone, i.e., a ratio of reproducing males to females of 9:1. Plants at all sites produced large flushes of leaves in the season immediately after fire, regardless of whether they subsequently reproduced or not, or indeed, were large and old enough to have been normally capable of reproduction.

The above data collectively suggests that frequently burnt female plants display some form of mechanism negating the strongly pyrogenic effect of fire on reproduction. This is not replicated to the same degree, if at all, in male plants, and, in being particularly operative in relatively small-sized female plants, may be viewed as a mechanism protecting plants from over exhaustion of resources through closely repeated episodes of coning. Conversely the fire-induced leaf flush, shown by all ages and sexes, may be viewed as generally advantageous, whether followed by reproduction or not, since it fosters a quick replacement of assimilatory potential destroyed by the fire and is thereby likely to generate sufficient photosynthate for renewed root growth, capture of nutrients released in the fire, and successful development of cones.

Coralloid Root Development and the Dependence of Plants on Symbiotic Nitrogen Fixation

Development and growth of coralloid roots on a selected set of 75 variously-aged plants with stem boles ranging from 2 - 20cm diameter were examined on an annual basis over the period 1977 - 86, using a non destructive observational technique. The study habitat at Medina had been previously used (Halliday and Pate, 1976) for assessment of symbiotic N_2 fixation, and comprised mixed *Banksia attenuata: Allocasuarina fraseriana* woodland, with *M. riedlei* as a dominant understorey component. The surface layers of sand around the stem bole of each plant were carefully removed at each time of examination to expose the sets of coralloid roots, and after observations and photographic records (e.g., Figs. 6 and 7) had been made, the sand was replaced. There was no evidence of damage to plants through such annual observations. Findings from the survey were as follows:

1. Seedlings had formed sets of coralloid roots usually within 2 or 3 years after germination from seed.

2. Each seedling typically developed a set of apogeotropic roots (see Pate, 1976) opposite each protoxylem pole of its diarch root and both sets usually became infected and developed basal clusters of coralloid roots (Fig. 6).

3. Among older plants it was common to encounter plants with only 1 set of actively growing coralloid roots.

4. Regular increments of tissue were added to a coralloid root cluster each growing season (e.g., see

Fig. 7) and dichotomous branching continued throughout the life of the cluster.

5. Most coralloid root masses persisted for the full nine years of the study, but over one fifth of those examined senesced and, in most cases, were replaced at some time during the study period.

6. Only 6 of the 45 older plants examined failed to carry at least one set of coralloid roots during every year of the study, indicating that plants were likely to have had continued access to fixed N.

Assessment of Proportional Dependence of Plants on Symbiotic N_2 Fixation

Study of N_2 fixation using the nitrogen isotope natural abundance technique was carried out at nine sites, five in open sand plain habitats, two in *Banksia* woodland and two in jarrah forests. Above-ground biomass was sampled from ten cycads (stem bole diameters 3 - 15cm) and from ten plants of each of a range (5 - 10 species) of co-habiting woody, non-N_2 -fixing, shrub or tree species, hereinafter referred to as reference plants. ^{15}N natural abundance values (** ^{15}n ‰) were determined for the total N of all samples. In seven of the habitats mean ** values of the reference plants - species presumably depending solely on soil N - lay within the range +2.7 to +5.3 ‰ and the corresponding values for the *Macrozamia* plant material were from +0.9 to +2.2‰. With a ** value of 0 for atmospheric N, this comparative set of samples accordingly indicated a 39 to 72% dependence of the *Macrozamia* on symbiotically fixed N (average for all site 59%). The other two sites gave inconclusive results, sine ** values for reference plants (+1.3 and +0.4‰) were not significantly different from the paired cycad material (+0.8 and -0.3f). Taken collectively the study suggested heavy dependence of young plants of this size range on atmospheric N - a result agreeing with our earlier demonstration using the acetylene reduction technique of substantial N_2 fixation rates in the species (Halliday and Pate, 1976) and the prevalence and persistence of coralloid roots or to have a much reduced proportion of their biomass as coralloid roots than young plants of this size range used in the N_2-fixation assays described here and by Halliday and Pate (1976). The symbiotic status of these older plants awaits further study.

Figures 1 - 7 : Biology of *Macrozamia riedlei*

Figure 1: Biology of *M. riedlei* in natural habitat S.W. Australia. Tall 3.1m mature plant of the large arboreal form of the species, Eneabba, W. Australia.

Figure 2: Examples of the smaller form of the species typical of understoreys of open woodland, Medina, W. Australia. All trunks of these mature specimens are buried below ground, the fronds of the largest species in the photograph are 1.5m in length.

Figure 3: Plastic-coated wire rings (arrowed) used to tag leaves in a long-term study of annual rates of leaf production in population of the large arboreal form at Eneabba (Fig. 1).

Figure 4: Technique used for counting leaf base numbers in above-ground trunks (e.g., Fig. 1 of *M. riedlei*)

Figure 5: Adult female plant of *M. riedlei* from the Medina population (Fig. 2) showing several age cohorts of seedlings (arrow heads) which have germinated close to the mother plants.

Figure 6: Partially excavated young seedling showing two age groups of apogeotropic roots, the older set in the mid-foreground (marked by star) blackened and dead after being burnt by fire, the younger sets still actively growing. The seedling proved tobe uninfected by the N_2 fixing cyanobacterium *Nostoc* (X 0.75).

Figure 7: *In situ* photographic of a still attached coralloid root mass exposed after removal of sand from the stem bole of a juvenile plant. Note tentacles on surface of coralloid root and dark cyanobacterial ring of tissue (arrow) visible where one of the lobes of the root mass has been cut (X 1.5).

Acknowledgments

I am indebted to Dr. D.S. Pate and Mr. T. Leary for assistance with the field work on age determination of cycads.

References

***Byrne, M.* and *James, S.H.* 1991.** Genetic diversity in the cycad *Macrozamia riedlei*. Heredity. (in press).

***Halliday, J.* and *Pate, J.S.* 1976.** Symbiotic nitrogen fixation by coralloid roots of the cycad *Macrozamia riedlei*: physiological characteristics and iconological significance. Aust. J. Plant Physiol. 3: 349 - 358.

***Ofori, F., Pate, J.S. and Stern, W.R.* 1987.** Evaluation of N_2 fixation and nitrogen economy of a maize:cowpea intercrop system using ^{15}N dilution methods. Plant Soil. 102: 149 - 160.

***Ornduff, R.* 1985.** Male-biased sex ratios in the cycad *Macrozamia riedlei* (Zamiaceae). Bull. Torrey Bot. Club. 112: 393 - 397.

***Pate, J.S.* 1976.** Transport in symbiotic systems fixing nitrogen. *In* Transport in plants. Encyclopedia of Plant Physiology, N.S., vol. 2: pt. B, Tissues and organs, pp. 278 - 303, L**utge, U., Pitma, M.G., eds. Springer, Heidelberg.

***Pate, J.S., Lindblad, P. and Atkins, C.A.* 1988.** Pathways of assimilation and transfer of fixed nitrogen in coralloid roots of cycad-*Nostoc* symbiosis. Planta. 176: 461 - 471.

Determination of Sexual Dimorphism in a Population of *C. mexicana* Brongn. (Zamiaceae) Based on Vegetative Characteristics

Maria Ydelia Sanchez-Tinoco, Maria Vazquez-Torres and Jorge Alberto Alejandre-Rosas
Centro de Investigaciones Biologicas
Universitad Veracruzana, Av. 2 Vistas S/N km. 2.5 Carret.
Xalapa-Veracruz, Apdo. Postal 294, Xalapa, Veracruz 91000 Mexico

Abstract

The starting point was the hypothesis that the apparent morphological characters of *Ceratozamia mexicana* have discrimination significance in the sexual dimorphism of the species. To determine its vegetative sexual differentiation, a search of variables having the maximal discriminant power was made. We found that the variables showing the highest levels of conficende for differentiation purposes between males and females were spine size and the number of spines present on the petioles of mature fronds in adult individuals. Hotelling's 'T-square' ($P < 0.0001$).

Introduction

Cycads and other related species make up a group of fossil plants, most of them provided with a dioecious system of reproduction (Chamberlain, 1935).

Ceratozamia mexicana Brongn. is a cycad distributed in Mexico (San Luis Potosi, Hidalgo, Puebla, Veracruz, Oaxaca, Tabasco and Chiapas), Guatemala and Belize (Moretti, et al. 1980). Morphologically, it shows intrapopulational variations in some of its structures mainly in terms of leaflet and petiole. These variations have permitted the segregation of the different species and varieties within the *Ceratozamia* genus (Stevenson, et al. 1986; Vovides et al. 1983), whose emphasis on the vegetative characteristics is due to the dioecism of the genus and to the fact that the reproductive structures are unfrequently found in the field or in herbarium specimens.

Sex is determined by the reproductive structure named strobilus; the term microstrobilus is often used for the male structures and the term mega or macrostrobilus for the female structures.

Interestingly, the presence of these structures is the only characteristic that can be used *in situ* to determine the sex of these plants, which happens in analogous way in the corresponding trees with stamens or ovules, being difficult to distinguish sex without the occurrence of sex organs. This difficulty has been approached by some researchers with the aim of distinguishing sex in various species of cycads making use of cytological methods (Alor, 1987; Norstog, 1980, 1981); but even considering the perspectives opened to new research, the difficulty demands complicated practices to achieve an adequate manipulation of data in the field.

The efforts to achieve sex separation in populations of cycads have been considered by Newell (1985), for *Zamia pumila* L., where she states that the number of leaflets per front is the only different variable between male and female individuals, taking into account the microenvironment in which each plant develops. In some other works on *Zamia latifoliolata* Prenl. (Britton & Wilson, 1926) and *Zamia debilis* L. f. (Lioger & Martorell, 1982), morphological variations had already been detected among individuals of one population.

Taking into account the morphological variations shown by individuals of *Ceratozamia mexicana* in the population under survey and the presence of dioecism, the aim of this research was to distinguish the sex of individuals in practical terms, taking morphological vegetative characteristics as a base.

Areas of Research

This survey was carried out on a hill located outside Coacoatzintla, Ver., 20km to the northeast of the city of Xalapa on the Banderilla-Misantla road. The zone is located at 19º 38' and 19º 39' N latitude 96º 57' W longitude at 1,550 metres above sea level (INEGI, 1984c).

The 'Coacoatzintla Mountain' is 150m tall with the summit at 1,700m. It is elongated in shape and it is oriented from SE to NW, which defines two slopes; one oriented to NE with little exposure to sunlight and high environmental humidity, and the other oriented to SW with more exposure to sunlight and drier environment. The hill belongs to the mountain system of the Eje Neovolcanico Transversal. The population of cycads is found on the north slope.

According to the climate classification of Garcia (1983) the climate is type C(fm)w"b(i)g, with an annual average temperature of 19ºC and annual average rainfall of 1,727mm.

Materials and Methods

In 100 individuals selected at random in the population, clear morphological characteristics that could be used to differentiate male and female individuals were sought. The characters were: stem length; stem diameter; number of fronds at the crown; frond length; petiole length; rachis length; number of leaflet pairs at the frond; length and width of leaflets; vein bifurcations in the number of leaflet; veins in the middle portion of the leaflet; distance from the base to the widest part of the leaflet; spine size in the petiole; and spine abundance in the petiole. All of these characters were statistically compared to analyse their discriminatory values. For this purpose, basic statistics of mean and standard deviation were calculated to obtain a preliminary comparison. For those selected characters, it was chosen to refine the statistical analysis by the Student's 'T' test for the null hypothesis that there were no differences in the means of the compared groups; multivariate analyses were also done by calculation of Hotelling's square 'T', Mahalanobis' square 'D', the 'F' value of Fisher's statistics, with their respective degrees of freedom and the correlation matrix was obtained to reduce the dimension of the data.

The results of the comparison showed that only some of the variables calculated, the leaflet characters and the spines present in the petioles, yielded high level of confidence. For this reason the procedure was used separately in two following analyses.

Experiment 1: Analysis of Leaflets

In a preliminary analysis, measurements of 50 specimens of each sex, selected at random in the population under survey were performed, recognising the sex of the individuals by the presence of a strobilus or the strobilus of the previous year. The opposite pair

Table 1

Description of variables considered for the Analysis of Leaflets

Variable	Description of Variable	Coding
2	Total length of the first leaflet of the pair.	LTOT1
3	Total of bifurcations of the first leaflet of the pair.	NBI1
4	Width of the first leaflet of the pair.	ANCHO1
5	Distance where is found the width of the first leaflet of the pair.	DISTA1
6	Distance where is found the last bifurcation of the first leaflet of the pair.	DISTU1
7	Amount of veins in the wider part of the first leaflet of the pair.	VENCH1
8	Amount of veins in the midpart of the first leaflet of the pair.	VENMD1
9	Total length of the opposite leaflet of the pair.	LTOT2
10	Total of bifurcations of the opposite leaflet of the pair	NBI2
11	Width of the opposite leaflet of the pair.	ANCHO2
12	Distance where is found the width of the opposite leaflet of the pair.	DISTA2
13	Distance where is found the last bifurcation of the opposite leaflet of the pair.	DISTU2
14	Amount of veins in the wider part of the opposite leaflet of the pair.	VENCH2
15	Amount of veins in the midpart of the opposite leaflet of the pair.	VENMD2

The column corresponding to coding refers to the abbreviated names of the variables when handled in the computer. Note that in the table variable 1 does not appear since this deals with a categorical type, which takes the value of 1 for the group of females and value 2 for the group of males.

of the central portion of a mature frond per each adult plant was taken, including 14 variables listed in Table 1.

Experiment 2: Analysis of Spines found in the Petioles of Adult Male and Female Individuals

From qualitative observations, it was evident that the spines present in the petioles of mature fronds and adult plant show clear differences in length and abundance in both sexes. Taking this criterion into account, 10 female individuals and 10 male individuals were randomly selected in the population. A front was chosen from each plant. The spines present in the petiole were counted; 10 spines were removed from each petiole and were measured. The results indicated that the conditions previously observed for the spines of adult female and male individuals were consistent. For this reason we considered to increase the size of the sample in order to gain more fidelity in the results. We collected 100 more individuals (50 males and 50 fe-

Table 2

Summary of Calculated Measures in relation to Recorded Variables at the Leaflets Pairs

Variable	Group	Mean	Std Dev	S.M.E.	Sample	Maximum	Z-Score	At Case	Minimum	Z-Score	At Case
2	Female	38.8180	3.7049	0.5239	50	45.4000	1.7766	30	31.400	-1.4624	9
	Male	36.4020	4.1918	0.5928	50	45.8000	2.2420	97	27.300	-2.1714	87
3	Female	6.7200	1.7617	0.2491	50	11.0000	2.4294	17	3.0000	-2.1116	38
	Male	5.8400	1.5854	0.2256	50	10.0000	2.6075	60	3.0000	-1.7801	61
4	Female	2.0160	0.2113	0.0299	50	2.5000	2.2909	42	1.6000	-1.9690	19
	Male	1.8840	0.2510	0.0355	50	2.9000	4.0477	81	1.4000	-1.9282	64
5	Female	9.3400	2.0106	0.2843	50	14.5000	2.5664	40	5.4000	-1.9596	10
	Male	9.1120	1.6345	0.2312	50	14.3000	3.1740	86	6.0000	-1.9039	100
6	Female	7.5400	3.7472	0.5299	50	18.1000	2.8181	33	2.3000	-1.3984	28
	Male	7.6900	3.9245	0.5550	50	23.4000	4.0031	71		-1.4753	61
7	Female	17.1400	2.1382	0.3024	50	23.0000	2.7406	42		-1.9362	3
	Male	16.2000	1.5119	0.2138	50	19.0000	1.8520	74		-2.7780	84
8	Female	15.4000	2.0102	0.2843	50	20.0000	2.2884	42		-2.6863	2
	Male	14.0400	1.4841	0.2099	50	17.0000	1.9945	74		-2.7223	61
9	Female	36.7440	3.7205	0.5262	50	27.1000	2.7835	30		-1.4095	24
	Male	36.5400	4.4614	0.6309	50	44.3000	1.7394	78		-2.5418	93
10	Female	6.6400	1.4813	0.3095	50	10.0000	2.2683	34		-2.4573	50
	Male	5.6800	1.5444	0.2184	50	9.0000	2.1496	73		-1.7353	55
11	Female	1.9940	0.2208	0.0312	50	2.5000	2.2920	39		-1.7847	18
	Male	1.9020	0.2503	0.0354	50	2.5000	2.3892	72		-1.6061	52
12	Female	9.0020	1.8201	0.2574	50	13.8000	2.6361	41		-1.8142	15
	Male	8.7940	1.9116	0.2703	50	14.0000	2.7234	65		-2.7695	63
13	Female	7.5460	4.9720	0.7031	50	33.1000	5.1396	31		-1.1959	29
	Male	7.2800	3.5655	0.5042	50	19.9000	3.5394	71		-1.4528	90

males), and we selected 3 petioles (from each frond per each individual). Spines found in the entire petiole were counted. 10 spines of each petiole were collected (1,500 spines per each group) whose length was determined.

To obtain the 'T' square statistics of one and two independent samples with dimension 3 vector, the processing of data was performed by computer, too; variables being 'PROMEL', 'CANT' and 'SEXO' ('PROMEL' is the name corresponding to the average of 10 measurements of length of the spines taken from 3 petioles per 50 individuals, 'CANT' corresponds to the number of the spines present in the petioles and 'SEXO' is the categoric variable for male and female plants).

Results

With the experiments described in the previous section the following results were obtained:

Experiment 1

Table 2 shows a summary of the calculated measurements of each of the variables relative to 2 leaflets per plant. The value of Hotelling's 'T' square to test the null hypothesis of equality of dimension 14 means per group, is significant. It is suggested that we can reject the null hypothesis at a level of 5% but not at a level of 1%.

This means that there must be at least one null hypothesis (among univariate 14 variables) that is rejected. The problem now is to determine which univariate hypotheses are rejected. As an illustration, the calculations additionally required for obtaining the standard difference of variable NBI1 are explicitly indicated, understanding that the other differences appearing in Table 3 can be obtained in a similar fashion. For example;

$$\text{Standardised difference Variable NBI1} = \frac{\text{Mean female - Mean male}}{\text{Standard Error difference}}$$

$$= (6.72\text{-}5.84)/((1.2491)^2+(0.2256)^2)^{1/2}$$

$$= 2.62$$

The question: Which variables have discriminating power in a univariate way to classify male or female plants? can be answered, by checking Table 3, that NBI1, VENDM1, VENCH1 y NBI2 can be used alone to detect sex. But they allude the need of being careful is one wishes to consider one of the variables for discriminatory purposes.

According to the results of the correlation matrices, we can see that there is a similar pattern in the results of measurements both for the male group and the female group with respect to their interrelations, (Tables 4, 5).

Table 3

Results of Standardises differences to 14 Variables of the Leaflets

Variable	Standardised Difference	Result
LTOT1	0.52	Non significant
NBI1	2.62	Significant
ANCHO1	2.84	Significant
DISTA1	0.62	Non significant
DISTU1	-0.19	Non significant
VENCH1	2.50	Significant
VENMD1	3.85	Significant
LTOT2	0.25	Non significant
NBI2	3.17	Significant
ANCHO2	1.95	Non significant
DISTA2	0.56	Non significant
DISTU2	0.31	Non significant
VENCH2	1.36	Non significant
VENMD2	1.16	Non significant

Experiment 2

The results of the computation for comparisons between variables 'PROMEL' and 'CANT', between spines from petioles of male and female plants are shown in Table 6.

From the inspection of the histograms for both groups, according to variables 'PROMEL' and 'CANT', it can be observed that they correspond to

Table 6

Summary of Calculated Measures pertaining the Variables 'PROMEL' and 'CANT'

	PROMEL	Variable Number 1		CANT	Variable Number 2	
Group	Male		Female	Male		Female
Mean	0.4231		0.2490	120.1400		40.3000
Std Dev	0.0881		0.0581	30.6780		16.1311
s.E.M.	0.0072		0.0047	2.5049		1.3171
Sample Size	150		150	150		150
Maximum	6.800		0.4300	203.000		85.0000
Z-score	2.9154		3.1164	2.7010		2.7710
Minimum	201		109	171		103
Z-score	0.2250		0.0700	66.0000		8.0000
At Case	-2.2484		-3.0815	-1.7648		-2.0023

	Statistics	P-Value	DF	Statistics	P-Value	DF
Separate T	20.21	0.0000	257.9	28.21	0.0000	225.5
Pooled T	20.21	0.0000	298	28.21	0.0000	298
Levene F For Variances	24.60	0.0000	1,298	66.23	0.0000	1.298

unimodal distributions, with greater dispersions for the group of male individuals (Fig's. 1A, B, C, D). Hotelling's 'T' square statistics had a value of 1,220.2896 and the associated 'F' value was 608.0973 with degrees of freedom of 2 in the numerator and 297 in the denominator, which corresponds to a very significant difference, with a probability of less than 0.0001 that the observed data yielded those extreme values in the case that the null hypothesis of equality of populational means was true. Therefore the multivariate analysis easily rejects the null hypothesis. In the corresponding univariete analysis of comparison of means for variables 'PROMEL' and 'CANT', the differences were highly significant, and specifically for variable 'PROMEL'.

Consistently with this discriminatory purpose, and given the reasonably large sample sizes of 150 in each case, we proceeded to adjust models of normal distribution with means and standard deviations calculated with the data indicated above to the variables. In order to state the regions of values, where a specimen is

Table 7

Classification probabilities and Frequencies of Classificated Cases according to variables 'PROMEL' and 'CANT'

(the cells 'male-male' and 'female-female' correspond to right classifications and the remaining ones to incorrect classifications)

Classification	Real Sex			
	Male		Female	
Variable 'CANT'				
Male	147	(97%)	6	(6%)
Female	3	(3%)	144	(94%)
Variable 'PROMEL'				
Male	125	(85%)	14	(8%)
Female	25	15%	136	(92%)

stated as belonging to the group of males, and if are falls located in a complementary zone, to the group of females. The critical value was obtained through the intersection of the curves of normal distribution corresponding to each group, which required a small series of algebraic manipulations corresponding to the solution of quadratic equations.

The classification criteria are as follows:

Classifying Criterion for Variable 'CANT':

We classify as 'female' every individual producing a measurement of less than 71.67. The probabilities of erroneous classification are calculated from the tables of normal distribution by means of the corresponding standardisation. We obtain -

Pr (classifying a female as 'male') = 0.571

Pr (classifying a male as 'female') = 0.262

Classifying Criterion for Variable 'PROMEL':

Classify as 'female' every individual producing a measurement of less than 0.3301 and as 'male' if the measurement is greater than or equal to 0.3301. The probabilities of erroneous classifications are as follows -

Pr (classifying a female as 'male') = 0.0823

Pr (classifying a male as 'female') = 0.1469

These classifications are summarised in Table 7, including the probabilities of correct classification for each variable.

Discussion

The morphology of each of the structures present among individuals of the population of *Ceratozamia mexicana* under survey show variations, some of which are related to sex.

By examining the variation in mature plants, we found that there are distinguishable characteristics in the morphology of leaflets which allow the separation of sexes in these plants, some characters such as the number of bifurcations of veins and identification, quantity of veins in the midpart and quantity of veins in the midwest part can be used singly to determine sex. But we learned that these characteristics have to be carefully dealt with, since the nature of the variable could result from microenvironmental conditions such

A

Class Interval (mm)	Frequency Group
0.04-0.11	92
0.12-0.17	35
0.18-0.23	490
0.24-0.29	373
0.30-0.35	435
0.36-0.41	59
0.42-0.50	16

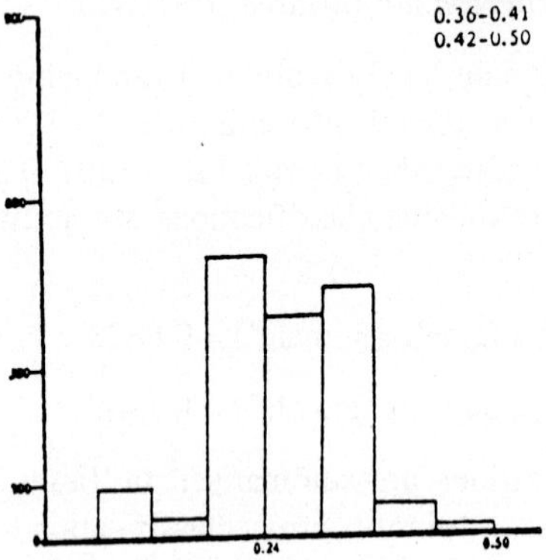

MEAN ($\bar{X}$) = 0.2490
STANDARD DEVIATION (S) = 0.0581
CONFIDENCE INTERVAL 95 % = $\bar{X} \pm$ 1.96 (S)
: UPPER LIMIT = 0.3948
: LOWER LIMIT = 0.1032
MINIMUM VALUE OBSERVATED = 0.04
MAXIMUM VALUE OBSERVATED = 0.50

B

Class Interval (mm)	Frequency Group
0.20-0.28	97
0.29-0.36	334
0.37-0.44	413
0.45-0.52	466
0.53-0.60	135
0.61-0.68	19
0.69-0.80	36

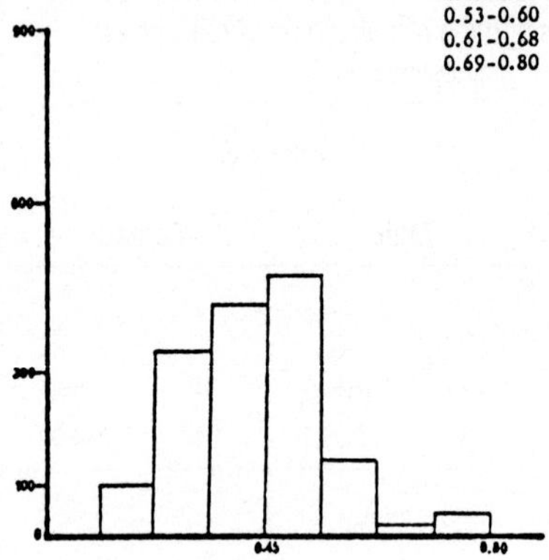

MEAN ($\bar{X}$) = 0.4231
STANDARD DEVIATION (S) = 0.1073
CONFIDENCE INTERVAL 95% = $\bar{X} \pm$ 1.96 (S)
: UPPER LIMIT = 0.6334
: LOWER LIMIT = 0.2128
MINIMUM VALUE OBSERVATED = 0.20
MAXIMUM VALUE OBSERVATED = 0.80

C

Class Interval	Frequency Group
8-19	10
20-30	30
31-41	42
42-52	34
53-63	20
64-74	11
75-85	3

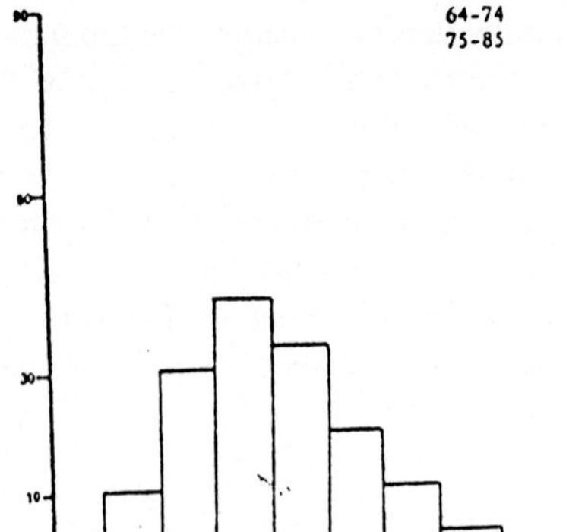

MEAN ($\bar{X}$) = 40.3000
STANDARD DEVIATION (S) = 16.1311
CONFIDENCE INTERVAL 95% = $\bar{X} \pm$ 1.96 (S)
: UPPER LIMIT = 71.9169
: LOWER LIMIT = 8.6831
MINIMUM VALUE OBSERVATED = 8
MAXIMUM VALUE OBSERVATED = 85

D

Class Interval	Frenquency Group
66-85	19
86-104	39
105-123	30
124-142	25
143-161	19
162-180	16
181-203	2

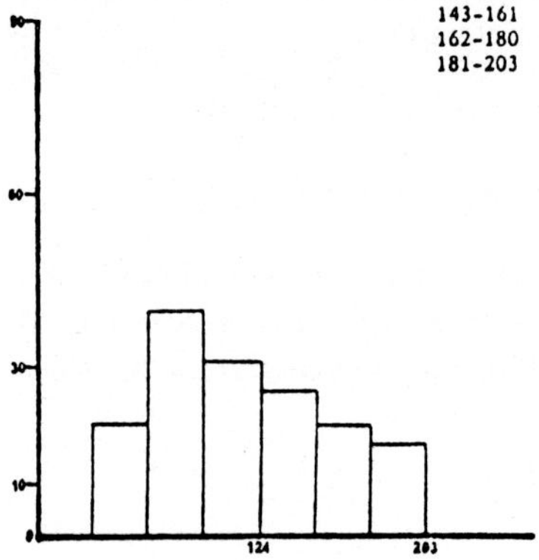

MEAN ($\bar{X}$) = 120.14
STANDARD DEVIATION (S) = 30.6780
CONFIDENCE INTERVAL 95% = $\bar{X} \pm$ 1.96 (S)
: UPPER LIMIT = 180.2689
: LOWER LIMIT = 60.0111
MINIMUM VALUE OBSERVATED = 66
MAXIMUM VALUE OBSERVATED = 203

as direct sun exposure of leaflets, the effects of temperature, or exposure to water.

The clear differentiation of sex in a practical methodological sense and with quite high margins of confidence; we found on counting and measuring the spines present in the petioles of mature fronds of adult plants. The counts and measures are greater in male individuals than in female individuals with respect to the variable length of spines. The results must be handled carefully since this variable is susceptible to greater variations with respect to environmental changes that could be imperceptible when taking samples. Apart from this, these features reveal high levels of confidence for discriminatory purposes.

Regarding the occurrence of vegetative characteristics for distinguishing sex in practical terms, a tool is provided that suggests the need of repeating the same analysis operation in other populations of the same species and being able to determine if it is a constant characteristic, and if this is the case, this would allow recognition of the sex of the whole population for a better understanding of the structure and dynamics in the ecology of plant populations.

The practical perspectives of the application of this finding could be focussed on dioecious species of biological importance (whether special or commercial), that determine their productivity in terms of a greater or lesser density of individuals of the given sex; with this implementation we would be able to search for characteristics until now ignored, through a systematic sampling criterion with the required statistical applications.

Acknowledgments

The authors would like to thank the contributions of Dr. Luis Cruz Kuri of the Instituto de Ciencias Básicas, Universidad Veracruzana, for his patient and devoted support in the statistical and computing work, as well as Dr. Sergio Sabato and Dr. Aldo Moretti of the Orto Botanico of the University of Napoles, for their fine suggestions and comments.

References

***Alor-Apango, R.C.* 1987.** Un estudio citológico en cicadaceas, satélites como indiciadores del sexo en *Ceratozamia hildae* Landry and Wilson y. *C. microstrobila* Vovides and Rees. X Congreso Mexicano de Botánica. Septiembre 27-Octubre 3. Guadalajara, Jalisco, Mexico.

***Britton, N.L. and Wilson, P.* 1926.** Descriptive Flora of Porto Rico and the Virgin Islands. Cycadales,. In Scientific survey of Porto Rico and the Virgin Islands. N.Y. Acad. Sci., N.Y. 6(3): 329 - 330.

***Chamberlain, C.J.* 1935.** Gymnosperms: structure and evolution. University of Chicago Press, Chicago. 184 - 216.

***Garcia, E.* 1983.** Apuntes de Climatologia. Universidad Autónoma de Mexico. Mexico, F.F.

***Liogier, A.H. and Martorell, L.F.* 1982.** Flora of Puerto Rico and Adjacent Islands. A systematic synopsis. Universidad de Puerto Rico. p. 28.

***Moretti, A., Sabato, S. and Vazquez-Torres, M.* 1980.** The distribution of *Ceratozamia mexicana* Brongn. (Zamiaceae). Delpinoa n.s. 21 - 22: 13 - 21.

***Newell, S.J.* 1985.** Intrapopulational variation in leaflet morphology of *Zamia pumila L.* in relation to microenvironment and sex. Amer. J. Bot. 72(2): 217 - 221.

***Norstog, K.* 1980.** Chromosome numbers in *Zamia* (Cycadales). Caryologia. 33: 401 - 409.

***Norstog, K.* 1981.** Karyotypes of *Zamia chiqua* (Cycadales). Caryologia. 34: 255 - 260.

***Stevenson, D.W., Sabato, S. and Vazquez-Torres, M.* 1986.** A new species of *Ceratozamia* (Zamiaceae) from Veracruz, Mexico with comments on species relationships, habitats and vegetative morphology in *Ceratozamia* Brittonia 38: 17 - 26.

***Vovides, A.P., Rees, J.D. and Vazquez-Torres, M.* 1983.** Zamiaceae. Flora de Veracruz. 26: 1 - 31.

←

Figure 1: Graphical description of (A) spine length of petiole in female plants; (B) spine length of petiole in male plants; (C) number of spines in petioles of female plants and (D) number of spines in petioles of male plants.

Heat and Odour Production in Cycad Cones and Their Role in Insect Pollination

William Tang
Fairchild Tropical Garden
Miami, Florida 33156 U.S.A.

Abstract

Heat production and a concomitant release of volatile compounds occurs in the male cones of a large number of cycad species during the pollen shedding phase of development. Heat production in cycad cones is cyclic and the time of peak heat production is consistent within a species. Gas chromatography-mass spectometry analyses of scents released by cones during heat production indicate that they are composed of complex mixtures of secondary compounds that have similarities with those in fragrances of angiosperm flowers. Preliminary choice experiments using a scent component isolated from *Zamia furfuracea* on its insect pollinator *Rhopalotria mollis* suggest that cone scent may be most attractive to insect pollinators at close range.

Introduction

Strong heat production in plants is rare, so when it does occur we want to ask: why? I conducted a survey of heat production in cones of 10 genera of cycads several years ago and found that heat production occurs in almost all the male cones of the species sampled (Tang, 1987a). Figure 1 is a graph of cone temperature over many days, demonstrating the rhythmic heating cycle found in most cycads. Heat production occurs even if the cone is removed from the plant. Figure 2 demonstrates heat production patterns in five Australian cycads. Australian cycads show the greatest variety of heat production patterns in cycad cones. *Bowenia spectabilis*, for instance, has the tendency to produce two peaks in heat production - one major peak early in the morning at about 1 - 2 a.m. and a secondary one in late afternoon. *Cycas media*, in contrast, produces heat weakly and continuously with no strong peaks. In species of *Lepidozamia, Macrozamia* section *Macrozamia*, (represented by *M. moorei*) and section *Parazamia* (represented by *M. lucida*) the pattern of heat production is that typically found in most cycads: a single daily peak. In massive cones like that of *Lepidozamia* the heating peak will be broad. In smaller cones the peak is brief and sharp. For the majority of species heat production peaks in the late afternoon or evening as in *M. moorei*. An exception to this pattern is *Macrozamia lucida*, which peaks shortly after noon time.

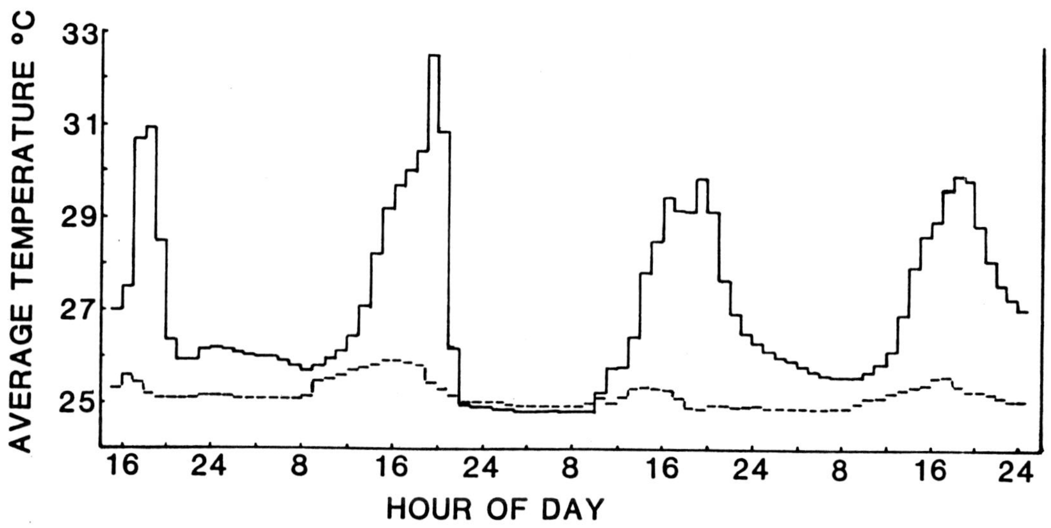

Figure 1: Temperature over a three and half-day period of a male cone of *Encephalartos ferox* removed from the plant and brought into the laboratory. Dashed line indicates ambient air temperature.

Large quantities of energy may be consumed in heat production and we find that those species that exhibit it have sporophylls filled with starch grains (Tang, 1987a). Quantitative measurements of starch in the cone before and after thermogenesis indicate that starch is metabolised during heat production (Tang, Sternberg and Price, 1987). Also, carbon isotope analysis of respired carbon indicates that, in *M. moorei* at least, lipids are also consumed in thermogenesis (Tang et al., 1987).

What is the function of heat production? It appears to be an adaptation. It occurs in a very orderly manner and the plants devote a great deal of energy to this activity. There are many possible functions of thermogenesis and these hypotheses are not mutually exclusive: (1) It may function to volatilise odours attractive to insect pollinators, (2) Heat may facilitate the movement of pollinators from cone to cone (see Heinrich, 1974), (3) It may facilitate the dehiscence of pollen sacs, (4) Heat may facilitate the rapid elongation of the male cone that is necessary for pollen shedding, and (5) It may be a byproduct of another metabolic activity; e.g., it may be a way for cycad cones to burn off excess cell sugars that may disrupt the osmotic balances. Thermogenesis may thus have more than one function. I will discuss only one of these in detail: its role in volatilising odours that attract insect pollinators.

Odour Production

All cycad cones that I have examined (male and female in over 40 species) release detectable odours. These odours are often strongest or detectable only during heat production (Tang, 1987a). By enclosing cones in bags with an absorbent for volatile organic compounds, these scents can be collected and analysed with gas chromatography-mass spectometry (Williams, 1983). Figure 3 demonstrates one of the simplest fragrances that has been analysed (Pellmyr, Tang, Bergstrom, Groth, and Thien, unpub. data). The

Table 1

Cone fragrance components of five species of cycad.
Only those components that comprise 5% or more of the fragrance are listed

M - Male, F - Female

Chemical	*Zamia furfuracea*	*Zamia pumila*	*Macrozamia moorei*	*Encephalartos altensteinii*	*Cycas rumphii*	
	M	M	M	M	M	F
β-Cycloheptenone	64.6			77.9		
Caryophyllene		8.2				
α-pentanone					12.3	3.6
Methyl-β-methyl butanoate					7.9	45.7
α-Pentyl acetate					14.8	
α-Penthanol					38.5	15.4
Limonene	0.2		6.0	2.7		
Alfa-terpinene			30.6			
p-Cymene	trace		31.2			0.4
α-Heptanol					10.6	7.7
Linalool	33.9		+			
Pentadecane		9.3				
Methyl Salicylate		53.4		1.4		

chromatogram represents the fragrance from the male cone of *Zamia furfuracea*. It is composed of two major components: one of these has been identified as linalool, a substance known to occur in some orchid flowers and believed to be attractive to euglosine bee pollinators (Hills, Williams and Dodson, 1968).

Cone scents also display sexual differences. Figure 4 compares the gas chromatograms of male and female cone scents of *Cycas rumphii*. Cone odours of this taxon are composed of complex mixtures of esters, alcohols and ketones not usually associated with the attraction of insect pollinators in flowering plants. There are sexual differences in chemical components

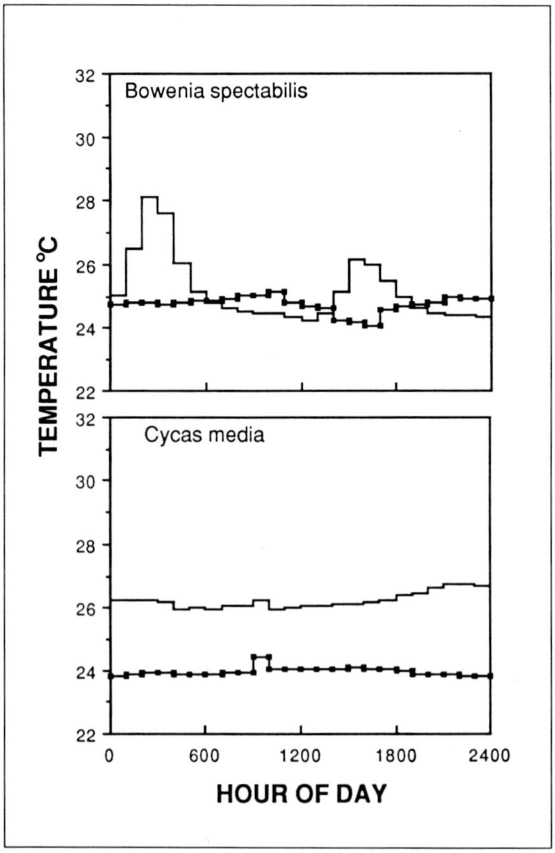

Figure 2, Part 1: Male cone temperatures of five Australian cycads over one heating cycle. Dotted lines indicate ambient air temperatures.

and in relative abundances of those components that are shared.

Table 1 summarises the major scent components in cones of five species of cycad. Many of these fragrance components are as yet unidentified. In *Z. pumila* cones the main scent component is methyl salicylate (oil of wintergreen). If you smell its cones during heat production you can detect a minty fragrance. Oil of wintergreen is a common component of orchid fragrances and known to be an effective general attractant of euglosine bee pollinators (Williams and Dodson, 1972). Many *Encephalartos* and *Macrozamia* have resiny odours and indeed they con-

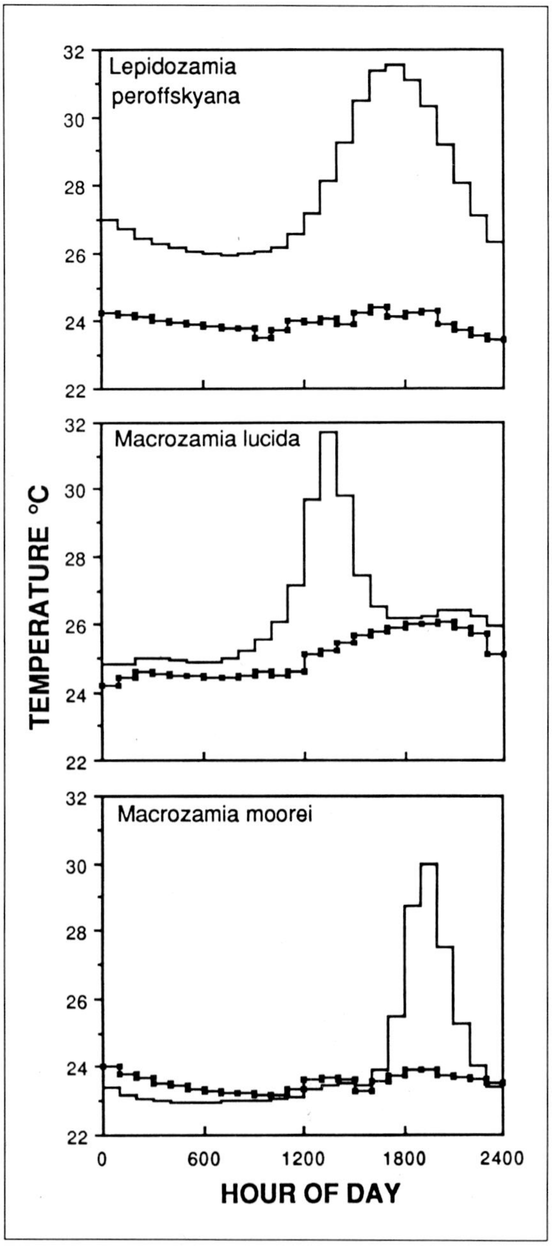

Figure 2, Part 1: Male cone temperatures of five Australian cycads over one heating cycle. Dotted lines indicate ambient air temperatures.

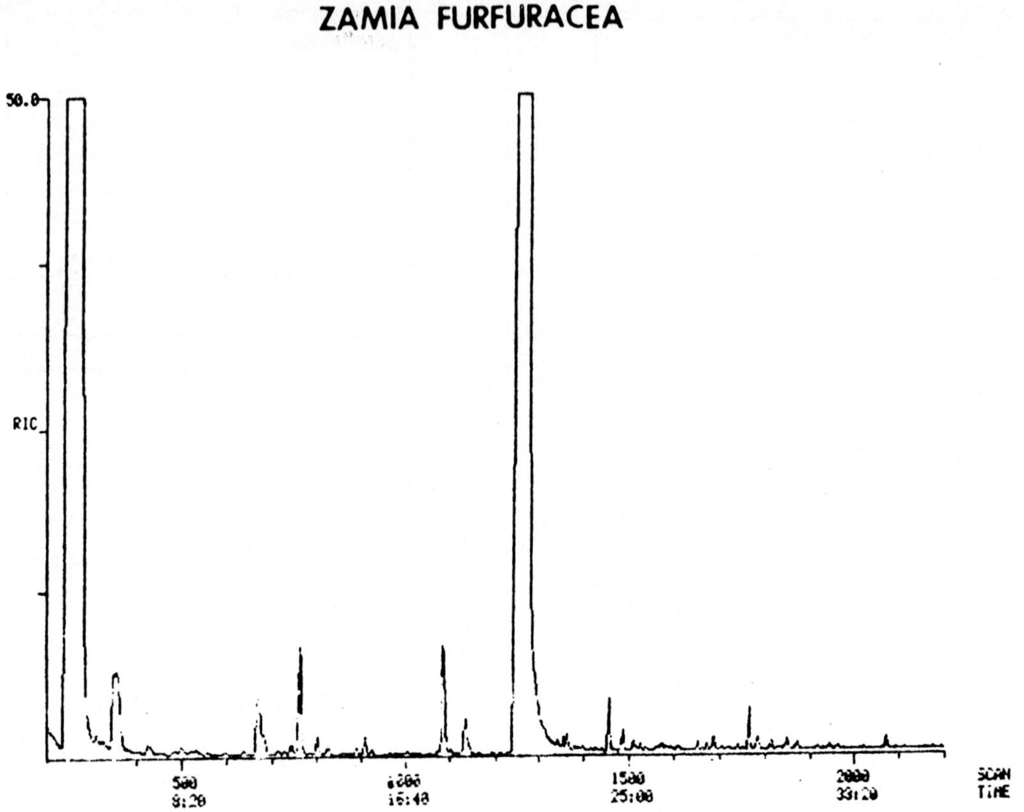

Figure 3: Gas-chromatogram of the male cone fragrance of *Zamia furfuracea*.

tain resiny smelling compounds like alfa-pinene and limonene. In *C. rumphii* there is a large sexual difference in relative proportions of fragrance constituents (e.g., 48% of the female sample consists of Methyl-β-methylbutanoate in males it makes up only 8% of the fragrance sample).

Insect Trials With Scents

To determine if these odours and the heat production that volatilises them function in insect pollination - the odours must be demonstrated to attract suspected pollinators. Two suspected pollinators of *Zamia pumila* in Florida are *Rhopalotria slossoni* and *Pharaxonotha zamiae* (Tang, 1987b). To test whether methyl salicylate, the major component of *Z. pumila*'s odour, attracts these beetles I placed petri dishes with this chemical in their pine forest habitat at the Fairchild Tropical Garden Research Centre over a 24hr period. No beetles were trapped or observed at these baited dishes. Bart Schutzman (pers. commun.) who has conducted his own independent research on cone

→

Figure 4: Gas-chromatograms of the male and female pollination organs of *Cycas rumphii*. These samples were taken in New Caledonia.

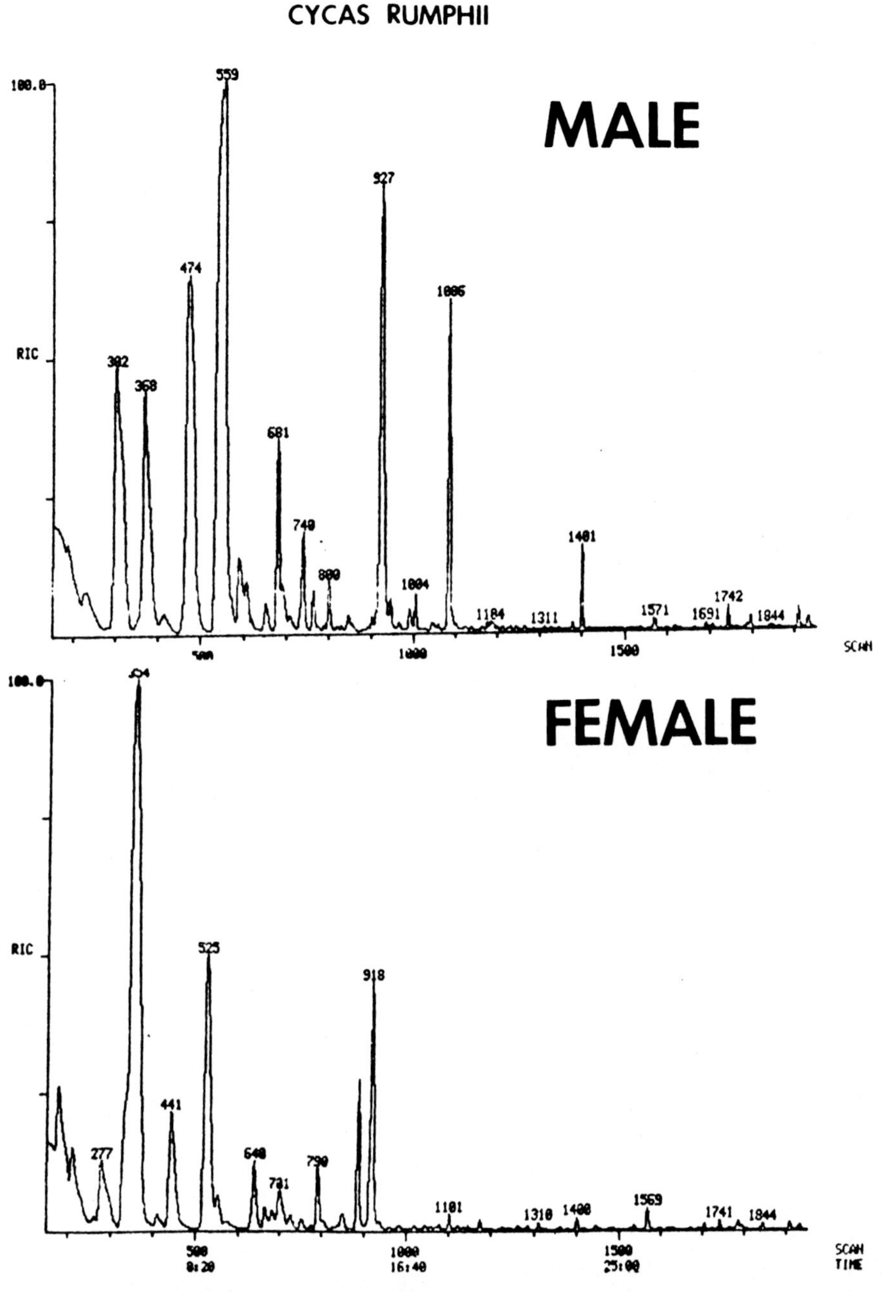
CYCAS RUMPHII
MALE
FEMALE
100.0
RIC
SCAN
TIME
500
1000
1500
8:20
16:40
25:00

Table 2

The number of Rhopalotria mollis that landed on filter paper baited with: (1) Linalool and Hexane solvent (2) Hexane control (3) Blank control

Run No.	Linalool + Hexane	Hexane	Blank
1	2	2	3
2	0	1	1
3	0	1	0

scents and field trails has, however, succeeded in trapping *Pharaxontha* overnight with this chemical bait using McPhail traps. Similar trials were conducted with linalool, a scent component of *Z. furfuracea* cones. Linalool in hexane solvent was presented in petri dishes for 30 minutes at distances of 1 to 10m from individuals of *Rhopalotria mollis* situated on cultivated plants of *Z. furfuracea* at the Fairchild Tropical Garden Research Centre. The pollination relationship between these species is well documented by Norstog, Stevenson and Niklas (1986) and Norstog and Fawcett (1989). Trials lasted only 30 min. due to the highly volatile nature of linalool sample. Again no beetles were observed to visit the bait.

Controlled experiments were conducted on afternoons using a scent component of *Zamia furfuracea* on *Rhopalotria mollis*. I collected these weevils and stored them in a refrigerator 1 - 3 hrs. prior to the experiment. I released ca. 60 weevils from a petri dish into a mesh-covered 30 gal. aquarium and provided them with three choices of filter paper: (1) soaked with linalool and hexane solvent, (2) soaked with hexane control, and (3) a blank control, and then counted the number of beetles that landed on each over a 30 min. period. The results (Table 2) show no particular attraction of the weevil to the scent component by itself. Energetic attempts by the weevils to escape the experimental enclosure during the experiment suggests that handling and the artificial conditions experienced by the weevils may have masked what attraction these weevils had for linalool.

Table 3

The number of Rhopalotria mollis that landed on jars enclosing: (1) A dehiscing male cone under a perforated lid (2) A dehiscing male cone sealed under an unperforated lid (3) An empty control in six separate trials

Run No.	Male cone in perforated jar	Male cone in closed jar	Empty jar
1	4	12	8
2	12	8	8
3	2	2	
4	3	3	
5	49	30	
6	26	14	

Using the same aquarium I attempted to test whether the beetles homed in on cones by vision or by a combination of vision and scent. I placed one dehiscing male cone in each of two transparent, sealed jars, one with 100 perforations in the lid to allow odour to escape. A third empty jar was used as a control. In most trials there seemed to be no attraction to cones, with or without scent (Table 3). Again the beetles were agitated and uncooperative.

There was one exceptional afternoon in which there was clear attraction to the cones, particularly to the cone in which weevils could detect odours. When these beetles landed on the lid with perforations they tried energetically to crawl into the holes toward the cone, suggesting that at close range the cone scents did have a strong attractive effect. I conducted similar trials with female cones. These were inconclusive (Table 4).

Table 4

The number of Rhopalotria mollis that landed on jars baited with:
(1) A dehiscing female cone under a perforated lid
(2) A dehiscing female cone sealed under an unperforated lid
in two separate trials

Run No.	Female cone in perforated jar	Female cone in enclosed jar
1	19	10
2	3	2

In conclusion, the findings on heat and odour production to date suggests that they are part of an insect-pollination syndrome in cycads. Preliminary tests on the attractiveness of the fragrance of one species. *Z. furfuracea*, to its pollinator suggest that it is attractive only under limited conditions and perhaps only at close range. In general, insect pollination in cycads has some similarities with nursery pollination in angiosperms like that employed by figs and yuccas, where pollinators reproduce on the pollination organ (Tang 1987b; Norstog and Fawcett, 1989). However, there appears to be no well-studied angiosperm pollination system that can serve as an appropriate model for cycads. Insect pollination in cycads may yet prove to be unique and distinct from that in angiosperms. The fine details of cycad-pollination mechanisms await more detailed research.

Acknowledgments

The preliminary results presented here on cycad cone fragrances were produced in collaboration with Olle Pellmyr (University of Cincinnati), Gunnar Bergstrom and Inga Groth (University of Gothenburg, Sweden) and Leonard Thien (Tulane University, New Orleans). I thank Bart Schutzman for his permission to cite his unpublished work and Terrance Walters for comments on a draft of this manuscript.

References

***Heinrich, B.* 1974.** Thermoregulation in endothermic insects. Science 185: 747 - 756.

***Hills, H.G., Williams, N.H. and Dodson, C.H.* 1968.** Identification of some orchid fragrance components. Am. Orchid Soc. Bull. 37: 967 - 971.

***Norstog, K., Stevenson, D.W. and Niklas, K.* 1986.** The role of beetles in the pollination of *Zamia furfuracea* L. fil (Zamiaceae). Biotropica 18: 300 - 306.

***Norstog, K. and Fawcett, K.S.* 1989.** Insect-cycad symbiosis and its relation to the pollination of *Zamia furfuracea* (Zamiaceae) by *Rhopalotria mollis* (Curculionidae). Amer. J. Bot. 76: 1380 - 1394.

***Tang, W.* 1987a.** Heat production in cycad cones. Bot. Gaz. 148: 165 - 174.

***Tang, W.* 1987b.** Insect pollination in the cycad *Zamia pumila* (Zamiaceae). Amer. J. Bot. 74: 90 - 99.

***Tang, W., Sternberg, L. and Price, D.* 1987.** Metabolic aspects of thermogenesis in male cones of five cycad species. Amer. J. Bot. 74: 1555 - 1559.

***Williams, N.H.* 1983.** Flora fragrances as cues in animal behaviour. Jones and R.J. Little (eds.), *In* C.E. Handbook of Experimental Pollination Biology. Van Nostrand Rheinhold Co. Inc., New York.

***Williams, N.H. and Dodson, C.H.* 1972.** Selective attraction of male euglossine bees to orchid floral fragrances and its importance in long distance pollen flow. Evolution 26: 84 - 95.

Seed Dispersal and Reproductive Effort in *Zamia pumila*

William Tang
Fairchild Tropical Garden
Miami, Florida 33156 U.S.A.

Abstract

A wild population of the cycad *Zamia pumila* in south Florida was studied over an 8-year period, revealing sharp differences in male and female reproduction. Males coned on average 2.3 times more frequently than females. Female cones were much larger than those of males and females invested on average 3.4 times more energy into reproduction than males over the study period. This dramatic sexual dimorphism in reproductive behaviour may have evolved in response to selection for seed dispersal in females. Dispersal success in this species is dependent on the size of the female reproductive effort. The number of seeds dispersed from a cone is positively correlated with the number of seeds it contains and the proportion of seeds dispersed from cones is positively correlated with seed size. Observations suggest that animal dispersers tend to deposit seeds underneath nearby shrubs and a sun/shade experiment indicates that seed germination and establishment is higher in shade than in sun. These findings indicate that there is selection in this species for females to produce large seeds and cones and that seed dispersal has played an important role in the evolution of dioecy and sexual dimorphism in cycads.

Introduction

Cycads are distinctive among seed plants in that the entire group is dioecious; male and female pollination organs are never found together on the same individual and are often strikingly different in size and shape. In this paper I will summarise two studies on the cycad *Zamia pumila* that explore differences in male and female reproduction: (1) a long-term study of reproduction in male and females in a wild population and (2) a study of seed dispersal of this species that examines how seed dispersal may account for differences in the reproduction of males and females.

Materials and Methods

Study Area

My two studies were conducted in subtropical pine forest in southeast Florida. This habitat, like that of other populations of *Z. pumila* throughout much of its range in Florida and the Greater Antilles, is characterised by a sparse tree canopy and a shrub layer of palms and hardwoods. This pine forest is a fire-climax habitat that burns at roughly 3 - 8 yr intervals (Wade et al., 1980). The subterranean stem of *Z. pumila* appears to be an adaptation to fire (Stevenson, 1980) and plants produce leaves and recover soon after burning.

1. Long Term Reproductive Study

The study site was a 7 x 114m transect through remnant pine forest at the Fairchild Tropical Garden Research Centre in the city of Miami. The study site burned in 1980. In 1982 I mapped all zamias in the plot and recorded the number of cones and leaves present on plants during the pollination season (December - January). Because female cones persist on plants for over a year female reproduction could be extrapolated back one additional year before the study commenced (Newell, 1983). During the years 1983 - 1989 seedlings that appeared on the plot were not included in the study. Male and female cones were collected in 1984 and 1988 from plants adjacent to the site and their dry masses were determined for a measure of energy allocation to cones. For further details on methods see Tang (1990).

2. Seed Dispersal Study

The study site was a 200 x 300m plot within a 8,000ha patch of forest at Everglades National Park. I marked and mapped 122 coning females in this population and counted the seeds on each from December, 1984 to February, 1985 as the cones began breaking apart. I took 1 - 3 seeds from each cone to determine average seed mass. Seed positions were mapped around a 4m radius of each mother plant in February. It was assumed that most seed dispersal would have been completed by this time. I monitored cones at 1 - 5 wk intervals thereafter until July when germination commenced and the period of dispersal was assumed to have ended.

Seeds dispersed by animals tend to be dropped in the shade of shrubs (see results). To determine if shade enhances germination and survival a sun/shade experiment was conducted. A group of 80 seeds was placed beneath each of a pair of 60 x 43cm wire mesh screens at the study site in July, 1985. One screen was shaded with 75% shade cloth the other left exposed. Four replicates were established. Germination, leaf production and herbivory were followed for 15 months. For details on methods see Tang (1989).

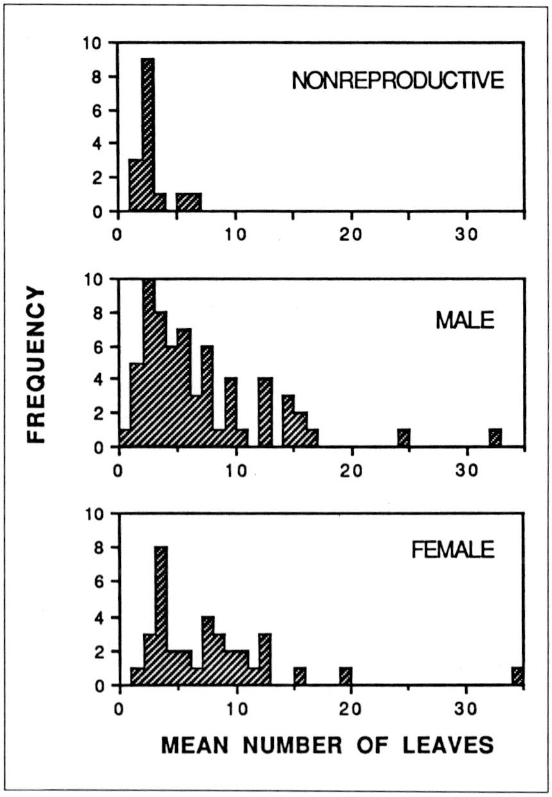

Figure 1: Frequency distributions of male, female, and non-coning plants by mean number of leaves over six seasons, an index of plant size. Reprinted from Tang (1990).

Results and Discussions

1. Long Term Reproductive Study

Of the 114 plants mapped 99 coned over the 8-yr period; of these 65% were males and 35% were females. Fifteen plants did not cone. No sex changes occurred. Using mean number of leaves over six seasons as an index of plant size, the majority of non-coning plants were juveniles (Fig. 1). The size

Table 1

Proportion of plants coning for 9 seasons in a South Florida Population of *Z. pumila*

The sex ratio for each season is also presented.
Data is from Tang (1990)

Season	81	82	83	84	85	86	87	88	89	Mean ± SD
Proportion coning										
Male	-	0.75	0.73	0.75	0.64	0.67	0.56	0.44	0.28	0.60 ± 0.17
Female	0.34	0.26	0.49	0.34	0.23	0.37	0.09	0.26	0.06	0.27 ± 0.14
Sex Ratio	-	5.3	2.7	4	5.1	3.3	12	3.1	9	5.6 ± 3.3

distributions of males and females were not significantly different (Kolmogorov-Smirnov test, $P > 0.05$). The sex ratio was significantly skewed toward males (male/female = 1.8; $X^2 = 8.49$, $P < 0.005$).

When plant size (as measured by mean leaf number) is plotted against coning frequency (Fig. 2A) three patterns are apparent: (1) Males cone at a smaller size, (2) male coning frequency increases faster with increasing plant size, and (3) males cone more frequently than females of equal size. Similarly, when plant size is plotted against number of cones produced per coning event (Fig. 2B) I find that males produce more cones than females within a given season.

At the population level I find that for several years after burning, approximately 75% of the male plants coned each season (Table 1). However, by the sixth

Table 2

Mean dry masses of male and female cones of *Z. pumila* with standard deviations and ranges.

Data are from Tang (1990)

	Total dry mass (g)	
Male cone	5.6 ± 2.4 (2.9 - 9.6)	11
Female cone		
at pollination	10.3 ± 3.2 (4.9 - 13.1)	5
ripe cone	79.2 ± 47.0 (16.4 - 158.5)	24
seeds	58.8 ± 41.4 (0.9 - 123.0)	24
support tissue	20.4 ± 8.0 (7.3 - 35.5)	24

Figure 2: Mean number of leaves (calculated from six seasons) vs. (a) proportion of seasons in which a plant coned, and (B) mean number of cones per coning event. r Spearman rank correlation coefficient, $^{*}P < 0.001$, $^{**}P < 0.05$. Reprinted from Tang (1990).

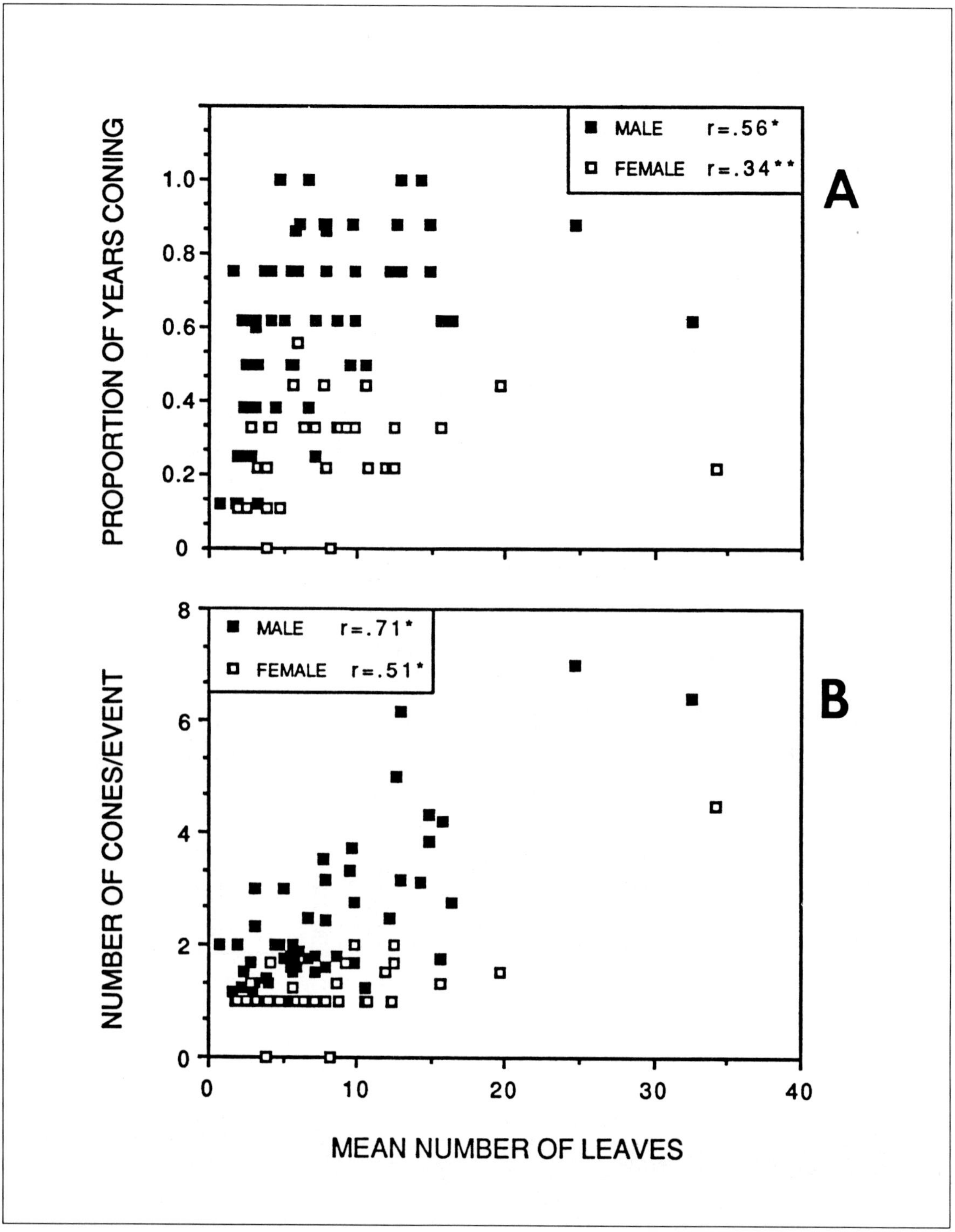
PROPORTION OF YEARS CONING
1.0
0.8
0.6
0.4
0.2
0
MALE r=.56*
FEMALE r=.34**
A
NUMBER OF CONES/EVENT
8
6
4
2
0
MALE r=.71*
FEMALE r=.51*
B
0
10
20
30
40
MEAN NUMBER OF LEAVES

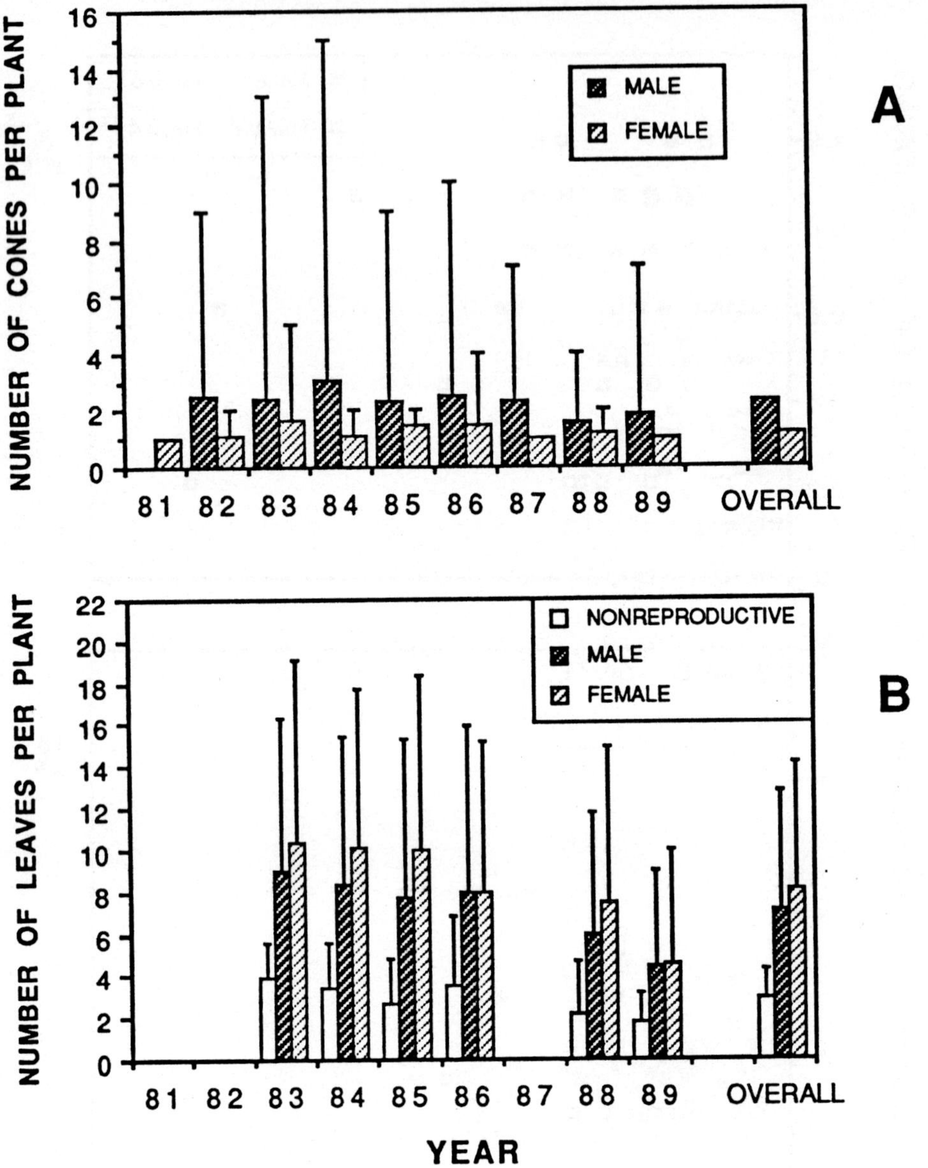

Figure 3: (A) Mean and maximum number of cones produced for those of male and female plants that coned during each of nine seasons. (B) Mean number of leaves of males, females and non-reproductive plants and standard deviations for six seasons. Reprinted from Tang (1990).

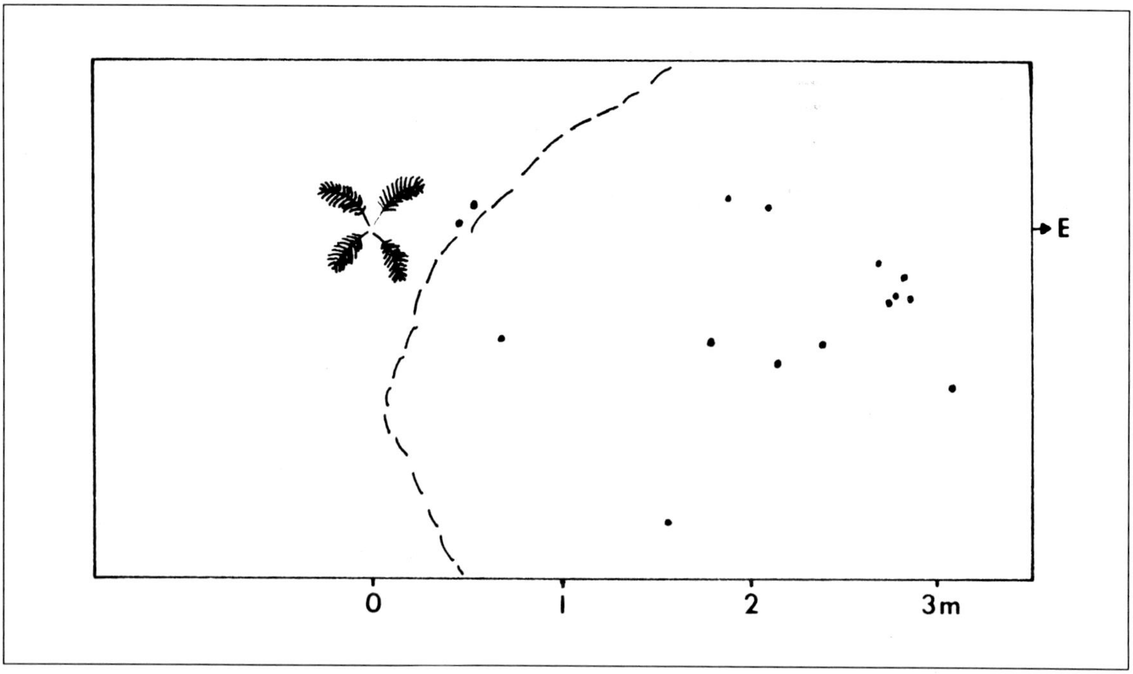

Figure 4: Map of the seeds dispersed within 4m of a mother plant of *Z.pumila* at the Everglades study site in February 1985. Dashed line denotes the edge of shrub cover east of the plant. All seeds have been dispersed from this plant.

year after fire the number of males coning each season begins to decline. By the tenth year only 28% are coning. Female coning in contrast is more erratic, fluctuating from year to year from as high as 49% of all females to only 6%. If we calculate the sex ratio for those plants that cone each season is calculated, it is found to vary from 2.7 - 12 male/female and that this variation depends mainly on the number of females that cone. In Fig. 3 plant size, as measured by mean leaf number, is plotted against the number of leaves and cones produced from season to season. There is a gradual decline in the production of both cones and leaves after the 6th season after fire. This decline in vegetative and reproductive vigour appears to be due in part to shading by shrubs recovering from fire. A second cause of decline may be an outbreak of the cycad butterfly *Eumaeus atala*, which destroyed 25% or more of the leaf area on 8 - 17% of the plants each year (Tang, 1990).

Table 2 shows the dry masses of male and female cones. On average a ripe female cone is 14 times more massive than a male cone. Taking into account that females produce fewer cones (mean: females 0.35 cone/yr, males: 1.45 cones/yr) and cone less frequently than males (frequency ± SD: females: once every 3.7 ± 1.2 yrs; males: once every 1.6 ± 0.4 yrs), females on average invested 3.4 times more energy into reproduction over the study period than males. Why is there such a big difference in reproductive behaviour between males and females? This study indicates that male and female reproduction are under different evolutionary constraints. There appears to be selective pressure on female plants to produce much larger cones than males. Since the amount of resources available for growth and reproduction is limited, a consequence is that females must cone less frequently than males.

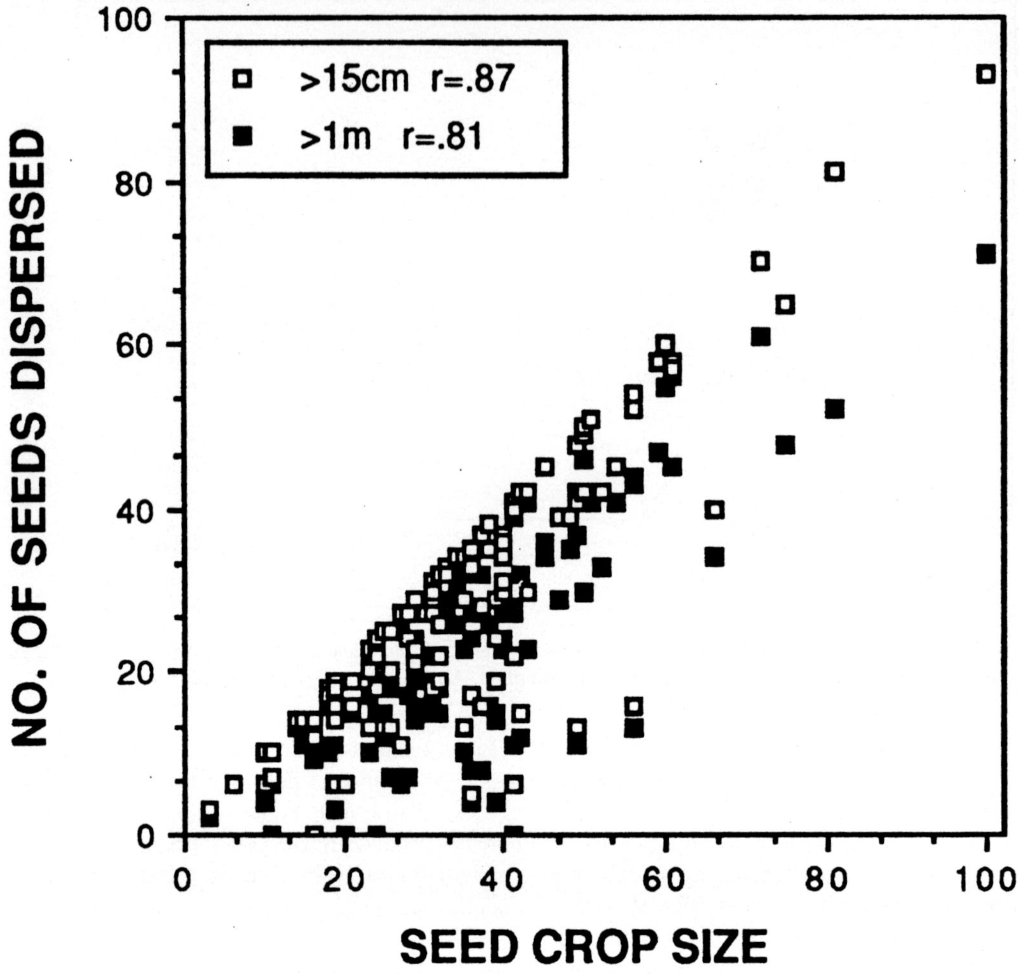

Figure 5: Seed crop size versus the number of seeds dispersed beyond 15cm and 1m of the cone for #122 plants at the Everglades study site in 1985.

Seed Dispersal Study

Figure 4 is a map of the seeds dispersed around a mother *Z. pumila* plant. Bite marks on the fleshy coats of dispersed seeds indicate that birds and rodents are the primary dispersers. In the study population seed dispersal away from each mother plant tends to be in a particular direction(s), especially toward and under nearby shrubs. This corresponds with observations of other vertebrate seed dispersers indicating that they often prefer to take fleshy fruits to cover before eating them (Howe, 1979; Fleming and Heithaus, 1981).

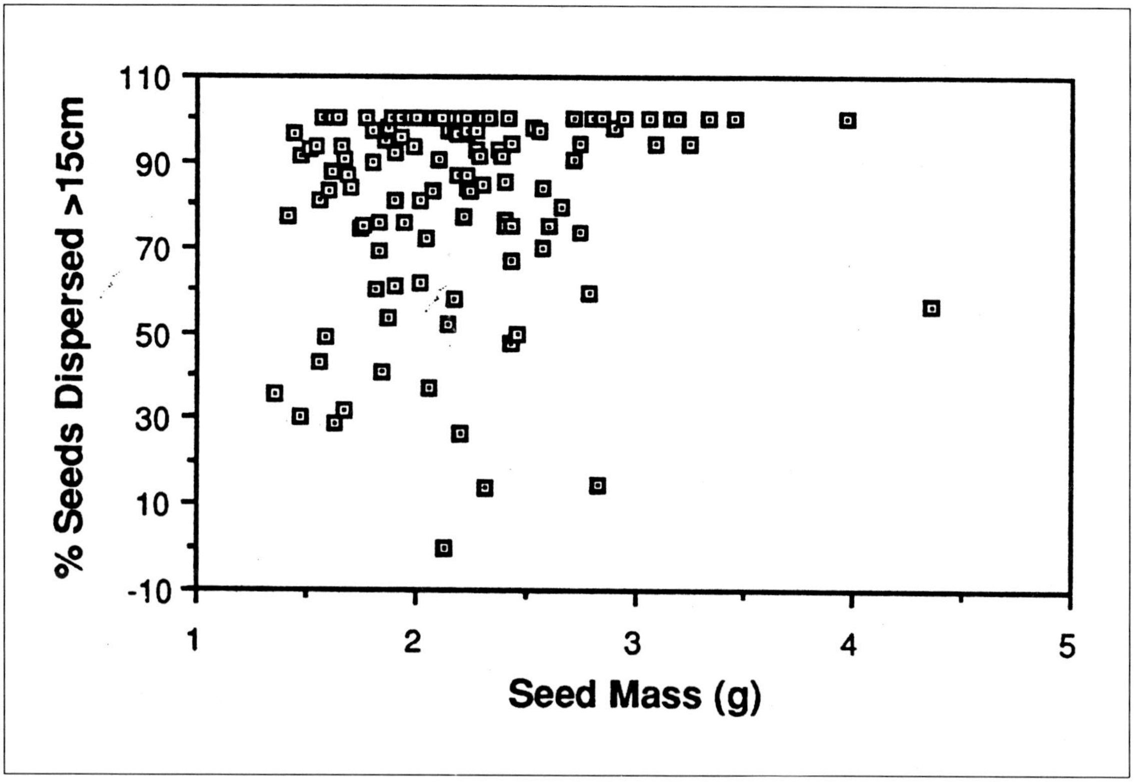

Figure 6: Seed mass versus the proportion of seeds dispersed beyond 15cm of the cone for 116 plants at the Everglades study site in 1985.

Shade is important in this seasonally dry habitat as the canopy is sparse and sunlight and heat are often intense. The shade beneath shrubs may be a safe site where seeds and seedlings have a better chance of survival.

There is a positive relationship between aspects of cone size and the amount of seeds that are dispersed. The greater the number of seeds in a cone the greater the number of seeds dispersed beyond 15cm and 1m of the cone (Fig. 5). These correlations are statistically significant ($P < 0.001$). Seed mass is also significantly correlated with the proportion of seeds dispersed beyond 15cm and 1m of the cone (Spearman rank correlation coefficient $r = 0.0234, 0.0232, P < 0.01$; Fig. 6).

After 5 months in the sun/shade experiment, by which time the period of seed germination had ended, seed germination success was 85% in shade and 53% in sun for all replicates combined (statistically different, *G*-test for heterogeneity, $P < 0.001$). After 15 months from the beginning of the experiment 22% of the seeds in the shade treatment had survived as seedlings while only 5% had survived in the sun treatment ($P < 0.001$).

Clumps of undispersed seeds and seedlings at the mothers' crown tend to suffer lethal predation by rodents and insects (Tang, 1989). If we look at a map of adult *Zamia pumila* in this habitat (Fig. 7) we find that plants do not form tight climbs; individuals are rarely closer than 1m apart. This is further evidence that dispersal away from the mother is crucial for the survival of a seed to adulthood.

In conclusion, this study provides evidence that seed dispersal is vital to successful reproduction in this species of cycad. Females with large seeds and large numbers of

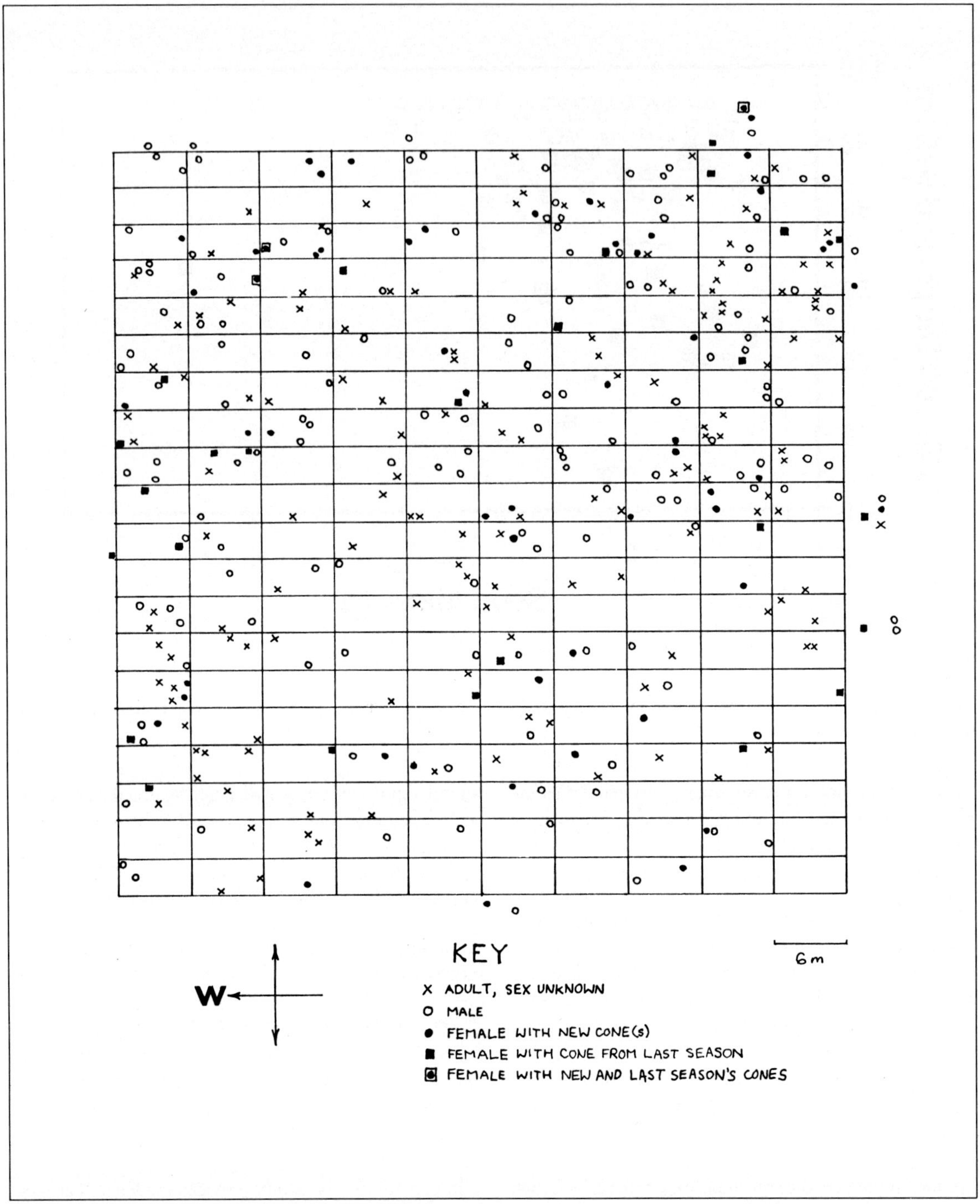

Figure 7: Map of adult plants at the Fairchild Research site in December, 1982. Reprinted from Tang (1985).

seeds receive better seed dispersal and reproduction. This suggests that selection is occurring in this cycad for the production of large cones and seeds. This may explain why females devote so much more energy into reproduction than males and why they cone less frequently. Other studies indicate that other species of cycads exhibit similar sexual differences in reproductive effort (Ornduff, 1985, 1989; Grobbelaar et al., 1989) suggesting that selective pressure for seed dispersal has helped mould the breeding systems of cycads throughout their long evolutionary history. The high expense of female reproduction has perhaps eliminated bisexuality as a viable reproductive option in this group of plants (Givnish, 1980).

Acknowledgments

I thank Terrence Walters for comments on a draft of this manuscript.

References

***Fleming, T. and Heithaus, E.R.* 1981.** Frugivorous bats, seed shadows, and the structure of tropical forests. Biotropica 13 (Suppl.): 45 - 33.

***Givnish, T.J.* 1980.** Ecological constraints on the evolution of breeding systems in seed plants: dioecy and dispersal in gymnosperms. Evolution 34: 959 - 972.

***Grobbelaar, N., Meyer, J.J.N. and Burchmore, J.* 1989.** Coning and sex ratio of *Encephalartos transvenosus* at the Modjadji Nature Reserve. S. African J. Bot. 55: 79 - 82.

***Howe, H.* 1979.** Fear and frugivory. Am. Nat. 114: 925 - 931.

***Newell, S.* 1983.** Reproduction in a natural population of cycads (*Zamia pumila* L.) in Puerto Rico. Bull. Torrey bot. Club 110: 464 - 473.

***Ornduff, R.* 1985.** Male-biased sex ratios in the cycad *Macrozamia riedlei* (Zamiaceae). Bull. Torrey Bot. Club 112: 393 - 397.

***Ornduff, R.* 1989.** Size distribution and coning behaviour in the Australian cycad *Lepidozamia peroffskyana*. Aust. J. Ecology 14: 241 - 245.

***Stevenson, D.W.* 1980.** Observations on root and stem contractions in cycads (cycadales) with special reference to *Zamia pumila* L. Bot. J. Linn. Soc. London 81: 275 - 281.

***Tang, W.* 1985.** Studies in cycad reproductive biology. M.S. thesis, University of Miami, Florida.

***Tang, W.* 1989.** Seed dispersal in the cycad *Zamia pumila* in Florida. Can. J. Bot. 67: 2066 - 2070.

***Tang, W.* 1990.** Reproduction in the cycad *Zamia pumila* in a fire-climax habitat: an eight-year study. Bull. Torrey Bot. Club 117: 368 - 374.

***Wade, D., Ewel, J. and Hofstetter, R.* 1980.** Fire in South Florida ecosystems. USDA For. Sev. Gen. Tech. Rep. SE-17. 125 pp.

Part Four

Physiology

Senescence Patterns in the Female Cones of *Encephalartos*

N. Grobbelaar
Department of Botany
University of Pretoria, 0002, South Africa

Abstract

The female cones of most *Encephalartos* species disintegrate spontaneously at maturity. The process is initiated by a decrease in the water potential of the seeds' sarcotesta which induces the seeds to enlarge as a consequence of enhanced water uptake. This sets up a longitudinal physical tension in the megasporophyll which induces the formation of an abscission layer near the megasporophyll base. The seeds subsequently abscise from the sporophylls. Sporophyll abscission is not induced in the case of *E. lanatus* and as a consequence the cone axis splits longitudinally before the cone as a whole dries out. The cones of *E. dolomiticus, E. dyerianus, E. eugene-maraissi* and *E. middelburgensis* simply dry out at maturity without any organs abscising or splitting.

Introduction

The female cones of most *Encephalartos* species, notably *E. altensteinii* Lehm. and *E. lebomboensis* Verdoorn which were used in this study, disintegrate spontaneously several months after the ovules become receptive to pollination. As can be deduced from the fact (see Fig. 1) that a cluster of detached seed-bearing megasporophylls often lifts up from the cone axis of a horizontally held disintegrating female cone, the disintegration of the cone is not simply the result of the abscission of the megasporophylls from the cone axis followed by the abscission of the ovules from the megasporophylls.

If the surrounding megasporophylls are artificially removed from one of the seed-bearing sporophylls of a female cone which is almost ready to disintegrate, it will be observed that the solitary sporophyll will not abscise from the cone axis (see Fig. 2) although the other remaining sporophylls will do so and the seeds will abscise from the sporophylls in all cases.

Abscission of Megasporophylls

The results of experiments carried out at the University of Pretoria (Grobbelaar, 1989) clearly suggest that a longitudinal physical tension has to be created in the megasporophyll in order to induce the formation of an

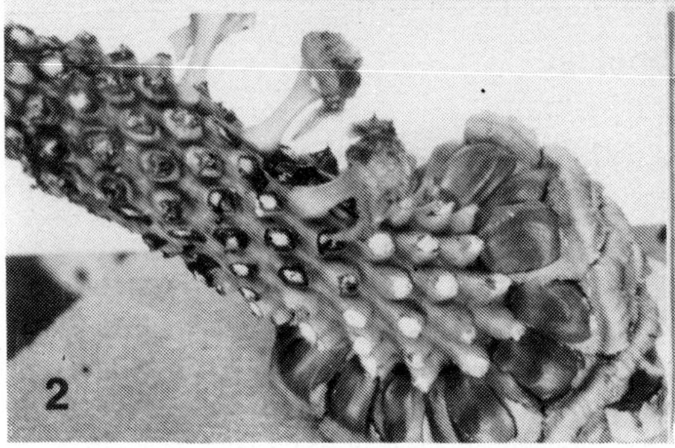
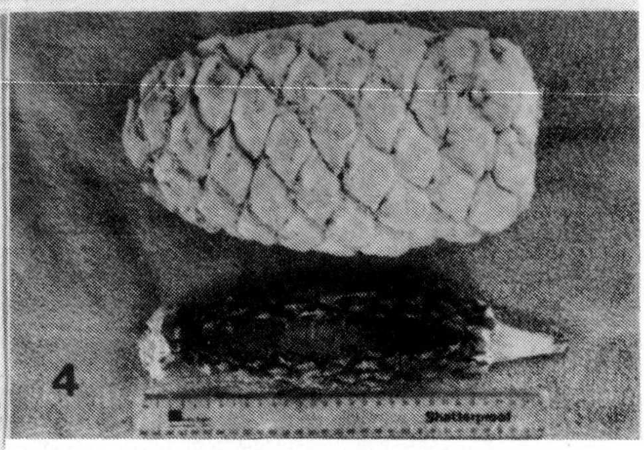

Figure 1: Horizontally held 550mm long female cone of *Encephalartos altensteinii* during early spontaneous disintegration. Note the raised position of the cluster of abscised megasporophylls near the cone apex.

Figure 2: Female cone of *Encephalartos lebomboensis* from which most sporophylls were removed at an early stage of cone disintegration. Solitary sporophylls in central part of cone did not form an abscission zone even after their seeds abscised and the mass of basal sporophylls to the right had also abscised.

Figure 3: Mature female cone of *Encephalartos lanatus* above with the longitudinally split axis of a similar cone below. Cone apices left; scale 300mm long.

Figure 4: Mature dry female cone of *Encephalartos eugene-maraisii*. Cone apex left; cone length 250mm.

abscission layer near the base of the megasporophyll. This tension is normally set up by a decrease in the water potential of the seed's sarcotesta, probably by the hydrolysis of macromolecular components of the sarcotesta. As a result of the drop in its water potential, the seeds absorb water from the remainder of the cone and increase in size. The swelling of the seeds necessitates an increase in the diameter of the 'seed circle' which is inhibited by the megasporophylls which are attached to the cone axis.

The megasporophylls abscise in a basipetal direction. The abscission of the sterile apical megasporophylls is induced by the acropetal pressure on them caused by the expanding apical seeds of the cone.

Exceptions

The female cones of all *Encephalartos* species do not disintegrate spontaneously. In the case of *Encephalartos lanatus* Stapf and Burtt Davy, the tension that develops in the megasporophylls during seed enlargement, does not induce the formation of an abscission layer at the sporophyll basis. Consequently the cone axis splits longitudinally (Fig. 3) whereafter the whole cone dries out without the seeds being liberated.

In the case of *Encephalartos dolomiticus* Lavranos & Goode, *E. dyerianus* Lavranos & Goode, *E. eugene-maraisii* Verdoorn (Fig. 4), the sarcotesta of the seeds do not undergo autolysis but simply dries out together with the remainder of the intact mature cone. The dry cones are brittle and readily break up on mechanical disturbance.

It has been reported in the paper by Grobbelaar (1989) that Oosthuyzen (pers. comm.) claims that the mature female cones of *E. friderici-quilielmi* Lehm., *E. humilis* Verdoorn and *E. laevifolius* Stapf & Burt Davy, also do not disintegrate spontaneously. After writing the paper, the author did however have occasion to witness the spontaneous disintegration of several female cones of *E. friderici-guilielmi*.

Acknowledgments

The financial assistance of the Foundation for Research Development and the University of Pretoria is gratefully acknowledged.

References

***Grobbelaar, N.* 1989.** Disintegration of *Encephalartos* megastrobili. S. Afr. J. Bot. 55: 581 - 585.

Specificity of the Cycad-Cyanobiont Symbiosis

N. Grobbelaar and J. Marshall
Department of Botany
University of Pretoria, 0002, South Africa

Abstract

The cyanobionts in the coralloid roots of cycads apparently are almost always species of *Nostoc* although species of *Anabaena* and *Calothrix* have also been encountered. Within the genus *Nostoc* at least six species have been found to be involved although *N. commune* appears to be by far the commonest. Different coralloid root clusters of a single plant can house different cyanobionts and different cyanobionts can also exist in the same coralloid root segment simultaneously although the latter situation does not appear to be common.

Cycad Cyanobacterial Cosymbionts

The cyanobacteria which can live in symbiosis with cycads, in all cases were found to be heterocystous filamentous forms. Although they have usually been identified as species of *Anabaena* or *Nostoc*, the latter genus has been recorded far more often than the former (Lindblad and Bergman, 1990). The only other cyanobacterial genus that has thus far been reported as a symbiont in cycad coralloid roots, is *Calothrix* from *Encephalartos hildebrandtii* (Grobbelaar et al., 1987; Huang and Grobbelaar, 1989). Thus Grobbelaar found that 40 of the 41 cyanobionts which they isolated from the coralloid roots of 38 plants belonging to 31 cycad species (30 *Encephalartos* species and *Stangeria eriopus*) were *Nostoc* species. During this study *Nostoc ellipsosporum* and *N. punctiforme* were isolated in different cases from *Encephalartos manikensis* individuals, whilst *Nostoc commune* and *N.paludosum* were isolated from different root clusters of the same *Encephalartos transvenosus* individual.

In a subsequent study (Marshall, Huang and Grobbelaar, 1989) the cyanobionts of ten *Encephalartos transvenosus* individuals growing in different habitats were isolated. One of the isolates was identified as a species of *Anabaena* whilst all the others were classified as *Nostoc* species. In one of the latter cases, a mixed culture of *N. paludosum* and *N. commune* was obtained from a single coralloid root transverse section by the method of Bowyer and Skerman (1968) and Grilli Caiola (1972). Although the ten *Nostoc* isolates were shown to differ from one another, only six of them could be identified to the species level.

Of the 52 cyanobiont isolates referred to above (Grobbelaar et al., 1987; Marshall et al., 1989), 50 were therefore identified as *Nostoc* species and one each as an *Anabaena* and a *Calothrix* species. The *Nostoc* endophytes that have been identified to the species level represent at least six species, namely: *N. commune, N. ellipsosporum, N. muscorum, N. paludosum, N. punctiforme* and *N. sphaericum.* Of these, *N. muscorum* has been found only in *Stangeria eriopus* whilst the others have been found only in the *Encephalartos* species that were investigated.

Of the five *Nostoc* symbionts of the *Encephalartos* species that were identified to species level, *Nostoc commune* was by far the commonest, occurring in 33 of the 47 coralloid root samples examined. *Nostoc punctiforme* occurred in five; *N. paludosum* in three; *N. ellipsosporum* and *N. sphaericum* in two each of the *Encephalartos* coralloid-root samples examined.

In the studies by Grobbelaar et al., (1987) and Marshall et al., (1989), a total of 21 cyanobiont isolates were made from the coralloid roots of *Encephalartos transvenosus.* This yielded 16 isolates that could be identified at the species level. Of these, 13 were assigned to *N. commune* and three to *N. paludosum.*

Experimental Infections

Marshall (1989) attempted to infect *Encephalartos transvenosus* seedlings under aseptic conditions with several isolated cycad cyanobionts. Unfortunately the experiment had to be terminated after 18 months due to poor growth of the seedlings. By this time, nearly all the plants had formed coralloid roots which, however, had not yet branched more than once. In five cases, the coralloid roots were found to be infected with a cyanobacterium which, after isolation and microscopic examination, appeared to be identical to the isolate that was used as inoculant. Thus *Nostoc commune* from *Encephalartos laevifolius; N. ellipsosporum* from *E. altensteinii; N. punctiforme* from *E. manikensis; N. sphaericum* from *E. natalensis* and a *Nostoc* sp. from *Cycas revoluta* were all found to be infective on *E. transvenosus.* In as much as the isolated cyanobiont of a cycad was successfully used to infect another cycad and was subsequently, after re-isolation, demonstrated to be identical to the inoculum, Koch's postulate was fulfilled. However, it must be pointed out that unfortunately none of the inoculants that were isolated from *E. transvenosus* originally, infected the *E. transvenosus* test plants by the time the experiment was terminated. Although *Nostoc muscorum* from *Stangeria* was also used as an inoculant, it also had not infected the *Encephalartos transvenosus* test plants by the time the experiment had to be terminated.

Specificity of Cyanobionts

Because only one isolation was made from *Stangeria*, nothing can be said about the specificity between it and its cyanobionts. It is clear, however, that the *Encephalartos* species are not very specific with regard to the heterocystous cyanobacteria with which it can live symbiotically. The available results reveal that all nine of the *Nostoc* species and an *Anabaena* species that were isolated from different *Encephalartos* species as well as a *Nostoc* species isolated from *Cycas revoluta* can act as cyanobionts for *E. transvenosus.* From the limited results available, it would appear that species from only three genera of heterocystous filamentous cyanobacteria (*Nostoc, Anabaena* and *Calothrix*) can enter into symbiosis with cycad coralloid roots. However, until many more cyanobacterial species have been tested for their ability to enter into a symbiotic relationship with cycads, one should be careful about drawing conclusions about the specificity of such relationships. The high frequency with which *Nostoc commune*, for instance, has been found in the coralloid roots of *Encephalartos* species could be due to an abundance of that cyanobacterial species in the soils on which the *Encephalartos* species that were examined happened to grow. On the other hand, all cyanobacteria-infected coralloid roots that have thus far been tested for nitrogen fixation, yielded positive results (Lindblad and Bergman, 1990). This suggests that the symbiosis between cycads and cyanobacteria is at least restricted to the nitrogen-fixing species of the cyanobacteria and apparently to its heterocystous filamentous forms.

Acknowledgments

The financial assistance of the Foundation for Research Development and the University of Pretoria is gratefully acknowledged.

References

***Bowyer, J.W. and Skerman, V.B.D.* 1968.** Production of axenic cultures of soil-borne and entophytic blue-green algae. J. Gen. Micro. Biol. 54: 299 - 306.

***Grilli Caiola, M.* 1972.** Cell morphology of the blue-green algae under culture conditions from *Cycas revoluta* isolated. I. Light microscopic observations. *Caryologia* 25: 137 - 145.

***Grobbelaar, N., Scott, W.E., Hatting, W., and Marshall, J.* 1987.** The identification of the coralloid root endophytes of the southern African cycads and the ability of the endophytes to fix dinitrogen. S. Afr. J. Bot. 53: 111 - 118.

***Huang, T.C., and Grobbelaar, N.* 1989.** Isolation and characterisation of endosymbiotic *Calothrix* (Cyanophyceae) in *Encephalartos hildebrandtii* (Cycadales). Phycologia 28: 464 - 468.

***Lindblad, P., and Bergman, B.* 1990**. The cycad-cyanobacterial symbiosis, pp. 137 - 159 *In* A.N.Rai (ed.), CRC Handbook of Symbiotic Cyanobacteria. CRC Press, Boca Raton, Florida.

***Marshall, J.* 1989.** Aspekte van die simbiose en stuifmeelmorfologie van seker lede van die Cycadales. M.Sc. thesis, Univ. of Pretoria.

***Marshall, J., Huang, T.C., and Grobbelaar, N.* 1989.** Comparative morphological and physiological studies on cyanobionts of *Encephalartos transvenosus*. S. Afr. J. Bot. 55: 574 - 580.

The Root Nodules and Nitrogen Fixation in *Cycas panzhihuaensis*

Liu Guo-fan, Deng Ting-xiu
Department of Pedology, Chengu Branch, Academia Sinica

Zhou Lin and Yang Si-Yuan
Department of Gardens, Panzhihua City
Peoples Republic of China

Abstract

Cycas panzhihuaensis is found in the hot, dry Jinshajiang River Valley of the tropical region of South China. Multiple branched coralloid roots containing cyanobacteria are found in all plants of this species. The coralloid roots are most numerous at 10cm below the soil surface. The active coralloid roots of seedlings total 8g of fresh weight per plant and those of 100 year old plants average 370g of fresh weight per plant. Nitrogen fixation activity ranges from 1.8 to 11.1 μmole C_2H_4/g fresh wgt/h in autumn and is obviously influenced by light and soil moisture. Within the range of daily variation, nitrogen fixation is much higher in the day than in the night. The amount of nitrogen fixed ranges from 0.64-18.68mg/plant. This suggests that *C. panzhihuaensis* may play an important role in the nitrogen cycle of its ecosystem.

Materials and Methods

Cycas panzhihuaensis grows in an area of 310 sq. km on the mountain slopes of the Jinshajiang River Valley near Panzhihua City. The climate is typical of the semi-arid river valleys of the South Asian tropics. The community has few trees and the woody vegetation is composed chiefly of small shrubs with an uneven distribution of plants of *C. panzhihuaensis*. Most of these plants are between 0.3-0.6m tall, with a few between 0.6-1.4m tall and from 40-100 years old. The soil is brown and derived from parent material consisting of limestone. Bare rock composes 50% of the coverage with large and small rocks scattered throughout. The roots of *C. panzhihuaensis* extend into the soil between these rocks. The soil is dry and slightly acid to neutral without a positive reaction to HCL tests. Simple plots were chosen on a south slope at 1,600m (Plot 1) and 1,700m (Plot 2) and on a north slope at 1,600m (Plot 3).

There are nearly 10,000 plants of *Cycas panzhihuaensis* of varying ages, grown from seed, in the botanical garden of the Panzhihua Horticulture Department. Most of these are samples transplanted to nutritive pots. Those plants over ten years old were transplanted to the garden from their native habitat. The soil used in transplanting was derived from a dry, red-brown sandstone with a 6.0-6.5 pH.

Nitrogen fixation activity was tested by the acetylene reduction method in the field and in cultivated plants in the botanical garden in order to determine daily fluctuations in the relationship between the age of the plant and the amount of nitrogen fixed.

Results and Discussions

1. Features of root nodules

Cycas panzhihuaensis has big nodule clusters with coral-like nodules with a diameter of 2mm that are either tightly or loosely organised into nodule clusters. The tight cluster looks like *Alnus cremastogyne*, with tight formation of nodules, and a cluster diameter of less than 10cm. The latter is more branched with 3 to 5 coral-like branches and a cluster diameter of 20cm. The colour of the nodule is aquamarine with an annular, columnar, green stripe very clear between the inside and outside cortices of the nodule. This is obviously related to pigment of the cyanobacteria and it is rather similar to other *Cycas* species. Furthermore, there is another coralloid root, thin, fine but not nitrogen fixing. Occasionally these non-nitrogen fixing coralloid roots are difficult to distinguish from the nitrogen fixing ones.

Root nodules are perennial. They can be divided into three nodule ages by their colour, cortices and branches:

(a) young nodules with light, delicate colour and smooth surface, with about 1cm long branches normally at the top of tight or loose nodule clusters;

(b) old nodules with dark colour, rough surface and many projecting nodular specks, and with approx. 2cm long branches which are normally found in the middle or bottom of loose nodule clusters; and

(c) robust nodules which are intermediate between the young and old nodules.

Many nodule clusters decline from their bases, and wrinkle, dry and rot longitudinally. The growth stages and relative ripeness can be told from the age of the root nodule. Natural or artificially cultivated *Cycas* plants that are 10 to over 100 years old have very few dead root nodules, demonstrating that their root nodules are long lived. This obviously tells us that it is much superior over other species. For example, many pulse root nodules decline in the current year, and other species, such as *Alnus cremastogyne*, also decline, mostly after 10 years. Information collected from 10 plants of *C. panzhihuaensis* showed 70% of the root nodules occurring on small pleurogenic roots within 30cm of the basal stem and within 10cm to the earth surface. On the same *Cycas* root, root nodules of different ages can be found.

2. The nitrogen fixation activity and relevant clusters are quite often restricted by many factors, causing numerous changes. It is mainly influenced by internal characteristics and the environment. The number of tested specimens for the activity is always greater than eight.

(a) *Influence to the activity by root nodule character and site conditions*

Root nodules at different ages have different nitrogen fixation activity. A test of a specimen taken from a plant about 100 years old in a natural forest at the bottom of a north slope, shows (Table 1) that a young nodule and a robust nodule have a nitrogen fixation activity comparable to the root nodule of *Alnus crematogyne*. In contrast, an old nodule has a significant reduction in nitrogen fixation activity (Table 1).

Some root nodules taken from another plant were soaked for about five minutes and then put into a test bottle. The activity of the soaking nodules was observed to be about two to three times more than the ones not soaked. This shows that dry soil causes some water loss of the root nodule, resulting in decreased nitrogen fixation activity and that appropriate water supply is favourable to the recovery of the activity. It is thus clear that nodule age and moisture are important factors affecting its nitrogen fixing activity.

Site condition comprehensively embodies water, heat and soil nutrients. Tests in natural communities and in cultivated Cycas plants with different site conditions, reveal an obvious trend (Table 2) that nitrogen fixation activity of cultivated *Cycas* plants is generally higher than that of natural ones which were tested at the same time. This may be due to careful artificial cultivation methods of spreading organic fertiliser and proper watering while the mountain slope remains dry and barren of organic materials.

(b) *Diurnal variation of nitrogen fixation activity*

Characteristics of diurnal variation of nitrogen fixation activity of *Cycas panzhihuaensis* are as follows: The sun rises at 8.00 am, nitrogen fixation activity reaches the highest point at 9.00 am. Dewdrops are found on leaves at this time, showing a high relative humidity. Until 3.00 pm the activity stays at a high level. After 4.00 pm it gets lower, and the lowest occurs from 7.00 pm to the next morning. It then gradually recovers and reaches its highest again at 9.00 am. This differs from *Alnus cremastogyne* and *Cajnus cajan*, also with high diurnal variations (11.1-1.1 Umole/g.h-1). The soil temperature, 5cm below ground, reaches its highest of 26ºC at 4.00 pm and the lowest of 21.5ºC after midnight, rising again when the sun is out. Accordingly co-ordination between light and conditions and relative temperature and moisture is rather important in affecting the diurnal variation of nitrogen fixation activity. It seems more susceptible to light conditions than *Alnus cremastogyne* and *Cajnus cajan*. This provides us with a scientific basis in estimating nitrogen fixing capability.

Table 1

Relationship between nodule character and nitrogen fixation activity

Specimen	N2-fixation activity (umole C_2H_4/g.f.w.h.)
Young nodule	4.45 + 1.00
Robust nodule	3.48 + 0.73
Old nodule	1.11 + 0.40
Soaking nodule	15.96 + 1.91
Comparison nodule (*Pinus crematogyne*)	4.24 + 0.51

3. Nodulation and nitrogen fixing capability

To a great extent, total nitrogen fixing capability is determined by the number of nodules and nitrogen fixation activity per nodule. The amount of nitrogen fixation of a plant in a period of time was studied for

Table 2

Relationship between Site Condition and Nitrogen Fixation Activity

Community	Topography	Soil	Sampling Time in mins.	No. of Repetitions	Activity
Natural	Slope	Brown	10-11 13-14 15-16	8 9 9	5.24 + 1.19 5.06 + 0.84 3.48 + 0.73
Artificial	Platform	Dry Red	9-11 11-15 15-18	13 16 14	9.85 + 1.78 7.88 + 2.09 5.45 + 1.48

Table 3

Nodule and Nitrogen Fixation Amount

Community	Sample Plot	Age	Number of Nodules		N_2 fix actiumole C_2H_4	N_2 fix amount mg N/Plant.h-1
			Living	Dead		
Natural	1 2 3	50 50-100 100	93 127 370	- - -	5.06 5.24 3.48	4.39 6.21 12.02
Artificial	1 4 10 25 35		8.0 18.0 19.5 198.0 205.5	- - - - 100.2	8.57 6.58 10.10 4.24 9.74	0.64 1.11 1.84 7.84 18.68

entire plants in order to obtain a relatively correct estimate. With available materials, we have made the first comparison with the nitrogen fixing ability of a single plant at different ages (Table 3). It shows that *Cycas panzhihuaensis* has a high number of nodules generally. They are infected by a cyanobacteria at the seedling period when the seedling is only 2.5cm in height and thickness in the first year, with a root nodule cluster weighing 8g. The weight of the root nodules increases according to tree age - 100 year old root nodules of natural *C.panzhihuaensis* weigh 370g without including dead nodules. In the artificial community, some root nodules over 25 years old decline or die. This is probably caused by injury of its root when transplanting, loosening the soil, and the spreading of fertiliser.

C. panzhihuaensis is a longevous species. There are many 100 year old trees in the surveyed areas, with roots possessing great numbers of living nodules and with good branching and tillering abilities. One of the trees of *C. panzhihuaensis* has 14 branches on its base, and new nodule clusters are formed continuously on the roots of these plants. There is no obvious relationship between nitrogen fixation activity and tree age and nodulation. To its best extent, the total nitrogen fixation amount per plant is decided by the number of nodules. Because the natural forest has only a sparse number of plants, the total amount of forest nitrogen fixation is not high. In a single plant, root nodulation nitrogen fixing ability increases steadily and through the life of a plant as more long lived nodules are produced. Plants of *C. panzhihuaensis* reinforce their nutrient nitrogen supply through nitrogen fixation and also effectively improve the soil in the intermediate area. The natural community is located in a dry, hot river valley which has dry seasons in winter and spring. With this poor site condition, the species still grows well. This shows that it has very strong resistance to aridity as related to nitrogen fixation of root nodules. Root nodules keep comparatively high nitrogen fixing activity as well as adaptability of symbiosis cyanobacteria to extreme environment. Its nitrogen fixation function plays a good role in nitrogen circulation of the whole ecosystem.

Cuticle, Epidermis and Anatomy of Pinnae in Some Species of *Cycas linn*

Divya Darshan Pant and Kabita Das
Department of Botany
University of Allahabad, Allahabad 211002, India

Abstract

The paper deals with the cuticle, epidermis and some features of the anatomy of eight species of *Cycas*, viz. *C. angulata*, R. Br., *C. armstrongii* Miq., *C. basaltica* C.A. Gardn., *C. calcicola* Maconochie, *C. kennedyana* F. Muell, *C. normanbyana* F. Muell, *C. pruinosa* Maconochie, *C. taiwaniana* Carruth. Additional anatomical features of the pinnae of *C. beddomei* are also described. As a result of the present work we now know such details for the pinnae of twenty one species of *Cycas* out of a total of about fifty four listed species.

The authors have enumerated the characters which are common or different between various species of *Cycas*. They point out that these are of taxonomic importance and they can help in distinguishing the genus from other genera of cycads and the various species of *Cycas* from each other.

Particularly interesting features noticed during the investigation include the occurrence of mucronatoid tips of *C. angulata, C. basaltica*, and *C. kennedyana*, a midrib, which falls short of the lamina at the extreme tip, in *C. pruinosa*, the presence of occasional stellate hairs in *C. calcicola*, amphistomatic pinnae in *C. armstrongii* (besides *C. comorensis* and *C. micholitzii* where such pinnae are already reported), three or two mucilage canals in the basal part of pinnae of *C. beddomei* and *C. taiwaniana*, respectively. The bundle sheaths around the midrib bundle show outer rings of thick-walled cells and inner chloroplasts. The structure of the bundle sheath is comparable with that of C_4 plants but C_4 character of *Cycas* can be confirmed only after someone has studied the photosynthesis of its leaves.

Introduction

Out of an up to date total of fifty seven named forms of *Cycas*, Pant and Nautiyal (1963) had earlier described the cuticles and epidermis of pinnae of thirteen species with occasional notes on their anatomy. The present account details with the details of the cuticle, epidermis and anatomy of the pinnae of eight additional species of the genus, viz. *C. angulata* R.Br., *C. armstrongii* Miq., *C. basaltica* C.A. Gardn., *C. cal-*

cicola Maconochie; *C. kennedyana* F. Muell., *C. normanbyana* F. Muell; *C. pruinosa* Maconochie, *C. taiwaniana* Carruth. Besides these, we are describing some additional details of the structure of pinnae of *C. beddomei* a species which had been investigated by Pant and Nautiyal (1963).

As a result of the present work the anatomical details of pinnae of twenty one species of *Cycas* have become available, eighteen of these species, including *C. angulata, C. kennedyana, C. normanbyana, C. pruinosa* and *C. taiwaniana*, among the newly investigated ones, have hypostomatic pinnae. The pinnae of the remaining three species, including those of *C. armstrongii* among the newly investigated ones, are amphistomatic.

The common megascopic foliar characters of *Cycas* include unipinnate leaves with the acroscopic side of their obliquely inserted pinnae subtending angles of about 35 to 120º with the rachis. The angles may differ in different species and sometimes in the same species or in the same leaf, e.g., in *C. basaltica*.

The pinnae are always sessile, having parallel sides. Their terminal portions taper steeply to end in pointed apices. In some species, e.g., *C. angulata, C. basaltica* and *C. kennedyana* the lamina is more abruptly contracted to give rise to a mucronatoid spinous tip. The lamina is traversed by a characteristic single median vein which extends right up to the apex. The only exception is *C. pruinosa*, where the midrib ends a little before the apex of the pinnae. In species having spine tipped pinnae, the vein extends into the apical spine.

The unusual repeatedly forked pinnae of *C. micholitzii* are also exceptional in having a median vein which forks with each bifurcation of the pinna so that a single midrib fork enters each pinna segment but even this species has a single median vein in each segment except at the junctions of the forks.

The pinnae are somewhat contracted a little above their basis but their extreme bases are once again more or less widened so that they become widely attached. The acroscopic margins of pinnae run straight to the rachis but their basiscopic margins become arched to make the lamina decurrent on that side.

The common microscopic characters of the cuticle and epidermis of all hitherto investigated species are:

(i) A thick cuticle on both sides of the pinnae but among these the cuticle on the upper side is usually somewhat thicker than the lower. The cuticles of pinnae are thinner over the domes of the epidermal cells and thicker over their anticlinal walls (see (III) below) and they show clearly marked outlines of the epidermal cells.

(ii) The anticlinal walls of epidermal cells are straight in all species except in *C. micholitzii* where, as described by Pant & Nautiyal (1963), their sides are sinuous.

(iii) The central portions of the surface walls of epidermal cells form shallow domes and are more or less bulged outwards and their peripheral portions are depressed (near their anticlinal walls).

(iv) Numerous simple pits occur in the surface walls of epidermal cells (Porsch, 1905; Karzel, 1908; Pant & Nautiyal, 1963). They usually tend to be arranged in rows but their shape and size may vary in different species.

(v) The cells of the epidermis over and below the midrib are longer than broad and arranged more or less in regular longitudinal rows.

(vi) The cells of the upper epidermis, on either side of the midrib are obliquely or transversely elongated and their linear arrangement is more or less disturbed. The upper epidermis often tends to show sporadic groups of two or more cells which have parallel sides and are of almost equal size in one director.

(viii) The lower epidermal cells are usually irregularly arranged and they show irregularly placed stomata between them. The frequency of stomata varies in different regions of a pinna, being highest in the middle region tending to diminish towards the base and apex whose extremities may be completely devoid of stomata.

(ix) The hair bases, rings of cells around them and papillate ends of subsidiary cells are highly cutinised as indicated by their taking up a deeper red safranine stain.

(x) The guard cells are always sunken. The stomata are sometimes monocyclic, with the guard cells lying below a single more or less regular ring of subsidiary cells but in other cases they are partly or wholly amphicyclic where the guard cells lie below one or more incomplete or complete additional rings of encircling cells.

(xi) The guard cells show well developed gymnosperm type of polar and lateral lignin lamellae.

(xii) A one to many layered hypodermis is present below the upper epidermis of a pinna. It may also be present over the lamina. As a rule the hypodermis is more than one layer of cells thick over the midrib. On either side o the midrib hypodermis may extend up to the lower side in the revolute or recurved margins. It is absent below the lower epidermis of lamina. In the basal parts of pinnae a continuous hypodermis lies below epidermis on the upper as well as the lower sides and at the margins. The number of layers of the hypodermis may vary in different parts of one and the same pinna or in pinnae of different species. The cells of the hypodermis are generally longer than the epidermal cells and they are invariably elongated longitudinally.

(xiii) The chlorophyllose mesophyll is differentiated into a typical palisade of dorsiventrally elongated cells on the upper side and there is a similar mesophyll of somewhat shorter palisade-like cells inside the lower epidermis.

(xiv) Between the two chlorophyllose mesophyll layers lie colourless cells of the accessory transfusion tissue whose transversely elongated cells run parallel to each other, stretching themselves between the midrib and margins.

(xv) The vascular bundle of the pinna has a typical cycadean diploxylic strand showing a wedge-shaped adaxial xylem strand in transections with its protoxylem at the apex on the abaxial (dorsal) side just inside the phloem. A few larger xylem cells lie in one or two patches on the inner side of cambium which gives rise to secondary phloem.

The vascular bundle is surrounded by a multilayered sheath whose outer cells particularly those on the abaxial side are frequently thicker-walled and crystalliferous. Some of the relatively thinner-walled cells lying next to the vascular elements show numerous chloroplasts.

Beside the above common features of *Cycas* leaves, the megascopic and microscopic specieswise peculiarities of the presently investigated species are mentioned in the accounts of the species (the characteristics mentioned for the various species incorporate not only our own observations on the material available to us but they also mention the ranges and details which have been given for the respective species by earlier authors like Miquel, 1969; Bailey, 1902; Lamb, 1923; Schuster, 1932; Coulter & Chamberlain, 1910; Johnson, 1961; Pant & Nautiyal, 1963; Pant, 1978; Maconochie, 1978).

Materials and Methods

Material of fresh leaves of *C. beddomei* Dyer was collected from plants growing locally, in the Roxburgh Botanical Garden of the University of Allahabad or from plants gorwing wild in the Cudduppah Hills. Material of the leaves of all other investigated species of *Cycas* was available only in the form of dried up herbarium specimens. Only a few pinnae or portions of leaves of *C. normanbyana* F Muell, and *C. taiwaniana* Carruth. were available for this work.

Cuticles of pinnae were obtained by macerating cut pieces of pinnae in Schulze's fluid (conc. HNO_3 + $KC1O_3$) and thereafter thoroughly washed in water and then treated with dilute NH_4OH. The cuticle coats were recovered in the maceration residues, were rewashed with water and mounted in safranine glycerine jelly.

Dried leaves were softened by soaking them for two to four hours in a mixture of equal parts of acetic acid and hydrogen peroxide or by boiling them in hot water before making epidermal peels or cutting free hand or microtome sections. Epidermal preparations of leaves were made by peeling the surface of pinnae with a sharp razor. Microtome sections were cut after paraffin embedding (Chamberlain, 1932) at thicknesses of 8-11μm.

Table 1

Key for the determination of 21 investigated species of *Cycas*

1. Pinnae forked Pinnae simple	*Cycas micholitzii* 2
2. Pinnae amphistomatic Pinnae hypostomatic	3 4
3. Substance of pinna thin Substance of pinna thick	*C.armstrongii* *C.comorensis*
4. Margin or lamina curved downwards (revolute) Margin and lamina flat	5 6
5. Lamina curved downwards on both sides of midrib Lamina flat but margin revolute	7 8
6. Pinnae with mucilage canals Pinnae without mucilage canals	*C.taiwaniana* 11
7. Pinnae copiously hairy on underside Pinnae scarcely hairy on underside	*C.calcicola* *C.cairnsiana*
8. Midrib falling short of extreme apex Midrib continued up to extreme apex	*C.pruinosa* 9
9. Pinnae with mucilage canals Pinnae without mucilage canals	*C.beddomei* 10
10. Stomata amphycyclic, encircling cells papillate forming a chimney above stomata Stomata monocyclic without a chimney formed by encircling cells	*C.revoluta* *C.siamensis*
11. Pinna apex mucronatoid Pinna apex tapering and acute	12 13
12. Undersurface of pinna scarcely hairy Undersurface of pinna copiously hairy	*C.angulata* *C.basaltica*
13. Midrib almost flush with the surface on both sides Midrib ridged	*C.seemannii* 14
14. Midrib raised only on one side Midrib ridged almost equally on both sides	15 16
15. Midrib raised on lower side only Midrib raised on upper side only	17 18

16. Upper palisade layer present over midrib and continuous from one margin to another, a lower palisade-like mesophyll also present.	*C.peetinata*
Upper palisade layer interrupted over midrib and no palisade-like mesophyll on lower side.	19
17. Hypodermis present on upper side of lamina up to margin but absent on lower side of lamina	*C.normanbyana*
Hypodermis present only over midrib and at margins	20
18. Upper palisade layer present over midrib and continuous from one margin to another	*C.media*
Upper palisade absent over midrib	*C.rumphii*
19. Hypodermis present only along margins and over midrib	*C.circinalis*
Hypodermis present below upper laminar epidermis but interrupted on two sides of midrib	*C.celebica*
20. Upper palisade layer present over midrib and continuous from one margin to another	*C.kennedyana*
Upper palisade absent over midrib	*C.thouarsii*

In the formulation of the above key, characters mentioned by Pant & Nautiyal (1963) have been utilised for distinguishing *Cycas beddomei*, *C. cairnsiana*, *C. celebica*, *C. circinalis*, *C. comorensis*, *C. media*, *C. micholitzii*, *C. pectinata*, *C.revoluta*, *C. rumphii*, *C. seemannii*, *C. siamensis* and *C. thouarsii* from each other and from other species.

Sections were stained in safranine-fast green combination and mounted in Canadian Balsam.

It may be mentioned that in spite of all precautions in making preparations of dried up leaves, their delicate structures like epidermal hairs or phloem were either partially lost and badly distorted or totally lost. However, the presence of hairs could still be inferred from cuticular preparations, epidermal peels or sections by the presence of hair bases.

Observations

Cycas angulata R. Br.
(Plate , Fig. 3; Text fig. 1, A-E)

Foliage leaves about 50-125cm long. Pinnae numerous (about 110 in our specimens), inserted on rachis at acroscopic angles of 50-70º. Pinnae ca. 6-25cm long x 3-7mm wide, straight or slightly arched towards base, apex acute and spine tipped. Lamina flat but midrib depressed on upper side and raised below. Pinnae essentially hypostomatic with only a small number of stomata (about 12-18) on upper side. Pinnae showing hairs on both surfaces but their frequency is higher on the upper side.

Upper cuticle of lamina thicker ca. 5µm thick over domes of epidermal cells. Upper cuticle of midrib thicker than that of lamina ca. 8µm thick over domes.

Lower cuticle of lamina ca. 2 µm thick over domes of epidermal cells. Lower cuticle of midrib thicker ca. 3µm over domes.

Upper epidermal cells ca. 24µm x 40µm, surface walls showing pits in rows, pits elongated oval to rounded ca. 3µm long x 1µm wide. Hair base cells triangular, oval or rounded ca. 37µm in diameter, having a warty surface. Upper epidermis over midrib non-stomatiferous, showing elongated cells in longitudinal rows, cells ca. 67µm long x 17µm wide. Surface walls of cells bulged out in the middle and pitted as in epidermal cells of lamina. Hairs and hair bases present but stomata absent.

Lower epidermis of pinna showing irregularly disposed ordinary epidermal cells over lamina up to margin, cells ca. 30µm x 22.5µm. Surface wall showing small, simple pits as in upper surface, hair bases like those in upper epidermis frequently present. Stomatal frequency ca. 94.1mm^2 and index ca. 21.05. Stomata generally all over lamina but absent near margins and a few stomata present below midrib. Stomata on the lower side of lamina surrounded by 5-9 prominently papillate subsidiary cells overarching guard cells but stomata below midrib having elongated polar subsidiaries and short laterals. Guard cells sunken in a pit about 11µm deep formed by overarching papillate subsidiaries, guard cells ca. 43µm long by 18µm wide. Lower epidermis over midrib showing cells like those of upper epidermis but slightly smaller, ca. 47.5µm long x 16.5µm wide, a few stomata present.

A one-layered hypodermis is present below the upper epidermis up to margin. Hypodermis becomes up to 4 cells high on either face of midrib, but is absent below lower epidermis of lamina. A continuous palisade layer of vertically elongated cells which occasionally becomes two cells high is present on the upper side of lamina below the hypodermis up to margins but absent in midrib region. The palisade cells are ca. 62µm long x 15µm wide. Banks of accessory transfusion tissue are three to four cells thick and a spongy mesophyll tissue is absent on the lower side of midrib.

Cycas armstrongii Miq.
(Plate 5, Fig 6; Text fig. 2A-H & A')

Foliage leaves about 62-87cm long and 12-42cm wide. Pinnae numerous (about 102 in our specimens), inserted on rachis at acroscopic angles of 35-60º ca. 12-27cm long x 4-8mm wide, widely attached apex acute and spine tipped, substance of lamina thin and flat but midrib depressed on upper side, raised below. Pinnae amphistomatic with stomata in basal portion of upper surface. Pinnae showing hair based on both surfaces but their frequency is higher on the upper surface.

Upper cuticle of lamina ca. 3µm thick over domes of epidermal cells. Upper cuticle of midrib thicker, ca. 5µm thick.

Lower cuticle of lamina thinner ca. 2µm thick. Lower cuticle of midrib ca. 3µm thick.

Upper epidermis over lamina showing stomata in basal region of pinna (text fig. 2F). Epidermal cells averaging at ca. 50µm x 40µm. Pits on surface walls of cells oval, ca. 3µm long x 2µm wide. Hair base cells occurring interspersed among ordinary epidermal cells. Hair based cells triangular, oval or rounded ca. 48µm in diameter. Stomata on upper surface like those in lower (see below). Upper epidermis over midrib non-stomatiferous showing elongated cells in longitudinal rows, ca. 88µm long x 23µm wide, hair bases and surface wall pits present as in upper epidermis of lamina.

Lower epidermis of lamina showing irregularly disposed ordinary epidermal cells up to margin, cells ca. 43µm x 31µm, hair bases and surface wall pits present as on upper surface. Stomata generally confined to surface of lamina. Stomatal freqency ca. 88.2mm^2 and index ca. 31.9. Stomata surrounded by 5 to 9 overarching prominently papillate subsidiary cells. Guard cells ca. 65µm long x 32µm wide, sunken in a 14µm deep pit. Lower epidermis over midrib non-stomatiferous and like upper epidermis of midrib.

A layer of hypodermal cells is present below the upper epidermis all over the pinna up to its margins in extreme apical part but their middle and basal portions show a hypodermis only at the margins and at the two sides of midrib. The hypodermis becomes 3 to 5 cells thick in basal portion of pinna. Below the hypodermis is present a continuous layer of palisade cells on the upper side of lamina towards the apex but the midrib lacks a palisade in the middle and the basal parts of a pinna. Palisade cells are ca. 78µm long and 22µm wide. The accessory transfusion tissue is 3 to 7 cells high. Apalisade like mesophyll of shorter cells is present on the lower side of lamina and midrib in the apical part of pinna but the layer is absent in the middle and basal parts of the pinna.

Cycas basaltica C.A. Gardn.
(Plate 1, Figs. 4-7; Plate 6, Figs 2-6; Text fig. F-K)

Leaves about 90cm long and 8-18cm wide, with about 130 pinnae per leaf, pinnae inserted on rachis on acroscopic angles of 55-120º. Pinnae ca. 2-19cm long x 4-8µm wide. Apices mucronatoid with a spinate tip. Midrib depressed on upper side and prominently raised below.

Pinnae pubescent on both surfaces but frequency of hairs higher on lower surface. Hairs 1 or 2 celled, usually thin-walled but occasionally thick-walled, hairs as a rule tapering and having pointed tips.

Upper cuticle of lamina 6µm thick over domes of epidermal cells. Lower cuticle of midrib thicker than lthat of lower side of lamina ca. 7μm thick over domes of epidermal cells.

Upper epidermis of lamina non-stomatiferous epidermal of cells ca. 35μm x 27.5μm. Surface wall of cells showing rows of numerous small elongated to rounded oval pits ca. 2μm long x 1.5μm wide. Hair bases interspersed among ordinary epidermal cells, oval or rounded, ca. 32.5μm diameter. Upper epidermis of midrib showing longitudinally elongated cells in longitudinal rows cells ca. 45μm long x 13μm wide, surface walls of cells bulged out in the middle and pitted as inepidermal cells of lamina. Hair and hair bases present but stomata absent.

Lower epidermis of pinna showing irregularly disposed ordinary epidermal cells over lamina up to margin, cells ca. 20μm x 19μm, surface walls showing small pits as in upper. Hair bases present as in upper epidermis but more frequent. Stomata generally confined to surface of lamina. Stomatal freqency ca. 105.8mm^2 and index ca. 21.6. Each stomata surrounded by 7-10 overarching prominently papillate subsidiary cells. Guard cells sunken in a 7μm deep pit, guard cells 32μm long x 15μm wide. Lower epidermis over midrib non-stomatiferous like upper epidermis of midrib.

A layer of hypodermal cells present all over lamina up to margins below epidermis. Hypodermis up to 4 cells high on either face of midrib but absent below lower epidermis of lamina. Acontinuous palisade layer, usually 1 but occasionally 2 cells high, is present on the upper side of lamina below the hypodermis. Palisade cells ca. 80μm high x 15μm wide. Banks of accessory transfusion tissue 6-8 cells thick. A palisade like spongy mesophyll is present on the lower side of pinna. Palisade and spongy mesophyll present up to margins but absent below midrib.

Cycas beddomei Dyer
(Plate 3, Fig. 1-4; Text fig. 3 A-F, B'-F')

Among its diagnostic characters, the genus *Cycas* is categorically distinguished from other genera of cycads on the basis of the absence of mucilage canals in its pinnae. However, Pant & Nortiyal (1963) reported that the pinnae of *C. beddomei* were exceptional in having single mucilage canals on the phloem side of their vascular bundles. A detailed reinvestigation of the structure of the pinnae of this exceptional species was therefore made by us to see the course of the mucilage canals. Our work has shown that the pinnae of this species may have three independent mucilage canals in theie basal parts, a median canal on the abaxial side of the midrib bundle and two laterals on its two sides (Plate 3, Fig.3; Text Fig. 3 F, F'). The laterals are short and end in the basal parts of the pinnae while the median canals are longer and they extend approximately as far as three-fourths of the length of the pinnae. It is now clear to us that Pant & Nortiyal had cut sections only above the basal parts of the pinnae and accordingly they found only a single median canal in their sections.

Cycas calcicola Maconochie
(Plate 2, Figs. 1-6; Plate 6, Fig 1; Text fig. 4, A-M)

Leaves dark green, about 60-115cm long and 9-27cm wide. Pinnae 150-300 per leaf, pinnae inserted on rachis at acroscopic angles of 70-110º. Pinnae straight to slightly arched towards base, about 7.2-12cm long x 2-4mm wide, pinna apices attenuate to mucronatoid, with a spinate tip. Midrib seen as a hairy furrow on upper side and prominently raised on lower face showing a few discontinuous longitudinal ridges and furrows on its surface, margins revolute, lower surface of pinna between raised midrib and revolute margins appearing like two wide longitudinal furrows on either side of midrib. Pinnae pubescent or glabrous on both surfaces but frequency of hairs higher on lower surface than on upper. Hairs usually one-celled, usually thin-walled, sometimes thick-walled, usually simple and tapering, sometimes branched and with two pointed ends. Hairs rarely stellate with tapering branches. Pinnae hypostomatic.

Upper cuticle of lamina ca. 8.5μm thick over domes of epidermal cells, upper cuticle of midrib ca. 10.5μm thick over the domes of epidermal cells.

Lower cuticle of lamina slightly thinner than upper, ca.7μm thick over domes of epidermal cells. Lower cuticle of midrib ca. 8μm thick over domes of epidermal cells.

Upper epidermis of lamina and its revolute margins showing cells ca. 74μm x 5.5μm, surface walls showing pits ca. 2.5μm long x 1μm wide. Single celled hair bases occurring interspersed among ordinary epidermal cells, hair based cells trianguloid, oval or rounded, about 26μm in diameter. Upper epidermis of midrib showing cells ca. 47μm long x 13μm wide, pits on surface walls of cells as in upper epidermal cells of lamina, hairs and hair bases present but stomata absent.

Lower epidermis of lamina showing irregularly disposed ordinary epidermal cells over the two grooves in the lamina in either side of midrib formed by the sides of midrib and the revolute margins, cells in grooves ca. 27.5μm x 13.5μm showing very frequent occurrence of rounded to oval hair bases, ca. 18μm, pits on surface walls of lower epidermal cells absent or obscure. Stomata present only in furrows between midrib and margins. Stomatal frequency ca. $100mm^2$ and index ca. 25. Stomata surrounded by 7-11 overarching prominently papillate subsidiary cells. Guard cells about 33μm long x 11μm wide, sunken in 4μm deep stomatal pit, lower epidermis of midrib non-stomatiferous and like upper epidermis of midrib.

A layer of hypodermal cells present below upper epidermis up to revolute margins. Hypodermis becoming up to 4 cells thick on either face of midrib but absent below lower epidermis of lamina. Acontinuous palisade layer of vertcally elongated cells which occasionally becomes two cells high is present on upper side of lamina below the hypodermis up to its revolute margins but absent in midrib region. The palisade cells on upper side are ca. 29.5μm long x 10μm wide. Often the lower ends of a few palisade cells are placed over common collecting cells. A 3-4 cell high accessory transfusion tissue is present. A palisade-like mesophyll of shorter cells present on lower side of the lamina up to its margins but it is absent below midrib.

Cycas kennedyana F. Muell
(Plate 5, Fig.7; Text fig. 5, A-I)

Foliage leaves about 49cm long and 22.5cm wide. Number of pinnae about 100 per leaf, pinnae inserted on rachis at acroscopic angles of 40-115º. Pinnae ca. 3-5mm wide by 6-13cm long, flat but midrib depressed on upper side and raised below. Apex mucronatoid and spine-tipped. Pinnae hypostomatic, showing hair bases on both surafces, hair frequency higher on upper surface.

Upper cuticle of lamina ca. 6μm thick over domes of epidermal cells. Upper cuticle of midrib ca. 8μm thick over the domes of epidermal cells.

Lower cuticle of lamina thinner than upper, ca.3μm thick over domes of epidermal cells. Lower cuticle of midrib ca. 4.5μm.

Upper epidermis of lamina non-stomatiferous, epidermal cells averaging at ca. 38μm x 31μm. Surface walls of cells showing ca. 2μm long x 1μm wide pits. Interspersed among ordinary epidermal cells, are seen triangular, or oval to rounded hair bases, ca. 32μm in diameter. Epidermal cells around hair bases forming rings of subsidiaries like those around stomata.

Upper epidermis of midrib also non-stomatiferous but hairs and hair bases present, epidermal cells ca. 70μm x 15μm, surface walls of epidermal cells showing pits like those in upper epidermal cells of lamina.

Lower epidermis of lamina showing irregularly disposed ordinary epidermal cells over lamina up to margin, cells averaging at ca. 40μm x 20μm, surface walls showing small pits as upper epidermal cells, hair bases frequently present Stomata generally confined to surface of lamina. Stomatal frequency ca. $111.7mm^2$ and index ca. 15.8. Stomata surrounded by 7-9 overarching prominently papillate subsidiary cells. Guard cells about 36μm long x 16.5μm wide sunken in 9μm deep pit. Lower epidermis of midrib non-stomatiferous like upper midrib epidermis.

A single layered hypodermis is present below epidermis on either face to midrib and pinna margins but it is absent elsewhere. However, the hypodermis becomes 4-5 cells thick in the midrib region as well as the lamina in the basal parts of pinnae. A continuous layer of palisade cells is present throughout the upper side, over the midrib and lamina up to the margins, but in extreme base of a pinna the palisade is absent over the midrib region. The palisade cells are ca. 56μm long and 13μm wide. The accessory transfusion tissue is 3-6 cells high in the pinnae but in their extreme basal regions these cells are not clearly differentiated. A palisade-like mesophyll of shorter cells is present on the lower side of lamina but absent below the midrib.

Cycas normanbyana F.Muell.

(Pl. 4, Fig. 5-10; Text fig. 6, A-H)

Foliage leaves are about 90-150cm long and 25cm wide. The numerous pinnae are ca 10-14cm long x 5-8mm wide, the lamina margins flat but midrib depressed on upper side and raised below. The apex is acute and spine tipped.

The pinnae are hypostomatic and hairy on both surfaces but the frequency of hairs is higher on upper side.

Upper cuticle of lamina ca. 8mm thick over domes of epidermal cells. Upper cuticle of midrib is ca 10mm thick over domes of epidermal cells.

Lower cuticle of lamina is ca. 4mm thick over domes of epidermal cells. Lower cuticle of midrib ca-5mm thick over domes of epidermal cells.

Upper epidermis of pinna shows cells which are often broader than long, averaging at ca. 43.5mm x 45mm, epidermal cells tend to be sporadically arranged in rows. Surface wall pits ca. 2mm long x 1.5mm wide. Hair base cells are ca. 34mm in diameter, showing triaguloid, oval or rounded outlines, interspersed among ordinary epidermal cells, hair bases showing a warty surface. Cells in upper epidermis of *midrib showing ca. 91mm long x 17.5mm wide cells, surface wall of cells bulged out in the middle and pitted as in epidermal cells of lamina. Hairs and hair bases present but stomata absent.

Lower epidermis over pinna showing irregularly disposed ordinary epidermal cells over lamina up to margin, cells ca. 54mm x 31.5mm, surface wall of cells without pits, hair bases frequently present. Stomata present over lamina but absent over midrib and margins. Stomatal frequency ca. 52.9mm and index ca. 13.8. Stomata surrounded by 7-10 overarching prominently papillate subsidiary cells. Guard cells sunken in 23mm deep pits, guard cells ca. 48mm long x 12mm wide. Lower epidermis of midrib non-stomatiferous and showing cells as in upper epidermis of midrib.

A layer of hypodermal cells is present below upper epidermis up to margins. Hypodermis becomes up to 5 cells thick on either face of midrib but is absent below lower epidermis lamina. A continuous palisade layer, which occasionally becomes two cells high, is present on the upper side of lamina below the hypodermis up to margins but is absent in midrib region. The palisade cells are ca. 94mm long x 25mm wide. Accessory transfusion tissue 5 to 7 cells high. Lower palisade-like spongy mesophyll layer of shorter cells is present up to the margin on the lower side of pinnae but it is absent below the midrib bundle.

Cycas pruinosa Maconochie.

(Pl. 5, Figs. 1-5; Text fig. 7, A-I)

Foliage leaves about 80cm long and 20cm wide. Number of pinnae lower leaf about 120 wide. Number of pinnae per leaf about 120-240, pinnae inserted on rachis at an acroscopic angle of 50º-65º Pinnae ca. 3-16cm long x 3-5mm wide. apex mucronatoid, spinous. Margins revolute, midrib depressed on upper side and raised below but median vein exceptional in falling short of extreme apex. Pinnae hypostomatic. Pinnae glabrous on both sides but hair bases very rarely visible in basal part of pinnae.

Upper cuticle of lamina unevenly thick ca. 5mm thick over domes of epidermal cells and upper cuticle of midrib ca. 7mm thick over domes.

Lower cuticle of lamina ca. 2mm thick over domes of epidermal cells and lower cuticle of midrib ca. 3mm thick over domes.

Upper epidermis of pinnae and their revolute margins shows cells which may be sporadically arranged in small groups of transversely or obliquely elongated almost equal narrow cells with parallel sides. These cell groups lie between irregularly arranged unequal cells elongated in different directions. The epidermal cells are in general greatly variable in length and width but their size averages ca. 46mm x 44mm. Surface walls of cells showing pits ca. 3mm long x 2 mm wide. Angular or oval hair bases, ca. 58mm in diameter, very rarely seen. Upper epidermis of midrib non-stomatiferous showing elongated cells in longitudinal rows, cells ca. 86mm long x 21mm wide. Surface wall puts as in upper epidermis of lamina.

Lower epidermis of pinna shows irregularly disposed ordinary epidermal cells over lamina up to the margin, cells ca. 37mm x 31mm, hair bases rare but their frequency is slightly higher in the lower epidermis than in upper. Stomata confined to the lower surface of lamina between midrib and margins. Stomatal frequency ca. 88.2 and index ca. 20.5. Stomata surrounded by 6-9 subsidiary cells. Guard cells are ca 56mm long x 23mm wide, sunken in a 14.5mm deep pit. The surface wall of epidermal cells shows pits as in upper epidermis. Lower epidermis of midrib is non-stomatiferous like its upper epidermis but hair bases are more frequent here than in lower epidermis of lamina.

A hypodermis of thick-walled cells is present on both sides of midrib and below the upper epidermis, it becomes up to 4 cells thick in midrib region but elsewhere it is generally only one or two cells thick. The hypodermis is absent below the lower epidermis of lamina but present below the lower epidermis of midrib although it is absent below the lower epidermis of midrib in its extreme apical portion. A continuous palisade layer is always present on the upper side below the hypodermis up to its revolute margins. The palisade cells are ca. 95mm long x 20.5mm wide. The accessory transfusion tissue is six to eight cells high. A palisade-like mesophyll of shorter cells is present on the lower side of the lamina up to its margins as well as the lower side of midrib but it is absent in the extreme basal portion of midrib.

Cycas taiwaniana Carruth
(Plate 2, Fig. 5; Text Fig. 8, A-1)

Leaves about 56cm long and 18cm wide, with about 60 pinnae per leaf, pinnae inserted on rachis at acroscopic angles of 35-60. Pinnae ca 1-13cm long x 2-6m wide, midrib of pinnae depressed on upper side and raised below, apex acute, spine-tipped, lamina flat. Pinnae hypostomatic, hairs present on both surfaces but frequency of hairs higher on upper side.

Upper cuticle of lamina thicker, ca 3µm over domes of epidermal cells and lower cuticle of midrib ca 3.5µm thick over domes.

Upper epidermis over lamina non-stomatiferous ordinary epidermal cells broader than long, ca 56µm x 27.5µm, surface wall pits, ca 3µm long x 2µm wide. Interspersed among ordinary epidermal cells are seen rounded to oval hair base cells ca 33µm in diameter. Upper epidermis of midrib shows elongated cells in longitudinal rows, cells ca 53µm long x 18µm wide. Hairs and hair bases present but stomata absent.

Lower epidermis of pinna shows irregularly disposed ordinary epidermal cells over lamina up to margin, cells ca 38µm x 30µm, with irregularly scattered hair bases as in upper epidermis. Stomata present and generally confined to surface of lamina but absent over midrib and near margins. Stomatal frequency 82.3mm^2 and index ca 23.7. Stomata amphicyclic with a ring of encircling cells raised to form a conical chimney, encircling cells having thicker darker stained walls with an underlying ring of 7 to 11 subsidiary cells mounted over guard cells. Subsidiary cells having walls and cuticles staining just like ordinary epidermal cells. Guard cells sunken in ca 24µm deep pit. Guard cells ca 51.5µm long x 26.5µm wide, radiating walls of subsidiaries usually continuous with those of encircling cells indicating thereby that the rings of subsidiary and encircling cells may have arisen by formation of tangential walls in common mother cells.

Below the upper epidermis is present a layer of thick-walled hypodermal cells all over the lamina. The hypodermis is up to 6 cells high on either face of midrib but absent below lower epidermis of lamina. Hypodermal cells are arranged in longitudinal rows. Palisade cells on upper wise ca 93µm long x 23.5µm wide. A four to eight cells high accessory transfusion tissue is present. A lower palisade-like mesophyll of shorter cells is present up to margins but absent below the midrib.

The expanded bases of pinnae with decurrent acroscopic margins show to mucilage canals, one on either side of midrib bundle, but these end in the constricted neck-like portion of the lamina.

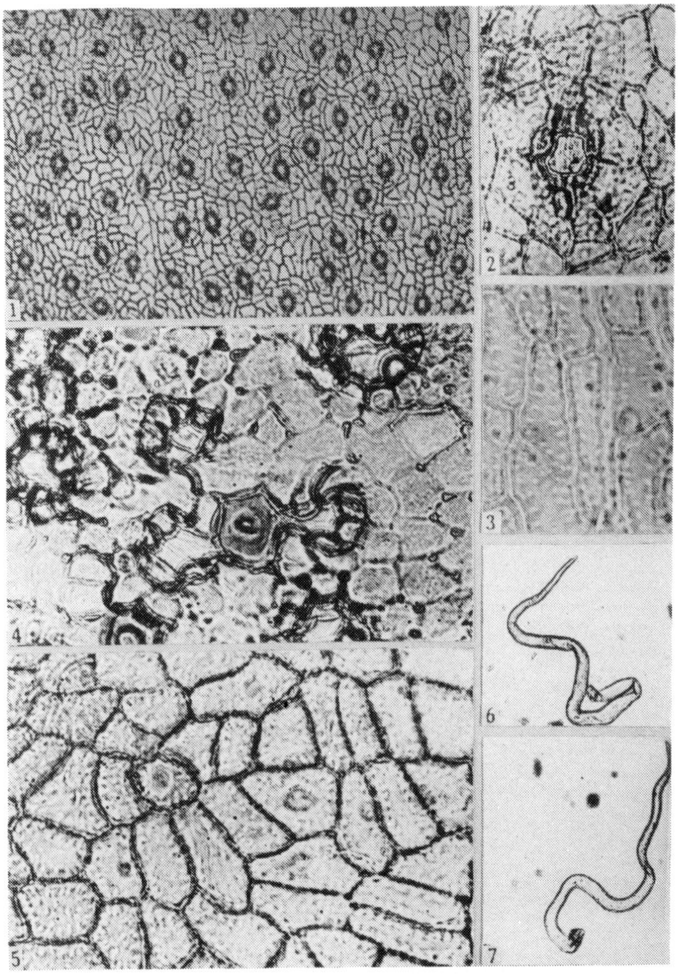

Plate 1

Cycas angulata

Figure 1: Lower epidermis of pinna. x 100.

Figure 2: A stoma from lower epidermis magnified to show surface wall pits in subsidiary and epidermal cells. x 400.

Figure 3: Upper epidermis of midrib showing surface wall pits. x 400.

Cycas basaltica

Figure 4: Lower epidermis showing stomata around a hair base. x 400.

Figure 5: Upper epidermis of lamina showing surface wall pits and sporadic rows of parallel equal cells and a hair base. x 400.

Figures 6-7: Single celled contorted hairs with pointed tips. x 100.

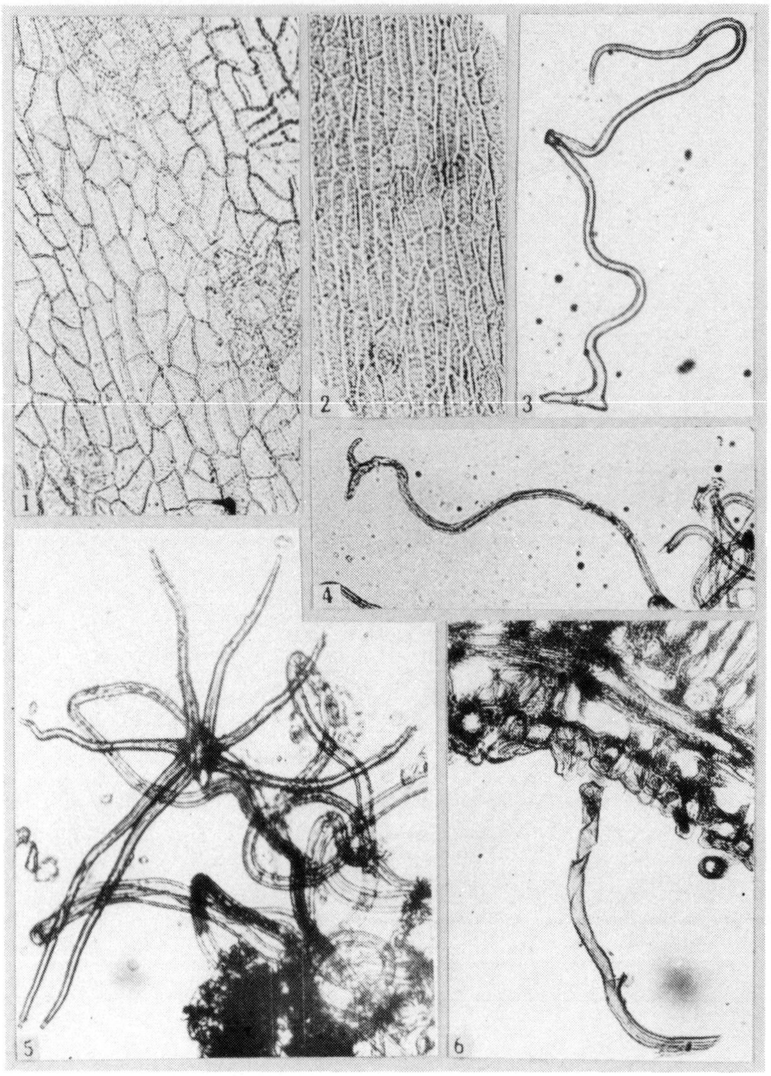

Plate 2

Cycas calcicola

Figure 1: Upper epidermis of lamina with occasional groups of parallel equal cells. x 200.

Figure 2: Upper epidermis of midrib. x 200.

Figures 3-4: Contorted hairs. Hair in Fig. 3 has a short arm near its base and hair in Fig. 4 is forked near apex. x 100.

Figure 5: A stellate hair lying over contorted hairs. x 200.

Figure 6: Portion of a T.S. of pinna showing an attached contorted hair. x 200.

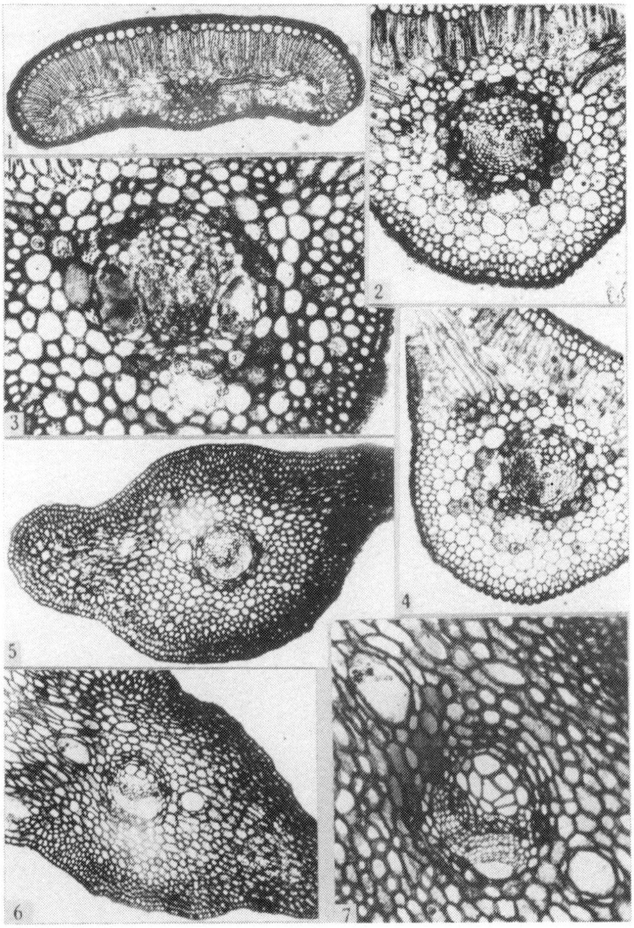

Plate 3

Cycas beddomei

Figure 1: T.S. of a pinna near apex without any mucilage canal. x 50.

Figure 2: T.S. of the middle part of a pinna with a single mucilage canal on abaxial side of the vascular bundle. x 100.

Figure 3: T.S. of extreme basal part of a pinna showing three mucilage canals. x 125.

Figure 4: T.S. of basal part of pinna showing two mucilage canals. x 100.

Cycas taiwaniana

Figure 5: T.S. of pinna showing single mucilage canal just above base (beyond the constricted portion of lamina). x 50.

Figure 6: T.S. of extreme base of pinna showing two mucilage canals. x 50.

Figure 7: Vascular bundle and two mucilage canals in Fig. 6 more magnified. x 100.

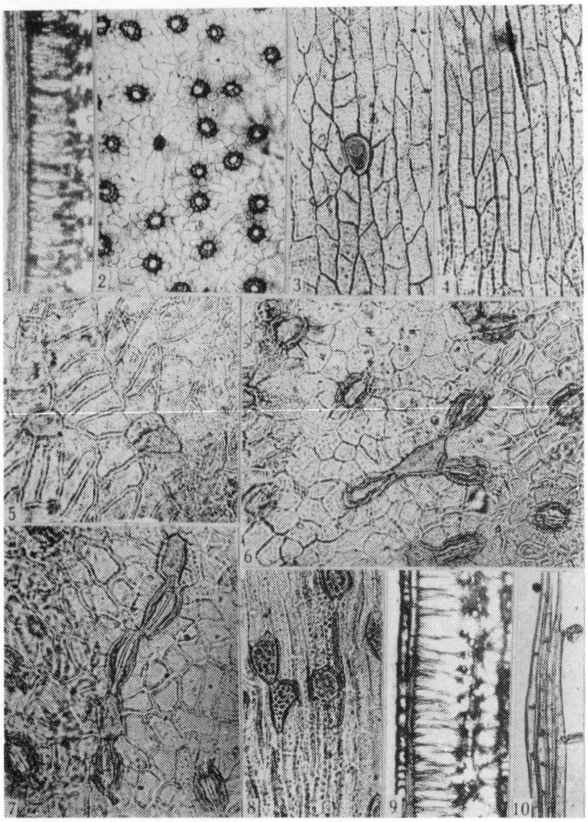

Plate 4

Cycas taiwaniana

Figure 1: Upper side of L.S. of pinna showing longitudinally elongated hypodermal cells between epidermis & palisade. x 100.

Figure 2: Lower epidermis of lamina showing stomata and hair base. x 100.

Figure 3: Upper epidermis showing a hair base and only a few groups of parallel, equal cells. x 200.

Figure 4: Upper epidermis of midrib. x 200.

Cycas normanbyana

Figure 5: Upper epidermis with single or paired hair base cells and groups of parallel equal cells. x 200.

Figure 6 & 7: Lower epidermis of lamina Fig. 6 shows three stomata around a hair base, and Fig. 7 shows two hair bases and three stomata (two of them are contiguous) arranged in a row. x 200.

Figure 8: Upper epidermis of midrib showing conspicuous surface wall pits and a few hair bases. x 200.

Figure 9: L.S. of a pinna showing longitudinally elongated hypodermal cells between epidermis & palisade cells. x 100.

Figure 10: Elongated hypodermal cells, separated by maceration of the substance of a pinna. x 100.

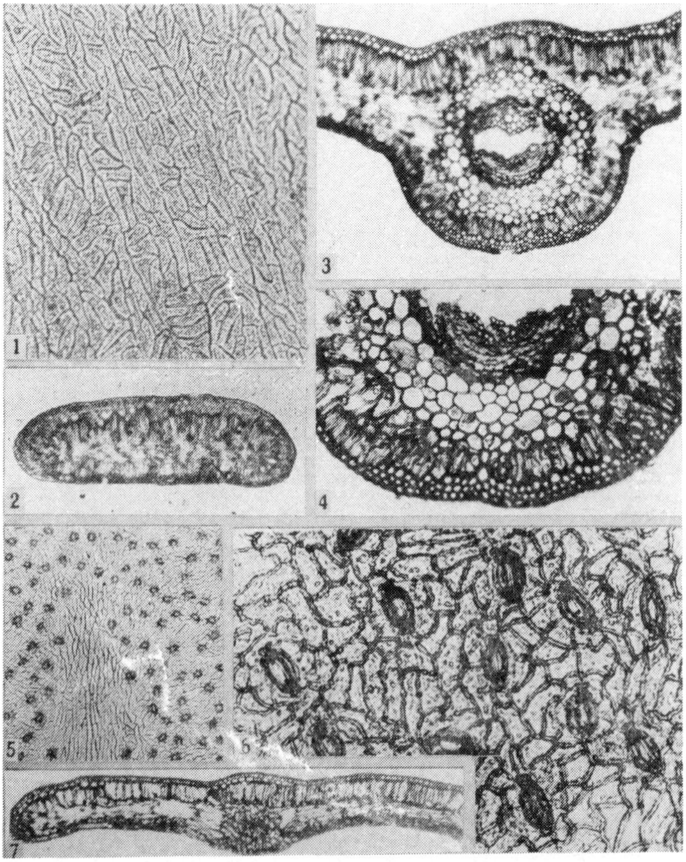

Plate 5

Cycas pruinosa

Figure 1: Upper epidermis of lamina showing groups of equal epidermal cells. x 100.

Figure 2: T.S. of acostate extreme apex of pinna. x 50.

Figure 3: Middle portion of T.S. of a pinna showing presence of palisade or palisade-like mesophyll on both faces of midrib. x 50.

Figure 4: More magnified view of palisade-like cells on lower side. x 100.

Figure 5: Lower epidermis of apical portion of a pinna showing presence of stomatiferous epidermis and the absence of differentiated elongated cells of midrib at extreme apex of a pinna. x 50.

Cycas armstrongii

Figure 6: Stomatiferous lower epidermis of lamina. x 200.

Cycas kennedyana

Figure 7: T.S. of apical part of pinna showing presence of hypodermis only above and below midrib and margin. x 50.

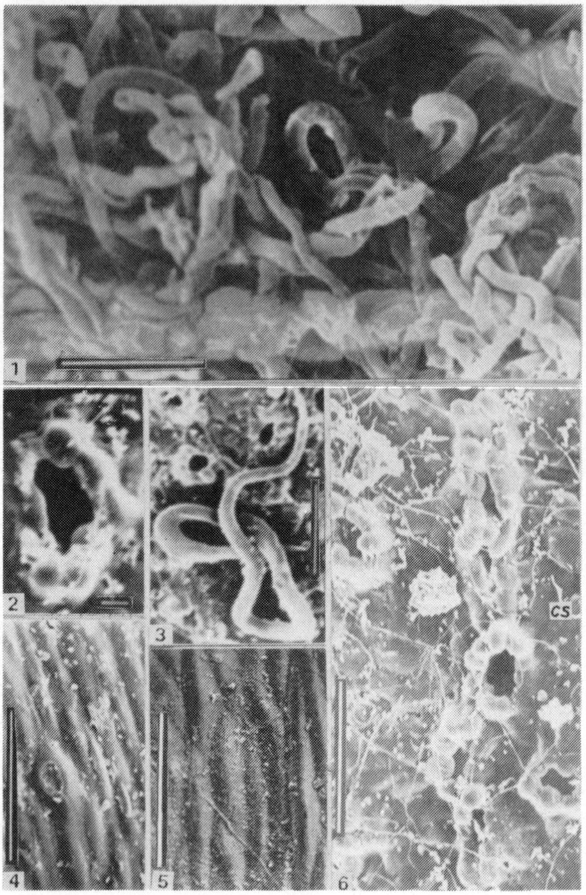

Plate 6

SEM photographs

C.calcicola

Figure 1: Lower surface of pinna showing matted hairs. Bar = 100μm

C.basaltica

Figure 2: Stometal pit bounded by popillate subsidiary cells. Bar = 100μm

Figure 3: Lower epidermis of pinna showing a twisted hair. Bar = 100μm

Figure 4: Lower surface of midrib showing epidermal cells an a hair base. Bar = 100μm

Figure 5: Cells of the upper surface of lamina. Bar = 100μm

Figure 6: Lower surface of pinna showing a row of contiguous stometa (cs). The fine threads are extraneous fungal hyphae and small or somewhat larger lobed white bodies are deposits of wax. Bar = 100μm.

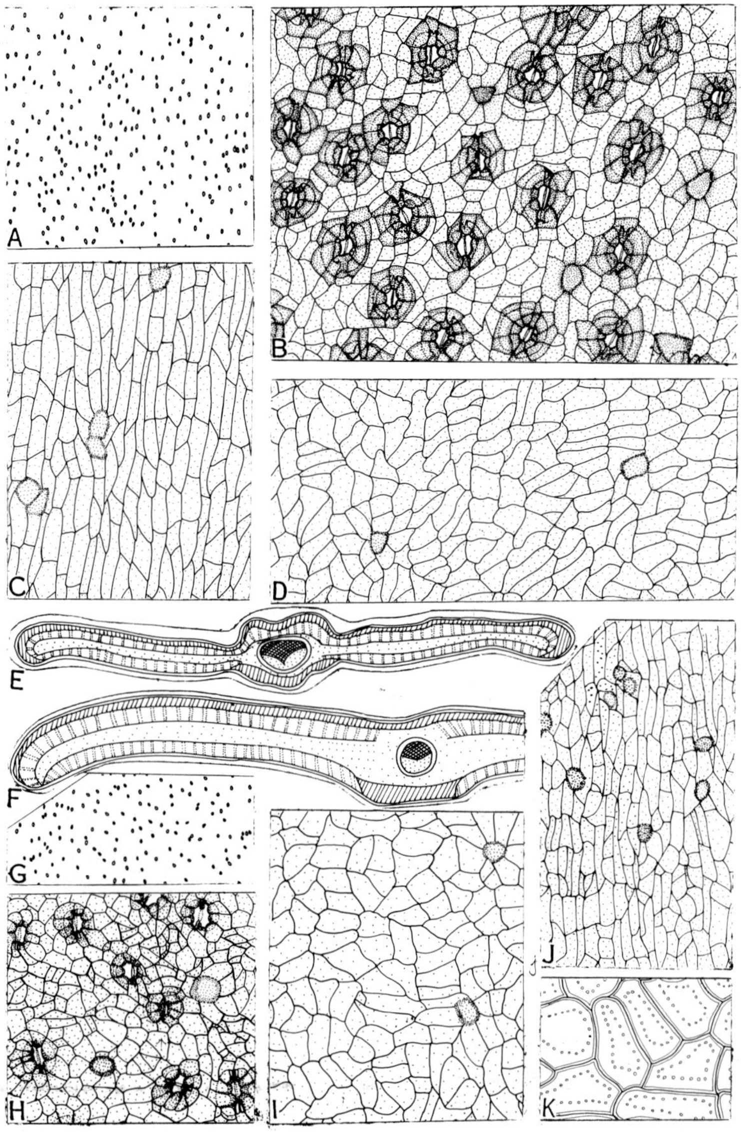

Text Fig. 1. A-E *Cycas angulata*, F-K *Cycas basaltica*. A, stomatal distribution pattern in lower surface of mid-region of pinna. Blower epidermis of pinna showing stomata and hair bases. C, upper midrib epidermis with a few hair bases. D, upper epidermis of pinna showing two hair bases among epidermal cells having sporadic groups of equally wide cells. E-F, transections of mid-regions of pinnae (topographic), G, stomatal distribution pattern in lower surface of mid-region of pinna. H, lower epidermis of pinna showing stomata and hair bases. I, upper epidermis of pinna showing epidermal cells with sporadic groups of equally wide cells and two hair bases. J, upper epidermal cells of midrib with a few hair bases. K, upper epidermal cells of lamina more magnified to show surface wall pits.

A, E-G x 25; B-D, H, I x 150; J x 250; K x 350.

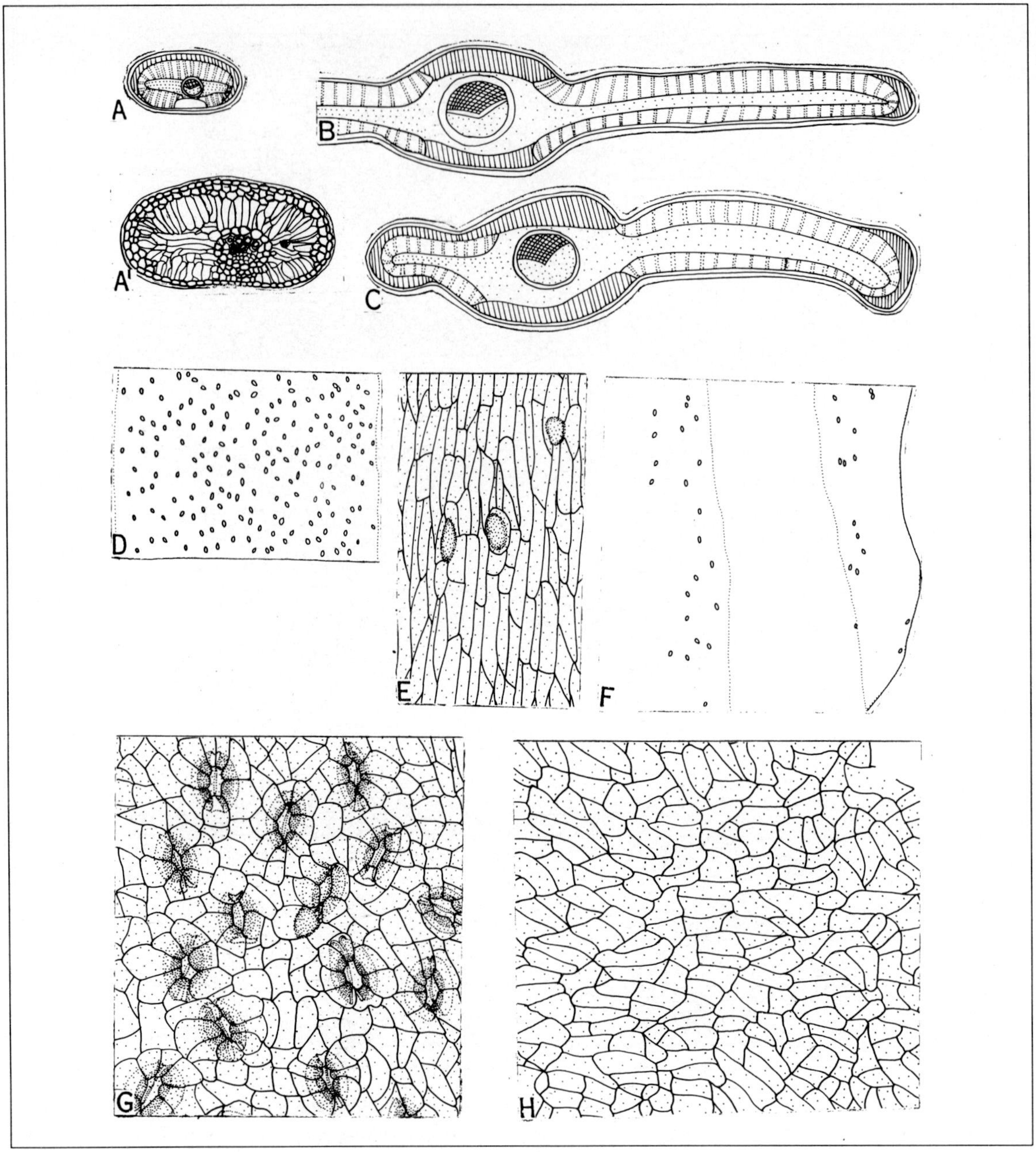

Text Fig. 2. *Cycas armstrongii*. A, B, C, transections of pinna (topographic). A, near apex. B, near middle. C, near base. A cellular details of section in A. D, lower surface of pinna showing stomatal distribution pattern. E, upper epidermal cells of midrib with a few hair bases. F, upper surface of pinna base showing stomatal distribution. G, lower epidermis showing stomata. H, upper epidermis of pinna showing sporadic rows of equally wide calls.

A-D, F x 25; A' x 50; E, G, H x 125.

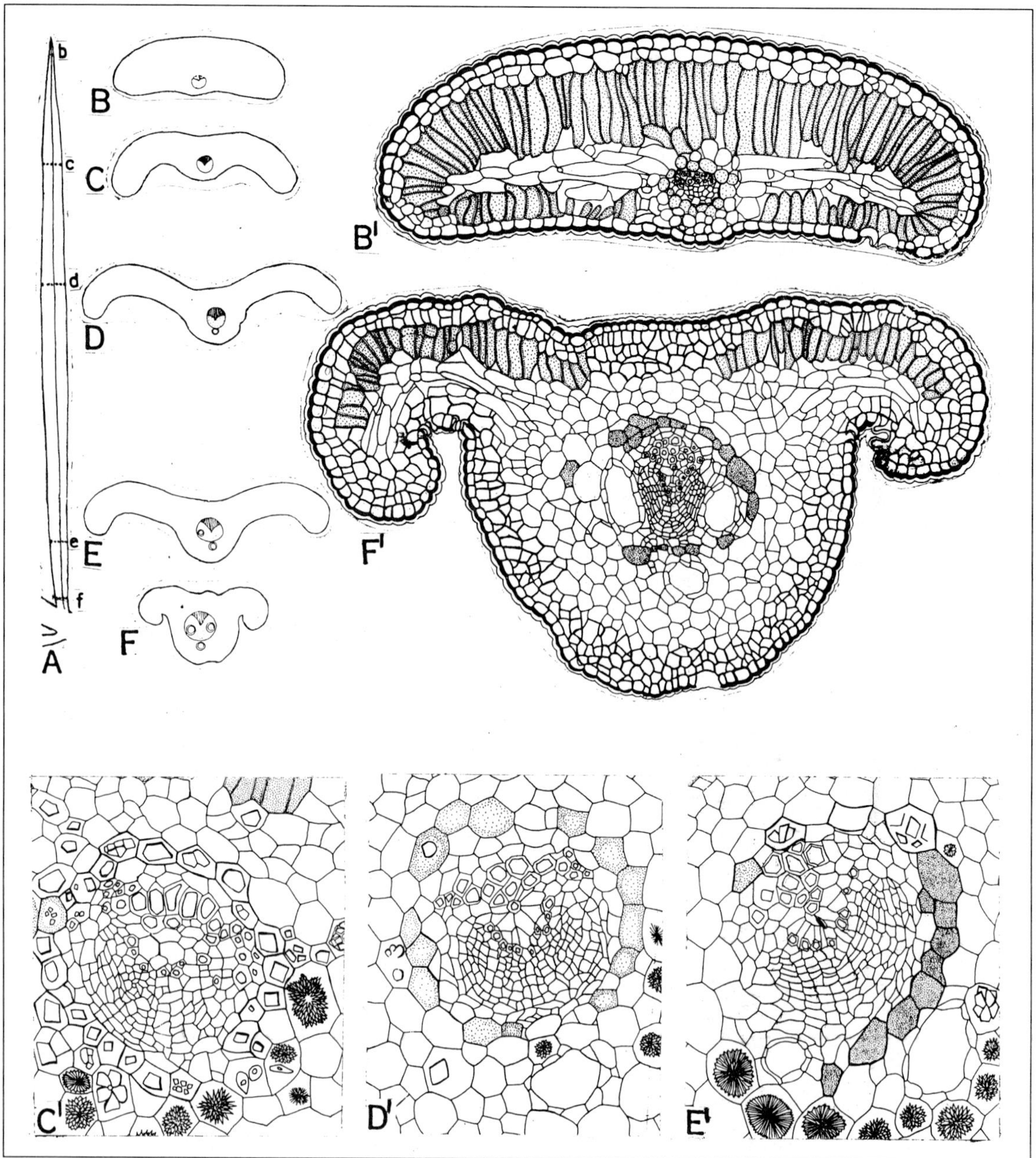

Text Fig. 3. *Cycas beddomei.* A, a pinna where dotted lines b to f indicate level of transections drawn in Topographic sketches, B to F, respectively. Note the absence of mucilage canals in B and C, one in D, two in E, and three in F. B' and F' shows cellular details of sections in B and F respectively. D', D' and E' are enlarged drawings of vascular bundles and mucilage canals in sections C, D and E respectively.

A x .5; F-F, 15 B', F'x 75; C', D', E' x 150.

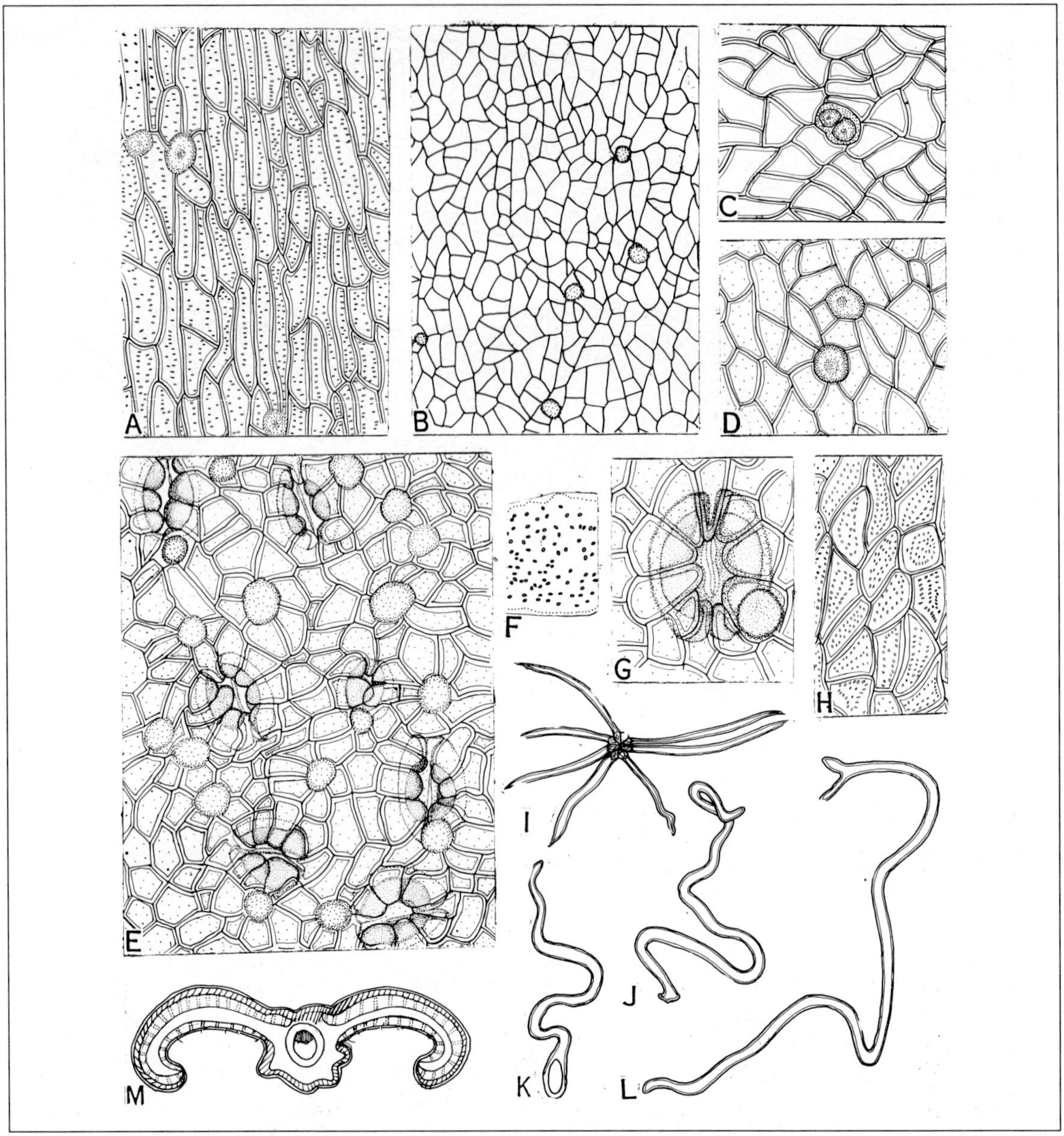

Text Fig. 4. *Cycas calcicola*. A, upper epidermal cells of midrib showing surface wall pits. B, upper epidermis of amina showing less obvious sporadic groups of equally wide epidermal cells. C-D portions of epidermis showing hair bases. E, portion of lower epidermis showing stomata and hair bases. F, stomatal distribution pattern on lower surface of a pinna. G, enlarged stoma with a prominent hair-like papilla base in a subsidiary cell. H, cells of upper epidermis further magnified to show surface wall pits. I-L, hairs from lower epidermis of pinna, hair in I is stellate; those in J, K, are simple; and that in L is forked. M, transaction of pinna (topographic).

A, E, H x 275; B x 125; C-D x 200; F, M x 25; I-L x 100; G x 375.

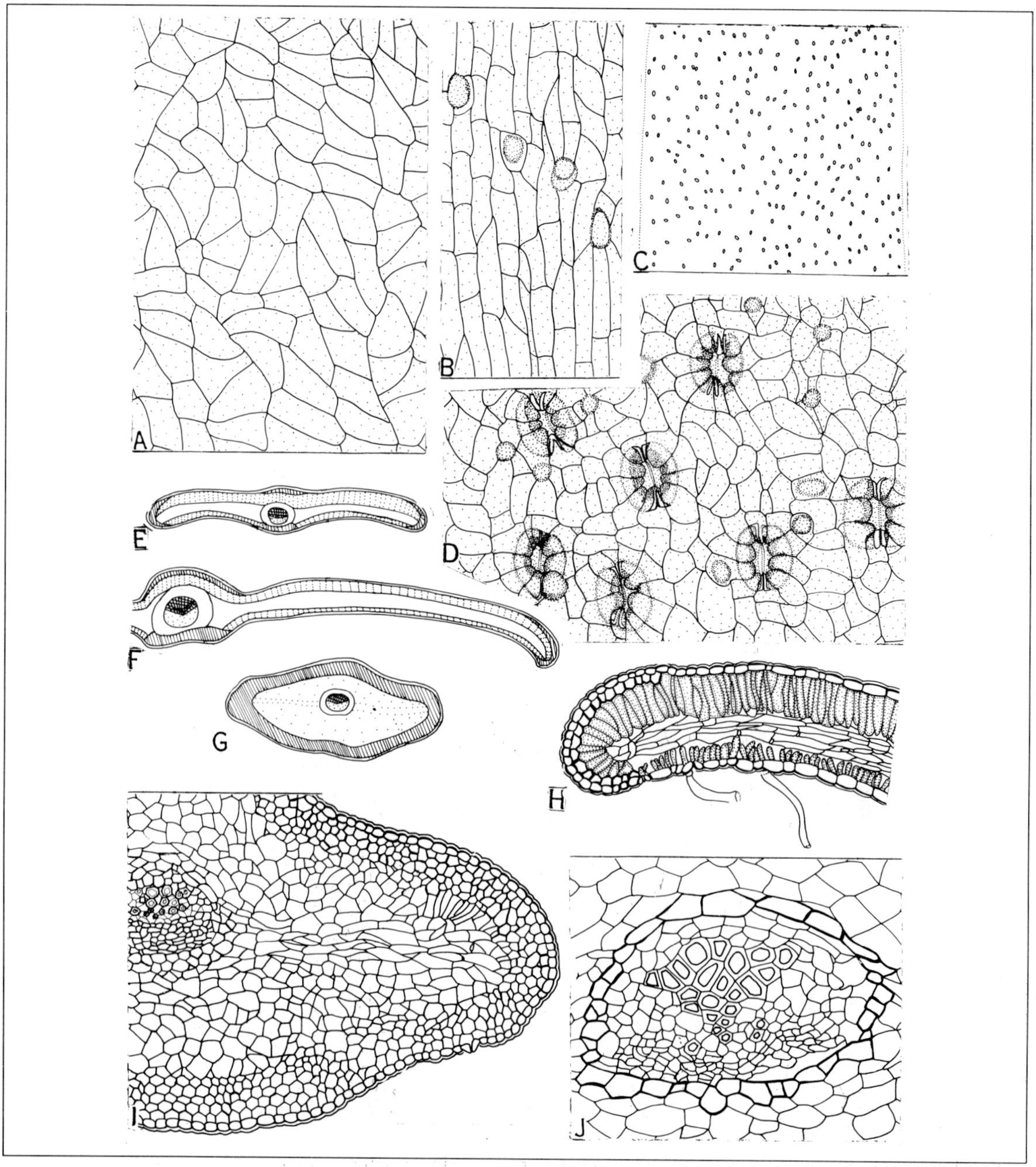

Text Fig. 5. *Cycas kennedyana*. A, upper epidermis of lamina showing sporadic groups of equally wide cells. B, upper epidermal cells of midrib showing hair bases. C, stomatal distribution pattern in lower surface of pinna. D, lower epidermis showing stomata and hair base. E-G, topographic transect details. I, part of the section in G to show cellular details and less elongated cells of accessory transfusion tissue. J, more magnified view of vascular bundle of transection in I.

A, B, D & J x 200; C, E-G x 25; H & I x 100.

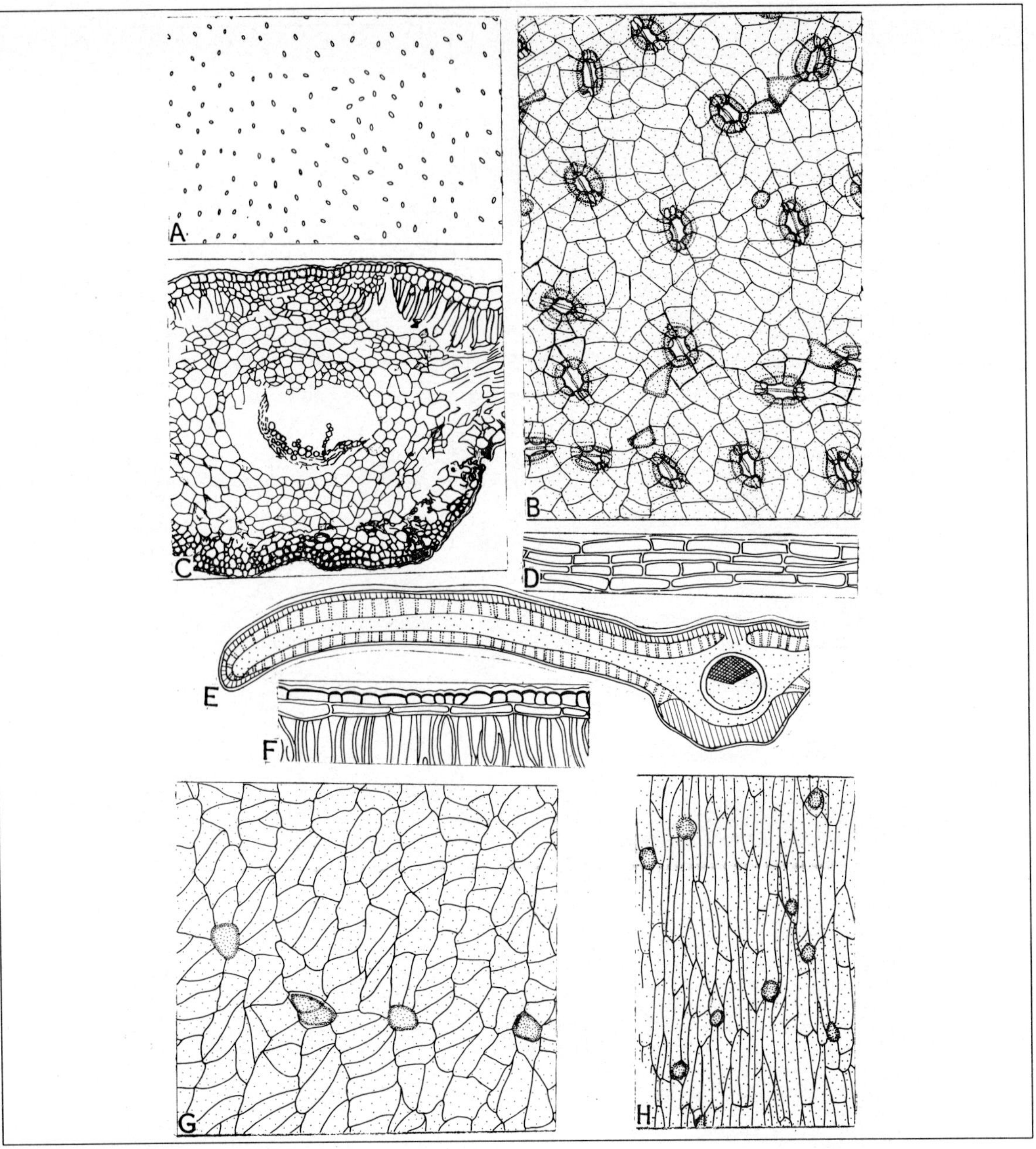

Text Fig. 6. *Cycas normanbyana*. A, stomatal distribution pattern in mid-region of pinna. B, lower epidermis of pinna, showing stomata and hair bases. C, part of transection showing midrib region with multilayered hypodermis on either side. D, macerated patch of hypodermal cells. E, transection of pinna from mid-region. F. portion of longisection of pinna showing epidermal, hypodermal and palisade mesophyll cell. G, upper epidermal cells of pinna showing sporadic rows of equally wide cells and a few hair bases. H, lower epidermis of midrib showing a few hair bases.

A x 35; B, G x 150; D, F x 100; C x 50; H x 125.

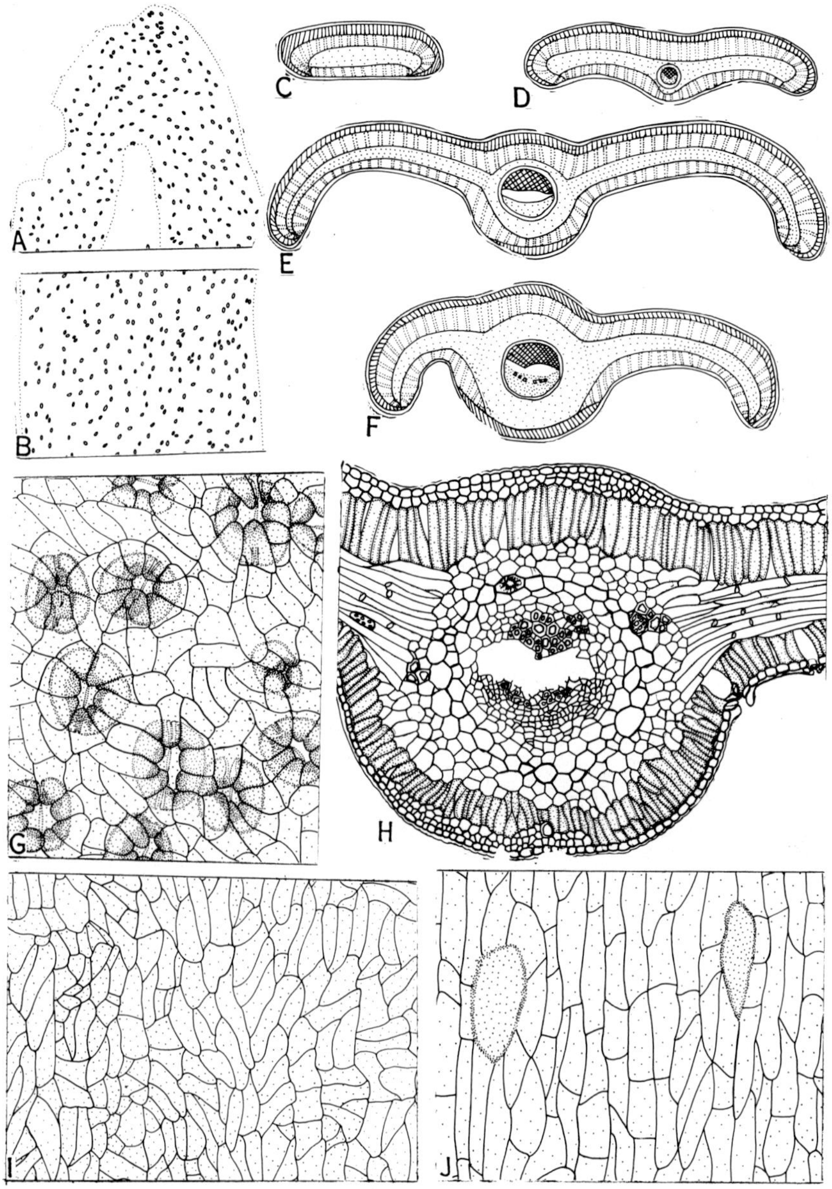

Text Fig. 7. *Cycas pruinosa*. A, stomatal distribution pattern on lower surface of apical region of pinna. Note the dome-like formation at apex due to midrib falling short of apex. B, stomatal distribution pattern in lower surface of mid-region of pinna. C-F, transactions of pinnae; C, through avascular region at apex. B, apical region with vascular bundle. C, mid-region. D, near basal region (all topographic). G, lower epidermis of lamina showing near midrib. I, upper epidermis of pinna showing sporadic rows of equally wide cells. J, upper epidermal cells of midrib.

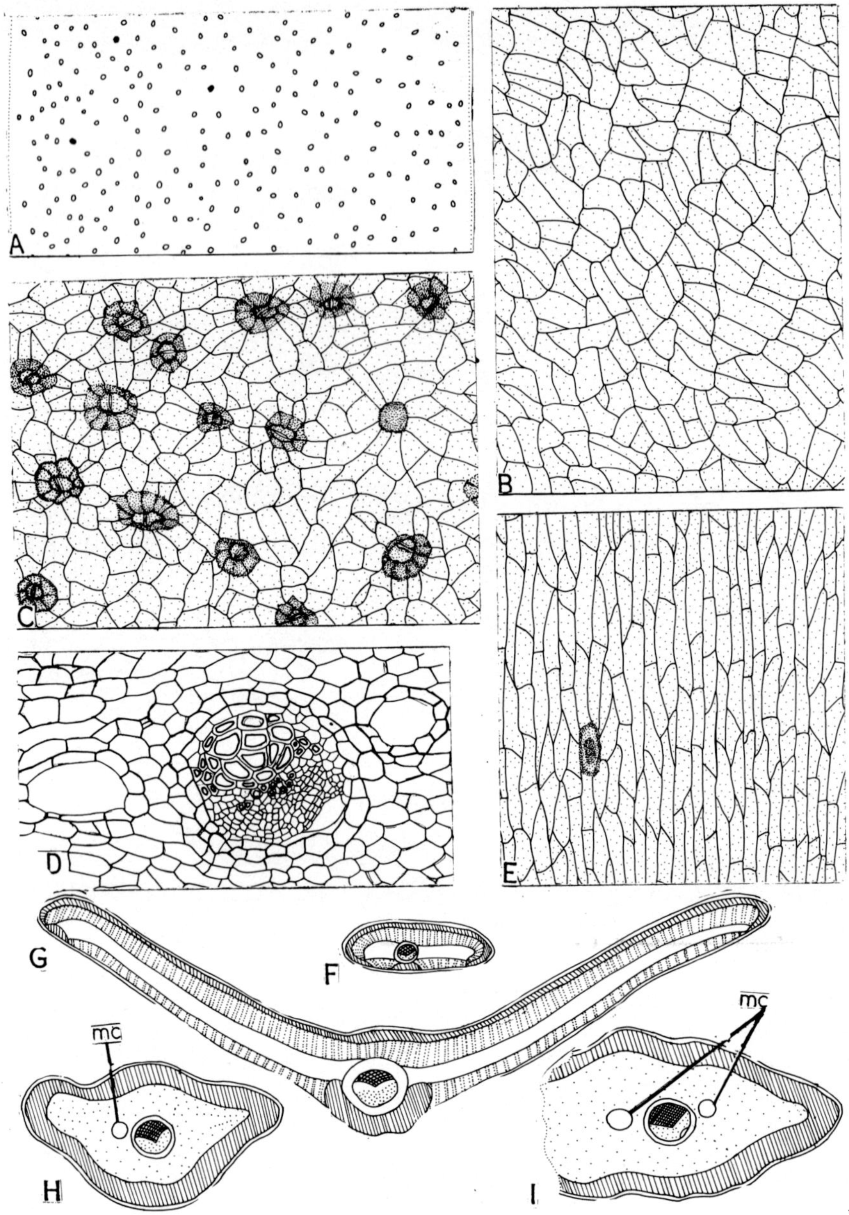

Text Fig. 8. *Cycas taiwaniana.* A, stomatal distribution in pattern in lower surface of mid-region of pinna. B, upper epidermis showing sporadic groups of equally wide cells. C, portion of lower epidermis showing stomata. D, portion of transection showing vascular bundle and two mucilage canals. E, upper epidermal cells of midrib showing a hair base. F-I, transection of pinna. F, near apex. G, near middle region. H, near base. I, near extreme base.

A, F-I x 25; B-E x 150.

Discussion

The epidermal characters of 13 species of *Cycas* have been dealt with by a number of previous workers (see Pant & Nautiyal, 1963 and references cited therein). The present study includes an account of cuticle and epidermis of eight additional species of the genus, viz. *C. angulata*, *C. armstrongii*, *C. basaltica*, *C. calcicola*, *C. kennedyana*, *C. normanbyana*, *C. pruinosa*, *C. taiwaniana* and their anatomical features. Besides these, we are also describing a few additional anatomical details of the pinnae of a ninth species, *C. beddomei*, which had been dealt with by Pant & Nautiyal (1963). In the present work we have at the outset listed the common epidermal and other anatomical characteristics of various species of *Cycas*. Among these, we find that:

1. Most species of the genus are hypostomatic, the only exceptions known earlier were *C. comorensis* and *C. micholitzii* (Pant & Nautiyal, 1963) and to these we have now *C. armstrongii* whose pinnae are amphistomatic. Besides these, *C. angulata* shows a few stomata on the upper surface of pinnae, although it is essentially hypostomatic.

2. The anticlinal walls of epidermal cells are generally straight but they are sinuous in *C. micholitzii* (Pant & Nautiyal, 1963). Among the newly investigated forms, the cells of the lower epidermis of *C. armstrongii* are slightly undulate.

3. A feature noticed by us for the first time is the sporadic but frequent occurrence of pairs or rows of a few equal upper epidermal cells lying between irregularly arranged unequal cells elongated in different directions. The common intervening walls of such equal cells may suggest that they were formed as a result of one or more parallel divisions in a protoderm cell while the epidermal cells were being formed.

4. The arrangement of encircling cells in stomata of *C. taiwaniana* suggested that they were formed by the formation of tangential walls in common mother cells which surrounded the guard cells or their mother cells to give rise to subsidiary cells and encircling cells as was shown by Pant & Mehra (1963) for *C. revoluta* occurrence papillate subsidiary cells in *C. taiwaniana* and *C. revoluta* is common to both species but *C. revoluta* has distinctly revolute margins while the margins of *C. taiwaniana* are flat and show two mucilage canals.

5. Hairs may occur on both sides of the pinnae but in most species they are more common on the upper side. Two species, *C. basaltica* and *C. calcicola*, are exceptional in having greater frequency of hairs on the lower surface. In the lower epidermis, the hair base cells which occur between stomata may be single or paired. The configuration of cells surrounding the hair bases resembles that of stomata and this may support the homologous nature of hairs and stomata (Pant, 1965).

6. The occurrence of a hypodermis of one or more layers of cells has been described by various workers in the pinnae of all species of *Cycas* whose anatomy has been worked out but a feature which no-one has noticed is the longitudinal elongation of its cells in the proximal distal direction of the pinnae. Another peculiar feature of the hypodermal cells is the occurrence of transverse septa in its cells. The longitudinal elongation may be resisting lengthwise bending of pinnae. The transverse septa may indicate transverse divisions in mother cells.

7. The pinnae of *Cycas* tend to be isobilateral with a palisade on the upper side and a palisade-like mesophyll of somewhat shorter cells on the lower side of the lamina. The midrib usually lacks an upper and lower palisade mesophyll.

8. The midrib of *C. pruinosa* is unusual in having an upper palisade as well as a lower palisade-like mesophyll. This species is also unique in having a midrib which falls short of its apex (Plate 5, Fig. 2 & 5; Text fig. 7A).

9. The midrib vascular bundle has multi-layered sclerenchymatous sheath with crystalliferous cells but a hitherto unrecorded feature of the bundle sheath in *Cycas* pinnae which we noticed in the present work is the occurrence of chloroplasts in the cells of the bundle sheath as is mentioned by Laetsch (1974) for grasses. This leads us to suspect that this genus may be a C_4 plant. However, so far no-one has reported the C_4

cycle in any cycads or for that matter in any of the present day gymnosperms, although Pant & Nautiyal (1984) had suspected that *Glossopteris talbragarensis* may have been a C_4 plant. Accordingly, it would be worthwhile examining *Cycas* and other cycads as well as other gymnosperms for the existence of a C_4 dicarboxylic cycle.

10. A character which distinguishes the genus *Cycas* from all other cycads i the usual absence of mucilage canals in its pinnae, although Pant & Nautiyal (1963) had reported an exceptional species, *C. beddomei*, where the midrib was accompanied by an abaxial mucilage canal. The present study has not only confirmed the observation of these authors but has shown that the pinnae of *C. beddomei* have instead of one three independent mucilage canals in their basal parts. Two of the lateral canals end gradually in the basal half of the pinna (at first one and then the other) but the third median abaxial canal ends only after it has traversed three-fourths of the length of a pinna. In addition, we found that *C. taiwaniana* also shows two lateral mucilage canals on either side of its vascular bundle. On the basis of epidermal characters, Pant & Nautiyal (1963) had grouped species of *Cycas* into three categories according to the size of the pits on their surface walls, we find that all the presently investigated species come under their third group, having short cells and in their fourth group having small surface wall pits. Besides enabling us to elucidate the common characters of various investigated species of the genus, the present work has further confirmed the importance of the form and anatomy of pinnae, particularly their epidermis and cuticle, in the taxonomic determinations or the confirmation of the determinations of various species of *Cycas*. A key based on pinna characters is mentioned below but we must add that the pinna and epidermal features may be used only to conform the identifications of the species of *Cycas* and not be regarded as the sold criteria for their determination.

Acknowledgements

Material of the pinnae of some species of *Cycas* used in this study was collected by the senior author and Prof. D.D. Nautiyal in 1982 at the Royal Botanic Garden and the national Herbarium, Sydney through the kind courtesy of the then Director, Dr L.A.S. Johnson. The authors are thankful to Dr Johnson for arranging to send them more material. They also thank Mr John L. Dowe, Mr Joe Perner and Mr Ken Hill of Australia for arranging to send them material of *Cycas* leaves. They are deeply indebted to Prof. D.D. Nautiyal for giving them laboratory facilities and his manifold help in their work. They thank too Dr D.K. Chauhan for his ready assistance in various ways.

References

***Bailey, F.M.*1902.** CXXIV. Cycadaceae. Queensland Flora, pp.1501-1502.

***Chamberlain, C.J.* 1932.** Methods in plants histology. Chicago, University of Chicago Press, 5th rev. ed.

***Coulter, I.M.* 1910.** "Morphology of Gymnosperms", Chicago.

***Chamberlain, C.J., Laetsch, W.M.* 1974.** The C_4 syndrome: A structural analysis. Ann. Rev. Plant Physiol, 25:27-52.

***Lamb, M.A.* 1923.** Leaflets of Cycadaceae. Bot. Gaz., 76 185-202.

***Maconochie, J.R.* 1978.** Two new species of *Cycas* from Northern Australia. J. Adelaide Bot. Gard., 1(3) 175-178.

***Miquel, F.A.W.* 1868-69.** Nieuwe bijdragen lot de kennis der Cycadeen/Verslagen Meded. Af. Natuurk. Kon. Akad. Wetensch. Ser., 2.3(1):1-57.

***Pant, D.D.* 1965.** On the ontogeny of stomata and other homologous structures. Plant Sci. Series Allahabad, 1:1-24.

***Pant, D.D.* 1987.** The fossil history and phylogeny of Cycadales. International Medal Award Lecture delivered on Nov. 28, at the Palaeobotanical Society, Geophytology, 17(2):125-162.

***Pant, D.D.* 1973.** *Cycas and Cycadales.* Central Book Depot, Allahabad.

***Pant, D.D. & Mehra, B.* 1963.** Development of stomata in leaves of three species of *Cycas* and *Ginkgo biloba*, L. J. Linn. Bot. Soc., 58:491-497.

***Pant, D.D. & Nautiyal, D.D.* 1963.** Cuticular and epidermal studies of some modern cycadean leaves, sporangia and seeds. Senck. Biol., 44(4):257-347.

1984. Notes on *Glossopteris talbragerensis* (Etheridge Jr.) comb. nov. and some other Australian glossopterids. Phyta, 4, 5:33-37.

***Schuster, J.* 1932.** Cycadaceae in A. Engler, Das Pflanzenreich, 4.1. Leipzig.

***Stevenson, D.W.* 1981.** Observations on ptyxis, phenology and trichomes in the Cycadales and their systematic implications. Amer. J. Bot., 68(8): 1104-1114.

An Exploratory Study of the In Vitro Propagation of Cycads

P.M. Resslar
Virginia Wesleyan College, Wesleyan Drive,
Norfolk, Virginia 23502 - 5599, U.S.A.

Abstract

Cycads, like other gymnosperms, have proved to be difficult plants to propagate using the techniques of tissue culture. Since foliar material is easily collected and, probably more importantly, its removal causes little damage to the plants, explants made from leaflets were used exclusively in this study. Foliar explants of several different species of cycad - *Bowenia spectabilis, Ceratozamia kuesteriana, Cycas circinalis, Microcycas calocoma, Stangeria eriopus, Zamia fischeri,* and *Z. integrifolia* - were placed on several different media. Calli were obtained for all species except *C. circinalis* and *S. eriopus*. Media that do not contain inorganic compounds with ammonium ions (e.g., White's Salt Base) appeared to give better results than media containing inorganic compounds with ammonium ions (e.g., Murashige & Skoög's). Because of the initial results, only media using White's Salt Base were used in recent cultures.

Preliminary results suggest that explants on a medium of White's Salt Base, sucrose, casein hydrolysate, inositol, and thiamine produce more and healthier calli than other media used. Present work is focused on varying concentrations of sucrose, vitamins, kinetin, and auxins; sources of nitrogen; and the addition of extra sodium phosphate to the medium.

Introduction

Since many species of cycads are rare, threatened, endangered, or extinct in their natural habitat and since they have recently become more popular in the horticultural trade, *in vitro* propagation of cycads, if developed, could be an important means of collecting and distributing germplasm among botanical gardens, and, also, providing larger numbers of plants and making more species available to gardeners and collectors. Osborne (1990) has reviewed the work done on the *in vitro* propagation of cycads.

This paper is a preliminary report on a continuing study to find a method or methods of *in vitro* propagation for cycads using explants that can be obtained by doing minimal damage to the plant.

Materials and Methods

Plants of *Bowenia spectabilis* Hook ex Hook. f., *Ceratozamia kuesteriana* Regel, *Cycas circinalis* L., *Encephalartos ferox* Bertol f., *Microcycas calocoma* (Miq.) A. DC., *Stangeria eriopus* (Kunze) Baillon, *Zamia fischeri* Miq., and *Z. integrifolia* L. F. (syn., *Z. floridana* A. DC. and *Z. pumila* L. ssp. *pumila*), grown in the greenhouse, were used in this study.

Leaflets, both juvenile and mature, were removed from the leaves and put through a rigorous process of disinfection. This process is outlined in Table 1. Sections of leaflet about 1 to 1.5cm long were cut from the disinfected leaflets. In species with broad leaflets, the explants were cut to a width of about 1cm.

Explants were placed in culture tubes containing 15ml of one of several media tested[1]. The tested media included Anderson's Rhododendron Starting Medium (Anonymous, 1986), Carrot Callus Medium (Anonymous, 1986), Lily Multiplication Medium (Stimart and Ascher, 1978), Murashige's Draceana Starting Medium (Miller and Murashige, 1976), Murashige's Syngonium Starting Medium (Miller and Murashige, 1976), Murashige and Skoög's Salt Base (Murashige and Skoög, 1962), Tobacco Callus Medium (Anonymous, 1986), and Woody Plant Medium (Anonymous, 1986). Explants were also placed on several media using White's Salt Base (White, 1964) with the addition of sucrose (5,300 - 30,000mg/1); inositol (100mg/1); thiamine (0.4 - 0.5mg/1); kinetin (1 - 15mg/1) or 2iP (20mg/1); 2,4-D (0.5 - 10mg/1), IAA (5 - 10mg/1), or NAA (5mg/1); sodium phosphate, monobasic [NaH_2PO_4 - H_2O] (143.5mg/1); and casein hydrolysate (400 - 500mg/1). All media, except for the Anderson's Starting Medium which had a pH of 4.9 and was a liquid, were adjusted to a pH of 5.7 and solidified with 8g of agar per litre of medium.

Table 1

Procedure used to Disinfect Leaflets

1. Leaflets washed in tap water and Tween 20 for 5 min.
2. Rinsed with tap water.
3. Rinsed with 70% ethanol or isopropyl alcohol for 30 sec.
4. Washed with 1.05% sodium hypochlorite* and a couple of drops of Tween 20.
5. Rinsed several times with sterile distilled water.
*The 1.05% solution of sodium hypochlorite was prepared by making a 20% solution of a commercial bleach that was 5.25% sodium hypochlorite.

Cultures were incubated in a growth chamber with the temperature maintained at 25ºC. Incubation in both a dark chamber and a chamber with a photoperiod of 18 hrs. of light and 6 hrs. of dark was tested. Cultures were observed regularly and those showing signs of contamination discarded.

Results and Discussion

These studies were initiated to develop techniques in *in vitro* propagation that would result in minimal damage to the maternal plant. For this reason, the use of explants from the stem or the root was rejected as being too destructive. Foliar material was determined to be the best part of the plant from which to make explants. Leaflets, and for that matter entire leaves, could be removed form mature plants with little or no long term damage to the maternal plant. Leaflets were chosen over petioles for two reasons: firstly, the petiole is extremely variable in diameter. In some species (e.g., *Z. fischeri*), the petiole has a diameter of a few millimetres, while others (e.g., *Encephalartos* spp.) have petioles with a diameter of more than 20mm. Secondly, the obdurate nature of the petiole, particu-

[1]The basic media used during this study were formulations made by Carolina Biological Supply Co., Burlington, North Carolina, U.S.A.

larly in the species with thicker petioles, makes it very difficult to dissect the petiole into small sections.

To date, cultures using White's Salt Base with a few additives (Table 2) placed in a chamber with a photoperiod of 18 hrs. of light and 6 hrs. of dark have given the best results. Callus visible to the naked eye formed in less than 3 weeks. Explants of *B.spectabilis, C. kuesteriana, M. calocoma, Z. fischeri*, and *Z. integrifolia* formed callus on media using White's Salt Base. Callus did not form on explants of *C. circinalis* or *S. eriopus*. The latter two species are to be retested to determine if they will form callus on the media containing White's Salt Base. The formation of callus was very poor or did not occur on the other media tested. One possible explanation could be the use of inorganic compounds with ammonium ions as a source of nitrogen. All of the media tested, with the exception of White's Salt Base, contain ammonium ions. It is possible that these ions are toxic to the cells of foliar explants. The results reported here using foliar explants differ from some of the earlier reports where explants from roots or stems were used. Koeleman and Small (1982) reported growth of callus from explants of stems and roots, taken from several species of *Encephalartos*, placed on media containing inorganic sources of ammonium, but they reported the best growth of callus on a modification of a medium designed N59. N59 (Norstog, 1967; Norstog and Rhamstine, 1967) is based on White's Salt Base. This study confirms some of the work reported by Norstog (Norstog, 1967; Norstog and Rhamstine, 1967) who also used a modification of White's medium and found that cultures of *Zamia integrifolia* grew better on media with reduced nitrogen. Present studies are also investigating the use of another salt base, Heller's (Heller, 1953), which like White's Salt Base, contains no inorganic compounds with ammonium ions.

The production of callus does not appear to be influenced greatly by the age of the leaflet. Leaflets taken from leaves of any maturity do form callus. Cultures made from leaflets of *Z. fischeri* that were at least one year old seem to produce callus as readily as leaflets that have just completed expansion. Whether the age of the leaflet influences how readily plantlets will form is, of course, not yet known.

Table 2

The formation of medium showing the best formation of Callus*

Whites Salt Base	**mg/1**
H_3BO_3	1.50
$Ca(NO_3)_2 . 4H_2O$	300.00
$Fe_2(SO_4)_3$	2.50
$MgSO_4 . 7H_2O$	720.00
$MnSO_4 . 4H_2O$	7.00
KC1	65.00
KI	0.75
KNO_3	80.00
$NaH_2PO_4 . H_2O$	16.50
Na_2SO_4	200.00
$ZnSO_4 . 7H_2O$	3.00
Addition to Salt Base	
Sucrose	5,000.00
$NaH_2PO_4 . H_2O$ (additional)	143.50
Inositol	100.00
Thiamine HC1	0.40
Kinetin	2.00
2,4-D	0.50
Casein hydrolysate	500.00
Agar	8,000.00

*The pH of the medium was adjusted to 5.7

Originally, media with relatively high concentrations of sucrose (30,000mg/1) were used. Later formulations of the media used much lower concentrations (5,000mg/1). Callus that formed on the media with lower concentrations of sucrose appeared healthier and grew more rapidly.

In general, the results of this study suggest that the medium or media that will be used in the successful culturing of cycads will be one with low concentrations of the necessary nutrients and, probably, one with little or no ammonium ions. Present work, using media prepared at one-half the usual concentration, is testing this hypothesis.

References

***Anonymous*, 1978.** Carolina Plant Tissue Culture Media Formulation Booklet. Carolina Biological Supply Co., Burlington, N.C.

***Heller, R.* 1953.** Recherches sur la nutrition minrale des tissus vgtaux cultivs *in vitro*. Ann. Sic. Nat. Bot. Biol. Veg. 14: 1 - 223.

***Koeleman, A, and Small, J.G.C.* 1982.** A note on callus formation by stem and root tissue of some *Encephalartos* species. S. Afr. J. Bot. 1(4): 165 - 166.

***Miller, L.R., and Murashige, R.* 1976.** Tissue culture propagation of tropical foliage plants. In Vitro 12: 796 - 813.

***Murashige, T., and Skoög, F.* 1962.** A revised medium for rapid growth and bioassay with tobacco tissue culture. Physiol. Plant. 15: 473 - 497.

***Norstog, K.* 1967.** Some characteristics of growth and development in cycad tissue cultures. Fifth Conference on cycad Toxicity, Miami Florida. IV - 1 - - IV - 8.

***Norstog, K., and Rhamstine, E.* 1967.** Isolation and culture of haploid and diploid cycad tissues. Phytomorphology 17: 374 - 381.

***Osborne, R.* 1990.** Micropropagation in cycads. Mem. N.Y. Bot. Gard. 57: 82 - 88.

***Osborne, R., and Ascher, P.D.* 1978.** Tissue culture of bulb scale sections for asexual propagation of *Lilium longiflorum* Thunb. J. Amer. Soc. Hort. Sci. 103(2): 182 - 184.

***White, P.R.* 1963.** The Cultivation of Animal and Plant Cells, 2. ed. Ronald Press Co., New York.

Symbiosis in Cycads With Special Reference to *Macrozamia communis*

Ian A. Staff and Charmian P. Ahern
Botany Department, LaTrobe University
Bundoora, Victoria 3083, Australia

Abstract

Cycads have several types of roots - taproots, lateral roots and special roots that often grow upwards and contain cyanobacteria (blue-green algae) in a symbiotic relationship. These roots may be papillose, with a covering of small scales, or coralloid, with a covering of cork and many lenticels. The cyanobacteria are situated in a ring in the cortex and their entry into the roots has been observed rarely. Details of the entry of cyanobacteria into roots of the Australian cycad *Macrozamia communis* and their migration through cortical cells and intercellular spaces are shown. Dramatic changes in the host cells in response to root invasion are described, including growth, extension, penetration and death. The ultrastructure of both host cells and cyanobacteria is illustrated. Nitrogen fixation is an important function carried out by heterocysts of the symbiont and a brief history of early studies of this process in cycads is included.

Introduction

The root system of cycads consists of a tuberous tap root, narrow branched lateral roots and special, swollen lateral roots which frequently grow upwards (Fig. 1). Apart from their distinctive appearance, they are unusual because they become invaded by nitrogen-fixing cyanobacteria (blue-green algae) mostly identified as species of *Nostoc* or *Anabaena*, and rarely *Chroococcus* (Plaut, 1910) and *Calothrix* (Grobbelaar et al., 1987). The cyanobacteria are generally confined to a specialised zone in the mid-cortex called the cyanobacterial zone (formerly algal zone) (Figs. 5, 7).

In seedlings of *Macrozamia*, these special roots are first produced endogenously within the short hypocotyl, and are always apogeotropic (growing upwards towards the soil surface) even before invasion by the symbionts and when grown in sterile culture (Lamont and Ryan, 1977), Such roots also develop from lower regions of the contractile tap root, and from normal secondary roots especially when in pot culture (Fig. 1). These roots have been given many names such as 'tubercles, nodules, apogeotropic roots, aerial roots, pneumatophores, and coralloids'. Most of the names used are not suitable; for example, 'apogeotropic roots' (coralloids are often not negatively geotropic -

Fig. 2), 'tubercles' (they are not small tubers), 'nodules' (they are modified lateral roots and not localised, spherical outgrowths of infected roots), 'pneumatophores' (they do not aerate water-logged plants as cycads are frequently located in dry habitats) and 'aerial roots' (they are subterranean and generally stop growing on reaching the soil surface).

The term 'coralloid' is the most appropriate name for these special roots since, following invasion by the cyanobacteria, they develop a coralloid appearance, i.e., they are brittle, with extensive dichotomous branching, and with thick periderm (cork) covered by whorled lenticels (Fig. 2). Prior to cyanobacterial invasion, these roots are negatively geotropic, dichotomously branched and often covered with yellow-brown membranous scales called papillae (Fig. 3). The papillae usually fall off before the coralloid stage is reached. In this paper, uninvaded roots will be referred to as 'precoralloids'. Although many precoralloids mature and degenerate without having reached the coralloid stage, the potential for invasion and coralloid development still exists.

Initiation of the precoralloid roots has been attributed to a variety of causes. The causal agents were believed to be bacteria by Life (1901), Bottomley (1910), Spratt (1915) and McLuckie (1922); while Schaede (1944) suggested that cyanobacteria were the necessary stimulants, and light and aeration were believed to be effective by Horejs (1910) and Spratt (1915). Lamont and Ryan (1977) and Webb (1983) showed that precoralloids can be produced by *Macrozamia* seedlings grown in sterile culture in the dark, confirming the original suggestion of Wittman, Bergersen and Kennedy (1965) that these roots are an inherent part of the plant. Subsequent experiments by Webb and Bildemback (1975) and Webb (1981, 1983) demonstrated that light was responsible for 'nodulation' (precoralloid formation) in *Zamia* and *Macrozamia*, and that in *Zamia* white light is the most effective stimulant, followed by blue, green, and then red light.

The occurrence of cyanobacteria in cycad roots was first described by Reinke (1872) who proposed that the symbiont entered through injured parts of the roots. Subsequently, various authors suggested alternative hypotheses for the mode of entry; e.g., by lenticels (Horejs, 1910; Spratt, 1915; Chaudhuri and Akhtar, 1931), the papillose sheath (Pearson, 1898; Schaede, 1944; Wittman et al., 1965) or breaks in the dermal layer (Schneider, 1894; Watanabe, 1924; Nathanielsz and Staff, 1975a; Milindasuta, 1975). The only clear evidence of a continuous channel from the epidermis to the cyanobacterial zone has been provided by Nathanielsz and Staff (1975a) and this initial invasion occurred on a precoralloid root several centimetres from the root tip at a place where no papillae or lenticels were present.

The function of coralloid roots has long been a matter of great interest. Life (1901) believed that these roots contributed to the aeration of the plant by means of their lenticels and associated air passages. The now well-established importance of coralloids as a source of nitrogen fixation originated with the growth experiments of Horejs (1910) and Schuster (1932) Horejs (1910) observed that the presence of intact leaves on plants was correlated with the presence of the live cyanobacteria in the roots, while damaged or incomplete leaves were associated with plants having no cyanobacteria in the roots. Schuster (1932) found that new leaf size was greatly reduced in plants whose coralloid roots had been removed.

Nitrogen fixation by isolated cyanobacteria was first demonstrated by Winter (1935) and confirmed by Schaede (1944) and Douin (1953). Watanabe and Kyohara (1963) showed the incorporation of $N_2 15$ in cultured cyanobacteria isolated from the coralloids, while Bergersen, Kennedy and Wittman (1965) and Renaut, et al., (1975) showed that coralloids containing cyanobacteria were able to fix $N_2 15$, and that the nitrogen so incorporated was transmitted to other parts of the plant. Coralloids devoid of the cyanobacteria could not fix $N_2 15$. Halliday and Pate (1976) demonstrated that physiologically significant levels of nitrogen were fixed by stands of *Macrozamia reidlei* in Western Australia. Hoare et al., (1971) observed dark heterotrophic growth of cyanobacteria (that remained green) in coralloid roots, and inferred that the cyanobacteria derived carbohydrates from their host plants. The symbiotic relationship thus involves provision by

B
A
B
1
L
2
3
PS
CZ
S
4
P
CZ
S
5
PS
6
7
RC
8

the cycad of carbon compounds which are utilised by the cyanobacteria, and provision by the cyanobacteria of nitrogen compounds which are utilised by the cycad.

Methods and Materials

Seeds and seedlings of *M. communis* were collected from Narooma and Kangaroo Valley in New South Wales, Australia, and were cultivated in a glasshouse. Fixation of specimens was carried out both in the field and the laboratory. Standard glutaraldehyde/osmium fixation was used followed by dehydration with either a graded acetone or ethanol-epoxypropane series. Spurr's low viscosity embedding medium (Spurr, 1969) was used for infiltration and embedding. Ultrathin sections for electron microscopy were mounted on formvar/carbon coated copper grids and stained with uranyl acetate and lead citrate and thiosemicarbazide (TSC) for acid mucopolysaccharides (Thiery, 1967). Stains for light microscopy included 1% toluidine blue, ruthenium red and alkaline hydroxylamine for the localisation of pectins, Sudan IV for suberin, acid phloroglucinol for lignin (Jensen, 1962), and alcian blue Gx for mucilages (Curran, 1964).

Results and Discussions

Figure 1, 2 and 3 illustrate the strikingly different morphology of normal lateral roots, precoralloids and coralloids of *M. communis*. The lateral roots are long and narrow, while the precoralloids and coralloids are short and swollen. A remarkable feature of precoralloids and coralloids is the specialised cyanobacterial zone (CZ) which forms a hollow cylinder in the middle of the cortex and extends to the root tip (Figs. 4, 5). In the precoralloids, the CZ is barely visible (Figs. 4, 6) but in the coralloids it forms a conspicuous part of the root (Figs, 5,7). It is interesting to note that this zone, which is destined to be colonised by the symbiont, is present in precoralloids long before invasion occurs, and is not found in normal lateral roots (Fig. 8).

The precise timing and location of invasion is unpredictable, and there are no visible external signs indicating the onset of invasion. It is apparently very rapid, and its location is masked by transformation of the root into a coralloid where the endophyte is confined to the mid-cortical CZ. Locating the site of the initial invasion is therefore very difficult.

From over 2,000 root branches of approximately 300 seedlings examined, only three specimens showed initial invasion. In all three specimens, the

←

Figures 1 - 8: Coralloid Roots

Figure 1: Seedling of *M. communis* showing special roots: (A) derived from the hypocotyl-root axis; and, (B) derived from a lateral root. Bar = 3cm.

Figure 2: Coralloid roots from the hypocotyl of a *M. communis* seedling. Note the positive geotropism of root tips, dichotomous branching and abundant lenticels - L. Bar = 1cm.

Figure 3: Papillose root tips of *M. communis* emerging from the soil surface near the leaf bases. Bar = 1cm.

Figure 4: Papillose precoralloid root - median longitudinal section (M.L.S.): PS - papillose sheath; CZ - undifferentiated cyanobacterial zone and S-stele. Bar = 300μm.

Figure 5: Coralloid root - MLS showing mature CZ containing darkly-stained mucilage and cyanobacteria: S-stele; P-periderm. Bar = 300μm.

Figure 6: Uninvaded precoralloid root - transverse section (T.S.) showing poorly differentiated CZ: papillose sheath. Bar = 300μm.

Figure 7: Coralloid root - TS showing well-differentiated CZ. Bar = 300μm.

Figure 8: M.L.S. normal lateral root having no CZ: RC-conical root cap. Bar = 200μm.

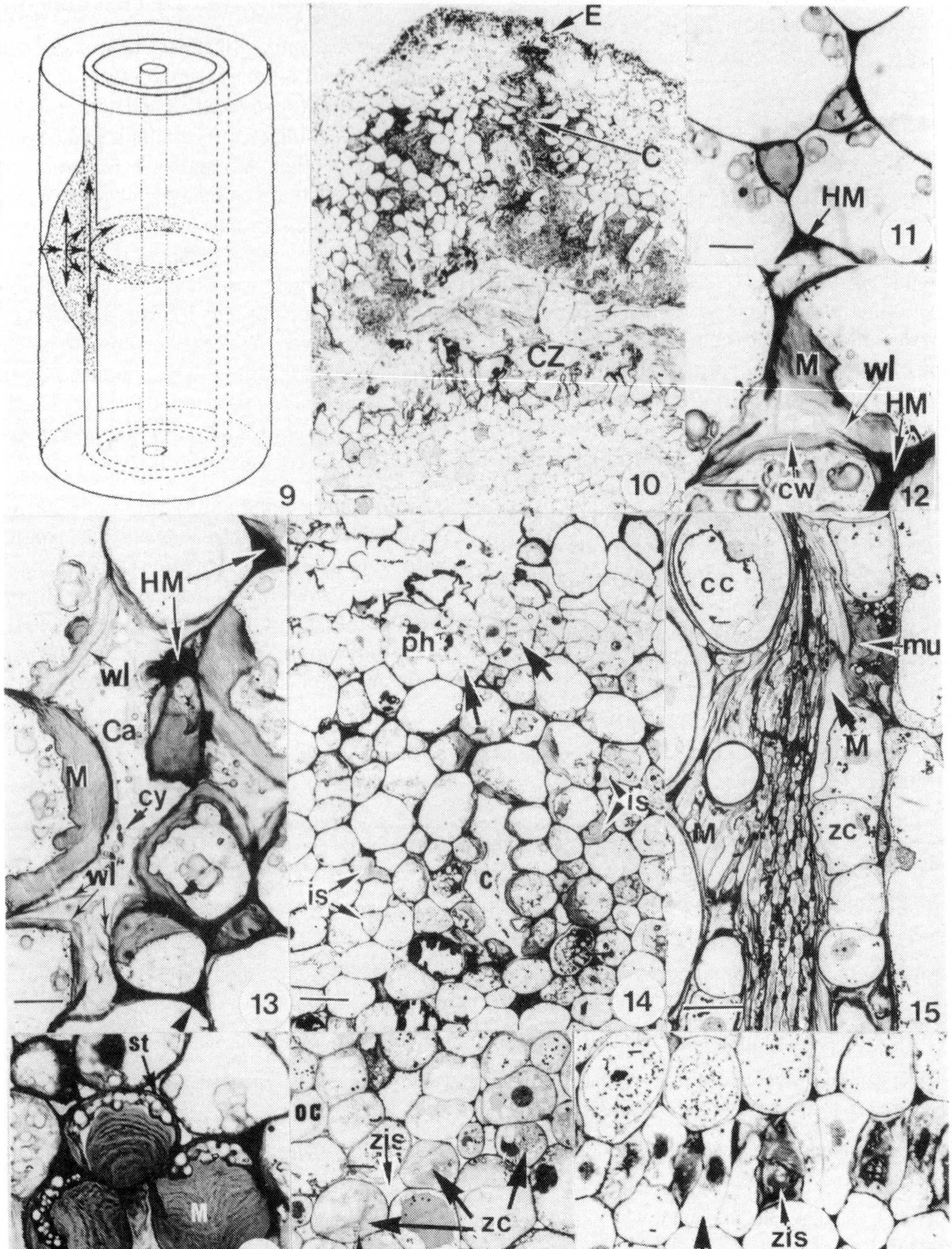

E
C
HM
11
CZ
M
wl
HM
cw
9
10
12
HM
cc
mu
wl
Ca
ph
M
M
cy
is
M
zc
is
wl
c
13
14
15
st
oc
zis
M
zc
zis
16
st
ic 17
zc
18

outer papillose sheath had been replaced by a think dermal layer. In all specimens examined, and in our previous study (Nathanielsz and Staff 1975a), initial invasion involved penetration through the dermal layer at a point on the root axis distal from the tip. Serial sectioning revealed that the endophyte, mucilage and host-cell response were restricted to a localised region of the cortex and CZ. Figure 9 is a diagrammatic reconstruction of the initial invasion zone showing the localised migratory pathway of the endophyte. Following entry through the dermal layer, the cyanobacteria migrate in all directions through the outer cortex to the hollow, cylindrical CZ. Here, they migrate upwards, downwards and laterally, but not inwards into the inner cortex. Examination of these roots confirmed similar findings by Nathanielsz and Staff (1975a) (Fig. 10).

During invasion, the middle lamella between the cortical cells dissolves and, as the cells separate, mucilage penetrates these enlarged intercellular spaces (Fig. 11). Cortical cell walls facing these mucilage-filled spaces swell as they themselves become impregnated with mucilage and split into layers which later rupture or dissolve (Fig. 12). As these processes continue, small cavities are formed (Fig. 13) which later merge to form one or more channels through the outer cortex (Fig. 14).

In the invasion zone, numerous intercellular spaces are filled with mucilage (stained blue and alcian blue Gx) forming a mucilage invasion front which precedes invasion by the endophyte (Fig. 14). Other intercellular spaces are filled with a matrix material, presumably of host origin, which has the same staining characteristics as the middle lamella (Figs 11 - 13). Mucilage may penetrate intercellular spaces that have not been filled with host matrix material. However, where this material is present, penetration by mucilage is apparently restricted. This supports our previous hypothesis

←

Figures 9 - 18: Root Invasion

Figure 9: Diagrammatic reconstruction of the initial invasion zone in *M. communis*. Arrows indicate the pathway of *Nostoc* migration.

Figure 10: T.S. of a precoralloid root showing initial invasion: C-channel; CZ-cyanobacterial zone; E-entry point. Bar - 10μm.

Figure 11: First stage of invasion showing mucilage-filled intercellular spaces in the outer cortex of the initial invasion zone; HM-host mucilage. Bar = 20μm.

Figure 12: Next stage of invasion: M-mucilage penetration of cell; CW-mucilaginous cell wall; wl-ruptured wall layers. HM-host mucilage. Bar = 20μm.

Figure 13: Later stage of invasion: Ca-cavity formed by separation and breakdown of cells; wl-wall layers; M-mucilage; cy-cyanobacteria, HM-host mucilage. Bar = 30μm.

Figure 14: Cortical invasion channel (C) in outer cortex; ph-phellogen; is-mucilage-filled intercellular spaces (mucilage invasion front) Bar = 50μm.

Figure 15: LS of CZ showing filamentous cyanobacteria in the enlarged intercellular spaces: cc-cortical cell; zc-zone cell; M-mucilage; mu-mucoid cell. Bar = 50μm.

Figure 16: Mucoid CZ cells; M-mucilage; st-starch. Bar = 20μm.

Figure 17: Uninvaded region of CZ in a newly invaded root; zis-zone intercellular space; oc-outer cortex; ic-inner cortex; st-starch. Bar = 50μm.

Figure 18: Invaded region of CZ showing radially elongated CZ cells; zis-zone intercellular space. Bar = 50μm.

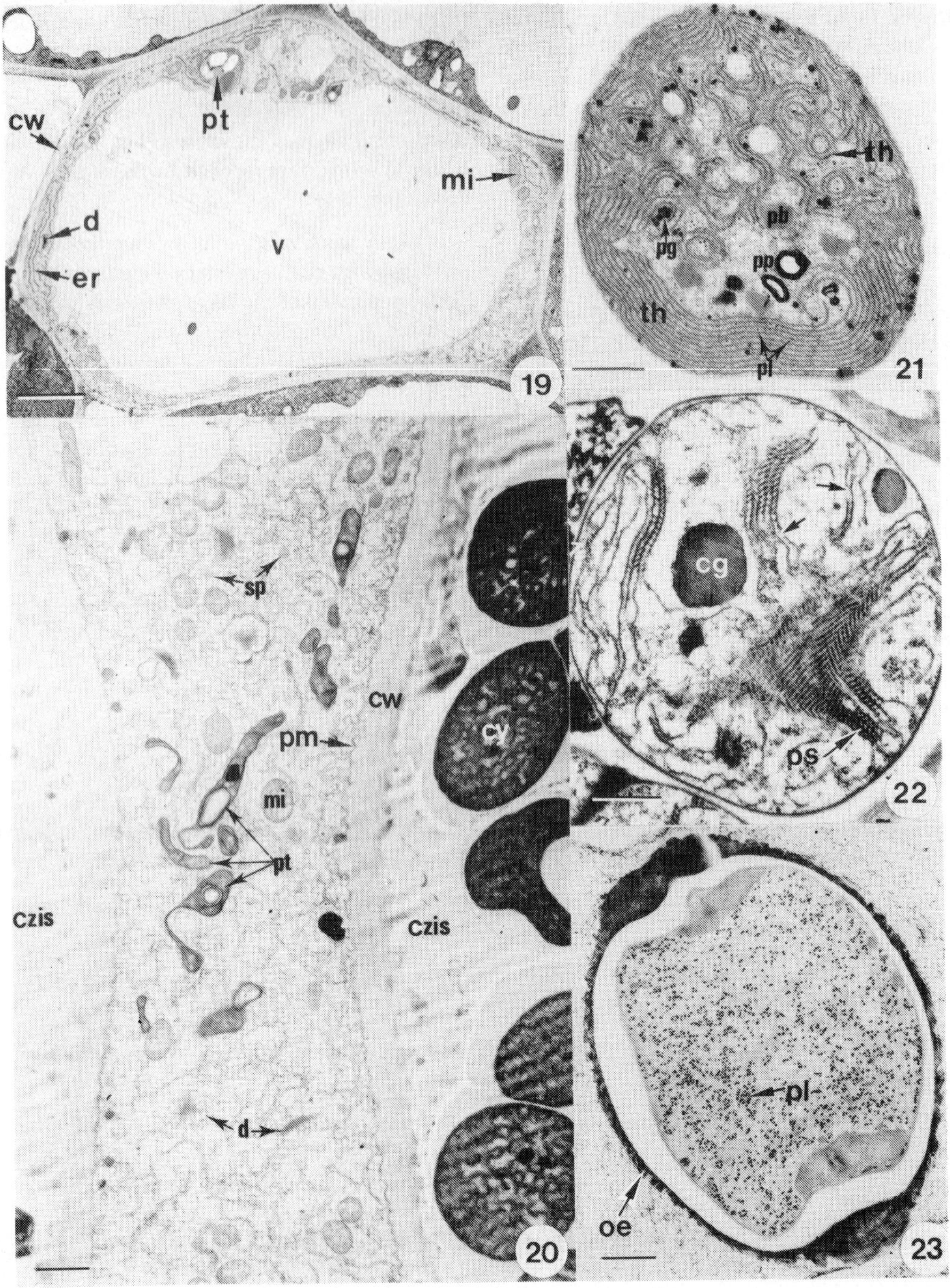
cw
pt
mi
d
v
er
19
th
pb
pg
pp
th
pl
21
sp
cw
pm
cy
mi
pt
czis
czis
d
20
cg
ps
22
pl
oe
23

(Nathanielsz and Staff, 1975a) that the host matrix material is produced by the plant as a defence mechanism against invasion.

On reaching the CZ, *Nostoc* cells migrate through the extensive intercellular space system of this zone (Fig. 15). In the initial invasion zone, extensive mucilage invasion of CZ cells produces characteristic mucoid cells (Fig. 16). Invasion of the CZ typically involves radial elongation of the host cells resulting in a greatly enlarged intercellular space system (Figs. 7, 17, 18).

Whereas the mature CZ in coralloids has been described by numerous workers since Reinke's original discovery (1872) see Nathanielsz and Staff (1975a), the undifferentiated CZ in papillose roots was first detected only recently by Wittman et al., (1965). The few early speculations on the invasion process were attempts to explain the change from an apparently uniform cortex (Fig. 6) to one bearing a mid-cortical zone of radially elongated cells (Fig. 7). It was generally accepted that migration of the endophyte occurred intercellularly via cavities formed by the dissolution of middle lamellae between cells (Meyer, 1966). Our observations show that invasion is a complex process involving a variety of host/endophyte reactions including dissolution of host cell walls and middle lamellae, penetration of cyanobacteria between and into host cells, mucilage penetration of cell walls and cells, cell destruction and cavity formation. As migration occurs away from the initial invasion zone, the CZ cells elongate radially, producing a wide cylinder with an extensive intercellular space system harbouring the abundant mucilage and cyanobacteria.

At the ultrastructural level, marked differences in host cell cytology are apparent between cells of the uninvaded and invaded zones. Cells of the uninvaded zone (Fig. 19) have a large vacuole with sparse organelles distributed peripherally. In mature host CZ cells, however, the presence of abundant mitochondria, Golgi dictyosomes, amyloplasts, tubular endoplasmic reticulum and convoluted plasmalemma (Fig. 20), suggests a greatly increased metabolic activity following invasion. The cell wall appears thickened and mucilaginous. These characteristics, together with their greatly increased surface area due to radial elongation, are typical of transfer cells which are involved in transfer of metabolites between the symbionts (Grilli, 1975; Obukowicz, Schaller and Kennedy, 1981; Lindblad et al., 1985).

Cyanobacteria occur as vegetative cells (Figs. 20, 21, 22) and heterocysts (Fig. 23) in the enlarged intercellular spaces of the CZ. Vegetative cells in the mature zone (Figs. 20, 21) are thin-walled, with abundant, often contorted thylakoids (lamellae). Heterocysts are known to be involved in nitrogen fixation, and in mature regions, heterocyst frequency is high. Ultrastructural features indicate that cyanobacterial metabolism is modified towards provision of nitrogen for the symbiosis and accumulation of carbohydrates through substrate oxidation of carbon available in the

---

←

Figures 19 - 23: Cells of Host and Phycobiont

Figure 19: Immature CZ cell in papillose root of *M. communis*: cw-cell wall; d-dictyosome; er-endoplasmic reticulum; mi-mitochondria; pt-plastid; v-vacuole. Bar = 2μm.

Figure 20: Mature CZ cell in coralloid root: cw-cell wall; cy-cyanobacteria; czis-CZ intercellular space; d-dictyosome; er-endoplasmic reticulum; mi-mitochondria; pm-plasmalemma; pt-plastid; sp=spherosome. Bar = 1μm.

Figure 21: Vegetative *Nostoc* cell from mature zone: pb-polyhedral body; pg-plastoglobuli; pp-polyphosphate granule; pl-polyglucoside granule; th-thylakoid. Bar = 1μm.

Figure 22: Vegetative *Nostoc* cell from initial invasion zone: cg-cyanophycin granule; ps-phycobilisomes; arrows-absence of polyglucoside granules. Bar = 1 μm.

Figure 23: Heterocyst from mature zone. Note reduced internal structure and thickened walls: oe-outer envelope; pl-polyglucoside granules. (TSC staining for acid mucopolysaccharides). Bar = 1μm.

host environment. This is evidenced as a reduction or absence of cyanophycin granules (nitrogen reserves) and phycobilisomes (associated with carbon fixation), but an increase in polyglucoside granules (storage sites of carbohydrates) and polyphosphate bodies (containing phosphorus and calcium, and linked with carbohydrate storage) see Figs. 20, 21 and Grilli Caiola (1974a; 1975a,b). In the initial invasion zone, the vegetative *Nostoc* cells have characteristics resembling free-living forms, namely large cyanophycin granules and numerous phycobilisomes, but few polyglucoside granules and polyphosphate bodies (Fig. 22). Thus it appears that the endophyte undergoes considerable changes associated with a symbiotic mode of life.

Ultrastructure of cultured and symbiotic cyanobacteria from numerous cycads has been studied extensively (Grilli, 1990). Lindblad et al., (1985) showed that cyanobacteria in *Zamia* have the same characteristics as free living organisms; i.e., phycobilisomes and carboxysomes. Obukowicz et al., (1981) found seasonal differences in nitrogen storage with much stored in winter and little in summer, and an association with many degenerating cells also in summer. Thus there is great variation in cyanobacterial ultrastructure that depends on position in root, season of sampling and the species of cyanobacterium.

As heterocysts are the cells in which nitrogen fixation occurs in cyanobacteria, Lindblad and Bergman (1986) sought and showed nitrogenase activity in heterocysts of symbiotic cyanobacteria with the aid of immunolabels. Lindblad (1990) showed maximum nitrogen fixation in the growing root tip, but maximum heterocyst numbers lower down in the root. He also showed assimilation of ammonia into citrulline and glutamine in some cycad species. Peters et al., (1978) suggested that the ammonia-assimilating enzymes are most likely located within the elongated host cells of the *Azolla-Anabaena* symbiosis and it is interesting to speculate that a similar situation may occur in the radially elongated cells of cycad roots. An in-depth treatment of metabolism in cycads appears in a later chapter (Pate, 1991)

References

***Bergersen, F.J., Kennedy, G.S., and Wittman, W.* 1965.** Nitrogen fixation in the coralloid roots of *Macrozamia communis* L. Johnson. Aust. J. Biol. Sci. 18: 1135 - 1142.

***Bottomley, W.B.* 1910.** The association of certain endophytic Cyanophyceae and nitrogen-fixing bacteria. Rep. Brit. Assoc. pp. 786 - 787.

***Chaudhuri, H., and Akhtar, R.* 1931.** The coral-like roots of *Cycas revoluta, C. circinalis* and *Zamia floridana* and the algae inhabiting them. Ind. J. Bot. Sci. 10: 43 - 59.

***Curran, R.C.* 1964.** The histochemistry of mucopolysaccharides. Int. Rev. Cytol. 17: 149 - 212.

***Douin, R.* 1953.** Sur la fixatioon de l'azote libr per les Myxophyces endophytes des Cycadaceies. C.R. Acad. Sci. Paris. 236: 956 - 958.

***Grilli, M., and Trabucchi, B.* 1963.** Ozzervazioni sulle strutture e la infrastrutture dell cellule dei tubercoli radiclai di *Cycas revoluta*. Annal. Fac. Agr. 3: 467 - 481.

***Grilli Caiola, M.* 1974.** A light and electron microscope study of the blue-green algae living either in the coralloid roots of *Macrozamia communis* or isolated in culture. Giorn. Bot. Ital. 108: 161 - 173.

***Grilli Caiola, M.* 1975b.** A light and electron microscopic study of blue-green algae living either in the coralloids of *Dioon edule* or in culture. Phykos 14: 29 - 34.

***Grilli Caiola, M.* 1975.** A light and electron microscopic study of the blue-green algae growing in the coralloid roots of *Encephalartos altensteinii* and in culture. Phycologia 14: 25 - 33.

***Grilli Caiola, M.* 1990.** Structural arrangements of phycobionts in coralloid roots of cycads. Mem. N.Y. Bot. Gard. 57: 94 - 103.

***Grobbelaar, N., Scott, W.E., Hattingh, W., and Marshall, J.* 1987.** The identification of the coralloid root endophytes of the southern African cycads and the ability of the isolates to fix dinitrogen. S. Afr. J. Bot. 53: 111 - 118.

***Halliday, J., and Pate, J.S.* 1976.** Symbiotic nitrogen fixation by blue-green algae in the cycad, *Macrozamia riedlei*: physiological characteristics and ecological significance. Austral. J. Plant Physiol. 3: 349 - 358.

***Hoare, D.S., Ingram, L.O., Thurston, E.L., and Walkup, R.* 1971.** Dark heterotrophic growth of an endophytic blue-green alga. Arch. Mikrobiol. 78: 310 - 321.

***Horejs, I.* 1910.** Symbiotiká rasa siná korenech u *Cycas revoluta*, Ceska Akad. Ved. A Umeni V. Paze 19 Tr. 11: 1 - 31.

***Jensen, W.A,* 1962.** Botanical histochemistry. W.H. Freeman and Co, San Francisco and London.

***Lamont, V.V., and Ryan, R.A.* 1977.** Formation of coralloid roots by cycads under sterile conditions. Phytomorphology 27: 426 - 429.

***Lewis, D.H.* 1973.** Concepts in fungal nutrition and the origin of biotrophy. Biol. Rev. 48: 261 - 278.

***Lewis, D.H.* 1974.** Micro-organisms and plants: the evolution of parasitism and mutualism. Symp. Soc. Gen. Microbiol. 24: 367 - 392.

***Life, A.C.* 1901.** The tuberlike rootlets of *Cycas revoluta*. Bot. Gaz. 31: 265 - 271.

***Lindblad, P., Bergman, B., Hosten, A.V., Hllbom L., and Nylund, J.E.* 1985.** The cyanobacterium-*Zamia* symbiosis: An ultrastructural study. New Phytol. 101: 707 - 716.

***Lindblad, P., and Bergman, B.* 1986.** Glutamine synthetase: activity and localisation in cycanobacteria of the cycads *Cycas revoluta* and *Zamia skinneri*. Planta 169: 1 - 7.

***McLuckie, J.* 1922.** Apogeotropic roots of *Macrozamia spiralis* and their physiological significance. Proc. Linn. Soc. N.S.W. 47: 319 - 328.

***Meyer, F.H.* 1966.** Mycorrhiza and other plant symbioses. *In* "Symbiosis" (Ed. S.M. Henry) Acad. Press. New York.

***Milindasuta, B.* 1975.** Developmental anatomy of coralloid roots in cycads. Amer. J. Bot. 62: 468 - 472.

***Nathanielsz, C.P., and Staff, I.A.* 1975a.** A mode of entry of blue-green algae into the apogeotropic roots of *Macrozamia communis* L. Johnson. Amer. J. Bot. 62: 232 - 235.

***Nathanielsz, C.P., and Staff, I.A.* 1975b.** On the occurrence of intracellular blue-green algae in cortical cells of the apogeotropic roots of *Macrozamia communis* L. Johnson. Ann. Bot. 39: 363 - 368.

***Obukowicz, M., Schaller, M., and Kennedy, G.S.* 1981.** Ultrastructure and phenolic histochemistry of the *Cycas revoluta-Anabaena* symbiosis. New Phytol. 87: 751 - 759.

***Pate, J.S.* 1991.** Biology of the S.W. Australian cycad *Macrozamia riedlei* (Fisch. ex Gaudich) C.A. Gardn. Proc. Int'l. Symp. Biol. Cycads (Cycad 90).

***Pearson, H.H.W.* 1898.** Anatomy of the seedling of *Bowenia spectabilis* Hook. F. Ann. Bot. 12: 475 - 490.

***Peters, G.A., Toia, R.E., Raveed, D., and Levines, N.J.* 1978.** The *Azolla-Anabaena azollae* relationship. New Phytol. 80: 583 - 593.

***Plaut, M.* 1910.** Untersuchungen zur Kenntnis der Physiologischen Scheiden bei den Gymnospermen, Equiseten und Bryophyten. Jahrb. Wiss. Bot. 47: 121 - 185.

***Renaut, J.A. Sasson,A., Pearson, H.W., and Stewart, W.D.P.* 1975.** "Nitrogen fixing algae in Morocco". *In* Nitrogen fixation by free living microorganisms. (Ed. W.D.P. Stewart) IBP Handbook No. 6. pp. 229 - 246, Cambridge Uni. Press, Cambridge.

***Reinke, J.* 1972.** Uber die anatomischen Verhaltnisse einiger Arten von *Gunnera*. Gott. Nachr. 57: 100 - 108.

Schaede, R. **1044.** Uber die korallen Wurzeln der Cycadeen und ihre Symbiose. Planta (Berl.) 34: 98 - 124

Schneider, A. **1894.** Mutualistic symbiosis of algae and bacteria with *Cycas revoluta*. Bot. Gaz. 19: 25 - 32.

Schuster, J. **1932.** Cycadaceae. *In* Engler, A., das Pflanzenreich 4.1, Leipzig.

Spratt, E.R. **1915.** The root-nodules of the Cycadaceae. Ann. Bot. 29: 619 - 626.

Spurr, A.R. **1969.** A low viscosity epoxy resin embedding medium for electron microscopy. J. Ultrastruct. Res. 26: 31 - 43.

Thiery, J.P. **1967.** Mise en evidence des polysaccharides surcoupes fines en microscopie electronique. J. Microscopie 5: 987 - 1018.

Watanabe, K. **1924.** Studien ber die Koralloide von *Cycas revoluta*. Bot. Mag. Tokyo 38: 165 - 187.

Webb, D.T. **1981.** Effects of light on root nodulation of seedlings in sterile culture of *Bowenia serrulata*. Phytomorphology 31: 122 - 124.

Webb, D.T. **1983.** Nodulation in light and dark-grown *Macrozamia communis* L. Johnson seedlings in sterile culture. Ann. Bot. 52: 543 - 547.

Webb, D.T., and Bildenback, D. **1975.** The formation of apogeotropic and nodular roots by *Zamia floridana* in sterile culture. Bot. Soc. Amer. Abstr. Corvallis, Oregon.

Watanabe, A., and Kiyohara, T. **1963.** Symbiotic blue-green algae of lichens, liverworts and cycads. *In* Studies on Microalgae and Photosynthetic Bacteria; Japanese Society of Plant Physiologists, eds. (Special Issue of Plant and Cell Physiology, pp. 189 - 196) Tokyo.

Wittman, W., Bergersen, F.J., and Kennedy, G.S. **1965.** The coralloid roots of *Macrozamia communis* L. Johnson. Aust. J. Biol. Sci. 18: 1129 - 1134.

Possible Role of Cycadales in Formation of Micro-Nuggets of Gold in Soil

Alan F. Wilson (Emeritus Professor)
Department of Geology and Mineralogy
University of Queensland, St. Lucia, Queensland, 4067 Australia

Introduction

Considerable geological and chemical research is trying to identify details of the role of certain plants in aggregating gold atoms into micro-nuggets of gold, or gold-bearing colloids within the upper layers of soils (Baker, 1973, 1978). In near-surface layers of plant litter in hardwood and conifer forests biochemical processes can result in traces of cyanide and thio-cyanate ions during decay of vegetation. These are capable of forming water-soluble gold complexes in aerated conditions (Curtin, Lakin and Hubert, 1970).

In the presence of nitrogen-fixing organisms it is envisaged that, if these could extract nitrogen from the gold cyanide complex rather than from the air, tiny particles of free gold should result. Once sub-micron nuclei of gold and suitable organic complexes of gold have developed, continued biochemical activity is expected to enlarge the gold particles by accretion. A significant feature of the micro-nuggets of very pure gold, which are common in some Australian soils formed above primary gold-bearing lodes or rocks, is their strange shape. They have grotesque or complex mamillated shapes which distinctly differ from the shape of gold particles in quartz veins (Wilson, 1984). However, these shapes are strongly reminiscent of shapes of nitrogen-fixing nodules of various plants (Wilson, 1987).

Possible Role of Cycads

The possible role of cycads in these phenomena is two fold. Firstly; cycads have very well established nitrogen-fixing symbiotic organisms within their prominent coralloid structures. It is postulated that these ought to be able to decompose auriferous nitrogen-bearing complexes. Secondly; the toxins within the cycad 'leaves' and seeds have chemical properties somewhat similar to those of the cyanides. Thus, the litter from cycads should be more effective than, say, the litter from North American conifer or hardwood forests which are known to produce notable amounts of nitrogen-bearing compounds capable of forming water-soluble gold complexes (Curtin et al., 1970).

During the Conference, i.e., Cycad 90, a progress report on the possible role of Cycadales was given. In Australia the most widespread cycads are the macrozamias. Although some exhibit above-ground caudices (trunks), such as *M. moorei* and some *M. riedlei*, most have large sub-surface caudices. Species common in some gold-bearing regions are *M. miquelii* in eastern Queensland, and *M. communis* in eastern New South Wales.

It is postulated that cycads, with long life on fairly level terrain and sub-surface caudices and their wide protective skirt of 'leaves', should be more efficient in conserving in the litter and soil any nitrogen-bearing compounds capable of forming water-soluble gold complexes.

It must be stressed that only soils which already contain free atoms of gold set free from decaying of (say) gold-bearing pyrite or arsenopyrite in certain rocks are candidates for accretion of micro-nuggets. Moreover, it must be noted that it is not yet established whether or not other factors, such as non-cycad related micro-organisms, could be involved in the postulated organic precipitation of discrete particles of gold.

A final word of warning is appropriate. Non-scientific persons commonly comment that the author digs up cycads and finds gold nuggets, much like finding potatoes on the 'roots' of the potato plant! In the interest of cycad conservation it must be stated that the several forms of micro-nuggets which are widespread in Australia are very small. Indeed, they are normally too small to be found by using a panning dish, and are best recognised by very careful study under very high-powered optical microscopes or electron microscopes.

References

***Baker, W.E.* 1973.** The role of humic acids from Tasmanian podzolitic soils in mineral degradation and metal mobilisation. Geochem. Cosmochim. Acta 37: 269 - 281.

***Baker, W.E.* 1978.** The role of humic acid in the transport of gold. Geochem. Cosmochim. Acta 42: 645 - 649.

***Curtin, G.C., Larkin, H.W. and Hubert, A.E.* 1970.** The mobility of gold in mull (forest humus layer) Geol. Surv. Research 1970, U.S. Geol. Surv. Prof. Pap. 700C, C127 - C129.

***Wilson, A.F.* 1984.** Origin of quartz-free gold nuggets and super-gene gold found in laterites and soils - a review and some new observations. Austral. J. Earth Sci. 31: 303 - 316.

***Wilson, A.F.* 1987.** Types of supergene gold to be expected in the Pacific Regions. Proceedings Pacific Rim Congress 87, Queensland, Australasian Inst. Mining and Metallurgy, Parkville, Victoria.

Part Five

Diversity

Molecular Biology in Cycad Systematics

Paolo Caputo, Cecilia Marquis, T. Wurtzel
Department of Biological Sciences
Lehman College, City University of New York
Bedford Park Boulevard West, Bronx, N.Y. 10458 U.S.A.

Dennis W. Stevenson
The New York Botanical Garden,
Bronx, N.Y. 10458 U.S.A.

Eleanore T. Wurtzel
(to whom correspondence should be sent)
Dipartimento di Biologia Vegetale,
Universita' di Napoli 'Federico II', 80139 Napoli. Italy

Abstract

A phylogenetic analysis of cycads was conducted by using chloroplast DNA restriction fragment length polymorphism analysis. *Bowenia spectabilis* Hook. ex Hook. f., *Ceratozamia mexicana* Brongn., *Chigua restrepoi* D. Stevenson, *Cycas revoluta* Thunb., *Dioon edule* Lindl., *Encephalartos natalensis* R.A. Dyer & Verdoorn, *Lepidozamia peroffskyana* Regel, *Macrozamia macdonnelli* (F. Muell. ex. Miq.) A. DC., *Microcycas calocoma* (Miq.) A. DC., *Stangeria eriopus* (Kunze) Baillon, *Zamia vazquezii* D. Stevenson ined. and *Zamia skinneri* Warsz. ex A. Dietrich were used. *Cycas revoluta* ,was used as an outgroup to the rest of cycads. A single most parsimonious phylogenetic tree was obtained in an exhaustive analysis, the topology of which only partly matches previous hypotheses. The exclusion of *Macrozamia* from the data set, on the basis of a possible polyphylesis of the genus, yielded a single most parsimonious phylogenetic tree that is almost completely congruent with previous hypotheses based on morphology.

Introduction

Cycads (Cycadales) are a primitive and isolated group of seed plants consisting of three families (Johnson, 1959; Stevenson, 1990a), 11 genera and approximately 155 species, discontinuously distributed in the pantropics and in subtropical areas around the world. Several morphological features, among which simple ovulate cones, coralloid roots, girdling leaf traces (Crane, 1985; Stevenson, 1990a) and the presence of unique secondary compounds (Siniscalco Gigliano, 1990) support the monophyly of the order. There has been a dearth of information on cycads for many years, due to their rarity and distribution (Stevenson, 1990a). Chamberlain (1935), in his monumental work on gymnosperms gave a full account of their morphology. Later Johnson (1959) attempted a classification, revised by Stevenson (1990a). Recently, Crane (1988) and Stevenson (1990a) undertook phylogenetic analyses of the cycads on a morphological basis, obtaining very similar results. However, the latter author could not objectively decide among three different phylogenetic trees, which differed mainly in the position of the genus *Dioon*.

The still partially unresolved intraordinal phylogeny of cycads, their role in the historical development of seed plants (Loconte and Stevenson, 1990 and references therein), and their endangered status were significant points suggesting that a further study of cycad phylogeny would be of general interest. Instead of relying on morphological features, which are weighted equally regardless of the number of genes involved, we chose a complementary approach utilising a direct analysis of DNA variation. The chloroplast DNA (cpDNA) has shown to be highly conserved on the basis of gene order and DNA sequence (Palmer, 1985 a,b) and consequently has been used extensively for systematic investigations (Palmer, 1987). Accordingly, we performed phylogenetic analyses of chloroplast DNA (cpDNA) restriction fragment length polymorphisms (RFLP) on neotropical cycads (Caputo et al., 1991), the goals of which were to assess the feasibility of a molecular approach to cycad phylogeny. Encouraged by the results obtained, we extended the analysis to all the genera of cycads.

Table 1

Taxa used in the present study

(the three letters abbreviation is the acronym used to label the true terminals)

(BOW)	*Bowenia spectabilis* Hook. ex Hook. f.
(CER)	*Ceratozamia mexicana* Brongn.
(CHI)	*Chigua restrepoi* D. Stevenson.
(CYC)	*Cycas revoluta* Thunb.
(DIO)	*Dioon edule* Lindl.
(ENC)	*Encephalartos natalensis* R.A. Dyer & Verdoorn.
(LEP)	*Lepidozamia peroffskyana* Regel.
(MAC)	*Macrozamia macdonnellii* (F. Muell. ex Miq.) A. DC.
(MIC)	*Microcycas calocoma* (Miq.) A. DC.
(STA)	*Stangeria eriopus* (Kunze) Baillon.
(ZFI)	*Zamia vazquezii* D. Stevenson ined.
(ZSK)	*Zamia skinneri* Warsz. ex A. Dietrich.

Materials and Methods

One representative for each cycad genus was chosen (Table 1). Samples of the already studied neotropical genera were included in the analysis as a control. All leaf samples used in this study came from plants growing at the New York Botanical Garden except that of the recently discovered *Chigua restrepoi* D. Stevenson (Stevenson, 1990b) which is in cultivation at the Fairchild Tropical Garden, Miami, Florida.

DNA was extracted from each taxon following the procedure used in Caputo et al. (Brittonia, submitted); briefly, 3g of leaflets were ground in liquid N_2, lysed in a grinding buffer, treated with solvents to remove proteins, and the resulting DNA concentrated.

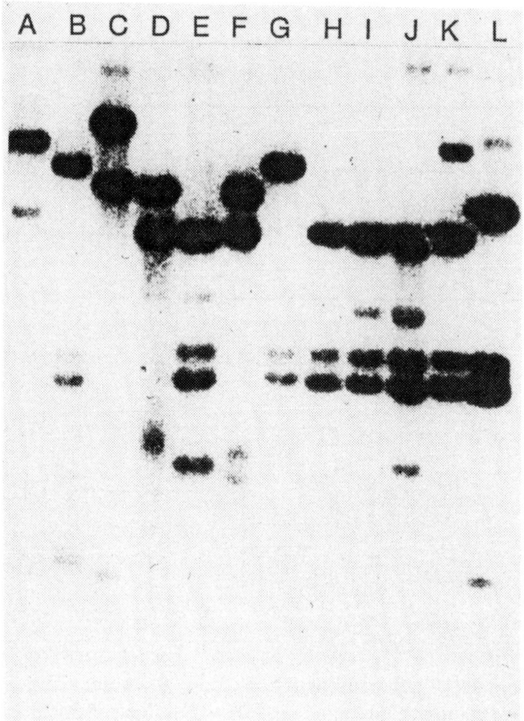

Figure 1: Autoradiogram of the DNAs in study, digested with *Eco*RI restriction endonuclease and hybridised against Ba-9a and Ba-9b clones. Taxa are identified by letters as follows: A, *Cycas revoluta*; B, *Stangeria eriopus*; C, *Bowenia spectabilis*; D, *Encephalartos natalensis*; E, *Macrozamia macdonnellii*; F, *Lepidozamia peroffskyana*; G, *Dioon edule*; H, *Ceratozamia mexicana*; I, *Zamia fischeri*; J, *Z.skinneri*; K, *Chigua restrepoi*; L, *Microcycas calocoma*

DNA samples were digested with several restriction endonucleases, bacterial enzymes which recognise and cleave DNA within a specific six base pair sequence. Digested DNAs were then subjected to agarose gel electrophoresis to separate restriction fragments on the basis of length. Following electrophoresis, the DNAs located within the agarose gel were denatured (i.e., made single-stranded) and transferred by capillary action to nitrocellulose filters (Southern, 1975). These filters were then hybridised against a set of denatured, radioactively labelled DNA probes isolated from clones of a tobacco chloroplast DNA library (Shinozaki et al., 1986). Experimental conditions were chosen in such a manner that tobacco DNA probes would bind to the filter-bound cycad DNA, forming a double stranded hybrid, only if the probe DNA and cycad DNA's shared at least 85% DNA sequence homology. Filters were then washed to remove the excess of unbound tobacco DNA probe and exposed to x-ray sensitive film.

On the resulting autoradiograms, hybridising restriction fragments were scored on a presence/absence basis, and the resulting data matrix was analysed using computer-aided phylogenetic methods. Phylogenetic systematics, or cladistics, is a systematic technique in which the basis tenets are: (a) relationships are implied through synapomorphies (shared derived characters) and (b) phylogenetic hypotheses are developed following the criterion of parsimony (to minimise the number of assumptions of independent gains or losses of characters necessary to support the data). Exhaustive searches for the most parsimonious cladogram(s) were made by using Hennig86 cladistic software (Farris, 1988). To identify which shared restriction fragments are derived characters, the polarity of mutations in restriction sites was determined by outgroup comparison (Stevens, 1980; Watrous and Wheeler, 1981). *Cycas revoluta* was used as an outgroup for the rest of the order. *Cycas* is regarded as the most primitive living cycad genus (Crane, 1988; Stevenson, 1990a) based on both morphology and the presence of the earliest unequivocal fossil among cycads (Zhifeng and Thomas, 1990).

Results and Discussions

Over 200 hybridising restriction fragments were found to be shared by two or more taxa; autapomorphies (fragments unique to one taxon) and symplesiomorphies (fragments shared by all the taxa, including the outgroup) were excluded from the analysis because they are uninformative in terms of assessing relationships. An example of the autoradiograms ana-

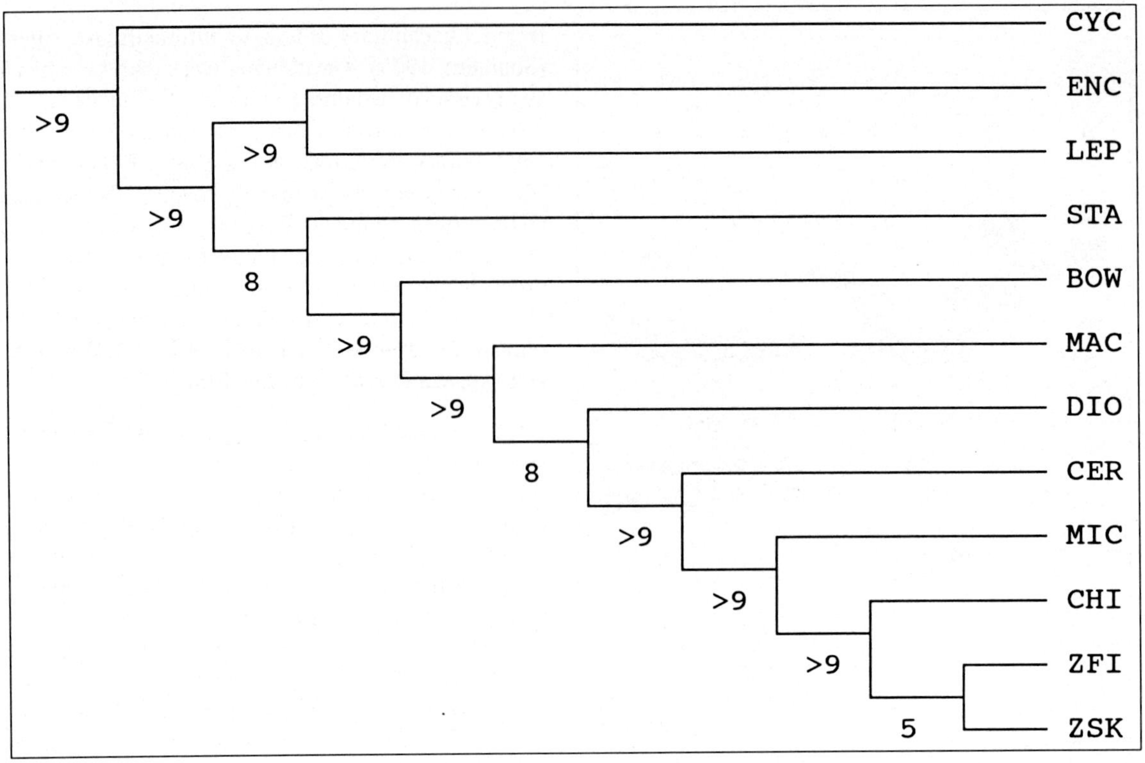

Figure 2: The most parsimonious phylogenetic tree for all the cycad genera (length 505, c.i. 0.50). Abbreviations for the taxa are found in Table 1. The numbers below each internode represent the number of shared derived characters that define the clade on the right.

lysed is shown in Fig. 1. By using *Cycas revoluta* as an outgroup for the rest of cycads, the analysis resulted in a single most parsimonious tree, 505 steps long and with a consistency index (Kluge and Farris, 1969) of 0.50 (Fig. 2). This phylogenetic tree is only partly congruent with previous phylogenetic hypotheses based on morphology. However, the phylogeny of neotropical cycads, the subject of a previous paper (Caputo et al., Brittonia, submitted), has been confirmed here and also with a wider choice of outgroups.

The main discordances between the present phylogenetic hypothesis and the previous ones based on morphological characters (Crane, 1988; Stevenson, 1990a) are the positions of *Encephalartos, Lepidozamia* and *Macrozamia*, from which the polyphylesis of Zamiaceae and the paraphylesis of Stangeriaceae (*sensu* Stevenson, 1990a) would be implied. It is difficult to say, at this point in the research, whether these results are dependent upon the choice of RLFP's as characters for our analysis, the analytical technique used, or the presence of some rearrangement in the chloroplast DNA of *Encephalartos, Lepidozamia*, or *Macrozamia*. Any major rearrangement in cpDNA would probably give rise to restriction fragment patterns dissimilar from those of the other plants in study. However, as the prerequisite for our analysis was that all cycad genera are monophyletic, and *Macrozamia* is questionable (Siniscalco Gigliano et al., in this issue), we attempted further analyses by excluding it from the data matrix. One single most parsimonious phylogenetic tree was obtained, shown in Fig. 3 (length 434, consistency index 0.51), matching the morphological evidence to a greater extend (Fig. 4a), in showing the monophylesis of both

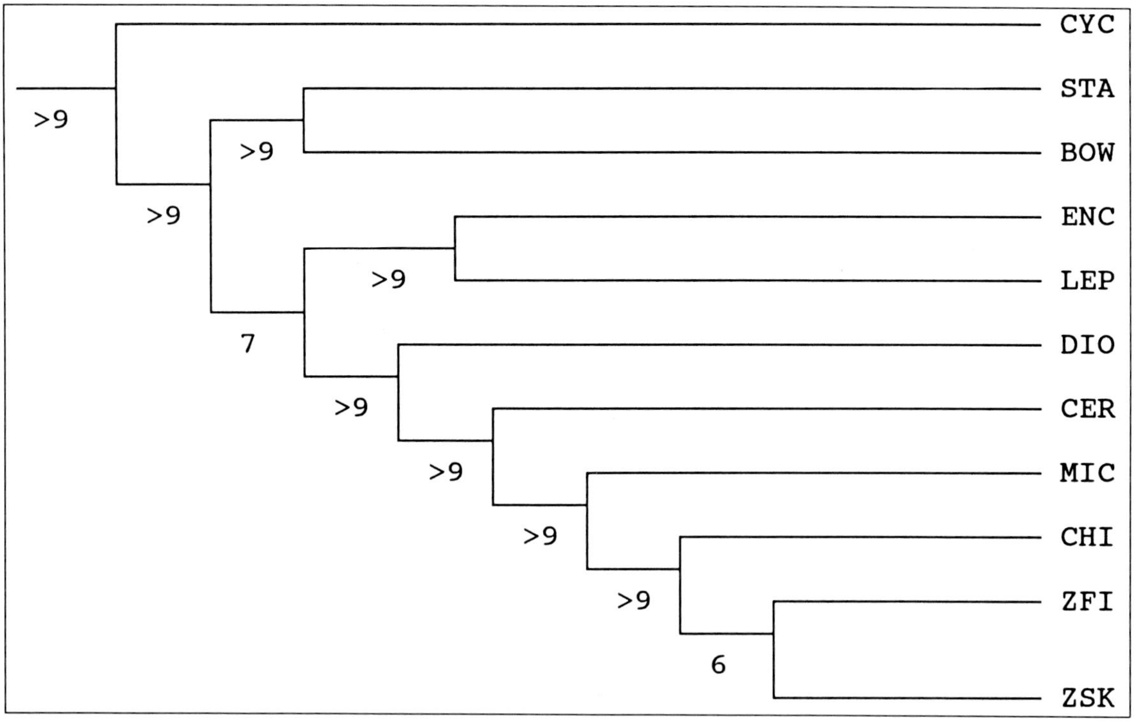

Figure 3: The most parsimonious phylogenetic tree for cycad genera after the exclusion of *Macrozamia* (length 434; c.i. 0.51). Abbreviations for the taxa are found in Table 1. The numbers below each internode represent the number of shared derived characters that define the clade on the right.

Stangeriaceae (*sensu* Stevenson, 1990a) and Zamiaceae, sister groups to each other. Moreover, the clade formed by *Encephalartos* and *Lepidozamia* is the sister group to the clade of the neotropical cycads, the phylogenetic relationships within which confirm our previous findings (Caputo et al., 1991). The position of *Dioon* as a sister group to the other American cycads is supported by geographical distribution, but not from morphological data, as shown in Fig. 4a, b (Stevenson, in this issue).

Therefore, the tree in Fig. 3 is strongly supported from external evidence, suggesting that it should be accepted as a working hypothesis for the phylogeny of cycads, awaiting further data that may allow a clearer understanding of the placement of *Macrozamia* within cycads.

Conclusions

This preliminary attempt of a phylogenetic analysis of cycads was aimed to test the usefulness of a molecular approach to cycad phylogeny, as well as to demonstrate the feasibility of RFLP analysis at the ordinal level.

The phylogenetic hypothesis inferred using molecular data is very similar to that obtained using morphological characters. Yet, some problems are still to be solved, mainly the position of *Macrozamia* (the cpDNA of which deserves further study *per se*) within Zamiaceae and the relationship of *Dioon* with *Encephalartos, Lepidozamia* and *Macrozamia*.

Further research is being carried out and analysis will be extended by choosing all the reasonable exter-

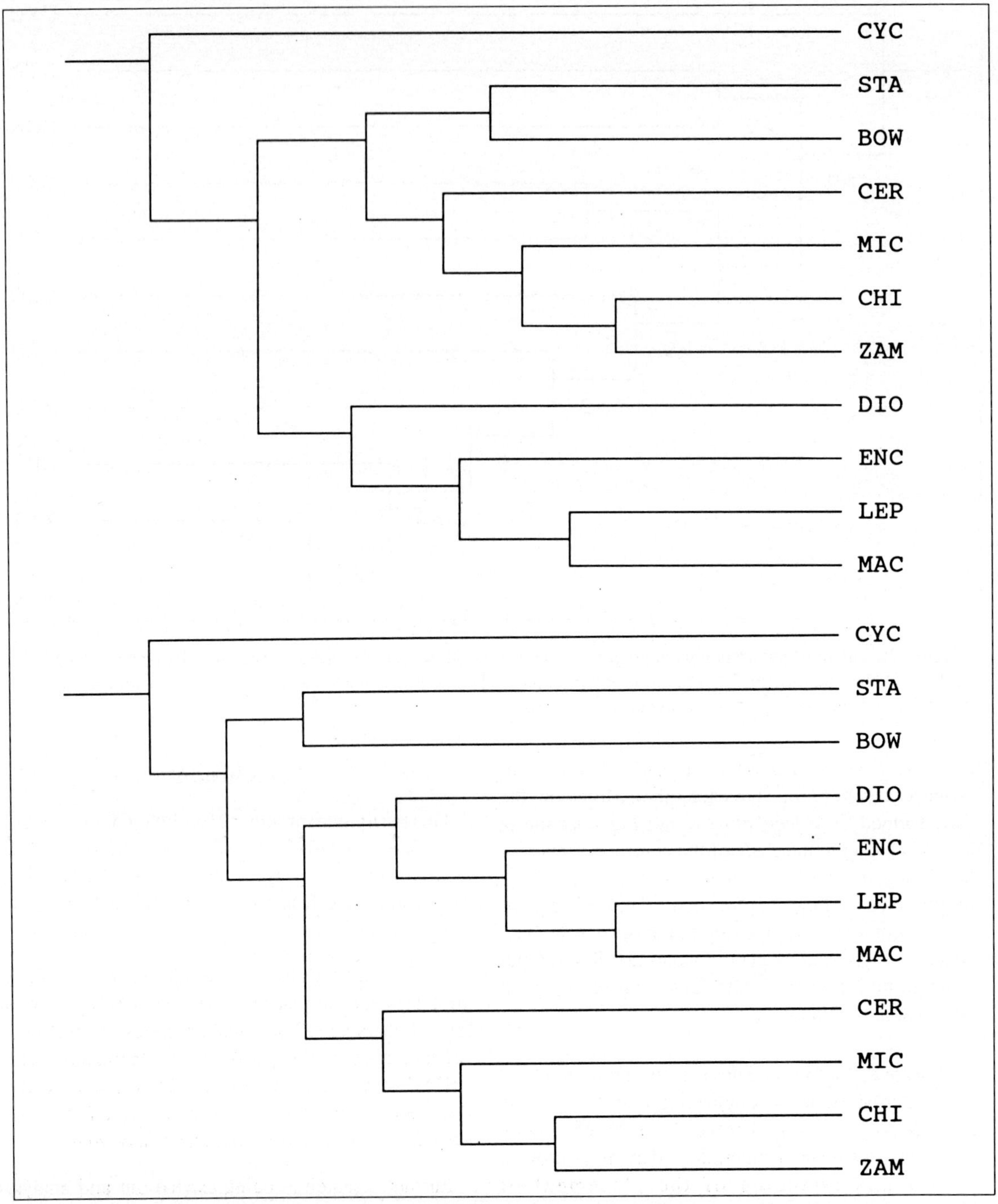

Figure 4a and b. The most parsimonious cladograms of cycads based on morphological data. Abbreviations for the taxa are found in Table 1, except for the genus *Zamia* (ZAM).

nal outgroups to our system as suggested by Loconte and Stevenson (1990), and by using other techniques of phylogenetic analysis in the hope of assessing the position of cycads in the phylogeny of seed plants.

Acknowledgment

The authors would like to gratefully acknowledge research support from the National Science Foundation, grants #RII-8805140 to ETW and #BSR-8796279 to DWS, the National Institutes of Health M.B.R.S. Program, grant #RRO8225 to ETW, and a Fulbright scholarship to PC.

References

***Caputo, P., Stevenson, D., Wurtzel, E.* 1991.** Brittonia 43(3): In Press.

***Chamberlain, C.J.* 1935.** Gymnosperms. Structure and evolution. University of Chicago Press. Chicago.

***Crane, P.R.* 1988.** Major clades and relationships in the 'higher' gymnosperms. *In*: C.B. Beck, editor. Origin and evolution of gymnosperms. Columbia University Press, New York.

***Farris, J.S.* 1988.** Hennig. Hennig86 Reference. Version 1.5.

***Johnson, L.* 1959.** The families of cycads and the Zamiaceae of Australia. Proc. Linn. Soc. New South Wales 84: 64 - 117.

***Kluge, A.G., and Farris, J.S.* 1969.** Quantitative phyletics and the evolution of anurans. Syst. Zool. 17: 1 - 32.

***Loconte, H., and Stevenson, D.W.* 1990.** Cladistics of Spermatophyta. Brittonia 42: 197 - 211.

***Palmer, J.D.* 1985a.** Comparative organisation of chloroplast genomes. Annual. Rev. Genet. 19: 325 - 354.

***Palmer, J.D.* 1985b.** Evolution of chloroplast and mitochondrial DNA in plants and algae. *In*: MacIntyre editor. Molecular evolutionary genetics, 131 - 240. Plenum Press, New York.

***Palmer, J.D.* 1987.** Chloroplast DNA evolution and biosystematic uses of chloroplast DNA variation. Amer. Naturalist 130: S6 - S29.

***Shinozaki, D., Ohme, M., Tanaka, M., Wakasugi, T., Hayashida, N., Matsubuyashi, T., Zaita, N., Chunwongse, J., Obokata, J., Yamaguchi-Shinozaki, K., Ohto, C., Torazawa, K., Meng, B.Y., Sugita, M., Deno, H., Kamogashira, T., Yamada, K., Kusuda, J., Takaiwa, F., Kato, A., Tohdoh, N., Shimoda, H., and Sugiura, M.* 1986.**
The complete nucleotide sequence of the tobacco chloroplast genome: its gene organisation and expression. EMBO J. 5: 2043 - 2049.

***Siniscalco Gigliano, G.* 1990.** Chemotaxonomic significance of MAM glycosides and mucilages in cycads. Mem. New York Bot. Gard. 57: 123 - 131.

***Southern, A.* 1975.** Detection of specific sequences among DNA fragments separated by gel electrophoresis. J. Mol. Biol. 98: 503 - 517.

***Stevens, P.F.* 1980.** Evolutionary polarity of character states. Annual. Rev. Ecol. Syst. 11: 333 - 358.

***Stevenson, D.W.* 1990a.** Morphology and systematics of the Cycadales. Mem. New York Bot. Gard. 57: 8 - 55.

***Stevenson, D.W.* 1990b.** *Chigua*, a new genus in the Zamiaceae with comments on its biogeographical significance. Mem N.Y. Bot. Gard. 57: 169 - 172.

***Watrous L., and Wheeler, Q.* 1981.** The out-group comparison method of character analysis. Syst. Zool. 30: 1 - 11.

***Zhifeng, G., and Thomas, B.A.* 1989.** A review of fossil cycad megasporophylls, with new evidence of *Crossozamia* Pomel and its associated leaves from the lower Permian of Taiyuan, China. Rev. Palaeobot. Palynol., 60: 205 - 223.

Studies of *Cycas* in Australia's Northern Territory

S. Chirgwin
Northern Territory University
Casuarina, N.T., Australia

Abstract

The existence of large undisturbed groups of *Cycas* in Australia's Northern Territory is not well known. Although descriptions of the genus in this location began in 1803, the work has been spasmodic and incomplete and there is confusion about species names and characteristics. Past descriptions of the genus in this geographical area are reviewed, and the reliability of some taxonomic characters questioned. Anatomical characters from both light microscope and Scanning Electron Microscope that have been found to be reliable and supportive in taxonomic work are also described.

Introduction

Australia's Northern Territory contains some of the largest remaining natural populations of *Cycas*, but information about species in this region is poorly and often incorrectly documented. Studies of the genus in the N.T. have been overlooked primarily due to the sheer volume of unfinished taxonomic work, although as far back as 1959 the need for an Australian wide review of the genus was acknowledged (Johnson, 1959).

Cycas occurs in the N.T. north of latitude 15ºS., usually as an understorey component in savanna woodland, where the dominant woody species are generally *Eucalyptus*. These plants occur in climatic conditions described as the 'wet/dry tropics', distinguished by consistently high temperatures throughout the year, rain from tropical storms and cyclones between October to May, and seasonal fires in the 'dry'.

Populations of *Cycas* can be found in areas virtually undisturbed by European settlement, since there has been no large scale land clearing for urban development or agriculture, and much of the total 134.62 million ha. of vegetation is still used for cattle grazing. Increasing amounts of land over the last decade have also returned to the control of traditional owners, with other areas (approximately 4%), converted to conservation parks (Leigh, Boden and Briggs, 1984).

Early Records of the Genus

The earliest record of the existence of *Cycas* occurs in Kakadu National Park, where cave paintings depict the use of the seed for food. European documentation

Figure 1: *C. calcicola*, near Bamboo Creek, Litchfield National Park. The pinnae of this species, described by Machonachie, are thickly covered in tomentum in localities where water is limited.

began with Robert Brown, who accompanied the British explorer Flinders on a voyage of circumnavigation around the Australian coast in 1803. Brown described the existence of two species - *C. angulata* and *C. media* along the coast between Cape York and Port Essington, the first unsuccessful settlement in the Darwin area (Brown, 1810).

Miquel (1861, 1863), reviewed collected samples and confirmed the existence of *C. angulata* and *C. media* and described a 'species dubia' from around Port Essington, which he felt was readily distinguished by 'a thornless petiole' (1863). He later proposed a name for this plant (1868).

In 1911 when the first Flora of the Northern Territory was published (Ewart and Davies, 1911), only one species - *C. media* was listed.

There was no further documentation of any species until 1958, when the information collected by the American-Australian expedition to Arnhemland in 1948 was published. This initially listed again *C. media*, but *C. angulata* was listed in an addendum (Specht, 1958).

Documentation From 1970 - 1980

In the early 1970s interest centred around *Cycas* species as a possible cause for the disease of cattle known as the 'zamia staggers' (Hooper, Best, and Campbell; 1974). The species used in all research in the Territory was listed as *C. media*. During this decade, however, the Officer in charge of the only government herbarium, John Machonachie, published descriptions of two new species of *Cycas* - one the

Figure 2: *Cycas* plants, Bynoe Harbour Road. These plants superficially resemble *C. armstrongii*, but differ in pinnae morphology and anatomy, and reproductive structures and timing.

spectacular blue-grey *C. calcicola*, endemic to the N.T. (Fig. 1), the other *C. pruinosa* from Western Australia (Machonachie, 1978). Machonachie is also credited with the renaming of the common N.T. species described as *C. media*. It had been recognised that this differed from the Queensland species of that name, so the specific epithet of the N.T. species reverted to *armstrongii*, the name given by Miquel to his 'species dubia' (Miquel, 1968). Thus by 1980 the documented species in the N.T. were *C. angulata, C. armstrongii* and *C. calcicola*, although many works after this time still continued to refer to *C. media*, instead of *C. armstrongii*.

Work From 1980 to the Present

Many local enthusiasts and collectors were aware that there were still large groups of plants that appeared different to documented species. Amongst these were the groups of plants in the Bynoe Harbour/Fogg Bay area (Fig. 2), those around Millingimbi and the Blythe River in Arnhem Land, the isolated group on Groote Eyelandt (Fig. 3), and those around Harriet Creek on the road to Kakadu National Park Fig. 4). A review of the morphology and anatomy of populations of the described species was carried out, and consistent, distinguishing morphological and anatomical characters determined. These were then used to establish the taxonomic status of accessible populations not previously documented.

Reliable Morphological Characters

Traditionally vegetative features of *Cycas* that have been used as taxonomic characters have included overall height of the plant, diameter of stem, and frond

Figure 3: *Cycas* plants on Groote Eyelandt. These plants have up to this point been described as *C. angulata*, but differ markedly in size of vegetative and reproductive structures and pinnae anatomy.

features. Whilst the maximum height is a useful guide, damage by seasonal fire and resulting branching in all N.T. species has led to marked variation in height and diameter values. The length and width of fronds exhibit considerable variation from area to area within a species, and even from season to season. Individual pinnae features such as colour (linked to tomentum and cuticle thickness), and number also vary.

Characteristics of the pinnae that are consistent within a species are amount of decurrence, gap between pinnae (measured at a consistent distance from the rachis), pinnae width, the prominence of the pinnae midrib both adaxial and abaxially, and the curvature of pinnae margins.

Supportive Anatomical Characters

Following the work of Pant and Nautiyal (1963), an investigation of transverse sections of pinnae with the light microscope revealed that some features were so distinctive that they could be used collectively to distinguish species. For example the presence or absence of stomata adaxially, the position of the midrib in relation to adaxial and abaxial surfaces, the presence or absence of hypodermis, and whether or not it is continuous, the dimensions of upper and lower photosynthetic tissue, and the presence or absence of a complete sheath around the vascular bundles.

Figure 4: *Cycas* plants at the base of a granite outcrop at Harriet Creek, off the Pine Creek to Kakadu Road. These plants have distinctive seagreen soft pinnae that are amphistomatic.

Surface scans of the pinnae with the Scanning Electron Microscope also enabled the shape and orientation of epidermal cells and the frequency, dimensions and position of stomata to be assessed. Trichome shape and dimension is also a useful character, as was the shape of the apex of the pinnae (Figs. 5 - 10).

Taxonomic Status of Present Populations

Up to this point in time there are still numerous populations that have not been visited or studied, generally in areas not accessible by road, many in areas requiring permits to visit. The establishment of

Figures 5-10: Scanning Electron Micrographs (SEM)

Figure 5: SEM of pinna apex of *C. angulata*, adaxial surface.

Figure 6: SEM of pinna apex of *C. armstrongii*, adaxial surface.

Figure 7: SEM of pinna apex of *C. calcicola*, adaxial surface.

Figure 8: SEM of pinna apex of *Cycas* sp from Bynoe Harbour, adaxial surface.

Figure 9: SEM of pinna apex of *Cycas* sp from Pine Creek, abaxial surface.

Figure 10: SEM of pinna apex of *Cycas* sp from Groote Eyelandt, abaxial surface.

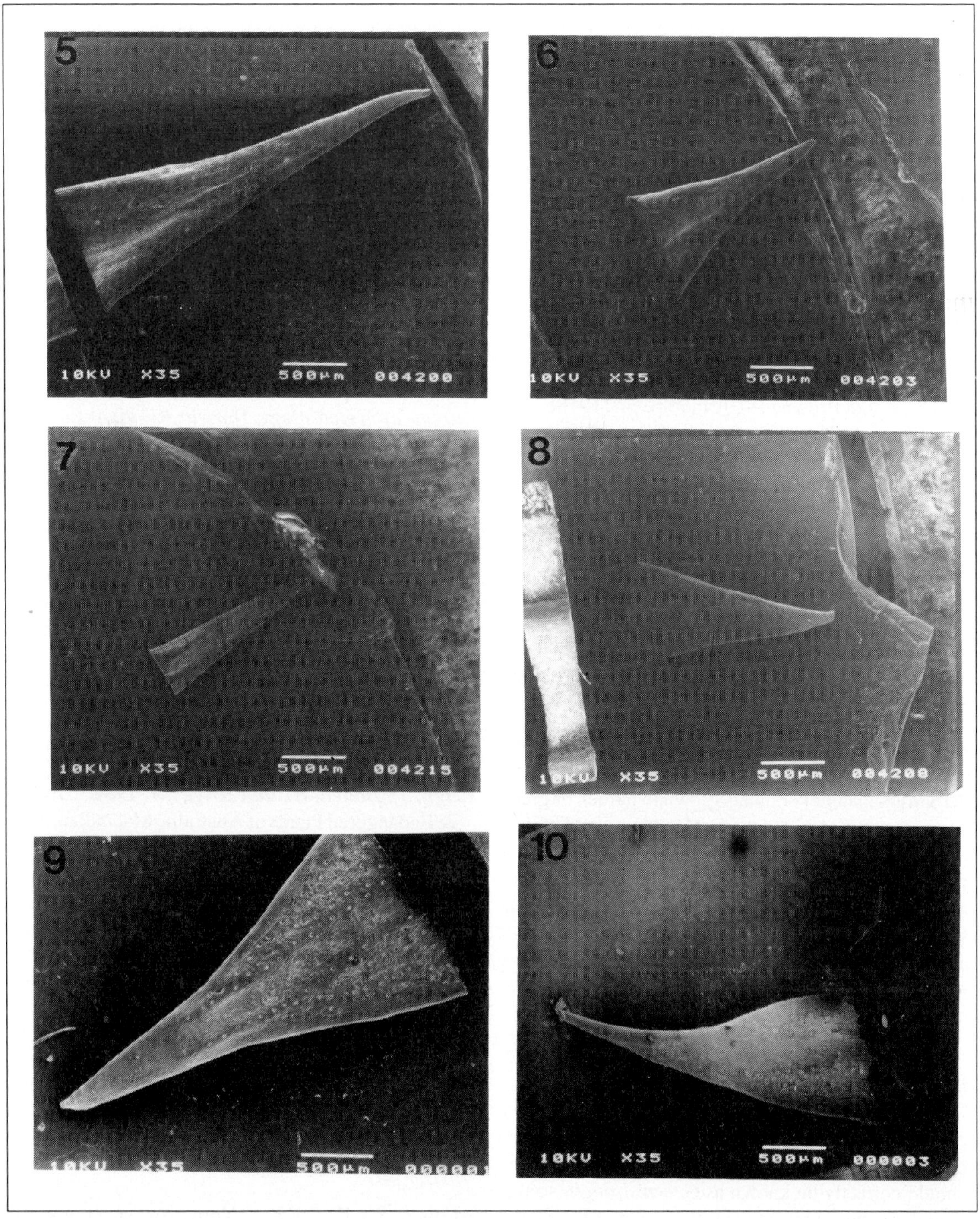
5
10KV X35 500µm 004200
6
10KV X35 500µm 004203
7
10KV X35 500µm 004215
8
10KV X35 500µm 004208
9
10
10KV X35 500µm 000003

a checklist of useful taxonomic characters has enabled the status of three large groups to be clarified. The name *C. conferta* (in ed.), is proposed for the group that exists in limited numbers in and around Harriet Creek, and in a small area in Kakadu National Park, adjacent to the Goodparla Road. It is interesting to note that this plant commonly (though not exclusively) exhibits 'thornless petioles'. The pinnae anatomy is however distinctive due to the presence of a few rows of stomata either side of the adaxial midrib, the lack of hypodermis, the absence of a sheath around the vascular bundle, and the reduced amount of lignified tissues. The isolated population on Groote Eyelandt which was previously documented as *C. angulata* by both Brown (1810) and Specht (1958), also exhibits markedly different morphological and anatomical features, as does the extremely large population (whose exact distribution is difficult to determine) in the Bynoe harbour/Fogg Bay area.

Conclusions

The N.T. offers the unique opportunity to collect information from large numbers of *Cycas* plants in habitats that have experienced little interference from European agricultural, industrial and urban activities.

In the past much of the information collected on this genus has relied to a large extent on small samples of herbarium material or plants grown in garden or green house conditions. Collection of large amounts of data from field plants has shown that in the N.T. there is variation within a species according to local resources and seasonal conditions, particularly for characters such as frond width and length, and megasporophyll and microsporangiate cone length. Comparison between field and cultivated plants has revealed that other features such as tomentum show variation.

The establishment of a checklist of reliable characters for the local members of the genus provides a working base for the identification of the taxon of a particular group. These characters provide further evidence that the N.T. plant formerly known as *C. media* should correctly be known as *C. armstrongii*, so that there are to this point in time three documented species - *C. angulata, C. armstrongii* and *C. calcicola*. A fourth species, again unique to the N.T. is proposed - *C. conferta* (in print). It has been also established that some populations that have been previously described as *C. angulata*, and *C. armstrongii* exhibit marked morphological and anatomical differences that are so consistent from field to greenhouse that these groups should also be afforded the rank of species.

References

***Brown, R.* 1810.** Prodromus florae Novae Hollandiae et Insulae Van Diemen. 1960. Facsimile by Englemen (J. Cramer) Wheldon & Wesley Ltd Weinhelm/Bergstr. Codicote Herts p. 348.

***Bureau of Meteorology, Darwin Regional Forecast Centre* (1989).** Climate of Northern Territory.

***Ewart, A.J., and Davies, O.B.* 1917.** A Flora of the Northern Territory. McCarron, Bird & Co., Melbourne.

***Hooper, P.T., Best, Susan M., and Campbell, A.* 1974.** Axonal Dystrophy in the Spinal Cords of Cattle consuming the Cycad Palm, *Cycas media*. Aust. Vet. Journ. 50. 146 - 149.

***Johnson, L.A.S.* 1959**. The Families of Cycads and the Zamiaceae in Australia. Proc. of Linn. Soc. New South Wales 84: 64 - 117.

***Leigh, J., Boden, R., and Briggs, J.* 1984.** Extinct & Endangered Plants of Australia. Macmillan, Aust.

***Machonachie, J.R.* 1978.** Two New Species of *Cycas* from Northern Australia. J. Adelaide Bot. Gard 1 (3) p. 175.

***Miquel, F.A.W.* 1842.** Monographia cycadearum. R. Natan, Utrecht.

***Miquel, F.A.W.* 1861.** Prodromus systematics cycadearum. C. v D. Post Amsterdam. C.G. v D. Post Jr. Utrecht.

***Miquel, F.A.W.* 1863.** Over de cycadeen in Neuw Holland. *In* Verslagen en medilingen Vol 15 363 - 376.

***Miquel, F.A.W.* 1868.** Nouveaux Matèriaux pour servir a la connaissaance des Cycadèes Premieretriosieme paties. *In* Archives Neerlandaises des Sciences exactes et naturelles 3: 235 - 236.

***Pant, D.D., and Nautiyal, D.D.* 1963.** Cuticular and Epidermal Studies of Some Modern Cycadean Leaves, Sporangia and Seeds Senck. Biol. 44(4) p. 267.

***Specht, R.L. and Mountford, C.P.* 1958.** Records of the American-Australian Expedition to Arnhem Land Vol. 3: Botany & Plant Ecology. University of Melbourne Press.

***Wilson, H.J.* 1982.** The Northern Territory Chronicle. History Unit, Dept of Chief Minister. N.T. Govt.

Utilization of Scanning Electron Microscopy in the Study of Surface Features in Cycadales

Bijan Dehgan, Bart Schutzman, and Fe Almira
Horticultural Systematics Laboratory,
Department of Environmental Horticulture
University of Florida, Gainesville, FL 32611 U.S.A.

Abstract

Advantages of scanning electron microscopy (SEM) in the study of various organs and/or their surfaces is well documented in the botanical literature. Amongst attributes studies in Cycadales, leaflets, sporangia, and pollen surfaces have provided several characteristics which are useful for inter- and infrageneric classifications. Leaflet surface features such as presence or absence of trichomes, topography of subsidiary and other epidermal cells, stomatal aperture shape, frequency and quantity of wax extrusions, etc., when distinct, assist in recognition of groups and often of specific taxa. In contrast, similarity of these features provide sufficient additional evidence that certain species such as *C. circinalis*, may be polymorphic. Further insights into inter- and infrageneric relationships have been realised with comparative study of microsporophylls and microsporangial surface morphology. Among other features, shape and configuration of sporangial sutures, cells shape, size and distribution, presence or absence of a discernable soral stalk and stomata, and number of sporangia per sorus, have proven systematic value. Distinctive pollen surfaces (fossulate, foveolate, and psilate) and pollen shapes (from narrowly to widely elliptical to suborbicular) have also contributed to understanding of relationships in cycad genera. In general, the result of comparative SEM studies, in concert with gross morphological features, indicate a striking resemblance between species of *Ceratozamia* and *Zamia*. Despite some minor differences, the alliance of these genera with *Microcycas* is also justified. Moreover, affinity of *Bowenia, Encephalartos, Lepidozamia* and *Macrozamia* is confirmed. *Dioon, Stangeria*, and *Cycas* possess features that are unique among all genera.

Introduction

Utilization of various surface features as taxonomic characters is not a new methodology. Several comprehensive studies of leaflet surface attributes have been published in the past 150 years, primarily because of their usefulness in paleobotany (see for example, Thomas and Bancroft, 1913, Greguss, 1966; Pant,

1973; Pant and Natiyal, 1963). However, despite use of scanning electron microscopy (SEM) in botanical studies for nearly 30 years and its obvious potential for the study of cycads, only a very limited number of investigations have been reported (for example, Baijnath, 1980, Dehgan and Yuen, 1983). In fact, contrary to the unfounded remarks that it was repeat of a previous work (Stevenson, 1990), the only comprehensive work thus far reported is that of SEM and TEM studies of whole cycad pollen grains from our laboratory (Dehgan and Dehgan, 1989). Only a limited number of genera were considered in earlier works (*c.f.* Audran and Masure, 1976).

The principal advantage of the scanning electron microscope (SEM) is that objects and surface characteristics are exhibited in three dimensions as opposed to two dimensions, as with the light and transmission electron microscopes. For a study to be thorough, however, observation of cleared leaves with light microscope is still indispensable. Other features, such as determination of pollen grain size, also require light microscopy. In contract, micromorphological features of sporophylls and sporangia can only be observed with SEM, because, unlike leaflets these cannot be cleared except perhaps in very early stages of development, prior to enlargement of ovules or pollen. Moreover, material examined with SEM provide considerable amount of information but do not require extensive preparation similar to that of light (LM) or transmission electron microscopy (TEM).

An overwhelming quantity of data on leaflet surfaces, and microsporophylls and microsporangia is currently available from our studies through a combination of SEM, TEM, and light microscopy. A detailed report of these studies will be published in the near future.

This paper is limited to a comparative overview of our findings on leaflet and sporangial surface morphology as they relate to the proposed revised classification of cycadales (Dehgan and Dehgan, 1989; Schutzman and Dehgan, in this issue of the proceedings) and is primarily concerned with genera and families with unique features (Fig. 1). Pollen characteristics are considered only briefly (Fig. 2). It should be noted that some of the observations mentioned below have also been reported by other workers using light microscopy.

I. *Stangeria, Dioon* and *Cycas*

A. Leaflet Surfaces

The only leaflet surface feature these genera have in common is that they are hypostomatic; that is, stomata occur only in their lower surface. They do, however, each have several unique attributes. For example, among unique features of *Stangeria* are the presence of radial cuticular lamellae and flat (not sunken) stomata in relation to other epidermal cells. In *Cycas* and *Dioon* the stomata are sunken. In *Cycas revoluta* Thunberg, *C.taiwaniana* Carruthers, and related species the subsidiary cells are raised above the epidermal cells, thus resembling a volcano or moon crater (Fig. 1A). In the *C. circinalis* L. complex, including *C. chamberlainii* Brown & Kienholz, *C. wadei* Merrill, and related taxa, the stomata are sunken but the subsidiary cells are flat in relation to other epidermal cells (Fig. 1B). The Australian *Cycas* species seem to be intermediate between the other two groups in leaflet surface morphology as well as other features (Dehgan and Schutzman, 1984). *Dioon* has a stomatal crypt, a unique feature among all cycads.

Although not directly related to the surface features, the very unique multiple vascular bundles in the midvein of *Stangeria* is noteworthy (Fig. 1C). There is only one vascular bundle in the midvein of *Cycas* (Fig. 1D), while *Dioon* lacks a midvein.

B. Sporophyll and Sporangia

There should be no difficulty recognising male cones of these genera. All three have palaceous (nonpeltate) microsporophylls. They are wide at the apex and attached to the cone axis by a relatively wide base. However, the prominent outgrowths from the distal edge of microsporophylls in *Stangeria* are unique. Another unique feature in *Stangeria* is the apparent solitary sporangia (*c.f.* Stevenson, 1990). The number of sporangia per sorus in *Dioon* and *Cycas* is three or more. The sori in *Dioon* are prominently stipitate,

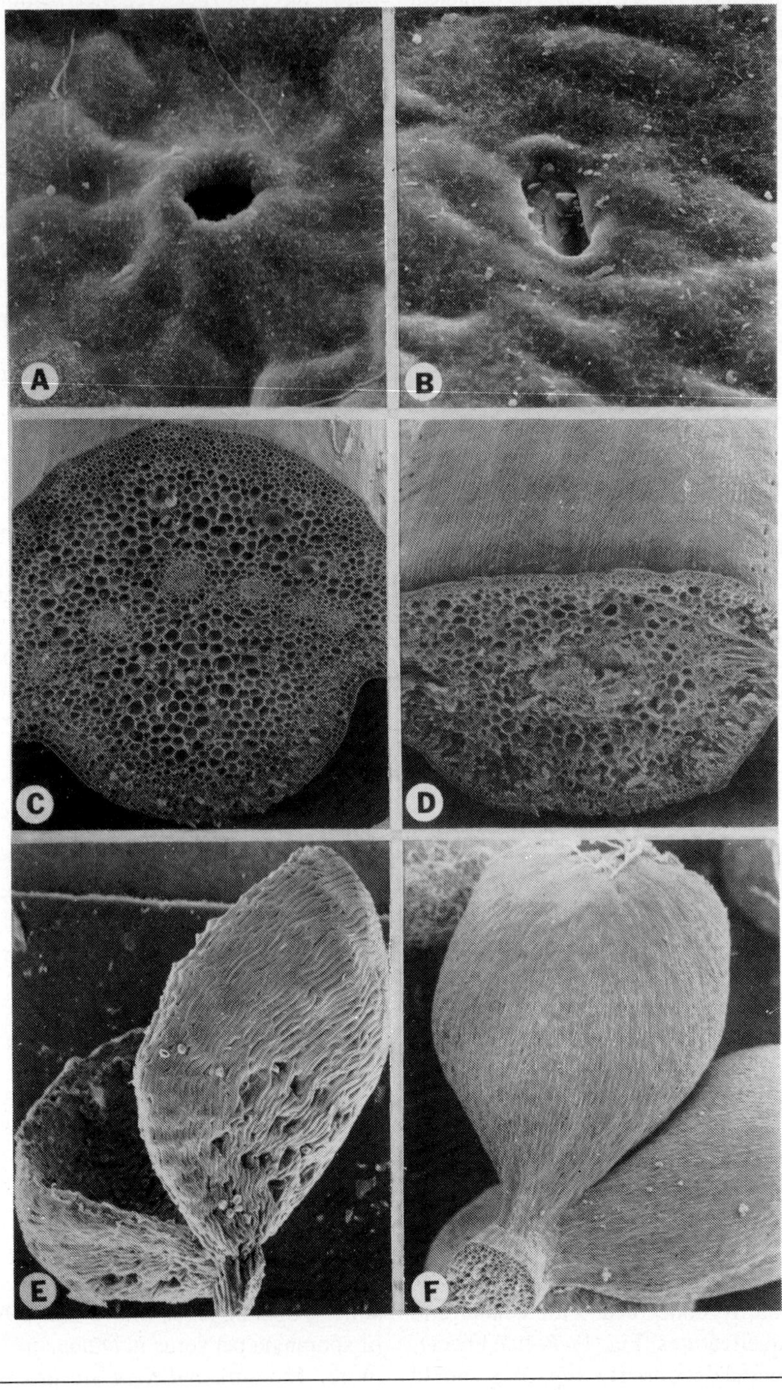
A
B
C
D
E
F

whereas, the stalk of *Cycas* microsporangia are embedded in the sporophyll tissue and reduced. Among these three genera, *Stangeria* is unique in having stomata at the base of the sporangia (Fig. 1E). *Cycas* (Fig. 1F) and *Dioon* both lack stomata on their sporangia.

The annulus along which sporangia dehisce is unique in both *Dioon* and *Stangeria*. *Dioon* lacks an organised line of cells that characterise all other cycads and the suture appears as a tear wherever there is a weakness in the cell walls, thus appearing irregular. In contract, the sporangial annulus of *Stangeria* is the most organised of all cycadales and has the appearance of two rows of interlocking cells with unusually thickened walls.

C. Pollen Grains

The shape of pollen in the three genera is rather similar. All three were classified as orbicular (Fig. 2A - C), having a width/length ratio of about 0.7 (Dehgan and Dehgan, 1989). With but two exceptions (*Ceratozamia hildae* Landry & Wilson and *Zamia skinneri* Warszewicz ex Dietrich) this is a feature unique to these genera. *Stangeria* and *Dioon* have a foveolate surface, whereas *Cycas* has a fossulate surface.

II. *Dioon* vs Other *Zamieae*

A. Leaflet Surfaces

All genera in the Zamieae more or less follow the same groundplan for leaflet surfaces; that is, they are mostly hypostomatic with well-organised stomata between the veins on the abaxial surface. In some *Zamias* and *Ceratozamia*, stomata with no particular organisation are found on the adaxial surface. Some features are unique to specific genera and allow their delimitation, such as the reticulate wax depositions around *Ceratozamia* stomata (which are also found to a lesser extent in a few species of *Zamia*). The typically angular epidermal cells and square-shaped stomata in *Microcycas* are also distinct features. In *Macrozamia* the wax extrusions occur in a reticulate pattern on the entire surface, not just around the stomata as in *Ceratozamia*. *Dioon* leaflet surface features more closely resemble those of *Macrozamia* and *Encephalartos* species than *Zamia* or *Ceratozamia*. The unique stomatal crypt of *Dioon*, with its elaborate arrangement of subsidiary and accessory cells, as noted above, is a feature not found in any other cycad.

B. Sporophyll and Sporangia

As was mentioned previously, the palaceous sporophylls of *Dioon* are unlike any in the Zamiaceae, which are generally more or less peltate. There is also

←

Figure 1: Examples of various features used in SEM studies.

A-B. Sunken stomata of leaflets.

A. *Cycas taiwaniana B.*

B. *C. media* Brown

Note the differences in topography and number of subsidiary cells.

C-D. Cross sections of the midvein.

C. Multiple vascular bundles of *Stangeria eriopus* (Kuntze) Nash.

D. Single vascular bundle of *C. revoluta L.*

E-F. Sori.

E. Single sporangium of *S.eriopus* (with numerous stomata).

F. Sporangia of *C. revoluta L.* (stomata lacking).

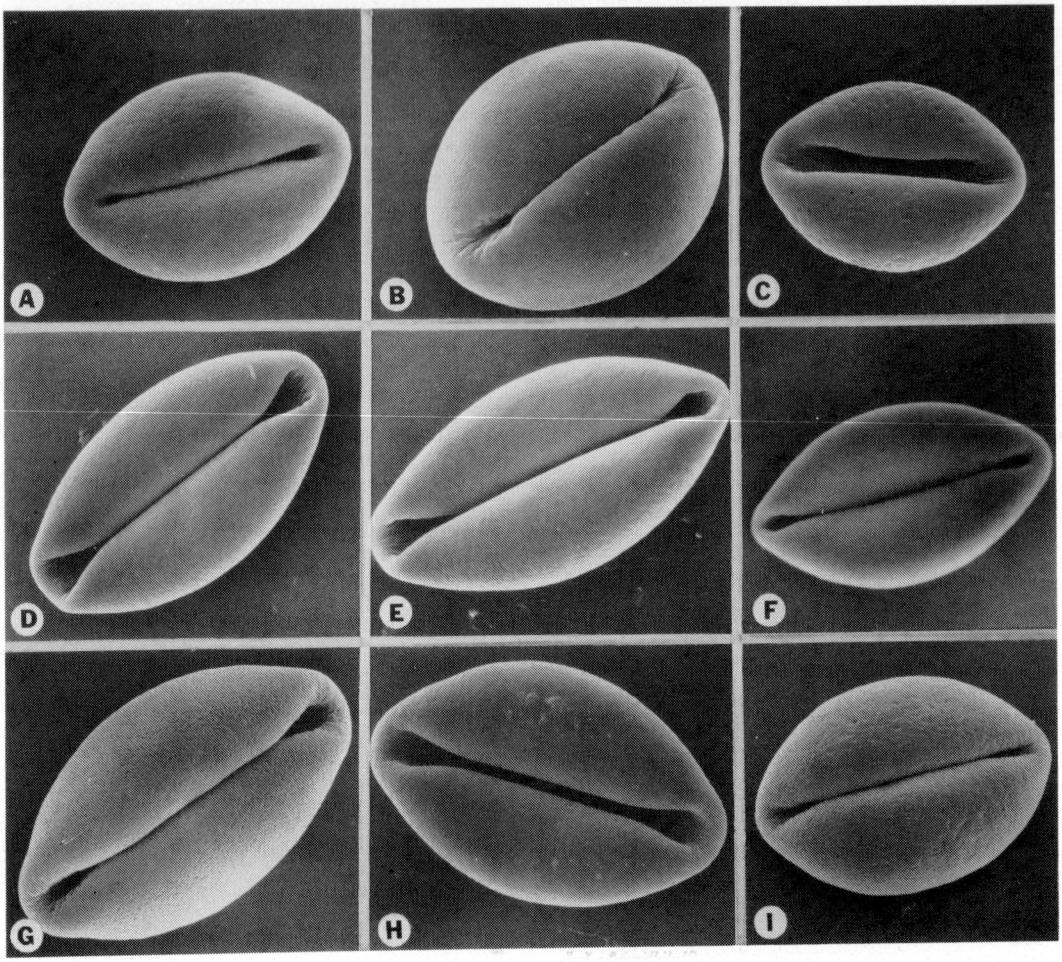

Figure 2: SEM micrographs of Cycadales pollen (from Dehgan and Dehgan, 1988).

A. *Stangeria eriopus.*

B. *Dioon edule* Lindley.

C. *Cycas revoluta*

D. *Encephalartos gratis* Prain.

E. *Lepidozamia peroffskyana* Regel.

F. Bowenia spectabilis Hook. ex Hook. f.

G. *Ceratozamia mexicana* Brongniart.

H. *Microcycas calocoma* (Miquel) De Candolle.

I. *Zamia acuminata* Oersted ex Thiselton - Dyer. Pollen of *Macrozamia* is similar to *Lepidozamia* and is therefore omitted.

a difference in the number of sporangia. The sporangia of *Zamia* are always in twos; those of *Microcycas* are mostly in two's, but probably as a result of abortion, exceptional solitary sporangia are occasionally observed. *Ceratozamia* is the only genus in the group with sori consisting of three or more sporangia and with a collectively short but horizontally elongated stalk, a feature which is lacking in *Dioon* and other Zamieae (*sensu* Johnson, 1959).

C. Pollen Grains

There is considerable difference in the pollen morphology of the Zamieae. Indeed, pollen characteristics in concert with other micro- and macro-morphological features originally prompted us to suggest a new taxonomic arrangement for cycads. With respect to SEM, the shapes of the pollen are generally variously elliptic with a few exceptions in *Zamia* and, of course, *Dioon* (Figs. 2D - E and 2G - I). Their surfaces, however, vary considerably. The Australian genera, including *Bowenia* (Fig.. 2F), are all psilate (smooth). Species of *Encephalartos* are variable; some are psilate while others are foveolate. In the American genera, *Dioon* is foveolate, *Zamia* and *Ceratozamia* are coarsely foveolate, and *Microcycas* is finely foveolate. Pollen surface features are significant for two reasons: they are conservative with respect to change over time; and their shape, size, and sculpturing were probably greatly influenced long ago by their mode of pollination.

III. *Stangeria* vs *Bowenia*

A. Leaflet Surfaces

The most obvious dissimilarity of the two genera is the absence of a midvein in *Bowenia*. While stomata of *Stangeria* are not sunken, those of *Bowenia* are partially sunken, and its stomatal opening is invariably filled with cuticular waxes (occluded). Cuticular outgrowths (thickenings) of the guard cells are more prominent in *Stangeria* and radial cuticular striations are lacking in *Bowenia*. *Stangeria* is hypostomatic while *Bowenia* is amphistomatic. Stomata in *Stangeria* are in random groups while those of *Bowenia*, similar to Zamiaceae, are in well-organised bands. *Stangeria* is mostly six subsidiary cells while *Bowenia* has only four. *Stangeria* has approximately 20 stomata per square millimeter while *Bowenia* has tow to three times as many, or about 65.

B. Sporophylls and Sporangia

Palaceous sporophylls of *Stangeria* has no resemblance to the peltate sporophylls of *Bowenia*. The prominent outgrowths from distal edges of sporophylls in *Stangeria* are also very unique. Moreover, as noted earlier, the seemingly solitary sporangia of *Stangeria* are unique in all Cycadales. The sporangial annulus of *Stangeria* is the most organised of all Cycadales, whereas that of *Bowenia*, similar to other Zamieae, consist of elongated, more or less disorganised cells.

C. Pollen Grains

The pollen grains of the two genera are as different as all the other features previously mentioned. The shape of *Stangeria* pollen is suborbicular (Fig. 2A), whereas that of *Bowenia* is elliptic (Fig. 2F). The surfaces are also different; that of *Stangeria* is distinctly foveolate while that of *Bowenia* is only finely foveolate and nearly smooth. There are also significant differences in pollen wall layer thickness of the two genera (Dehgan and Dehgan, 1989).

It should be clear by now that other than superficial resemblances, these two genera have little in common. The only general statement one can make is that they are both cycads. Considering the differences noted by Schutzman and Dehgan in these proceedings, Stevenson's (1990) contention that these genera are related is completely unfounded. Indeed *Bowenia* shares more characteristics with the other Australian-African cycads (*Encephalartos, Macrozamia*, and *Lepidozamia*) than it does with *Stangeria*, a fact born out by the cladistic results discussed by Schutzman and Dehgan in this volume.

Conclusions

Good classifications are based on the totality of characteristics, and reflect as many similarities and differences among organisms as possible. Attempts at simplification and use of inaccurate information, as has been noted is some recent publications (see, for example, Crane, 1988; Stevenson, 1990), merely ignore significant evidence and result in artificial classifications. In cycads, superficial resemblances and minor differences, particularly those observed on herbarium sheets, can be very deceiving. Much more evidence is needed before we can have an adequate classification, one that may indicate probable affinities and represent phylogenetic lines. Scanning electron microscopy of various surfaces confirms earlier findings and provides additional support for a new classification proposed by in a recent paper on pollen grain morphology (Dehgan and Dehgan, 1989) and by Schutzman and Dehgan in these proceedings. Other than some changes in the levels of hierarchy, our classification does not significantly alter that of Johnson (1959). Indeed, as noted by Professor Johnson (pers. comm.), had he had such data available, he would probably have considered the same arrangement as that which we are proposing.

On the basis of similarities and differences observed in surface features of leaflets, sporangia, and pollen characteristics, as well as the cladistic analysis discussed by Schutzman and Dehgan in these proceedings, the most plausible classification of cycads should recognise the genera *Cycas*, *Stangeria*, and *Dioon* as distinct monogeneric families. *Bowenia*, pending additional information from DNA studies, should be recognised as a distinct subfamily (as opposed to its current familial status) in a family which would also include *Macrozamia*, *Lepidozamia*, and *Encephalartos*. These latter genera, as noted previously, are sufficiently distinct from *Zamia*, *Ceratozamia*, and *Microcycas* to be considered as a distinct family. Zamiaceae will then be restricted only to the latter three American genera. Nonetheless, one has to be cognizant of the more profound differences often found in some angiosperm genera than in the entire cycadales. The only justification for suggesting a revised classification of the order lies in the fact that we are dealing with an ancient group of plants with but meager knowledge of their connecting links. The analogy is that of a jigsaw puzzle with numerous missing pieces.

It therefore seems prudent to treat the genera of groups of genera that exhibit distinct features as discrete entities until such time as sufficient fossil or other convincing evidence is (are) presented to otherwise justify their union. The karyotypic data recently presented by Moretti (1990, and in this issue), provides such additional data and in general, supports the suggested groupings.

Acknowledgment

Experiment Station Journal Series No. R-01425. We thank Drs. Alan Meerow and Dennis McConnell for review of this manuscript.

References

***Audran, J.C., Masure, E.* 1976** Precisions sur L'infrastructure de l'exine chez les cycadales (Prepermaphytes). Pollen & Spores 18: 5 - 26.

***Baijnath, H.S., Naidoo, Ramcharan, S.* 1980.** Leaf surface characters of selected genera of cycadales. Proc. Elect. Microsc. Soc. S. Afr. 10: 35 - 36.

***Crane, P.R.* 1988.** Major clades and relationships in the 'higher' gymnosperms. In Origin and Evolution of Gymnosperms (pp. 218 - 272). C. B. Beck, ed. Columbia University Press, New York.

***Dehgan, B.*, Dehgan, N.B. 1989.** Comparative pollen morphology and taxonomic affinities in cycadales. Amer. J. Bot. 75: 1501 - 1516.

***Dehgan, B., Yuen, C.K.K.* 1983.** Seed morphology in relation to dispersal, evolution, and propagation of *Cycas* L. Bot. Gaz. 144: 412 - 418.

***Johnson, L.A.S.* 1959.** The families of cycads and the Zamiaceae of Australia. Proc. Linn. Soc. N.S.W. 84: 64 - 117.

***Greguss, P.* 1966.** Xylotomy of the Living Cycads, with a description of their leaves and epidermis. Akad. Kiado, Budapest.

***Moretti, A.* 1990**. Karyotypic data on North and Central American Zamiaceae (Cycadales) and their phylogenetic implications. Amer. J. Bot. 77: 1016 - 1029.

***Pant, D.D.* 1973.** Cycas and the cycadales, 2. ed. Central Book Depot, Allahabad, India.

***Pant, D.D., Natiyal, D.D.* 1963.** Cuticle and epidermis of recent Cycadales. Leaves, sporangia and seeds. Senckenberg. Bio. 44: 257 - 348.

***Stevenson, D.W.* 1990.** Morphology and systematics of the Cycadales. Mem. N.Y. Bot. Gard. 57: 8 - 55.

***Thomas, H.H., Bancroft, N.* 1913.** On the citicle of some recent and fossil cycadean fronds. Trans. Linn. Soc. London (Bot.) 8: 1545 - 20.

Origin, Evolution, and Growth Patterns of Cycads

T. Delevoryas
Department of Botany
University of Texas, Austin, Texas 78713-7640

Introduction

The traditional concept of the evolution of the cycad megasporophyll considers that among extant forms, that of *Cycas revoluta* represents the most primitive type, with progressive reduction of the lamina resulting in a peltate structure with two ovules directed toward the cone axis. An alternative hypothesis would have the primitive megasporophyll as an entire, strap-shaped leaf with the ovules borne along the edge of the lamina. Two developments made the latter hypothesis less attractive. One is that it is now known that the Permian entire-leaved sporophyll, *Phasmatocycas*, had ovules borne not on the margin, but along a flange-like structure attached to the abaxial side of the sporophyll. The second is the discovery of megasporophylls in the Permian of China that had dissected distal laminae like those in *Cycas*. More evidence is presented to support the suggestion that early cycads had slender stems. Most are thought to have been deciduous. As among some extant cycads, some ancient forms shed individual pinnules. Although cycads and cycadeoids belong to different divisions, there are parallisms in habit and pattern of leaf drop.

Megasporophylls

Until relatively recently, the fossil record of cycads did not provide evidence concerning the origin and evolution of the megasporophyll. Suggestions concerning primitive vs. specialised sporophylls were derived solely from comparative morphology. It was generally felt that a megasporophyll such as that of *Cycas revoluta*, with its pinnately divided distal lamina, was an example of the most primitive type, resembling most closely a progenitor that was considered to have had compound leaves. *Cycas media* (Fig. 1) and *C. circinalis* were considered to be somewhat reduced, with a smaller distal portion with remains of pinnae scarcely evident. Further testament of the primitive situation in *Cycas* is the fact that the megasporophylls are not aggregated into a strobilus, but are borne in a loose cluster at the apex of the stem. As the sporophylls mature, the stem apex at the top of the sporophyll cluster begins to produce vegetative leaves.

All other genera of extant cycads have megastrobili. The megasporophyll of *Dioon* is considered to be

---

Figure 1: Megasporophylls of *Cycas media*.

Figure 2: Megasporophyll of *Dioon edule*.

Figure 3: Young megastrobili of *Macrozamia communis*.

Figure 4: Megastrobili of *Encephalartos altensteinii*.

1
2
3
4

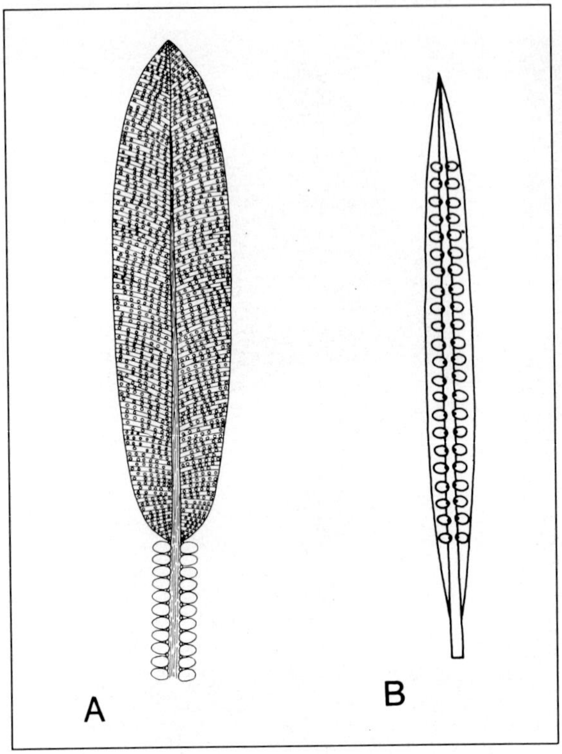

Figure 5: Reconstruction of megasporophyll of *Phasmatocycas.*

A. Mamay, 1976.

B. Gillespie and Pfefferkorn, 1986.

somewhat less primitive than that of *Cycas*. The megastrobilus consists of an axis with helically arranged megasporophylls. Each sporophyll has a slender stalk that is attached at right angles to the strobilar axis. Then the sporophyll bends upward at about a right angle, with the distal portion broad and leaf-like. It is covered with a dense mat of hair, but sometimes the bladelike portion of the megasporophyll turns green. On the inner face of the expanded blade, near the bottom portion, are attached two ovules, with the micropyles directed inward (Fig. 2). *Macrozamia* has a compact ovulate strobilus with the megasporophylls quite reduced and packed tightly in the cone. A distal pointed projection of the megasporophyll extends upward (Fig. 3). *Lepidozamia* also had compact strobili, with megasporophylls closely aggregated, but here, too, there is a fleshy bladelike portion at the end of the sporophyll. In genera such as *Zamia* and *Encephalartos*, the megasporophyll is not recognisable as a modified leaf (fig. 4). At the surface of the strobilus, the megasporophylls have the appearance of closely aggregated polygons - hexagonal in some; rhombic in others.

Before the appearance of Mamay's (1969, 1973, 1976) papers there were very few seed-bearing organs attributed to cycads. Those most frequently cited are *Cycadospadix, Dioonitocarpidium, Palaeocycas*, and *Beania*. Mamay (1976) summarised the uncertainties of interpretation of *Dioonitocarpidium* and *Palaeocycas*. No seeds had been found attached to *Cycadospadix*, which has been synonymised with *Crossozamia* (see reference to paper by Gao and Thomas, 1989, below). *Beania*, therefore, seemed to be the only bona fide cycad megasporophyll. It resembles a megastrobilus of a member of the Zamiaceae except that the cone axis is elongated and the cone was thought to have been lax and probably pendant.

The appearances of Mamay's papers (1969, 1973, 1976) initiated a renewal of activity concerning the origin of cycad megasporophylls. Because fossil sporophylls were so rare, palaeobotanists were desperate to find fossils that could be interpreted as possible cycad precursors. Mamay's suggestion that the primitive cycads had leaves and megasporophylls with undivided laminae was innovative and based on interesting new finds. Two genera, *Phasmatocycas* and *Archaeocycas*, were named by Mamay (1973) for ovule-bearing structures. *Phasmatocycas*, as originally described, consisted of an axis, thought to represent the base of a leaf, with two lateral rows of ovules. Associated with the fertile remains were taeniopterid leaf fragments. On the basis of glandlike structures on the fertile axis that corresponded with similar glandlike structures on the taeniopterid laminae, Mamay suggested that the parts probably be-

Figure 6: Accumulation of leaves of *Otozamites hespera* from the Jurassic of Oaxaca, Mexico.

longed to the same fertile leaf, and his reconstruction shows a proximal rachis with two rows of ovules and a distal expanded taeniopterid blade (Fig. 5A). Mamay, in 1976, presented a hypothetical series of diagrams indicating how such a megasporophyll could have evolved from a *Spermopteris*-type of fertile leaf that is essentially a taeniopterid leaf with seeds borne on the abaxial surface of a blade near the margin. Subsequent dissection of the distal taeniopterid lamina of a *Phasmatocycas* megasporophyll would have led to the evolution of a megasporophyll such as that of *Cycas revoluta*.

Gillespie and Pfefferkorn (1986) demonstrated that in *Phasmatocycas* seeds are indeed borne on a taeniopterid leaf. Attachment of seeds, however, is not as Mamay has visualised it, but they were borne along a flange-like structure suspended along the abaxial side of the leaf (Fig. 5B).

As described by Mamay (1969, 1973, 1976), *Archaeocycas* is a genus based on ovules borne on the abaxial surface of a lamina that partially overlaps the seeds. No distal lamina was observed, but Mamay postulated one on the basis of a slight expansion of the seed-bearing leaf just distal to the seed-bearing zone. Mamay suggested that most likely the distal bladelike portion was entire and that subsequent dissection could have produced a *Cycas*-like megasporophyll.

Recently Leary (1990) reported on an entire leaf, *Lesleya*, from Lower Pennsylvanian deposits. The entire leaf, which resembles *Taeniopteris*, bore ovules along the rachis. Leary felt that this discovery adds more weight to Mamay's hypothesis that cycad megasporophylls arose from ancestors with entire leaves. Preservation is poor, however, and, obviously, every entire, strap-shaped leaf need not be considered ancestral to cycad sporophylls.

Delevoryas (1982) presented a simpler, alternative hypothesis, suggesting that cycad megasporophylls probably arose from compound seed-bearing leaves of pteridosperms. He cited, among known forms, *Eremopteris zamioides* (Delevoryas and Taylor, 1969) and *Tinsleya texana* (Mamay, 1966) as having leaf forms from which cycad fronds and megasporophylls could have evolved. Gillespie and Pfefferkorn (1986) still felt that an ancestor with an entire lamina was more likely because "there are no intermediate forms to substantiate the Delevoryas model." Certainly the absence of an 'intermediate form' is no more serious than the absence of an intermediate form between Gillespie's and Pfefferkorn's reconstruction of *Phasmatocycas* and a *Cycas* megasporophyll.

With the discovery of cycad megasporophylls from the Lower Permian of Taiyuan, China (Gao and Thomas, 1989), it would seem that all previous arguments are moot. *Crossozamia* megasporophylls (variously referred to previously as *Cycadospadix, Cycalacis, Norinia*, and *Primocycas*) are as old as, or older than, *Phasmatocycas* and *Archaeocycas*, and closely resemble megasporophylls of *Cycas*, with ovules borne laterally along a rachis and a flared, divided distal lamina. Hence, cycad-like megasporophylls had evolved by the time of appearance of *Phasmatocycas* and *Archaeocycas*; it is inconceivable that the latter two could have been ancestral to the cycad situation when they were situated on the other side of the world when *Crossozamia* was in existence.

Gao and Thomas found megasporophylls arranged loosely in helical fashion around a slender axis they considered to be a strobilar axis. They found it difficult to explain how the situation in *Cycas*, in which megasporophylls are not arranged in a strobilus, could have developed from one in which the primitive type of megasporophyll was supposed to have been borne on a strobilus. In light of the frequent occurrence of slender-stemmed cycads in the early Mesozoic or before (Delevoryas and Hope, 1971; Kimura and Sekido, 1985; Zhang and Liang, 1985; Zhang and Wang, 1986), I would prefer to consider the megasporophylls of *Crossozamia minor* as having been

Figure 7: Pinnately compound cycadophyte leaves resembling *Pterophyllum* but with cycadeoidalean epidermal features. Upper Triassic, North Carolina, U.S.A.

Figure 8: Detached leaflets of *Zamites lucerensis*. Jurassic, Oaxaca, Mexico.

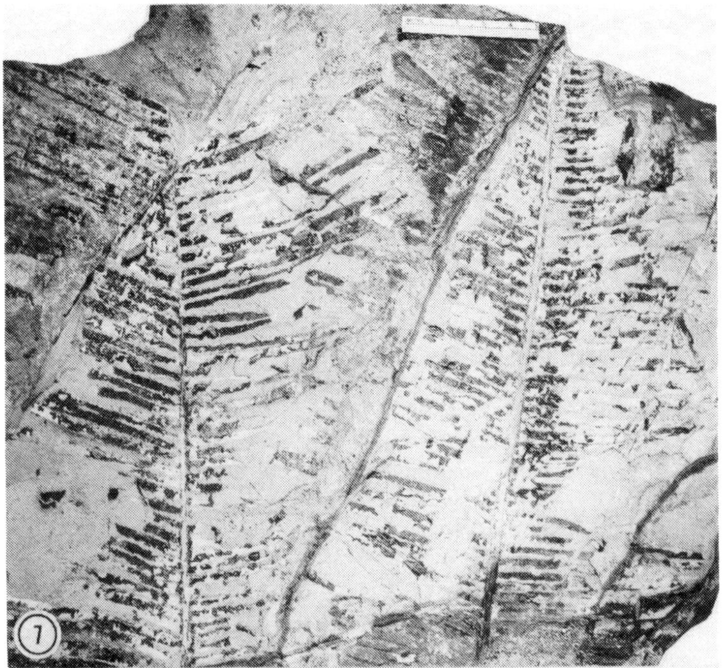
7

8

borne directly on the stem in a fertile zone; there is no strobilus. Then it would be simple to envisage the arrangement of megasporophylls in *Cycas* as a result of telescoping the stem.

Leaves

One of the pieces of evidence used by Mamay to support the 'entire leaf hypothesis' for cycad origins is that "considerable indirect evidence favours the assumption that all the known pre-Jurassic cycads, with the exception of the Upper Triassic *Leptocycas*, bore entire-margined taeniopterid foliage" (Mamay, 1976). In my 1982 paper (Delevoryas, 1982) I cited a number of pre-Jurassic cycad fronds that were pinnately compound. To that list may be added others. For example, Ash (1991) described from the Chinle Formation the genus *Aricycas*, with pinnately divided foliage. Another example is still undescribed foliage from the Upper Triassic of North Carolina that superficially resembles *Pterophyllum*, a cycadeoidalean genus, but that has cycadalean epidermal features (Fig. 6). One can go even farther back into the fossil record to see pinnately compound cycad leaves. *Yuania striata, Y. chinensis* (Du and Zhu, 1982; Gao and Thomas, 1989) and *Tianbaolinia circinalis* (Gao and Thomas, 1989) are early Permian leaves from China. (According to Gao and Thomas [1989] *Tianbaolinia circinalis* could represent the immature stage of *Yuania chinensis*). Perhaps, as I indicated in my 1982 paper, certain cycads may have arisen from entire-leaved ancestors. In light of the newer interpretation of *Phasmatocycas* by Gillespie and Pfefferkorn (1986), and in view of the early Permian existence of cycad megasporophylls with divided distal laminae, however, strength of that argument is considerably weakened.

Growth Habit and Patterns

Since our suggestion that the primitive growth habit for cycads was a plant with a slender stem (Delevoryas and Hope, 1971) there has been more evidence to support that viewpoint. Kimura and Sekido (1975), Delevoryas and Hope (1976), Zhang and Mo (1981), Zhang and Liang (1985), and Zhang and Wang (1986) all support the concept of a slender growth habit for early cycad stems. Stevenson (1990) criticised the original suggestion presented by Delevoryas and Hope (1971) and in subsequent papers, but really on mistaken premises. His interpretation of our work is that we implied that early cycads such as *Leptocycas* were leptocaulous and not pachycaulous as are extant cycads. Actually, we made no reference to leptocauly vs. pachycauly; the implication was, simply, that the stems were erect and slender. There is no way to discern the nature of the internal structure, of course. It would not be incongruous if, indeed, early cycads were demonstrated to have been pachycaulous. Possible pteridosperm precursors (e.g., *Medullosa*) were certainly pachycaulous, yet their stems were not short and squat.

No matter the ancestry or time of appearance, there is no question of the dominance of cycadophytes during the Mesozoic, particularly during the Triassic and Jurassic periods. In 1975 (Delevoryas, 1975), I remarked about the abundance of cycadophyte foliage in Mesozoic sediments and made suggestions concerning their frequency. One suggestion was that cycad leaves were periodically abscised. This does not imply that cycads were deciduous, but some plants that are typically evergreen drop their leaves at certain times of the year. The excellent preservation of many fossil cycad leaves, with no evidence of shrivelling or breaking, suggests a much different pattern of leaf loss than is found among most extant cycads. Furthermore, fossil cycad leaves frequency occur in dense mats

⟶

Figure 9: Leaf of *Zamia integrifolia* broken off by wind.

Figure 10: Leaf of *Dioon spinulosum* broken off by wind.

Figure 11: *Microcycas calocoma*. This species has articulated pinnae, but they shrivel before articulation.

Figure 12: *Zamia furfuracea*. In this species pinnae are articulated, but drop off before shrivelling.

9
10
11
12

(Fig. 6). Most frequently among living cycads, leaves die on the plant and shrivel while they are still attached (e.g., *Dioon, Lepidozamia*) (Fig. 11). If such leaves were to have fallen off and become preserved as fossils, they would not have the appearance of many of the almost intact leaves in the fossil record. This suggests periodic abscission; there is often a clean break at the base of the petiole in the fossils. Leaves that abscised on a seasonally periodic basis would result in an accumulation of many leaves on the ground, and this would explain the frequent occurrence of fossil cycad leaves in dense mats.

One factor that should be considered is the role of wind in separating whole leaves from the plants. Illustrated (Figs. 9, 10) are leaves of *Zamia intergrifolia* and *Dioon spinulosum* broken off by wind. In both of these examples the leaves are relatively unscathed and could have produced fossils in an excellent state of preservation. However, I do not believe that wind alone would have accounted for such concentrations of cycadophyte leaves in the fossil records. The detached leaves figured were from unprotected cultivated plants, kept indoors for part of the year, in Austin, Texas. Naturally occurring cycads are more hardened to the elements and often grow adjacent to other, protecting plants. One seldom observes masses of detached leaves in the vicinity of extant cycad plants.

The occurrence of closely packed separated pinnae in the fossil record (Fig. 8) suggests that some kinds of cycadophytes did not drop their leaves entire, but had articulated pinnae. This feature is common among extant cycads (e.g., *Zamia, Microcycas*) (Stevenson, 1990, shows a number of cycad leaves with articulated pinnae) (Fig. 12). (*Microcycas* [Fig. 11] has articulated pinnae, but they tend to shrivel before they drop off). The same phenomenon appeared to have occurred in certain Mesozoic cycadophytes.

Cycads (Division Cycadophyta) and the extinct cycadeoids (Division Cycadeoidophyta) have many superficial similarities. There are almost identical leaf types in both groups (except for epidermal features). Method of leaf drop was the same in both groups among the Mesozoic taxa. In some fossil forms entire leaves fell off and were preserved as fossils. Also, in both groups, were leaves with articulated pinnae. Whether the similarities imply a natural relationship cannot be determined at present.

References

***Ash, S.R.* 1991.** A new pinnate cycad leaf from the Upper Triassic Chinle Formation of Arizona. Bot. Gaz. 152: (in press).

***Delevoryas, T.* 1975.** Mesozoic cycadophytes. In K.S.W. Campbell (Ed.), Gondwana geology. Papers from the Third International Gondwana Symposium Canberra, Australia, 1973. Australian National University Press, Canberra, A.C.T., Section 2, Gondwana flora 15: 173 - 191.

***Delevoryas, T.* 1982.** Perspectives on the origin of cycads and cycadeoids. Rev. Palaeobot. Palyn. 37: 115 - 132.

Delevoryas, T. and Hope, R.C. A new Triassic cycad and its phyletic implications. Postilla 150: 1 - 21.

***Delevoryas, T. and Hope, R.C.* 1976.** More evidence for a slender growth habit in Mesozoic cycadophytes. Rev. Palaeobot. Palynol. 21: 93 - 100.

***Delevoryas, T. and Taylor, T.N.* 1969.** A probable pteridosperm with eremopterid foliage from the Allegheny Group of Northern Pennsylvania. Postilla 133: 1 - 14.

***Xianming, D. and Jianen, Z.* 1982.** The emendation of a cycad genus *Yuania* and the discovery of *Y. chinensis* sp. nov. Mem. Beijing Nat. Hist. Mus. 17: 1 - 6 (in Chinese with English Abstract).

***Zhifeng, G. and Thomas, B.A.* 1989.** A review of fossil cycad megasporophylls, with new evidence of *Crossozamia* Pomel and its associated leaves from the Lower Permian of Taiyuan, China. Rev. Palaeobot. Palyn. 60: 205 - 223.

***Gillespie, W.H., and Pfefferkorn, H.W.* 1986.** Taeniopterid lamina on *Phasmatocycas* megasporophylls (Cycadales) from the Lower Permian of Kansas, U.S.A. Rev. Palaeobot. Palyn. 49: 99 - 116.

***Kimura, T., and Sekido, S.* 1975.** *Nilssoniocladus* n. gen (Nilssoniaceae n. fam.), newly found from the early Lower Cretaceous of Japan. Palaeontographica 153B: 111 - 118.

***Leary, R.L.* 1991.** Possible early Pennsylvanian ancestor of the Cycadales. Science 249: 1152 - 1154.

***Mamay, S.H.* 1966.** *Tinsleya*, a new genus of seed-bearing callipterid plants from the Permian of north-central Texas. U.S. Geol. Surv. Prof. Rap. 523E: E1 - E15.

***Mamay, S.H.* 1969.** Cycads - fossil evidence of late Paleozoic origin. Science 164: 295 - 296.

***Mamay, S.H.* 1973.** *Archaeocycas* and *Phasmatocycas* - new genera of Permian cycads. U.S. Geol. Surv. Jour. Res. 1, 6: 687 - 689.

***Mamay, S.H.* 1976.** Paleozoic origin of the cycads. U.S. Geol. Surv. Prof. Pap. 934: 1 - 48.

***Stevenson, D.W.* 1990.** Morphology and systematics of the Cycadales. Mem. New York Bot. Gar. 57: 8 - 55.

***Shanzhen, Z. and Xianyuan, L.* 1985.** Note on the discovery of cycadophytes with slender growth habit in the Permian of N. China with special reference to the reconstruction of cycadophytes. Jour. Northwestern Univ. 47(2): 76 - 80 (in Chinese with English abstract).

***Shanzhen, Z. and Zhuangguan, M.* 1981.** On the occurrence of cycadophytes with slender growth habit in the Permian of China. Geol. Soc. Amer. Spec. Pap. 187: 237 - 242.

***Shanzhen, Z. and Oingzhi, W.* 1986.** On the growth of habit of fossil cycadophytes. Invest. Stud. Nat. Mus. Hist. Nat. Shanghai 6: 5 - 19 (in Chinese with English abstract).

PGM Polymorphism in Cycads

Gaudio, L. and Esposito, E.
Dipartimento di Genetica, Universit di Napoli

Siniscalco Gigliano, G., Caputo, P., Muoio, R., and Moretti, A.
Dipartimento di Biologia Vegetale, Universit di Napoli

Abstract

An electrophoretic analysis on phosphoglucomutase has been carried out on cycads, to assess the systematic value of such an enzyme in these plants.

Methods

Leaflets were collected from several species of all genera but *Chigua*. The authors analysed 1 specimen of *Bowenia*, 3 specimens of *Ceratozamia*, 35 of *Cycas*, belonging to 13 species, 11 specimens of *Dioon*, belonging to 9 species, 53 specimens of *Encephalartos*, belonging to 30 species, 2 specimens of *Lepidozamia*, belonging to 1 species, 3 specimens of *Macrozamia*, belonging to 2 species, 1 specimen of *Microcycas*, 2 of *Stangeria*, and 21 specimens of *Zamia*, belonging to 7 species.

Total proteins were extracted on Tris-HC1 buffer, pH 8.0, purified through successive steps of ammonium sulphate precipitation and dialysed. Crucial steps towards a successful extraction were the presence of high quantities of several protease inhibitors and dialysis in presence of EDTA. Electrophoresis was carried out on starch gels, and samples stained with a specific enzymatic reaction.

Results

All the species examined showed only the presence of cytoplasmic PGM activity and no chloroplast specific PGM activity was detected. Our study allowed the separation of cycads into two groups (Fig. 1), according to the behaviour of PGM: *Bowenia, Ceratozamia, Stangeria* and *Zamia* showed one allozymic form, whereas all the others showed two isoenzymatic forms (due to the single sample analysed, authors cannot place *Microcycas* in any of the two abovementioned groups).

Within the first group, *Ceratozamia* and *Zamia*, which have proven to be a closely knit unit from a morphological point of view, show a very similar electrophoretic pattern from which the presence of the same alleles in the two genera can be inferred.

In the second group a thorough analysis has been carried out for *Encephalartos* and *Dioon*. These two

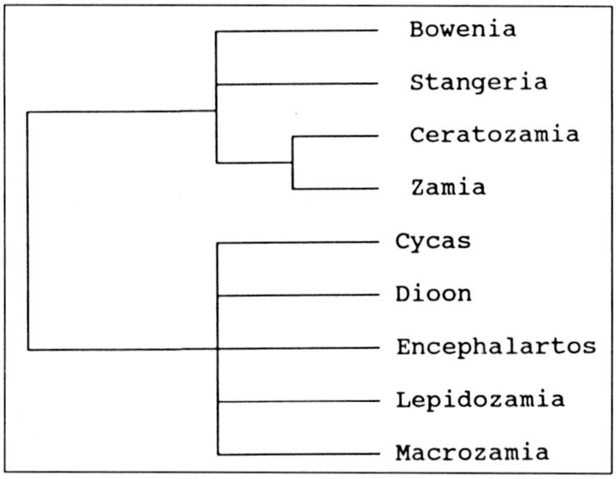

Figure 1: Phonogram of the studies taxa according to the number of PGM isozymes. Data for *Microcycas* did not allow its placement in the phonogram.

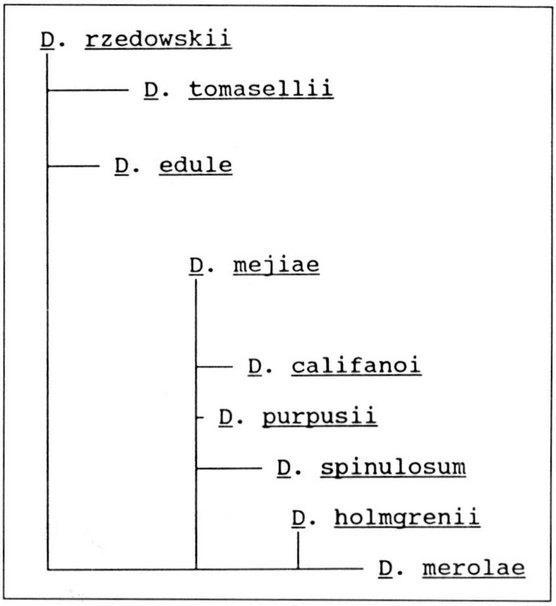

Figure 2: Phonogram for the studied species of the genus *Dioon*, using the Fitch and Margoliash method (1967).

genera too are regarded as similar from a morphologic point of view, and their electrophoretic patters are similar in showing comparable mobility. The presence of more electromorphs in *Dioon* than in *Encephalartos* implies that the former genus has more allelic variability than the latter, in spite of the higher number of species in this last genus. This unusual event may be explained by the wider ecological range of *Dioon; Encephalartos*, in fact, although distributed in a very wide geographical range, is more constant in its ecological requirements. In addition, *Dioon* shows polymorphisms for both the isoenzymatic forms, whereas *Encephalartos* shows polymorphism in only one locus.

The wider allelic variability in *Dioon*, allowed us to analyse the data pertaining to this genus further: a similarity coefficient was defined as the ratio between the number of shared bands and the total number of bands for each couple of taxa within the genus; a distance matrix was then prepared and analysed by using the Fitch and Margoliash method (1967) as implemented in the FITCH program in the PHYLIP software package (Felsenstein, 1985). The resulting phonogram is shown in Fig. 2. The phonetic pattern purported by our data does not match previous data on the genus (Sabato and De Luca, 1985). However, a formal phylogenetic or phonetic analysis is not yet available. Because of the discrepancies between the present phonogram and previous data, authors are currently analysing the genus by using molecular approaches (Siniscalco Gigliano, et al., this issue).

Conclusions

This preliminary report shows the feasibility of the study of enzymatic polymorphisms in cycads, in spite of the very low number of samples that it is possible to analyse. Further research is being presently carried out, in the hope that this approach can help elucidate the affinities among the genera.

References

***Felsenstein, J.* 1985.** Confidence limits on phylogenies: an approach using the Bootstrap. Evolution 39: 783 - 791.

***Fitch, W.M., and Margoliash, E.* 1967.** Construction of phylogenetic trees. Science 155: 279 - 284.

The Phylogeny of Some Cycad Genera: A Molecular Approach

Siniscalco Gigliano G., De Luca, P., Caputo, P.,
Moretti, A., Cozzolino, S.
Dipartimento di Biologia Vegetale
Universit di Napoli

Gaudio, L., and Tat, R.
Dipartimento di Genetica
Universit di Napoli

Abstract

A Restriction Fragment Length Polymorphism (RFLP) analysis on chloroplast DNA (cpDNA) was performed on several genera of cycads in order to obtain information on phylogenetic relationships within genera.

Methods & Results

Twenty species, belonging to all genera but *Chigua*, were analysed, namely *Bowenia serrulata, Ceratozamia hildae, C. matudae, Cycas circinnalis, C. media, C. revoluta, C. rumphii, Dioon edule, D. mejiae, D. spinulosum, Encephalartos altensteinii, E. villosus, Lepidozamia peroffskiana, Macrozamia communis, M. macdonnellii, M. secunda, Microcycas calocoma, Stangeria eriopus, Zamia furfuracea*, and *Z. paucijuga*. CpDNA-enriched total DNA was extracted from leaflets, digested with five restriction endonucleases (*Bam*HI, *Eco*RI, *Eco*RV, *Hind*III, *Pst*I), electrophoresed, transferred to nylon filters, prehybridised and hybridised (50ºC, 0.45M NA^+) against several radiolabelled fragments of a cpDNA library from *Petunia hybrida*. Filters were then washed (the most stringent conditions were 50ºC, 1xSSC, 30') and exposed to X-ray sensitive film. Hybridising restriction bands were scored for presence/absence and analysed by using an exhaustive strict parsimony algorithm with Hennig86 software. All genera, except *Macrozamia*, proved to be strictly monophyletic, being identified by unique restriction patterns. Infrageneric variability has proven to be very narrow, although variability in morphological characters has been reported in some of these genera. However, infrageneric differences provide some useful information about the

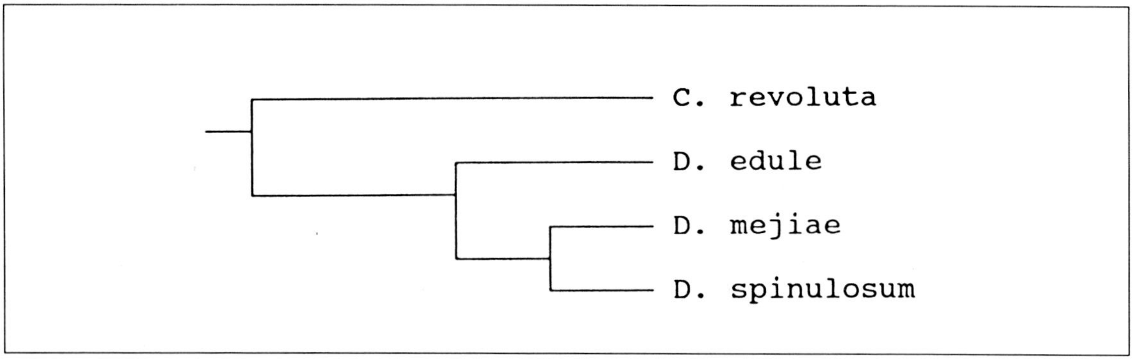

Figure 1: Cladogram showing the phylogenetic relationships in the species of the genus *Dioon* used in this study. *Cycas revoluta* was used as an outgroup.

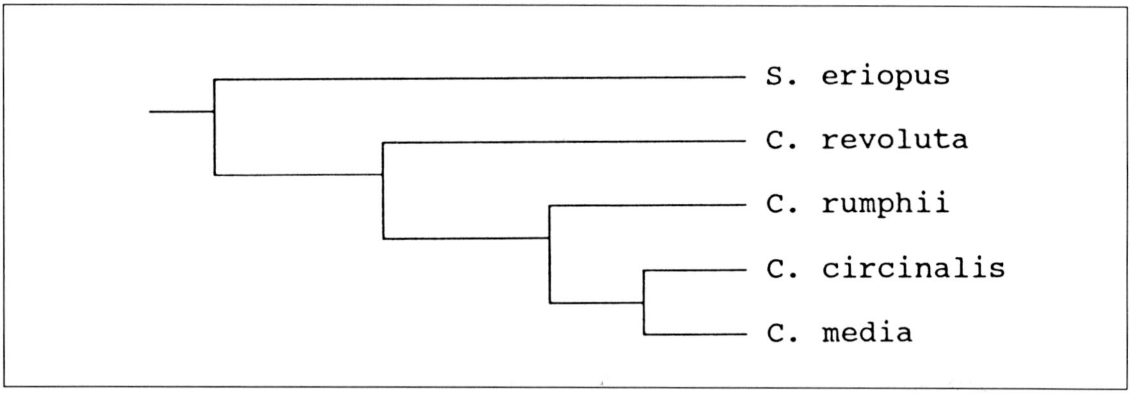

Figure 2: Cladogram showing the phylogenetic relationships in the species of the genus *Cycas* used in this study. *Stangeria eriopus* was used as an outgroup.

phylogenetic relationships among the species of *Dioon* and *Cycas*. For *Dioon* (by using *Cycas revoluta* as an outgroup), the phylogenetic pattern shows (Fig. 1) that *D. mejiae* is a sister group to *D.spinulosum*, and that *D.edule* is primitive.

Conclusions

The former data supports the morphological relationships reported between the two species, but the latter does not confirm the derivativeness of *D. edule*, suggested on morphological bases. For *Cycas* (by using *Stangeria* as an outgroup), our data show (Fig. 2) that *C. revoluta* is the most primitive among the examined species, and *Cycas circinnalis* and *C. media* are derived.

For some of the other genera, problems are still to be solved. For example, although *Encephalartos* is clearly the sister group for part of the genus *Macrozamia*, the later genus appears to be polyphyletic. Authors do not know yet whether this unusual result depends upon the small size of our data set or on some real problems concerning the origin of *Macrozamia*.

Completion of the analysis with all the available probes will probably clarify this and other controversial points.

History of the Taxonomy of Australian Cycads

K.D. Hill
National Herbarium of New South Wales
Royal Botanic Gardens, Sydney NSW 2000, Australia

Abstract

The history of the discovery and naming of the Australian cycads is discussed, with comments on the major contributors to their taxonomy. Currently accepted Australian species are listed, with their synonyms, and details of types and type localities for all Australian names.

Taxonomic History

The cycads were not known to the very early European herbalists and naturalists who provided the base for our present systems of plant classification. This is because none occurred in areas familiar to them. With the Renaissance and age of European exploration and expansion, innumerable new and exotic peoples, plants and animals became known to the Europeans. Among these were the first cycads. In particular, *Cycas circinalis* was of considerable economic significance as a 'Sago palm' of Southeast Asia. This plant was described by Hendrik Adriaan van Rheede tot Draakestein, who was a Dutch colonial administrator and representative of the Dutch East India Company. He recorded it in his 'Hortus Indicus Malabaricus' in 1862 under the local Indian name of 'Todda panna'.

Modern botanical classification and nomenclature was established by Linnaeus with his monumental 'Species Plantarum' in 1753, in which he built on and adapted the works of earlier naturalists. He recognised two genera of cycads, *Cycas* from Southeast Asia, and *Zamia* from the New World. Each was described with a single species. *Cycas circinalis* was based in part on Rheede's illustration, but also included other early illustrations which are now considered to represent other species.

The genus *Zamia* was used as an 'umbrella' category by subsequent workers to include species not only from the New World, but also from Africa and Australia. These latter are now placed in other genera such as *Encephalartos* and *Macrozamia*.

The first Australian cycad recognised was what is now called *Macrozamia spiralis*. This was described in 1796 by the British botanist Richard Salisbury, in an account of plants grown at Chapel Allerton. He based it on material collected by David Burton at Port Jackson, and placed it in the genus *Zamia*. His material does not appear to have survived. The identity of this species was subsequently greatly confused until unravelled by the landmark work of Lawrie Johnson (1959, 1961).

Even greater confusion surrounds *Zamia tridentata*, named by the German botanist Carl Ludwig Willdenow in 1805 in his revised 'Species plantarum'. This was named from a juvenile plant in cultivation, the seed coming from either South Africa or Australia. This may never be resolved, since it is almost impossible to determine juvenile *Macrozamia* plants to species. In fact, it is often difficult to separate early juvenile *Macrozamia* from *Encephalartos*.

The British botanist Robert Brown (1810) then recognised several more species, although he confused *Macrozamia spiralis* with two species now known as *Macrozamia communis* and *Macrozamia reidlei*. This confusion continued until resolved by Johnson in 1959. Brown also recognised two *Cycas* species, *Cycas media* and *Cycas angulata*. These he collected during the circumnavigation of Australia with Matthew Flinders in the years 1802 - 1803.

Several more species were described in the next decades, but most were based on limited material and little field knowledge. The genus *Encephalartos* was established in 1834 by the German botanist Johann Georg Christian Lehman. It included African and Australian segregates from *Zamia*. He transferred to this new genus the two Australian species now known as *Macrozamia spiralis* and *Macrozamia riedlei*.

The Dutch botanist Freidrich Anton Wilhelm Miquel was successively director of botanic gardens at Rotterdam, Amsterdam and Utrecht. He studied the cycads over a thirty-year period (1840 to 1870; see Read & Solt, 1986). His was a major and outstanding contribution, although he was known to change his mind at times. He described the genus *Macrozamia* in 1842 to cover Australian species, but later reconsidered, placing *Macrozamia* and the then recently described *Lepidozamia* back in *Encephalartos* as sections (1863). At the same time, he created the section *Parazamia*. In 1868, he returned to his previous stand, restoring *Macrozamia* as a genus with a section *Lepidozamia*.

During this period, Miquel named various species: *Macrozamia macleavi* (now *M. miquelii*), *M. fraseri* and *M. oldfieldii* (now *M. riedlei*) and *M. macdonnelli* (taken from a manuscript from Ferdinand von Mueller). He also named *Cycas armstrongii* and *Cycas gracilis* (now *C. media*).

The famous Australian botanist Ferdinand von Mueller of the Melbourne Botanic Gardens published a number of notes and new species in the period 1858 to 1889 (see Read & Solt, 1986). His work was, however, limited by little field knowledge and frequent changes of mind. He alternated freely between *Macrozamia* and *Encephalartos* as the appropriate generic position for the Australian species. This has led to a confusing number of synonyms in these genera. To his credit, Mueller stimulated a 'Golden Age' in Australian botany, creating great interest and leading to extensive field collection by others, not only in cycads but in all plant groups.

Mueller described *Macrozamia moorei, M. miquelii, M. pauli-guilielmi, M. diplomera* (as a variety of *M. spiralis*), *M. denisonii* (now *Lepidozamia peroffskyana* and *M. dyeri* (now *M. reidlei*). He also named *Cycas cairnsiana, C. kennedyana* and *C. normanbyana*.

The middle 1800s was a pinnacle of systematic botany as a science, with very many extremely able people active in the discipline. Techniques and concepts remained practically unchanged form that time to the 1970's, when the new molecular techniques and new philosophies of evolutionary biology developed. Most of the botanists worked from Europe on herbarium specimens, often without the benefit of field examination of the groups they studied.

Workers in that period who contributed to cycad knowledge included the Kew botanists, William Jackson Hooker, his son Joseph Dalton Hooker, and George Bentham. Joseph Dalton Hooker (1863) described the genus *Bowenia*, with one species. The name *Bowenia* had earlier been proposed by his father but not published. George Bentham wrote the entire Flora of Australia without leaving Kew, based on often inadequate herbarium material. In many groups, including cycads, his treatment has stood up better than the work of Mueller who had the opportunity to study the plants in the field. Bentham's work on cycads (1873) was a compilation, with no new species described.

The German-born Russian botanist Eduard August von Regel contributed valuable studies during this period (see Read & Solt, 1986). His work was mostly based on extensive collections of living material that he accumulated while scientific director of the Imperial Botanic Gardens of St. Petersburg. He described the genus *Lepidozamia* (1857), which was included in *Macrozamia* by following workers until recognised again by Johnson in 1959.

Also of this era was the Swiss botanist Alphonse de Candolle of Geneva. He treated the cycads in the Prodromus (1868), a major work intended to catalogue and classify the world's flora, which was edited by him, his father and his son at different times. This treatment largely followed Miquel's earlier work.

The works of Bentham and de Candolle were on a grand scale. Australian botanists of the late 1800s also published on the cycads, although their works were much more modest. Charles Moore, a director of the Botanic Gardens of Sydney, studied *Macrozamia* in the field and in cultivation in Sydney. He published these studies in 1884, recognising a number of new species and showing a good understanding of the species. He had, however, followed the confused nomenclature of earlier workers. New species he described were *Macrozamia secunda, M. fawcettii, M. heteromera* and *M. flexuosa* (now a subspecies of *M. pauli-guilielmi*).

Frederick Manson Bailey, government botanist of Queensland treated the cycads in his descriptive catalogues and flora of Queensland (see Read & solt, 1986). He also studied the plants in the field. He named *Macrozamia platyrachis* and *M. mountperriensis.*

The monograph published in Das Pflanzenreich by Julius Schuster in 1932 is almost breath-taking in its confusion. Schuster was primarily an historian, starting as curator of various scientific archives and becoming professor of the history of biology at the University of Berlin. In his taxonomic work, he ignored the basis rules of priority in nomenclature, and created infraspecific hierarchies with subspecies, varieties and forms that bear no apparent relationship to reality. He placed closely-related species in widely separated categories with obviously unrelated taxa. For example, he treats *Cycas armstrongii* as a synonym of *Cycas circinalis*, while regarding *Cycas lanepoolei* (now regarded as a synonym of *C. armstrongii*) as a variety of *Cycas media.*

The lowest point of Schuster's work on the Australian cycads was his treatment of *Macrozamia.* Subsequently, this has been clearly resolved by Dr Lawrie Johnson of the Royal Botanic Gardens of Sydney, in a paper published in 1959. Schuster's *Cycas* study is equally confused, but it is easier to resolve the Australian species in that group. Johnson did not study that group, and to date no adequate monographic treatment of *Cycas* exists. The late John Maconochie had begun work on the Australian species, and had resolved some of the problems (1978). He unfortunately died in a motor accident in Somalia before he could tackle the difficult *Cycas media* complex.

The nomenclatural confusion in *Macrozamia* and its eventual resolution by Johnson illustrate clearly the necessity for following the rules of nomenclatural priority and the importance of the type specimen in defining species. Finding the type specimen is not always easy with many early names. Even when located, early specimens are frequently found to be totally inadequate in distinguishing a taxon from later segregates. This then necessarily leads to locating and visiting type localities to study the populations from which early type specimens were collected.

Current Taxonomic Status - Cycad Families

All cycad species were by general consensus placed in the one family *Cycadaceae* until the critical studies by Lawrie Johnson. Non-conformists had earlier suggested division into more than one family, but had been largely ignored. The German botanist Heinrich Gustav Ludwig Reichenbach, director of the Dresden Botanic Garden, studied cycads. He placed all genera except *Cycas* in the family *Zamiaceae*, in 1837 in his 'handbuch des naturlichen Pflanzensystems'. This was however followed by very few botanists. Johnson's work clearly indicated that the cycads fell into three groups, which he recognised as the families *Cycadaceae, Stangeriaceae* and *Zamiaceae*. This has been almost universally followed, with the studies of Dennis Stevenson later illustrating the separate nature of *Boweniaceae* (1981). Family relationships are becoming clearer with recent morphological, cladistic and macromolecular studies, although apparently some DNA studies do not show congruent patterns with morphological studies.

Australian Cycads

Boweniaceae

Boweniaceae is a monogeneric Australian endemic family, with 3 species in Queensland.

Pinnule morphology varies, and each species has been erroneously recorded from within the range of another when identification has been solely on the basis of this character. The two named species are *Bowenia spectabilis*, occurring in wet forest country between Cooktown and Cardwell, and *Bowenia serrulata*, occurring in the ranges northeast of Rockhampton. An undescribed taxon occurs in the Tinaroo district of the Atherton Tableland.

Zamiaceae

The *Zamiaceae* is a large and diverse family with 8 genera and over 100 species, occurring in Africa, Australia and Central America. Two endemic genera with about 20 species occur in Australia.

The Australian genera are distinguished on the basis of frond and female cone morphology. The Australian *Zamiaceae* was in a state of taxonomic confusion virtually since the description of the first species in 1796, until Lawrie Johnson resolved much of the confusion with his work published in 1959 and 1961. His work stands largely unaltered.

The first genus is *Lepidozamia*, with two species occurring disjunctly in wetter areas of eastern Australia. *Lepidozamia peroffskyana* grows in wet sclerophyll forests of southern Queensland and northern New South Wales, from near Rockhampton to the Manning River. *Lepidozamia hopei* is found in rainforests of northern Queensland, roughly between Cooktown and Cardwell.

The second genus is *Macrozamia*, with 16 species occurring across mainly temperate regions of Australia. Two sections are recognised, sections *Macrozamia* and *Parazamia*.

The first section, *Macrozamia*, is distinguished by the presence of mucilage canals in the leaves. Plants also tend to be large and palm-like, with heads of numerous fronds and sometimes an erect aerial caudex. The fronds often have reduced spine-like pinnae towards the base (except in *M. mountperriensis*), and veins within pinnae are not raised. Species in this are typically associated with areas recognised as refugia, such as the southwest corner of Western Australia, the MacDonnell Ranges in Central Australia and the Carnarvon Range in Central Queensland.

Section *Parazamia* is distinguished by the absence of mucilage canals from the leaves. Plants are small, usually with a small number of leaves in the crown (as few as 2 in *Macrozamia secunda*). The caudex is small and subterranean, and veins are often prominent in the pinnae. Lower pinnae on fronds are not reduced to spines.

The section shows interesting distribution and inferred evolutionary patterns. The cycads are generally accepted as an ancient group, and many taxa worldwide show clearly relictual distributions. Sections *Parazamia*, however, shows a close geographic replacement pattern consistent with relatively recent expansion and speciation. This is typical of distribution patterns displayed by many modern plant groups that are actively evolving and speciating. The conclusion is that this section is a vigorous and actively evolving group and not a dead-end relic as many would regard the cycads.

Cycadaceae

The *Cycadaceae* is a monogeneric family. The genus *Cycas* comprises over 20 species occurring mainly in Australia and Southeast Asia, extending to southern China, Polynesia, Madagascar and East Africa. Species delineation is very difficult, and many more names exist than species. This is probably the group in the cycads that is in most need of critical taxonomic study. About 14 species occur in Australia, 11 of them endemic.

Evidence is mounting that there are weak fertility barriers between cycad species (Vorster 1986, Norstog 1990). This is also shown by the numbers of natural hybrids being discovered. The lack of pollinator specificity when combined with the apparently weak inherent fertility barriers means that geographic separation is the only real reproductive barrier between *Cycas* species in nature. If this hypothesis is correct, two predictions can be made. The first is that natural *Cycas* species will be geographically distinctly separated, and the second is that hybrid and intermediate zones will be found where different species have spread to within pollination range of each other.

Naturally occurring hybrids and intergrades have been recorded from N.T., where 4 species grow in relatively close proximity. Here, hybrids are reported between *C. armstrongii* and *C. conferta, C. calcicola* and *C. armstrongii*, and *C. calcicola* and an undescribed species. In Queensland, the Irvinebank cycad population is morphologically and geographically intermediate between *C. media* and an undescribed species, and also shows the considerable variability to be expected in hybrid swarms.

Breakdown of breeding barriers would also appear to have contributed to the complexity of the '*C. media* complex' in eastern Queensland. Here, it seems that several discrete species have arisen at some time in the past, probably when the range of an ancestral taxon was fragmented by climatic change. More recently, their ranges have expanded to meet and overlap. This has resulted in the present almost continuous variation in this complex in Queensland.

Accepted Species, Synonyms and Type Localities

Accepted names are numbered sequentially and listed in **bold**, synonyms are listed under the appropriate accepted names.

Bowenia

1. ***Bowenia spectabilis*** Hook. ex Hook. f., Bot. Mag., 89 (1863), sub t.5398. Type: the illustration, Bot. Mag., t.5398 (1863).

2. ***Bowenia serrulata*** (W. Bull) Chamberlain, Bot. Gaz. 54: 419 (1912); *Bowenia spectabilis* Hook. ex Hook. f., var. *serrulata* W. Bull, Catal., 4, t.5 (1878). Type: the illustration, W. Bull, Catal., t.5 (1878).

One undescribed species also occurs in northern Queensland.

Lepidozamia

1. ***Lepidozamia hopei*** Regel, Gartenfl., 6 (1876). Type: Based on a living plant grown by Haage and Schmidt.

Cetakidozamia hopei W. Hill, Gard, Chron., 1107 (Nov., 1865); *Macrozamia hopei* W. Hill ex Bailey, Syn. Queensland Fl., suppl. 1:52 (1886). Type: 'Daintree and Johnstone Rivers' (holo BRI)

Macrozamia denisonni C. Moore & F. Muell. var. *hopei* (W. Hill) Schuster, Pflanzenr. 4(1): 101 (1932). Type: 'from tropical Eastern Australia'.

2. ***Lepidozamia peroffskyana*** Regel, Bull. Soc. Imp. Nat. Mosc. 1: 184 (1857). Type: From a living plant in the Botanic Garden, St. Petersbourg.

Macrozamia denisonii C. Moore & F. Muell., in F. Muell., Fragm. 1: 41 (1858). Type: Manning River, N.S.W., Stephenson (lecto MEL; isolecto N.S.W., fide Johnson 1959).

Macrozamia

1. ***Macrozamia moorei*** F. Muell, Austral. Chem. Drugg. 4: 84 (1881).

Encephalartos moorei (F. Muell) F. Muell., Fragm. 11: 1225 (1881). Type '... Nogoa River ... Springsure ... J.G. Macdonald; P.A. O'Shanesy.' (syntypes MEL).

2. ***Macrozamia riedlei*** (Fisch. ex Gaudich.) C. Gardner, Enum. Pl. Austral. Occid., 3 (1930).

Cycas riedlei Fisch. ex Gaudich., in Freycinet, Voy. Aut. Monde, Bot., 434 (1826). Type: "In Novae Hollandiae ora occidentali Port du roi George" (holo P).

Macrozamia fraseri Miq., Monogr. Cycad., 37 (1842); *Encephalartos fraseri* (Miq.) Miq., Versl. Meded. Koninkl. Akad. 15: 368 (1863). Type: Geographe Bay, W.A. Fraser (holo U).

Macrozamia preissii Lehm., Nov. Stirp. Pug. 8: 31 (1844).; *Encephalartos preissii* (Lehm.) F. Muell., Quart. J. Pharm. Soc. Victoria 2: 90 (1859). Type: Western Australia, Preiss 1313 (type not seen).

Encephalartos oldfieldii (Miq., Versl. Meded. Koninkl. Acad. 15: 370 (1863); *Macrozamia oldfieldii* (Miq.) A.DC., Prodr. 16 (2): 535 (1868). Type: Swan River, W.A., Oldfield (holo U).

Encephalartos dyeri F. Muell., Austral. Chem. Drugg. 8: 12 (1855); *Macrozamia dyeri* (F. Muell.) C. Gardner, Enum. Pl. Austral. Occid., 3 (1930); *Macrozamia preissii* Lehm. ssp. *dyeri* (F. Muell.) Schuster, Pflanzenr. 4 (1): 102 (1932). Type: Coast near Esperance, W. Knight & W. Webb (holo MEL).

3. ***Macrozamia macdonnellii*** (F. Muell. ex. Miq.) A.DC., Prodr. 16 (2) 537 (1868).

Encephalartos macdonnelli F. Muell. ex. Miq., Versl. Meded. Koninkl. Akad. 15: 376 (1863). Type: "... flumen Neales ... Macdonnell-range ... J.M. Stuart". Collected by the explorer John McDouall Stuart in the MacDonnell Ranges. The species does not occur near the Neales River as it is presently known.

4. ***Macrozamia miquelii*** (F. Muell.) A.DC., Prodr. 16 (2): 535 (1868).

Encephalartos miquelii F. Muell., Fragm. 3: 38 (1862). *Macrozamia tridentata* (Willd.) Regel ssp. *mountperriensis* (Bailey) Schuster var. *miquelii* (F. Muell.) Schuster, in Engl., Pflanzenr. 4 (1): 90 (1932). Type: Rockhampton, Qld, Thozet (lecto MEL, fide Johnson 1959).

Encephalartos spiralis (Salisb.) Lehm. var. *major* Miq., Versl. Meded. Koninkl. Akad. 15 368 (1863). Type: three types are cited, from the Richmond River (Moore), Moreton Bay (Hill) and Fitzroy River (Thozet). All specimens are *M. miquelii.*

Macrozamia macleayi Miq., Arch. Neerl. 3: 250 (1868). Type: Moreton River Region, Qld., Macleay (holo U?).

Macrozamia tridentata (Wild.) Regel var. *oblongifolia* Regel, Acta Hort. Petrop. 4: 320 (1876). Type: possibly the type of *M.miquelii* (collected at Rockhampton by Thozet).

Macrozamia mackenzii Hort. ex Mast., Guard. Chron. n.s. 7: 665 (1877); *Macrozamia tridentata* (Wild.) Regel ssp. *mountperriensis* (Bailey) Schuster var. *mackenzii* (Hort. ex Mast.) Schuster, i.c. 90. Type: Rockhampton, J.L. Boorman (holo MEL).

Macrozamia tridentata (Wild.) Regel ssp. *mountperriensis* (Bailey) Schuster var. *miquelii* (F. Muell.) Schuster forma *milkaui* Schuster, 1.c. 90. Type: Rockhampton, L. Diels 8249, 11.5.1902 (holo B).

Macrozamia cylindrica C. Moore, J. & Proc. Roy. Soc. New South Wales 17: 119 (1884); *Macrozamia tridentata* (Wild.) Regel ssp. *cylindrica* (C. Moore) Schuster, 1.c. 91. Type: Upper Richmond River, N.S.W. (holo N.S.W. 40610).

5. ***Macrozamia douglasii*** W. Hill ex Bailey, Syn. Queensland Fl., 500 (1883). Type: Fraser Island Qld, J. Douglas (holo BRI).

Encephalartos douglasii F. Muell., Chem. Drugg. Austral. 4: 80 (1883); *Macrozamia tridentata* (Wild.) regel ssp. *mountperriensis* (Bailey) Schuster var. *douglasii* (F. Muell.) Schuster, 1.c. 90: Type: Fraser Island, Qld, J. Douglas (holo MEL).

6. ***Macrozamia mountperriensis*** Bailey, Syn. Queensland Fl., Suppl. 1, 50 (1886). Type: Mount Perry, Qld. (holo BRI).

Macrozamia tridentata (Willd.) Regel ssp. *mountperriensis* (Bailey) Schuster, 1.c. 89.

7. ***Macrozamia communis*** L. Johnson, Proc. Linn. Soc. New South Wales 64 (1): 98 (1959). Type: About

three miles west of Mossy Point, near Mogo, N.S.W., J.W. Vickery N.S.W. 43071, 4.1958 (holo N.S.W.).

Macrozamia tridentata (Wild.) Regel ssp. *cylindrica* (C. Moore) Schuster var. *corallipes* (Hook. f.) Schuster forma *wallsendensis* Schuster, in Engl., Pflanzenr. 4 (1): 93 (1932). Type: Wallsend, J.L. Boorman 10.1899 (holo MEL).

Macrozamia tridentata (Wild.) Regel ssp. *mountperriensis* (Bailey) Schuster var. *miquelii* (F. Muell.) Schuster forma *oblongifolia* (Regel) Schuster, 1.c. 90. Type: Port Jackson, F. Mueller 1881 (holo MEL).

8. ***Macrozamia diplomera*** (F. Muell.) L. Johnson, Proc. Linn. Soc. New South Wales 84: 98 (1959).

Encephalartos spiralis (Salisb.) Lehm. var. *diplomera* F. Muell., Fragm. 5: 172 (1866).

Macrozamia spiralis (Salisb.) Miq. var. *diplomera* (F. Muell.) A.DC., Prodr. 16: 535 (1868).

Macrozamia tridentata (Wild.) Regel. ssp. *cylindrica* (C. Moore) Schuster var. *pungens* (Ait.) Schuster forma *diplomera* (F. Muell.) Schuster, in Engl., Pflanzenr. 4 (1): 94 (1932). Type: Castlereagh River, N.S.W., W. Carron (lecto MEL, fide Johnson 1959).

9. ***Macrozamia platyrachis*** Bailey, Queensland Agric. J. 3: 356 (1898). Type: Range near Planet Downs, Qld, H.C. Brock-Hollingshead (holo BRI, iso N.S.W.).

10. ***Macrozamia lucida*** L. Johnson, Proc. Linn. Soc. New South Wales 84: 102 (1959). Type: Southern side of Ngungun, abt. 400ft. alt., Glass House Mountains, Qld, L.A.S. Johnson N.S.W. 40668, 13.6.1951 (holo N.S.W.). N.B. All records of *Macrozamia lucida* from New South Wales refer to juvenile plants of *Macrozamia moorei*.

11. ***Macrozamia spiralis*** (Salisb.) Miq., Monogr. Cycad., 36 (1842).

Zamia spiralis Salisb., Prodr. Stirp., 401 (1796); *Encephalartos spiralis* (Salisb.) Lehm., Pugill. 6: 13 (1834). Type: none preserved.

Macrozamia corallipes Hook. f., Bot. Mag 98: sub t. 5943 (1872); *Macrozamia spiralis* (Salisb.) Miq. var. *corallipes* (Hook. f.) Benth., Fl. Austral. 6: 252 (1873); *Macrozamia tridentata* (Wild.) Regel. ssp. *cylindrica* (C. Moore) Schuster var. *corallipes* (Hook. f.) Schuster, in Engl., Pflanzenr. 4 (1): 93 (1932). Type: Cult. W. Bull, the holotype is the plate, Bot. Mag. 98: t. 5943 (1872).

Macrozamia tridentata (Wild.) Regel. ssp. *cylindrica* (C. Moore) Schuster var. *corallipes* (Hook. f.) Schuster forma *dielsii* Schuster, 1.c. Type: Mt Dangar, Gungal, J.L. Boorman, 11.1904 (holo MEL).

12. ***Macrozamia secunda*** (C. Moore, J. & Proc. Roy. Soc. New South Wales 176: 122 (1884).

Macrozamia spiralis (Salisb.) Miq. var. *secunda* (C. Moore) Maiden & Betche, Census, 9 (1916); *Macrozamia tridentata* (Wild.) Regel. ssp. *cylindrica* (C. Moore)(Schuster var. *secunda* (C. Moore) Schuster, in Engl., Pflanzenr. 4 (1) 91 (1932). Type: Reedy Ck., near Mudgee, N.S.W., C. Moore, 1858 (lecto MEL, fide Johnson 1959).

Macrozamia spiralis (Salisb.) Miq. var. *secunda* Benth., Fl. Austral. 6: 252 (1873). Type: Reedy Creek, N.S.W., C. Moore (holo K).

13. ***Macrozamia heteromera*** C. Moore, J. & Proc. Roy. Soc. New South Wales 17: 122 (1884).

Macrozamia spiralis (Salisb.) Miq. var *heteromera* (C. Moore) Maiden & Betche, Census, 9 (1916). Type: Warrumbungle Ranges, N.S.W., N.S.W. 40720; lecto: N.S.W., fide Johnson (1959).

Macrozamia heteromera C. Moore var. *glauca* C. Moore, J. & Proc. Roy. Soc. New South Wales 17: 122 (1884). Type: Near Narrabri, N.S.W., E. Betche (holo N.S.W.).

Macrozamia heteromera C. Moore var. *tenuifolia* Schuster forma *harmsii* Schuster, in Engl., Pflanzenr. 4 (1): 96 (1932). Type: Pilliga Scrub, J.H. Maiden (syntype N.S.W.); Narrabri, J.L. Boorman, 6.1907 (syntype MEL); Warrumbungle-range, E. Betche, 1.1883 (syntype MEL).

14. ***Macrozamia stenomera*** L. Johnson, Proc. Linn. Soc. New South Wales 84: 106 (1959). Type: Above Coryah Gap, Nandewar Mtns., N.S.W., L.A.S. Johnson & E.F. Constable N.S.W. 32204 (holo N.S.W.).

Macrozamia heteromera C. Moore var. *tenuifolia* C. Moore, J. & Proc. Roy. Soc. New South Wales 17: 122 (1884); *Macrozamia heteromera* C. Moore var. *tenuifolia* Schuster, in Engl., Pflanzenr. 4 (1): 96 (1932), excl. forma *harmsii*. Type: Near Tamworth, N.S.W., E. Betche (holo N.S.W.).

Macrozamia heteromera C. Moore var. *dicranophylloides* Schuster, l.c. Type: Bingara, J.L. Boorman, 9.1907 (holo MEL).

15. ***Macrozamia pauli-guilielmi*** W. Hill & F. Muell., Fragm. 1: 86 (1859).

Encephalartos pauli-guilielmi (W. Hill & F. Muell.) F. Muell., Quart. J. Pharm. Soc. Victoria 2: 91 (1859). Type: Vicinity of Moreton Bay, Qld, W. Hill (holo MEL).

15b. ***ssp. plurinervia*** L. Johnson, Proc. Lin. Soc. New South Wales 84: 108 (1959). Type: "Reedy Creek" Station, near Bonshaw, N.S.W., J. Leader N.S.W. 40958, 4.1956 (holo N.S.W.).

15c. ***ssp. flexuosa*** (C. Moore) L. Johnson, Proc. Linn. Soc. New South Wales 84: 109 (1959).

Macrozamia flexuosa C. Moore, J. & Proc. Roy. Soc. New South Wales 17: 121 (1884).

Macrozamia flexuosa C. Moore, J. & Proc. Roy. Soc. New South Wales 17: 121 (1884).

Macrozamia spiralis (Salisb.) Miq. var. *flexuosa* (C. Moore) Maiden & Betche, Census, 9 (1916). Type: Limeburner's Creek, N.S.W., E. Betche N.S.W. 40951, 1.1883 (holo N.S.W.).

16. ***Macrozamia fawcettii*** C. Moore, J. & Proc. Roy. Soc. New South Wales 17: 120 (1884).

Macrozamia spiralis (Salisb.) Miq. var. *fawcettii* (C. Moore) Maiden & Betche, Census, 9 (1916). Type: Upper Richmond River, N.S.W., C. Fawcett N.S.W. 40653 (holo N.S.W.).

The population near Grafton previously regarded as *M. moorei* is now thought to represent another species, and *M. pauli-guilielmi* is recognised as a complex possibly involving two species and four or five subspecies. These are to be discussed in forthcoming taxonomic publications.

Cycas

1. ***Cycas calcicola*** Maconochie, J. Adelaide Bot. Gard. 1: 175 (1978). Type: 16km N. of Katherine, N. Terr., J.R. Maconochie 1314, 10.6.1971 (holo: NT; iso: BRI, CANB, K, L, PERTH).

2. ***Cycas pruinosa*** Maconochie, J. Adelaide Bot. Gard. I: 177 (1978). Type: Teronis Gorge, Durack Ranges, W. Aust (17º 25'S, 127º 20'E), D. Symon, 6.1975 (holo: NT; iso; BRI, CANB, K, L, N.S.W., PERTH).

3. ***Cycas cairnsiana*** F. Muell., Fragm. 10: 63 (1876). Typ: "Ad montes Newcastle-Range; Armit." (holo MEL 68032). Recent collection form O'Briens Creek match the type.

4. ***Cycas armstrongii*** Miq., Arch, Neerl. 3: 235 (1868).

Cycas media R.Br. forma *inermis* Miq., Linnaea 19: 412 (1842); *Cycas media* R.Br. var. *inermis* A.DC., Prodr. 16 (2): 528 (1868). Type: "Ad Port Essington (Armstrong 380 in h. Hook.)" (holo U 028107). This species does not occur at Port Essington, the specimen probably being gathered in the hinterland some distance to the southeast.

Cycas lane-poolei C. Gardner, For. Dept. Western Austral. Bull. 32: 30 (1923); *Cycas media* R.Br. var. *lane-poolei* (C. Gardner) Schuster, in Engl., Pflanzenr. 99: 78 (1932). Type: "The type is No. 144 of the Forests Department Herbarium." Collected by Gardner near Mt Hann, W. Aust. (holo PERTH).

Cycas papuana F. Muell., Descr. Notes Papuan Pl. 1 (4): 71 (1876). Type: "On the Fly-River; D'Albertis" (holo MEL 68056). Recent collections from the Western Province of Papua New guinea match the type, and also match *C. armstrongii*.

5. ***Cycas basaltica*** C. Gardner, For. Dept. Western Austral. Bull. 32: 31 (1923); *Cycas media* R.Br. var. *basaltica* (C. Gardner) Schuster, in Engl., Pflanzenr. 99: 78 (1932). Type: "The type is 1490, of the Forests Department Herbarium." Collected by Gardner, from basalt hills near the Lawley River, W. Aust. (holo PERTH).

6. ***Cycas furfuracea*** W. Fitzg., J. & Proc. Roy. Soc. Western Austral. 3: 108 (1918).

Cycas media R.Br. var. *furfuracea* (W. Fitzg.) Schuster, in Engl., Pflanzenr. 99: 78 (1932). Type: "Summits of Mts. Herbert, Broome, and Bold Bluff (W.V.F.). "Collected by Fitzgerald (syntypes PERTH).

7. ***Cycas angulata*** R.,Br., Prodr., 348 (1810). Type: "(T.) v.v.". Probably collected on 3-4 Dec. 1802 from Bountiful Island, near Mornington Island in the Gulf of Carpentaria, fide Groves & Moore 1989 (holo BM, Bennet 3016 p.p.). Recent collections from the same island match the type.

8. ***Cycas media*** R.Br, Prodr., 348 (1810): Type: "(T.) v.v.". Probably collected on 16 Oct 1802 from Calder Island in the Cumberland Group off Mackay, fide Groves & Moore 1989 (holo BM, Bennet 3016 p.p.).

Cycas gracilis Miq., Versl. K. Acad. Wet. Amsterdam 15: 366 (1963). Type: "... in regione fluminis Burdekin, prope Prom. Upstart: F. MUELLER" (holo U 028108). The specimen was annotated "Burdekin Expedition/probably from Cape Upstart" by Mueller. The Burdekin Expedition was led by J.W. Smith in 1860 and included E. Fitzalan as botanical collector.

Cycas normanbyana F. Muell., Fragm. 8:169 (1874), *Cycas rumphii* Miq. ssp. *normanbyana* (F. Muell.) Schuster, in Engl., Pflanzenr. 99: 75 (1932). Type: Near the Burdekin River estuary, Qld., E. Fitzalan (holo MEL).

Cycas gracilis Miq. var. *glauca* Regel, Act. Hort. Petropol. 4: 282 (1876). Type: "C.media h. Van Houte. - C.Boddami h. Haage et Schm. - Macrozamia latifrons h. Bull. - C.Normanbyana h. Belg." This was from cultivated plants from the growers cited.

Cycas gracilis Miq. var. *virdis* Regel, Act. Hort. Petropol. 4: 282 (1876). Type: "C.gracilis h. Turic. Habitat in Nova Hollandia tropica." This was from cultivated plants from the two growers cited.

Cycas kennedyana F. Muell., Austral. Chem. Drugg. 4(47): 85 (1882). Type: "In the Normanby-Ranges, near Port Denison; Eugene Fitzalan, Esq." (holo MEL).

9. ***Cycas circinalis*** *L.*, Sp. Pl. 1: 1188 (1753); Gen. Pl. 5th edn, 495 (1754). Type: The plates, Rheede (1682), t. 13-21.

Seven undescribed taxa also occur in Australia, three in the Northern Territory and four in Queensland. These are to be described in forthcoming taxonomic treatments.

References

***Brown, R.* 1810.** Prodromus Florae Novae Hollandiae et Insulae Van-Diemen (Taylor: London).

***Bentham, G.* 1873.** Cycadaceae, in Flora Australiensis. (L. Reeve: London).

***De Candolle, A.P.* 1868.** Cycadaceae, in Prodr. Syst. Naturalis & Reg. Veg. 16(2): 522 - 548.

***Groves and Moore* 1989.** Proc. Linn. Soc. New South Wales 111: 65 - 102.

***Hooker, J.D.* 1863.** Bowenia spectabilis. Bot. Mag. 98: t. 6008.

***Johnson, L.A.S.* 1959.** The families of cycads and the Zamiaceae in Australia. Proc. Linn. Soc. New South Wales 84(1): 64 - 117.

***Johnson, L.A.S.* 1961.** Cytological and taxonomic notes on Zamiaceae. Contr. New South Wales Natl Herb. 4(4): 235 - 240.

***Lehman, J.G.C.* 1834.** Novarium et minus congnitarum stirpium pugillus.

***Maconochie, J.* 1978.** Two new species of Cycas from northern Australia. J. Adelaide Bot. Gard. 1(3): 175 - 178.

***Miquel, F.A.W.* 1842.** Monographia Cycadearum (Utrecht).

***Miquel, F.A.W.* 1863.** Over de cycadeen in Nieuw-Holland. Vers. Meded. Afd. Natuurk. Kon. Akad. Wetensch. 15: 363 - 376.

***Miquel, F.A.W.* 1868.** Noveaux materiaux pour servir a la connaissance de Cycadees. Arch. Neerl. Sci. exact. nat. 3: 193 - 254.

***Moore, C.* 1884.** Notes on the genus *Macrozamia*. J. Roy. Soc. New South Wales 17: 115 - 122.

***Norstog, K.* 1990.** Studies of cycad reproduction at Fairchild Tropical Garden. Mem. New York Bot. Gard. 57: 63 - 81.

***Read, R.W.,* and *Scott, M.L.* 1986.** Bibliography of the living cycads (annotated). Lyonia 2(4): 33 - 199.

***Regel, E.* (1857).** Zwei neue Cycadeen. Bull. Soc. Imp. Nat. Moscou 30(1): 161 - 191.

***Reichenbach, H.G.L.* 1837.** 'Handbuch des naturlichen Pflanzensystems' (Dresden).

***Schuster, J.* 1932.** Cycadaceae, in Engler, Das Pflanzenreich 99(IV.I): 1 - 168.

***Stevenson, D.W.* 1981.** Observations on ptyxis, phenology and trichomes in the Cycadales and their systematic implications. *Amer. J. Bot.* 68(8): 1104 - 1114.

***Vorster, P.* 1986.** Hybridism in Encephalartos. Excelsa 12: 101 - 106.

The Interpretation of the Stem Lineage of the Cycadales

Henry Loconte
New York Botanical Garden
Bronx, New York 10458, U.S.A.

Abstract

In Hennigian systematics, a clear distinction is made for the treatment of extant and fossil organisms (Hennig, 1966; Ax, 1985, 1987, 1989). Extant taxa are subordinated into a hierarchy of sister-group relationships, which forms the basis for character polarisations by outgroup comparison. In contrast, fossils are excluded from the subordination of sister-group relationships between extant organisms; subsequently, fossils are excluded from outgroup comparisons. Instead, fossils are sequenced as parts of stem lineages. A stem lineage represents the extinct common ancestry of an extant monophylum. Fossil taxa can be classified as part of the stem lineage of an extant monophylum, if the fossils exhibit one or more autapomorphies of that monophylum. Therefore, evidence from the fossil record can be utilised as a secondary criterion for character polarisation that is capable of reversing a polarisation based on outgroup comparison, or resolving an equivocal polarisation (Loconte, 1990).

Introduction

The Cycadales occupies a unique phylogenetic position as the most primitive extant seed plants (Fig. 1). As a result, character polarisation by outgroup comparison can be problematic. The sister group of the Cycadales is [Ginkgoales-(Coniferales-(Gnetales-Angiospermae))], which was named as the Cladospermae by Loconte and Stevenson (1990). The second outgroup of the cycadales is the Pteridophyta (Loconte, in prep.), which can be summarised as [Ophioglossales-(Marratiales-(Osmundales-(Schizaeales-(Gleicheniales-(Cyatheales-(Polypodiales-Pteridiales))))))]. The Equisetales and Lycopodiales represent the putative third and fourth outgroups of the cycadales, respectively. The first two outgroups of the Cycadales are inclusive clades and, therefore, character-state optimisations are indirect. Additionally, the second outgroup of the Cycadales, Pteridophyta, has no comparison for characters that are unique to seed plants (secondary growth, pollen grains, seeds, etc.,). Both of these aspects of the Cycadales contribute to the ineffectiveness of character polarisation by outgroup comparison.

The fossil Lyginopterids and Medullosans were Paleozoic seed plants, which are referred to as pteridosperms because of their resemblance to ferns (Bierhorst, 1971; Gifford and Foster, 1988). The fossil Medullosans can be interpreted as part of the stem

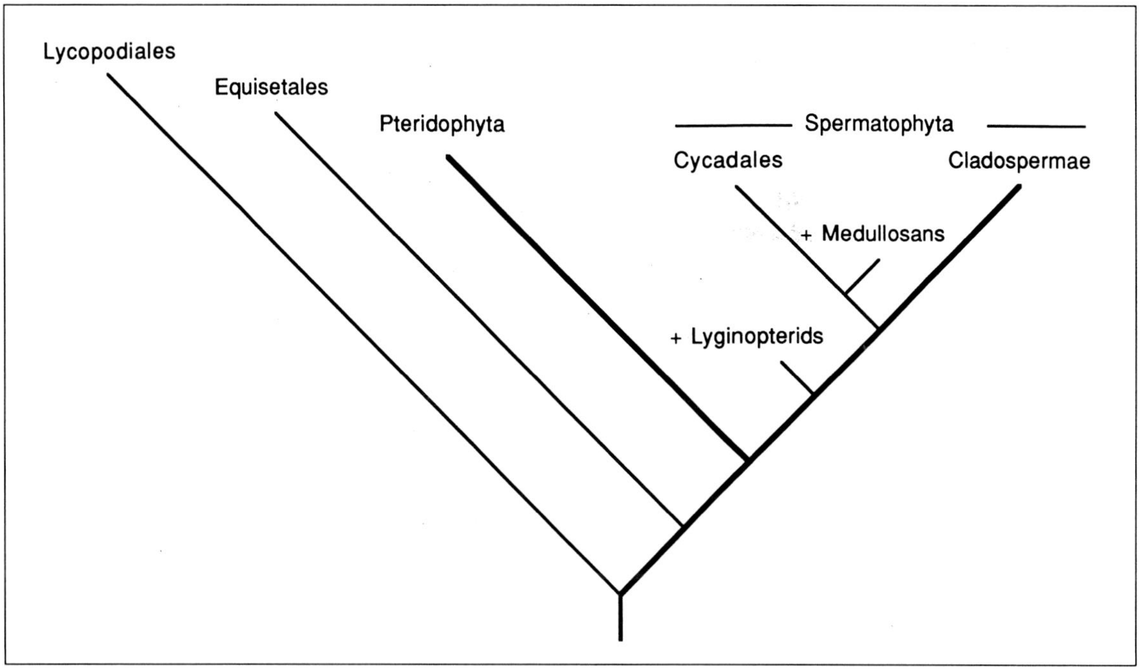

Figure 1: Superordinate relationships for the Cycadales.

lineage of the Cycadales, and the fossil Lyginopterids can be considered part of the stem lineage of Spermatophyta (Fig. 1). Apomorphic agreements between the Cycadales and the fossil Medullosans include multiple leaf traces and the double vascular system of the seeds.

Cycadales exhibit either platyspermic seeds, as in *Cycas*, or radiospermic seeds in all other genera (Stevenson, 1990). The platyspermic seeds of *Cycas* have been considered a plesiomorphic character-state for the Cycadales (Crane, 1988; Donoghue and Doyle, 1989); this conclusion is supported by outgroup comparison with the platyspermic Cladospermae. However, the fossil Medullosans and Lyginopterids have radiospermic seeds, thereby supporting the conclusion that radiospermic seeds are plesiomorphic for the Cycadales.

References

Ax, P. **1985.** Stem species and the stem lineage concept. Cladistics 1: 279 - 287.

Ax, P. **1987.** The Phylogenetic System: the systematisation of organisms on the basis of their phytogenesis. J. Wiley, Chichester.

Ax, P. **1989.** The integration of fossils in the phylogenetic system of organisms. Abh. Verh. Naturwiss. Vereins Hamburg 28: 10 - 26.

Bierhorst, D.W. **1971.** Morphology of Vascular Plants. The Macmillan Company, New York.

Crane, P.R. **1988.** Major clades and relationships in the higher gymnosperms. *In* C.B. Beck (ed.). Origin and Evolution of Gymnosperms. Columbia University Press, New York.

***Donoghue, M.J., and Doyle, J.A.* 1989.** Phylogenetic studies of seed plants and angiosperms based on morphological characters. *In* B. Fernholm, K. Bremer and H. Jornvall (ed's.). The Hierarchy of Life. Elsevier Science Publishers, Amsterdam.

***Gifford, E.M., and Foster, A.S.* 1988.** Morphology and evolution of vascular plants. W.H. Freeman, New York.

***Hennig, W.* 1966.** Phylogenetic Systematics. University of Illinois Press, Urbana.

***Loconte, H.* 1990.** Cladistic classification of Amniota: a response to Gauthier, et al. Cladistics 6: 187 - 190.

***Loconte, H., and Stevenson, D.W.* 1990.** Cladistics of the Spermatophyta. Brittonia 42: 197 - 211.

***Stevenson, D.W.* 1990.** Morphology and systematics of the Cycadales. *In* D.W. Stevenson (ed.). The Biology, Structure, and Systematics of the Cycadales. Memoirs of the New York Botanical Garden 57: 8 - 55.

Karyotypes of New World Cycads

Moretti, A., Caputo, P., and Cozzolino, S.
Dipartimento di Biologia Vegetale
Universit di Napoli, Via Floria 223, 80139 Napoli, Italy

Gaudio, L.
Dipartimento di Biologia generale e Genetica
Universit di Napoli, via Mezzocannone 8, 80134 Napoli, Italy

Abstract

Known and newly reported karyological data on neotropical cycads are discussed in relation to Robertsonian changes involved in the diversification of karyotypes and to phylogenetic implications. *Zamia* has interspecific chromosome variation and a diversified series of karyotypes ranging from 2n = 16 to 2n = 28. Several species of *Zamia* also have infraspecific chromosome variation. In contrast, *Ceratozamia* and *Dioon* are karyotypically uniform in their genera, with all species of the former being 2*n* = 16 (12M+2S+2T) and all species of the latter being 2*n* = 18 (8M+8A+2T). The monotypic genus *Microcycas* is 2*n* = 26. (2M+2S+2A+2OT), and *Chigua restrepoi*, the only species of this genus available in cultivation, is 2n = 18 (10M+2S+2A+4T). Comparative analysis of data shows that the karyotypes of *Chigua restrepoi* and *Microcycas* also occur in *Zamia*, and that the karyotype of *Ceratozamia* is very similar to those species of *Zamia* that are 2n = 16 (12M+2S+2A). The karyotype of *Dioon* (2n = 18 with 8M+8A+2T) is very distinct from the karyotypes of any other neotropical cycad. Suggested karyotypic similarities and differences have correlations with phenotypic characters.

Introduction

Five cycad genera in the Zamiaceae are endemic to America: *Zamia* L., *Chigua* D. Stevenson, *Microcycas* A. DC., *Ceratozamia* Brongn., and *Dioon* Lindl. *Zamia* is composed of 40 - 45 widespread species in Florida, Mexico, the West Indies, and Central and South America; *Chigua* has two species in Colombia; the monotypic *Microcycas* is endemic to Cuba; *Ceratozamia* is composed of eleven species in Mexico, Belize, and Guatemala; and *Dioon* has ten species in Mexico and Honduras (Eckenwalder, 1980; Balduzzi, De Luca and Sabato, 1982; Vovides, Rees and Vazquez Torres, 1983; Sabato and De Luca, 1985; Stevenson, Sabato and Vazquez Torres, 1986; Stevenson, 1990a, b).

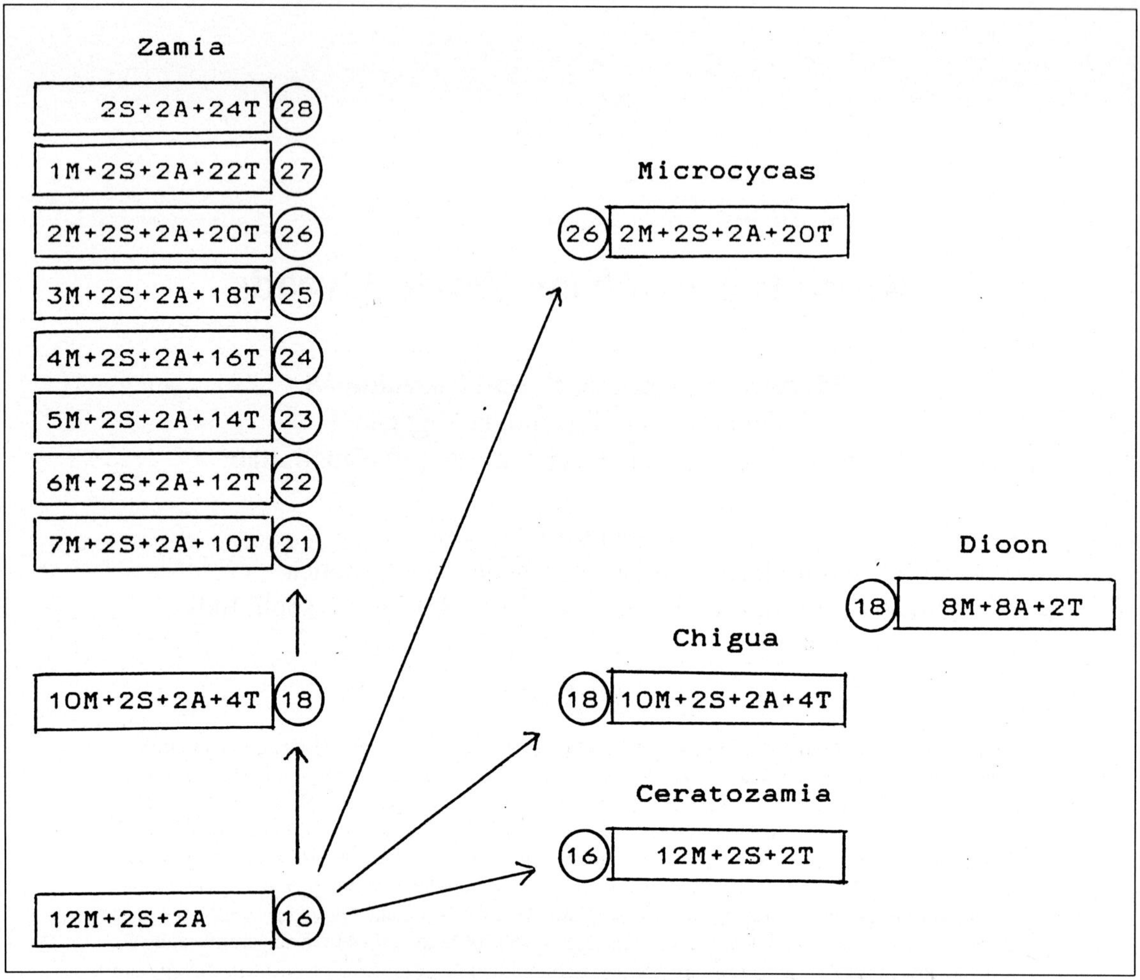

Figure 1: Chromosome numbers (circled) and karyotypes (in rectangles) of New World cycad genera are arranged in this scheme to indicate their karyological relationships. *Zamia, Chigua* and *Microcycas* karyotypes are shown to be originated by means of a mechanism of centric fission of metacentric chromosomes, starting from the lowest numbered *Zamia* karyotype (2n = 16). This karyotype gives rise also to the *Ceratozamia* karyotype by deletion of short arms of its acrocentric chromosomes. Alternative karyotypic diversifications explaining the origin of karyotypes in the four genera above are discussed in the text. No karyological relationships are suggested between *Dioon* and other genera. The *Dioon* karyotype is very distinct from any karyotype in the other American cycads in lacing submetacentric chromosomes and showing a high number of acrocentric chromosomes.

In the last ten years, several karyological studies have been carried out on these genera (Norstog, 1980, 1981; Vovides, 1983, 1985; Moretti and Sabato 1984; Schutzman, Vovides and Dehgan, 1988; Moretti, 1990a, b; Moretti, et al., in press). These studies have filled up a gap in the information on the karyology of American cycads. However, karyological data are not completed for *Zamia* and not available for *Chigua*. Besides, mechanisms involved in the karyotypic evolution of American cycads are still debated.

In the present paper, all known karyotypic data for American cycads, and new ones acquired recently for some species of *Zamia* and *Chigua* (Moretti, in preparation), are listed and discussed in relation both to the mechanisms involved in the karyotypic diversification and to their phylogenetic implications.

Discussion

Known and newly reported karyological data on New World cycads are listed in Table 1. In order to give uniformity to data in the literature, we present our personal interpretations of the chromosome morphology of some of the karyotypes reported in Table 1. The compiled data show that *Zamia* has interspecific and infraspecific karyotypic variation, whereas *Chigua, Microcycas, Ceratozamia* and *Dioon* have infrageneric karyotypic uniformity. In addition, *Zamia* shows all the chromosome numbers found in the other genera (Table 1). On the basis of these observations, comparative analysis between karyotypes of *Zamia* and karyotypes of the other American cycads could likely provide useful information on karyological relationships among the genera. Karyological affinities and diversities resulting from this analysis are discussed below and schematised in Fig. 1.

Zamia

In Fig. 1, all karyotypes of *Zamia* spp. but *Z.roezlii* are put in a sequence that can be interpreted as the direct consequence of the occurrence of a centric-fission mechanism of metacentric chromosomes. This mechanism produces an increase in chromosome number, two telocentric chromosomes being formed for each metacentric chromosome that undergoes the fission. The centric fission was proposed for *Zamia* spp. by Moretti and Sabato (1984), Moretti (1990b), and Moretti, et al., (in press) (Table 1). These authors observed derived characters, such as morphological adaptation to xeric conditions, in plants with higher chromosome numbers. In contrast to these authors, Norstog (1980, 1981) explained the occurrence of infraspecific chromosome variation in *Z. roezlii* (previously reported as *Z. chigua*) (Table 1) as a result of chromosomal fusion. This mechanism produces a decrease in chromosome number, a metacentric chromosome being formed for each pair of telocentric chromosomes that undergoes the fusion. Norstog (1981) drew support for his interpretation by observing in this species primitive morphological and reproductive characters in plants with a higher number of telocentric chromosomes. Cytotypes of *Z. roezlii* are not reported in the scheme of Fig. 1, where the arrangement of *Zamia* karyotypes is drawn on the basis of the fission mechanism only. The two contrasting suggestions on the mechanisms involved in the karyotypic evolution in *Zamia* represent one of most intriguing aspect in the karyological study of cycads (Marchant, 1968; Norstog, 1980, 1981; Moretti, 1990b). In the attempt to explain this controversial aspect, two hypotheses are suggested here. First, the evolution of karyotypes in *Zamia* could have proceeded in two distinct way, one by centric fission, the other by centric fusion. Due to the antiquity of the genus (Balduzzi, De Luca and Sabato, 1982), many relatives of living species would have been lost, so that a part of the memberships that occurred in the past in each lineage, as well as the putative link between the two lineages, would be lacking today. However, a second hypothesis, either a fission or fusion mechanism exclusively, also would have worked in *Zamia*. Moreover, in the course of the long evolutionary history of these plants, several biological characters could have evolved independently and at the different levels, so that both derived. However, a second hypothesis, either a fission or fusion mechanism exclusively, would have worked in *Zamia* and primitive characters would occur today in the same species. This would mean that in *Zamia*, as in all cycads and other ancient groups (Ehrendorfer, 1976; Moretti, 1990b) the karyotypic diversification would not necessarily parallel the morphological specialisation. Broader studies on the karyology as well as on the morphology of cycads are needed to clarify which of two hypotheses is correct or to establish whether some other biological event has to be taken into consideration as an alternative to those suggested here.

Chigua

Karyological data for *Chigua* rather to *Chigua restrepoi* only. Very few plants of this species are in

Table 1

Chromosome Numbers and Karyotypes in New World Cycads

Species	2n	Karyotype	Source
Zamia amblyphyllidia	16	12M+2S+2A	Moretti (1990b)
Z.angustifolia	16		Norstog (1980)
Z.boliviana	16		Moretti (in preparation)
Z.cremnophila	16		Schutzman et al. (1988), Moretti (1990b)
Z.fischeri	16		Marchant (1968), Moretti et al. (in print)
Z.inermis	16		Moretti (in preparation)
Z.integrifolia	16		Marchant (1968), Moretti (1990b)
Z.lecointei	16		Moretti (in preparation)
Z.lindenii	16		Norstog (1980, Moretti (in preparation)
Z.portoricensis	16		Norstog (1980), Moretti (1990b)
Z.pumila	16		Moretti (1990b)
Z.pygmaea	16		Norstog (1980), Moretti (1990b)
Z.pseudoparasitica	16		Norstog (1980), Moretti (in preparation)
Z.soconuscensis	16		Schutzman et al. (1988)
Z.splendens	16		Schutzman (1985), Moretti (1990b)
Z.wallisi	16		Moretti (in preparation)
Z.fischeri	18	10M+2S+2A+4T	Moretti et al. (in print)
Z.furfuracea	18		Norstog (1980), Moretti (1990b)
Z.loddigesii	18		Norstog (1980), Moretti (1990b)
Z.manicata	18		Moretti (in preparation)
Z.skinneri	18		Norstog (1980), Moretti (in preparation)
Z.spartea	18		Vovides (1983), Moretti (1990b)
Z.sylvatica	18		Moretti (1990b)

Z.picta	21	7M+2S+2A+10T	Moretti et al. (in print)
Z.picta	22	6M+2S+2A+12T	Moretti et al. (in print)
Z.polymorpha	22		Moretti et al. (in print)
Z.roezli	22	4M+4S+2A+12T	Norstog (1981)
Z.paucijuga	23	5M+2S+2A+14T	Moretti and Sabato (1984), Moretti (1990b)
Z.acuminata	24	4M+2S+2A+16T	Norstog (1980), Moretti (in preparation)
Z.muricata	24		Norstog (1980), Moretti (in preparation)
Z.paucijuga	24		Moretti (1990b)
Z.roezlii	24	2M+4S+2A+16T	Norstog (1981)
Z.paucijuga	25	3M+2S+2A+18T	Moretti and Sabato (1984), Moretti (1990b)
Z.roezlii	25	1M+4S+2A+18T	Norstog (1981)
Z.paucijuga	26	2M+2S+2A+20T	Moretti and Sabato (1984), Moretti (1990b)
Z.polymorpha	26	2M+2S+2A+20T	Moretti et al. (in print)
Z.ulei	26		Moretti (in preparation)
Z.roezlii	26	4S+2A+20T	Norstog (1981)
Z.paucijuga	27	1M+2S+2A+22T	Moretti and Sabato (1984), Moretti (1990b)
Z.paucijuga	28	2S+2A+24T	Moretti and Sabato (1984), Moretti (1990b)
Chigua restrepoi	18	10M+2S+2A+4T	Norstog (pers. comm), Moretti (in preparation)
Microcycas calocoma	26	2M+2S+2A+20T	Moretti (1990b)
Ceratozamia spp.	16	12M+2S+2T	Vovides (1983, 1985), Moretti (1990b)
Dioon spp.	18	8M+8A+2T	Vovides (1983), Moretti (1990b)

cultivation and no plants of *C. bernalii*, the other species in the genus, are available in cultivation. The single plant of *C. restrepoi* cultivated at the New York Botanical Garden is 2n = 18 and its karyotype seems to be 10M+2S+2A+4T (Moretti, in preparation). A similar karyotype is reported for a plant cultivated at Fairchild Tropical Garden in Miami, Florida (Norstog, pers. comm). These data suggest that the karyotype of *C.restrepoi* is the same as those of 2n = 18 species of *Zamia* (Table 1). Thus, *Chigua* and the 2n = 18 *Zamia* spp. could have originated from a common ancestor. A 2n = 16 *Zamia* is proposed here as the ancestry both of *Chigua* and of 2n = 18 *Zamia* spp. (Fig. 1), the 2n = 16 *Zamia* having undergone centric fission of two metacentric chromosomes to produce a 2n = 18. As a result, karyotype of such metacentric fission were formed in *Chigua*.

Microcycas

Microcycas calocoma is 2n = 26 with 2M+2S+2A+20T (Moretti 1990b). This karyotype corresponds to that described in some plants of *Zamia paucijuga* and *Z. polymorpha* (see Table 1). Both *M.calocoma* and the two *Zamia* species could have originated from a common ancestor with a lower numbered karyotype by means of repeated centric fissions of metacentric chromosomes. Among all putative ancestral karyotypes, 2n = 16 the lowest occurring in *Zamia* is suggested here (Fig. 1).

Ceratozamia

All species of *Ceratozamia* are 2n = 16 with 12M+2S+2T (Vovides, 1983, 1985; Moretti, 1990b). This karyotype is related to the 2n = 16 karyotype with 12M+2S+2A described in some *Zamia* spp. (see Table 1). The Robertsonian change proposed to interpret the karyotypic origin of *Ceratozamia* is that of proceeding from *Zamia* to *Ceratozamia* through deletion of short arms of two acrocentric chromosomes of *Zamia* (Moretti, 1990b). As a result, in *Ceratozamia* two telocentric chromosomes are formed. However, another Robertsonian change could account for a derivation proceeding in the opposite direction, that is from *Ceratozamia* to *Zamia*. In this case, a pericentric inversion may have occurred on two telecentrics in *Ceratozamia* giving rise to two acrocentrics of *Zamia*. Finally, the additional hypothesis of a derivation of both genera from a common extinct ancestor is perhaps probable. Among the three possible derivations, only the first one is shown in Fig. 1.

Dioon

All species of *Dioon* are 2n = 18 with 8M+8A+2T (Vovides, 1983; Moretti 1990b). Karyotypes of *Dioon* appear very distinct from those of *Zamia, Chigua, Ceratozamia* and *Microcycas* (Table 1). *Dioon* lacks submetacentric chromosomes, seen in all species belonging to other neotropical cycad genera, and has eight acrocentric chromosomes, a very high number as compared to two acrocentric chromosomes in species from the other genera (Table 1). The origin of *Dioon* karyotypes from chromosomes of any species or common ancestor of *Zamia, Chigua, Microcycas* and *Ceratozamia*, would indeed be very difficult to interpret. Thus, in the scheme presented in Fig. 1, no karyological relationships are proposed between *Dioon* and the other genera.

Correlation between Karyological and Phenotypic Characters in New World Cycads

Inferred phenotypic characters in American cycads support the karyological relationships proposed in Fig. 1.

The similarity between karyotypes of *Chigua restrepoi* and some *Zamia* spp. is supported on the similarity in the general morphology (Stevenson, 1990a) and phytochemistry (Siniscalco Gigliano, pers. comm.) of the two genera. Comparative analysis of *Zamia* and *Microcycas* also shows that the two genera are related. *Microcycas* is scarcely differentiable from *Zamia* in morphology, the only characters that consistently distinguish the two genera being embryological ones (Norstog, 1977; Eckenwalder, 1980; Stevenson 1990a). *Zamia* and *Ceratozamia* have also a very high degree of morphological similarity. Many species of the two genera are only distinguishable on the basis of strobilar structure comparison (Stevenson, 1990a). In *Dioon*, typical vegetative, reproductive, and ecological characters confirm its karyotypic diversification (Balduzzi De Luca and Sabato, 1982; Dehgan and Dehgan, 1988; Stevenson, 1989a). Due to its distinct features, Stevenson (1990a) has attributed *Dioon* to the subfamily Encephalartoideae and separated it from *Zamia, Chigua, Microcycas* and *Ceratozamia*, all included in the subfamily Zamioideae.

References

***Balduzzi, A., De Luca, P., and Sabato, S.* 1982.**
A phytogeographical approach to the New World cycads. Delpinoa 24: 185 - 202.

***Dehgan, B., and Dehgan, N.B.* 1988.** Comparative pollen morphology and taxonomic affinities in Cycadales. Amer. J. Bot. 75: 1501 - 1516.

***Eckenwalder, J.E.* 1980.** Taxonomy of the West Indian cycads. J. Arnold Arbor. 61: 701 - 722.

***Ehrendorfer, F.* 1976.** Evolutionary significance of chromosomal differentiation patterns in gymnosperms and primitive angiosperms. In Beck, C.B. (ed.). Origin and Early Evolution of Angiosperms. 200 - 240. Columbia University Press.

***Merchant, C.J.* 1968.** Chromosome patters and nuclear phenomena in the cycad families Stangeriaceae and Zamiaceae. Chromosoma 24: 100 - 134.

***Moretti, A.* 1990a.** Cytotaxonomy of cycads. Mem. N.Y. Bot. Gard. 57: 114 - 122.t

***Vovides, A.P.* 1983.** Systematic studies on the Mexican Zamiaceae. I. Chromosome numbers and karyotypes. Amer. J. Bot. 70: 1002 - 1006.

***Vovides, A.P.* 1985.** Systematic studies on Mexican Zamiaceae. II. Additional notes on *Ceratozamia kuesteriana* from Tamaulipas, Mexico. Brittonia 37: 226 - 231.

***Vovides, A.P., Rees, J.D., and Vazquez Torres, M.* 1983**. Flora de Veracruz. 26: 1 - 31.

Spermatogenesis In *Microcycas*: Evolutionary Significance of Male Gametes of Seed Plants

Knut J. Norstog
Fairchild Tropical Garden, Miami, Florida U.S.A.
Missouri Botanical Garden, St. Louis, Missouri, U.S.A.

Abstract

The Cuban cycad, *Microcycas calocoma*, has pollen tubes containing up to 16 sperm rather than the usual two of most other cycads. Multispermy, as exhibited always by *Microcycas* and, rarely, by *Ceratozamia* is often considered evidence that these are the more primitive cycads now in existence. However, it also is possible that multispermy in these two cycads is an advanced evolutionary character. In this report evidence is presented that multispermy in *Microcycas* is a primitive character and that its development appears to exhibit a retrogressive trend in the direction of diplospermy. Evolution toward diplospermy probably occurred independently in several lineages of cycads.

Introduction

Two living genera of cycads, *Ceratozamia* and *Microcycas*, exhibit multispermic pollen tubes (i.e., having more than one pair of male gametes); the remainder are diplospermic (having one pair of male gametes). *Microcycas* also differs from other cycad species in producing a variable number of spermatozoids (i.e., motile sperm), ranging from 8 to 16 (Caldwell, 1907), and *Ceratozamia*, although usually diplospermic, occasionally produces four spermatozoids in a single pollen tube (Chamberlain, 1912). One may ask, therefore, if multispermy in these two cycads represents a departure from an ancient general plan, or, conversely, a general multispermic ancestral state for members of the Cycadales and, possibly, for all other gymnosperms as well.

Nine of the 11 genera of living cycads are represented by at least one member in which spermatozoids have been described, and it can be assumed that all cycads have them. However, with the sole exception of *Ginkgo biloba* all other seed plants have non-motile male gametes. The spermatozoids of *Ginkgo* are similar to those of cycads but this is believed not to indicate a close relationship but rather to reflect the general occurrence of zooidogamy (fertilisation by motile sperm) among the various ancestors of modern gymnosperms (Foster and Gifford, 1974).

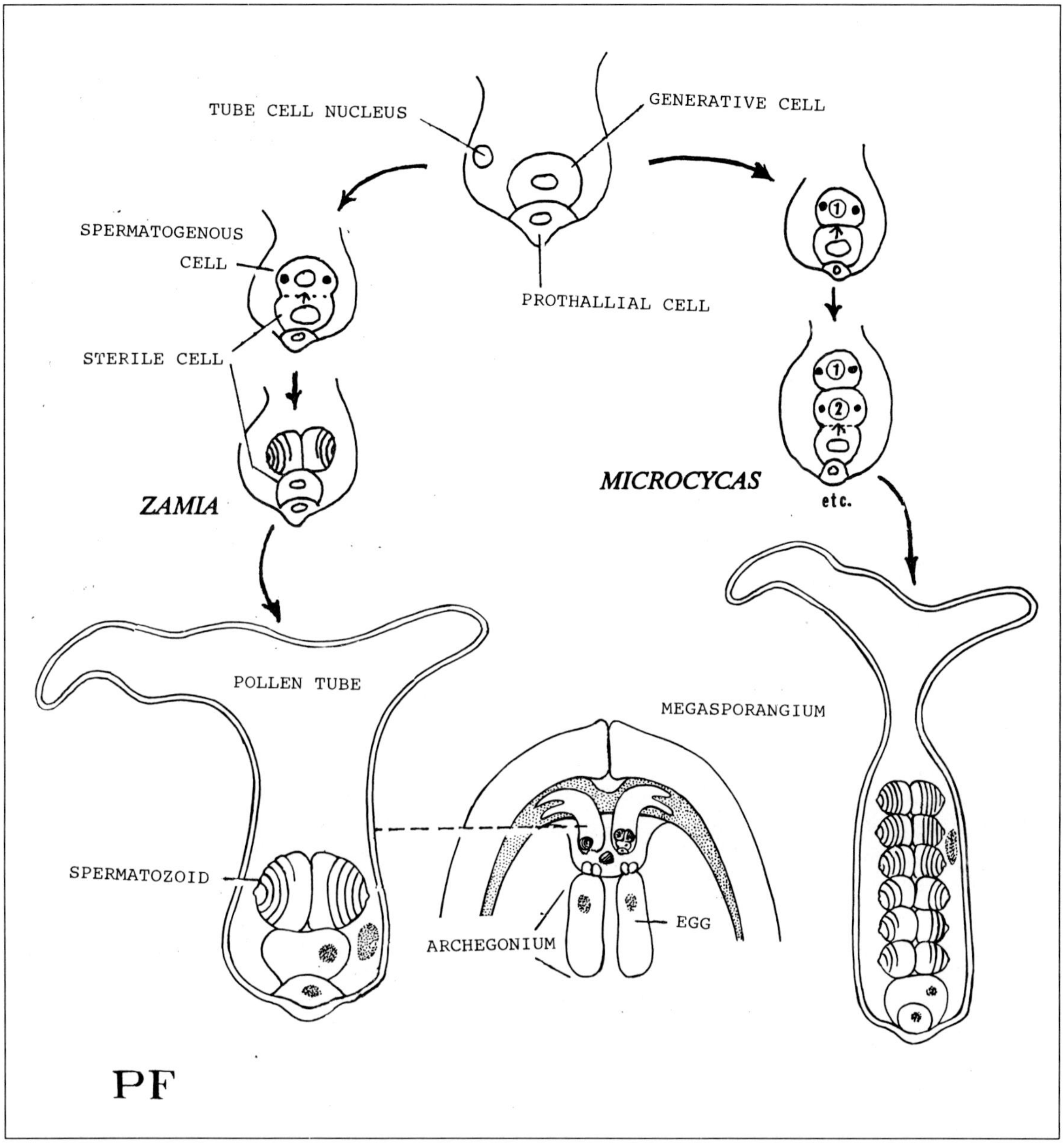

Figure 1: Male gametophytes of *Zamia* and *Microcycas* compared diagrammatically. (A) Division of sterile cells of *Zamia* and *Microcycas* compared. The sterile cell of *Microcycas* divides repeatedly as many as 11 times to produce a file of spermatogenous cells. (B) Pollen tubes compared. Note that the pair of spermatozoids nearest the tip of the pollen tube of *Microcycas* has fewer gyres of the spiral band. (C) Archegonia of *Zamia* are located beneath a depression in the female gametophyte; *Microcycas* (not shown) is similar but lacks the depression and has more archegonia.

Multispermy

The male gametes of a majority of algae are spermatozoids, as is true also of the so-called 'green-line' charophycean algae which may have given rise to all higher land plants (Pickett-Heaps, 1975). However, diplospermy which is so characteristic of pollen tubes of nearly all seed plants, is not characteristic of green-line algae; most of them are multispermic. Nor is diplospermy found in mosses, ferns and fern allies (Smith, 1955; Carothers, 1975). Therefore, because seed plants evolved from such ancestors, it has been suggested that multispermy also was characteristic of the earliest seed plants (Chamberlain, 1935; Sterling, 1963). This assumption owes much to the occurrence of multispermy in *Microcycas* (Caldwell, 1907; Downie, 1928), but *Ceratozamia* also is occasionally multispermic (Chamberlain, 1912). Unfortunately there is at present no firm evidence in the fossil record that male gametophytes of primitive seed plants were sometimes, often, or exclusively either diplospermic or multispermic. This is not surprising. Such diminutive male sexual stages are poorly represented in the fossil record. Their minute and delicate natures mitigate against fossilisation. Nevertheless, male gametophytes have been identified in certain fossil seed ferns, and at least some of these seem to have been multispermic as well as endosporic, that is, the sperm developed within the pollen grain instead of a pollen tube (Sterling, 1964). Chamberlain (1935) proposed that endosporic male gametophytes of this kind, were drawn into the ovules of ancient cycadophytes where without ever forming pollen tubes they released numerous, actively motile spermatozoids. This reasoning is the basis for thinking that primitive cycads were multispermic.

Multispermy in Cycads

If the remote ancestors of seed plants had multispermic male gametophytes, how may the almost universal occurrence of diplospermy in modern seed plants be accounted for? To answer that question, evidence for a multispermic origin of modern forms must be marshalled if one is to argue convincingly that modern diplospermic genera and species once were multispermic. Even though evidence supporting this hypothesis may be obtained from study of sperm development in *Microcycas*, and to a lesser degree *Ceratozamia*, it still must be shown that multispermy is a retained primitive character in these two examples rather than a recent evolutionary event.

Spermatogenous Cell

At the time of pollination, a cycad pollen grain is already a three-celled male gametophyte (tube cell, prothallial cell, generative cell, Fig. 1). When a pollen grain germinates within an ovule, it produces a male gametophyte which invades the megasporangium (nucellus) and some months later produces a pollen tube which projects into a space between the megasporangium and female gametophyte in which archegonia are located (Fig. 1). At about this time, the generative cell divides to produce a sterile cell and a sperm-forming or spermatogenous cell. The sterile cell ordinarily does not divide again (hence 'sterile') but, as the cycad male gametophyte approaches maturity, the spermatogenous cell than simultaneously develops a pair of multicentriolar bodies, the blepharoplasts, which are the eventual source of the flagella of the spermatozoids (Figs, 1, 2). After differentiation and development of its pair of blepharoplasts, the spermatogenous cell divides to form two spermatozoids, each of which receives one blepharoplast. The blepharoplast of each spermatozoid then rapidly produces flagella and the result is a spermatozoid.

Multispermy in *Microcycas*

The only two studies of sperm development in *Microcycas* are complementary. Downie (1928) noted that as many as 11 spermatogenous cells were present

Figure 2: Blepharoplasts of *Microcycas* compared. A last-formed blepharoplast is superimposed over a first-formed one to show that they are of the same size (8.0μm) and organisation. Blepharoplasts are composed of a vacuolate core covered by a single layer of centrioles (C). Centrioles are the flagella-forming structures of motile cells, X9250.

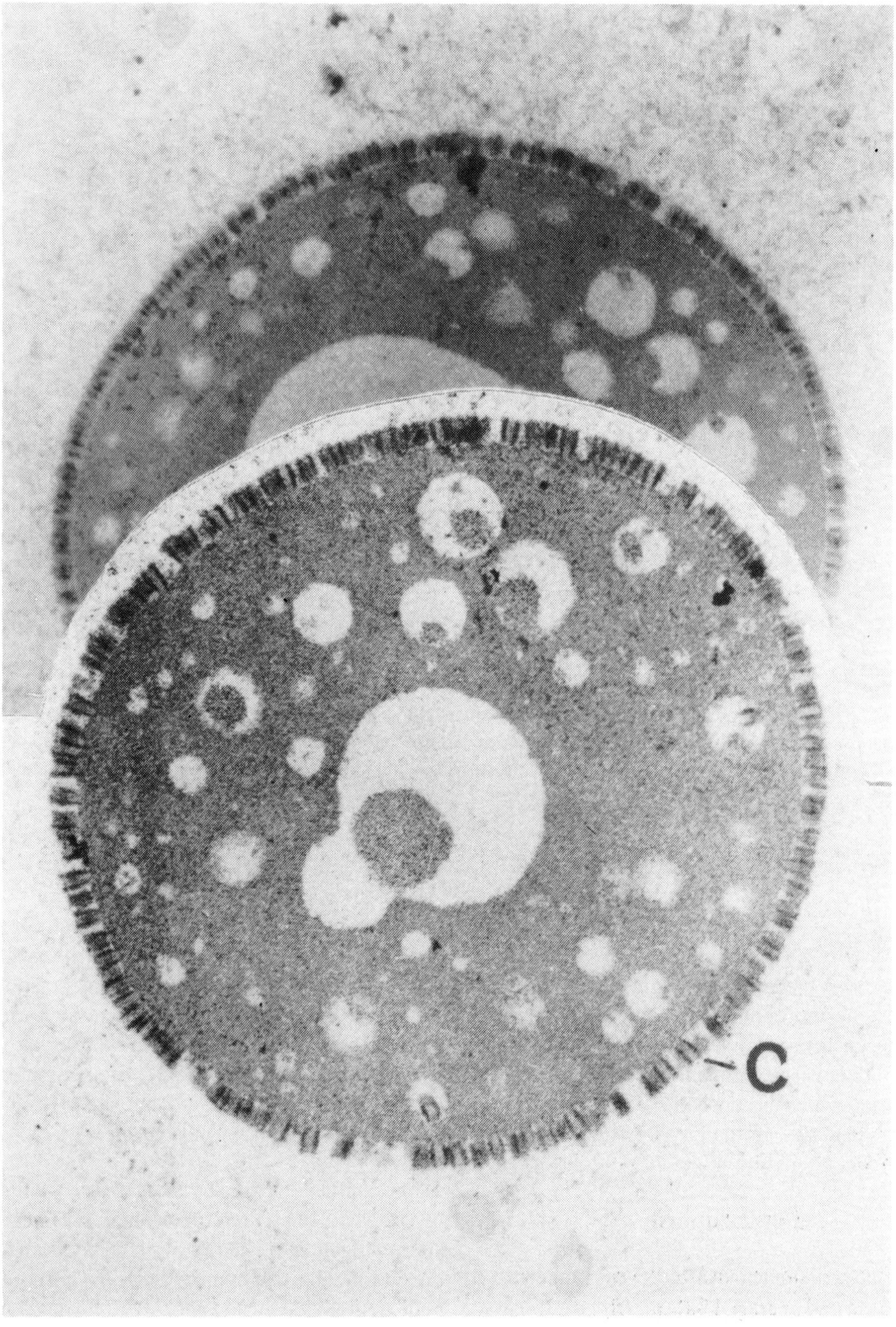
C

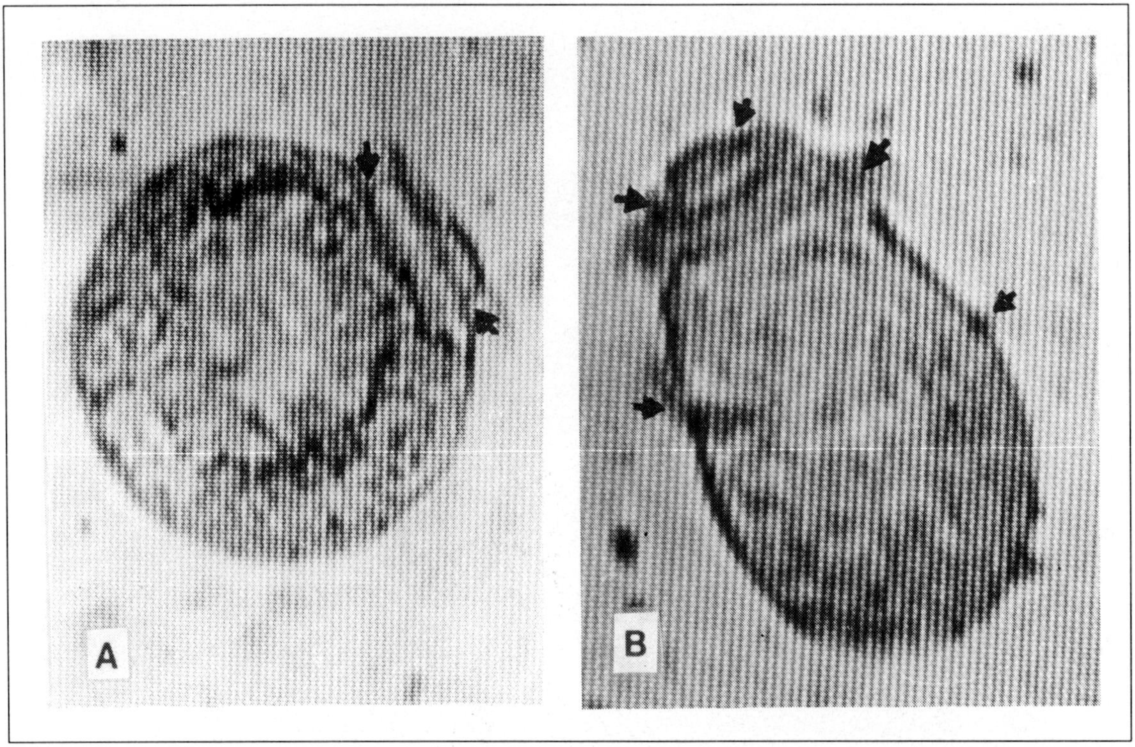

Figure 3: Videotape images of living spermatozoids of *Microcycas*. Arrowheads indicate positions of gyres of their flagellated bands. (A) spermatozoid typical of a last-formed cell in having spherical form and a short flagellated band of ca. 2 gyres, X1160. (B) spermatozoid typical of a first-formed cell in having ovoid form and a flagellated band of about 5 gyres, X1125.

in a single male gametophyte. She said they were produced serially by repeated divisions of a sterile cell, and also observed that, although spermatogenous cells are produced progressively, their blepharoplasts all develop synchronously. Caldwell (1907) observed development of spermatogenous cells to mature spermatozoids, of which he observed as many as eight pairs per pollen tube. In the present study, however, sperm numbers ranged from 8 to 12 per tube.

Blepharoplast

Blepharoplast diameters in *Microcycas* ranged from 5.7 to 12.8μm, and averaged 9.0μm, (fig. 2). Relative positions, hence relative age, of nine blepharoplasts were determined and it was found that size was not correlated with position. Blepharoplasts of last-formed spermatogenous cells, therefore, appear to be about the same-size as those in earlier-formed ones (Fig. 2). The significance of this observation is that variations in size and structure of spermatozoids observed in this study could not have resulted from variations in blepharoplast complexity.

Following the last division of the sterile cell of *Microcycas*, the pollen tube contains a linear array of spermatogenous cells, all at about the same stage of development, as shown in Fig. 1. In short order, these cells then divide synchronously to form a double row of up to 16 sperm (Caldwell, 1907), and the chronol-

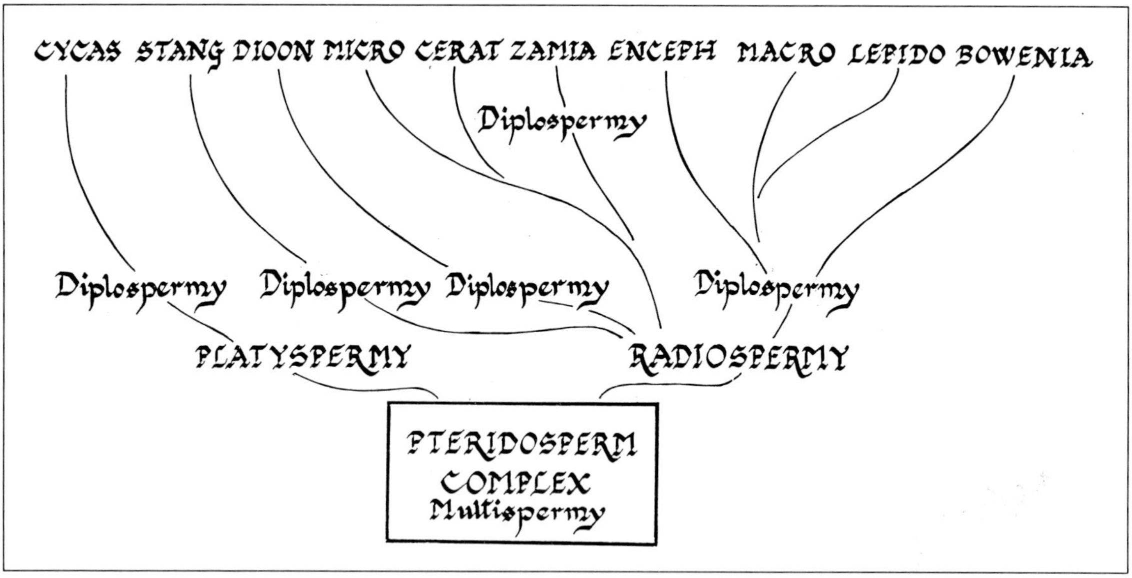

Figure 4: Relationships of cycad genera. In this scheme, the basic concept is that cycads originated, more-or-less independently from several pteridosperms (seed ferns) as, for example, a form having platyspermic (flattened) seeds as does *Cycas*. The remainder of modern genera have radiospermic (circular in cross-section) seeds and may share a common seed pteridosperm ancestry. Several lines apparently have evolved separately and it is unlikely that any present-day genera are very closely related to one another, although *Microcycas, Ceratozamia* and *Zamia* probably have a relatively recent common origin (i.e., are monophyletic) as are *Encephalartos* and *Macrozamia*. Probably all cycad had multispermic ancestors and evolution toward diplospermy may have occurred more recently in some lines than in others (e.g., *Microcycas* and *Ceratozamia* vs. *Cycas*). Among the radiospermic lines, it is considered plausible on biogeographic grounds that *Macrozamia* and *Encephalartos* became diplospermic relatively early although neither genus has yet evolved a mechanism for aborting an unfertilised female gametophyte; something all other cycads have achieved. *Dioon* also is considered to be a relic of an ancient line not closely related to other New World cycads. *Ceratozamia* and *Microcycas* appear to be the more closely related of New World taxa. In addition to multispermy, both have similar foliage and both bear large cones with truncated, more or less bifid sporophyll apices.

ogy of cell initiation is still identifiable. It is possible, therefore, to compare ultrastructural features of 'earlier' and 'later' spermatozoids (Norstog, 1990).

Spermatozoid

Light-microscope observations of living spermatozoids of *Microcycas* revealed two distinct types of motile cells: (A) slowly moving spherical cells bearing short flagellated bands of about 1½ to 2 turns (gyres), and (B) rapidly moving egg-shaped cells bearing flagellated bands of 4 - 5 gyres (Fig. 3). The difference in shape probably results from the degree of development of the internal 'skeleton' in the two kinds of cell. This skeleton, more specifically a "cytoskeleton", encircles the anterior of the cell, giving it an ellipsoidal shape. In the less-active spermatozoids, a less-well-developed cytoskeleton does not support the cell very well. The result is that surface-tension produces a spherical cell.

Examinations of spermatozoids in electron micrographs, shows that differences in the cytoskeleton are directly related to the position of the spermatozoids in the pollen tube (Norstog, 1990). It is the last-formed spermatozoids in a pollen tube (i.e. distal-most) that are less well-developed than the earlier-formed ones (Fig. 1). First-formed spermatozoids are remarkably similar both in overall form and in ultrastructure to the considerably larger spermatozoid of *Zamia pumila* (Norstog, 1974). Last-formed spermatozoids tend to be smaller and less complex ultrastructurally than first-formed ones.

For a time, it was thought that last-formed cells were mature and, in fact, in some features this is the case. However, it was subsequently observed that such cells were active when discharged in sugar water (Fig. 3) and further examination by electron microscopy shows that they have a completely developed flagellar system.

Evolutionary Significance of Multispermy

Because multispermy is characteristic of lower forms of plant life from which seed plants arose, the occurrence of multispermy in *Microcycas* led both Caldwell (1907) and Downie (1928) to conclude that the genus is the most primitive member of the order. Other botanists have concurred (Chamberlain, 1919; Pant 1973; Bold, Alexopoulos and Delevoryas, 1987).

Although it is possible to assume that multispermy in *Microcycas* is not a primitive character but rather is a derived condition and that ancestors of this and other cycads always were diplospermic, other evidence favouring a multispermic ancestry seems stronger. If multispermy in *Microcycas* were a character of secondary derivation, as outlined above, then it should be possible to explain it in terms of some real or apparent selective advantage. The concept of evolutionary economy or parsimony, a salient feature of cladistic analysis, seems to argue against consideration of multispermy as simply a duplication of little or no selective importance; one ought to be able to demonstrate some reproductive advantage to offset the energy expended in producing a dozen or more gametes per pollen tube.

Presumably, the first-formed spermatozoids of *Microcycas*, which are the more active swimmers, are those most likely to locate and enter archegonia. It seems rather doubtful that the poorly developed, slow-swimming, last-formed sperm could be competitive or even capable in fertilisation. Thus, they may represent vestigial cells exemplifying a parsimonious evolutionary trend toward fewer spermatozoids. The pollen tube of *Microcycas* is, therefore, not only primitive but it also demonstrates a trend which in other cycads has attained the evolutionary end-point of diplospermy.

Although multispermy is rare among modern cycads, its occurrence in *Microcycas* and *Ceratozamia* should not be viewed as exceptional features peculiar only to the two genera. Rather, evolution from multispermy to diplospermy should be considered as a general evolutionary tendency among cycadophytes and also among other ancient lineages of spermatophytes. Figure 4 presents the concept that reduction of spermatozoids from multispermy to diplospermy probably occurred independently in the ancestors of all modern cycad groups. *Ceratozamia*, which commonly is diplospermic but occasionally quadrispermic represents a later stage in this reductional trend than does *Microcycas*.

The Origins of Diplospermy

Diplospermy, *per se*, probably is neither more nor less efficient than, say, monospermy or trispermy, were these to occur; the ovules of most seed plants are entered by several to many pollen tubes so that the number of sperm per pollen tube many not be a critical factor in fertilisation. In gymnosperms, single fertilisation is the rule and one sperm may suffice to produce a fertile seed. However, several to many pollen tubes commonly are found in a single ovule and competition probably occurs among them to effect fertilisation.

One may speculate that seed plants, having achieved a closed system of fertilisation as a consequence of the evolution of the ovule, evolved further in the direction of siphonogamy (i.e., pollen tubes with non-motile sperm) by elimination of the flagella from male gametes. *Ginkgo biloba*, the only gymnosperm other than members of the Cycadales, in which spermatozoids are found appears to have evolved further in the direction of spermatozoid reduction than have cycads. *Ginkgo* spermatozoids (Fig. 5) are markedly like the relatively non-functional last-formed spermatozoids of *Microcycas* (Fuji, 1899; Lee, 1955; Li, Wang and

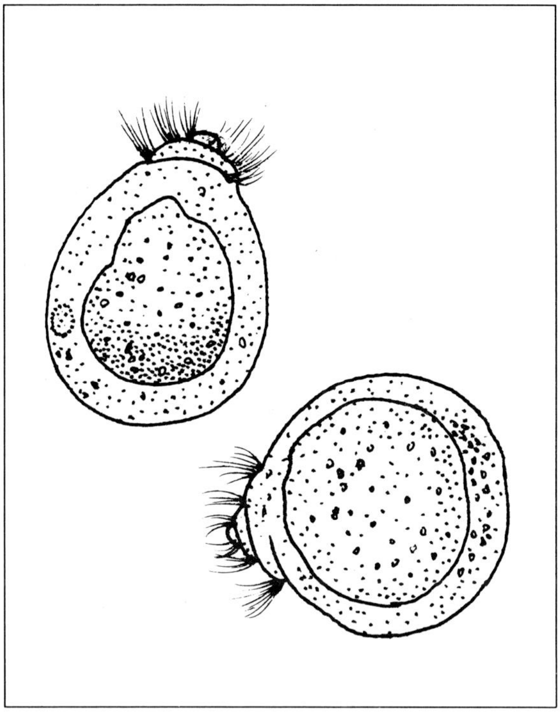

Figure 5: Spermatozoids of *Ginkgo biloba* (re-drawn from Fuji, 1899). Note that the flagellated band, of about 1½ gyres, is confined to the cell's apex (cf. Fig. 3), X600.

Knox, 1989). If other gymnosperms, such as the conifers with which *Ginkgo* is allied, had zooidogamous ancestors, one can suppose that their pollen tubes and spermatozoids passed through stages resembling those of *Ginkgo*. In all plants having motile sperm a spermatogenous cell possesses a pair of blepharoplasts, and the function of its division is the separation and distribution of one such flagellum-forming organelle to each spermatoid. Therefore, diplospermy can be viewed in those gymnosperms as well as in all flowering plants as a consequence of evolution from ancestors having blepharoplasts and motile sperm.

In angiosperms, where siphonogamy and double fertilisation occur, one can explain diplospermy as a necessity for fertilisation of the egg and polar cell of the embryo sac, but there is little doubt that diplospermy predated double fertilisation. Conifers, which are considerably more ancient in origin than flowering plants, also are siphonogamous and diplospermic, but their fertilisation involves only one of the two sperm in a pollen tube. In that case, what is the explanation for diplospermy and, for that matter, why is it so universal in seed plants?

It is quite plausible that it is simply the consequence of the original necessity of segregation of a pair of blepharoplasts and their distribution to spermatids. Once sperm cells lost their flagella in the transition from motile sperm to non-motile sperm, the requirements for blepharoplasts disappeared and there now are other reasons for the survival of diplospermy, such as double fertilisation in flowering plants.

Acknowledgments

This study was supported by National Science Foundation grant BSR-8604221. I wish to express my appreciation of Zane Carothers and Robert Ornduff for their helpful comments on portions of the manuscript.

References

***Bold, H.C., Alexopoulos, C.J., and Delevoryas, T.* 1987.** Morphology of Plants and Fungi, ed. 5. Harper & Row, New York.

***Caldwell, O.W.* 1907.** *Microcycas calocoma*. Bot. Gaz. 44: 118 - 141.

***Carothers, Z.B.* 1975.** Comparative studies on spermatogenesis in bryophytes. Biol. J. Linn. Soc. London 7 (suppl. 1): 71 - 85. Academic Press, London.

***Chamberlain, C.J.* 1912.** Morphology of *Ceratozamia*. bot. Gaz. 53: 1 - 19.

***Chamberlain, C.J.* 1919.** The Living Cycads. University of Chicago Press, Chicago.

***Chamberlain, C.J.* 1935.** Gymnosperms Structure and Evolution. University of Chicago Press, Chicago.

***Downie, D.G.* 1928.** Male gametophyte of *Microcycas calocoma*. Bot. Gaz. 85: 437 - 450.

***Foster, A.S., and Gifford, Jr, E.N.* 1974.** Comparative Morphology of Vascular Plants, ed. 2. W.H. Freeman and Co., San Francisco.

***Fuji, K.* 1899.** Observations on the morphology of the pollen tube and the spermatozoid of *Ginkgo* (in Japanese). Bot. Mag. Tokyo 13: 28 - 30.

***Lee, C.L.* 1955.** Fertilisation in *Ginkgo biloba*. Bot. Gaz. 117: 79 - 100.

***Li, Y., Wang, F.H., and Knox, R.B.* 1989.** Ultrastructural analysis of the flagellar apparatus in sperm cells of *Ginkgo biloba*. Protoplasma 149: 57 - 63.

***Norstog, K.* 1974.** Fine structure of the spermatozoid of *Zamia*: the *vierergruppe*. Am. J. Bot. 61: 449 - 456.

***Norstog, K.* 1977.** The spermatozoid of *Zamia chigua*. Bot. Gaz. 138: 409 - 412.

***Norstog, K.* 1990.** Spermatozoids of *Microcycas calocoma*: ultrastructure. Bot. Gaz. 151: 275 - 284.

***Pant, D.D.* 1970.** *Cycas* and the Cycadales, ed. 2. Central Book Depot, Allahabad, India.

***Picket-Heaps, J.D.* 1975.** Structural and phylogenetic aspects of microtubular systems in gametes and zoospores of certain green algae. Biol. J. Linn. Soc. London 7 (suppl. 1): 37 - 44.

***Smith, G.M.* 1955.** Cryptogamic Botany. Volume II. Bryophytes and Pteridophytes. McGraw-Hill Book Co., New York.

***Sterling, C.* 1963.** Structure of the male gametophyte in gymnosperms. Bot. Ev. 38: 167 - 203.

A Numerical Phenetic Study of the Genus *Encephalartos lehm*

R. Osborne, N. Grobbelaar
University of Natal, King George V Avenue
Durban, 4001 South Africa

P.L.D. Vincent
PO Box 15357, Lynne East, 0039 South Africa
University of Witwatersrand, PO Wits, 2050 South Africa

Standard techniques of numerical phenetics (Rohlf et al, 1971; Sneath and Sokal, 1973) were used to access infragenic relationships in *Encephalartos* Lehm. The matrix subjected to the multivariate statistical analysis comprised data from 86 characters for 52 OTUs. Vegetative, reproductive, anatomical, biochemical and other measurements and observations on living and herbarium specimens amplified data extracted from literature descriptions to constitute the character set while OTUs were restricted to defined species. The raw data were standardised and used to construct a matrix of taxonomic distances which were compared by an unweighted pair-group method, the output being represented in the from of a phenetic tree diagram (Fig. 1). All but four taxa fell within five groups which we have named the ALTENSTEINII, LEHMANNII, VILLOSUS, EUGENE-MARAISII and CYCADIFOLIUS groups. In many respects, our groupings are consistent with geographic distributions, intuitive assessments based on gross morphology and with published taxonomic opinion. This study lays the foundation for a cladistic analysis of the genus.

References

***Rohlf, F.J., Kishpaugh, J. and Kirk, D.* 1971.** NT-SYS. Numerical taxonomy system of multivariate statistical programmes. Technical Report of the State University of New York, Stony Brook, New York.

***Sneath, P.H.A. and Sokal, R.P.* 1973.** Numerical Taxonomy. W.H. Freeman, San Francisco.

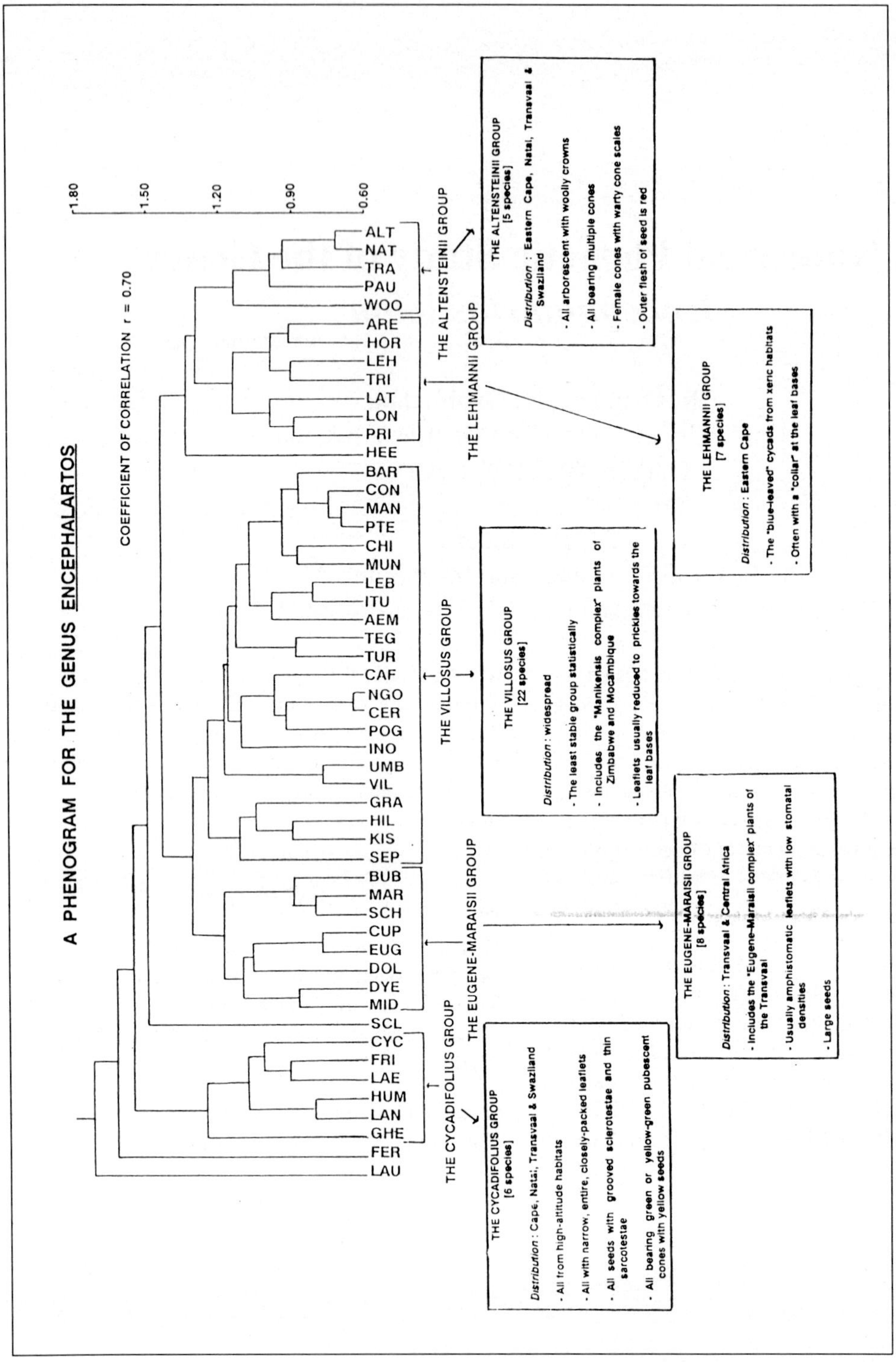

Figure 1: A phenogram for the genus *Encephalartos* based on taxonomic distances obtained from 86 characters from 52 species. The three-letter codes represent the first three letters from each species' name.

Computer-Assisted Systematics in the Cycadales

Bart Schutzman and Bijan Dehgan
Horticultural Systematics Laboratory,
Department of Environmental Horticulture
University of Florida, Gainesville, FL 32611 U.S.A.

Abstract

Systematic studies in our laboratory utilise computer programs and the resolving power of phonetic and cladistic analyses. Our software programs are written to aid cycad systematic research, with emphasis on current work in *Zamia*, *Cycas*, and the cycadales as a whole. Fourier analysis of shapes aided by the MORPHOS program have enabled fine distinctions, greater objectivity, and repeatability in the description of leaf shapes and other outlines whose shapes can now be objectively quantified and compared statistically. Principal component analysis (*PCA*), which produces a heuristic tool for taxonomists to encapsulate their data in two or three dimensions, is enhanced by the *PCAPLOT* program to create two-dimensional and two-point perspective renderings of three-dimensional *PCA* scattergrams. Current work with cycad hybrid species detection and validation, as well as infrageneric species comparisons, are productively diagrammed using this software, which allows zooming and identification of individuals (observations) in the scattergrams. In the realm of phylogenetic analysis, studies are enhanced by computer programs such as PAUP 2.4.1, CLADOS (Beta Test Version) and HENNIG86 1.5. In particular, cladistic analyses using Hennig86 have supported ideas first propounded by Dehgan and Dehgan (1988) that a reclassification of the cycadales was warranted. Cladistic analyses have also required careful consideration of morphological features in an attempt to discover parallel evolution, convergence of characters, and para- or polyphyletic taxa. These analyses, which are sensitive to differing interpretations of characters, focus our attention on areas which require further study. The end result of the use of new, more sophisticated software will be more robust classifications at all levels of the taxonomic hierarchy.

Introduction

One of the more puzzling aspects of cycad systematics is the characterisation of taxa, which have the potential for extremely wide morphological variation in vegetative as well as reproductive features. Overlap is not only possible, but seems the rule rather than the exception.

As we look more at cycads, we are finding that their descriptions reflect topological thinkings, i.e., that all plants of a species are expected to conform to some preconceived stereotype rather than to be understood as a variable entity that can be characterised. Developing the methodology to characterise and describe cycads is a continuing challenge in our laboratory rather than a boring taxonomic necessity. It is our goal to find a consistent technique for identification, not to present yet another jigsaw puzzle for others to piece together. Computers have become an important tool to us in our studies of taxonomic problems, which range from the species level through the revision of the order Cycadales.

Nowadays, systematic studies attempt to use as many features as possible to characterise taxa, but this creates a problem because it is difficult for us to mentally process this much data simultaneously. Instead, we create a data matrix of characters for a group of taxa and analyse it with one or more multivariate statistical techniques using a computer. The purpose of this paper is to detail the hardware and software, written in our laboratory or elsewhere, that is currently being used to enhance systematic studies, both phonetic and phylogenetic, of the cycads, at all levels of the taxonomic hierarchy.

Phonetic Analysis

Phonetic analysis attempts to find groupings of taxa without inferring evolutionary relationships. This is in contrast with Phylogenetic Analysis (see next section), in which evolutionary hypotheses are synthesised.

In order to perform numerical analyses with principal component analysis (*PCA*) or other multivariate statistics, one first needs to quantify data, assigning numerical values to biologically significant features. Then, a data matrix is constructed from these values and analysed. One very important feature in cycad identification is leaflet shape. However, taking variation into account is crucial, especially in cycads. Thoughtless characterisation of shape has resulted in topological thinking, such as the work of Schuster (1932), in which *Zamia* species were defined based entirely on leaflet widths.

Fourier Transformation

The Fourier transform is a crucial part of today's music recording and playback technology, known as compact disc. The details are little understood by the general public, but the digitisation of music waveforms and their storage in binary form on optical disk is reliant on the Fourier transform. The musical waveforms, as you might observe on an oscilloscope, are translated into values which are stored on optical disk. When they are played, the values are read by lasers from disk and translated back into waveforms. The outlines of objects can be similarly converted to numerical values with the same methodology (not necessarily stored on optical disk, however), and these values may be used to compare shapes in an objective way. There are those who would criticise the replacement of a visual shape by a series of numbers, claiming the numbers themselves have no biological interpretation (Bookstein et al., 1982). However, in phonetic studies, now commonly encountered in the systematic literature, characters are coded by researchers who are most likely completely unaware of the relationship of those characters to gene expression and other underlying phenomena, which have resulted in those characters. Comparisons of nucleic acid and protein sequences are similar in that each nucleotide or amino acid cannot be related to a visible phenomenon. Thus, we frequently deal with unknowns but continue to make comparisons, because the characters we work with can be quantified and do have significance within an analytical framework. This is the major benefit of fourier analysis; the shapes that are commonly characterised by subjective nomenclature can be given numerical values which we can average, display graphically or use in multivariate statistical analyses.

The arguments of Bookstein et al. (1982) and others are not fallacious; it is always preferable to work with numbers that can be easily interpreted, but until the state of knowledge advances to this point, we do not believe it sensible to discard useful data.

The Fourier transform can be applied to a large number of objects whose outlines we can capture; cells, leaves, cone outlines, sporophylls, etc. The first author of this paper has been developing a computer

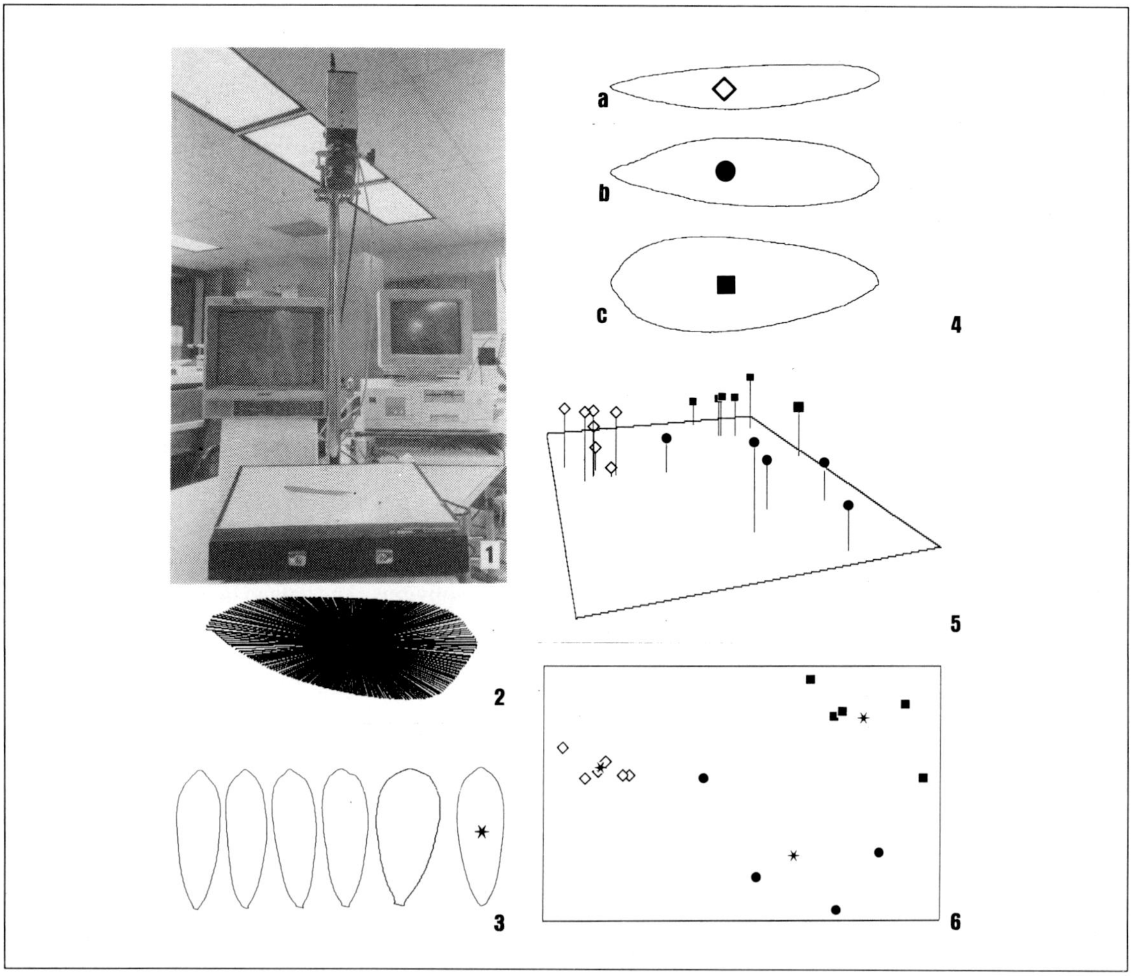

Figures 1 - 6: Analyses of Leaflets.

Fig. 1: Hardware configuration for image analysis system consisting of high resolution black and white television camera on copy stand, high resolution monitor, and 80386 AT-compatible computer with frame grabber.

Fig. 2: Example of a leaflet outline (*Zamia furfuracea* L.f.) used to calculate Fourier descriptors. It consists of 512 lines extending from the leaflet centroid which are measured, converted to a periodic waveform, and used as input to the Fourier transform.

Fig. 3: Example of the use of fourier descriptors to determine an average shape. Five leaflets of *Zamia furfuracea* L.f. are shown with their average, the starred leaflet to the right.

Fig. 4: Symbolism for principal component scattergrams in Figs. 5 and 6. (a) *Zamia loddigesii* leaflet, (b) natural hybrid of *Zamia furfuracea* and *Zamia loddigesii*, and (c) *Zamia furfuracea*.

Fig. 5: Example of *PCAPLOT* program output, a three-dimensional scattergram in two-point perspective using samples of the three taxa shown in Fig. 4.

Fig. 6: Two-dimensional version of the same scattergram shown in Fig. 5, with calculated average leaflet shapes indicated by stars. The averages may be found roughly in the center of each cluster.

program, Morphos, to compute fourier transforms of outlines. Cycad leaflets have been the focus of this study, but as mentioned before, any outline which we can digitise, or reduce to points in a Cartesian coordinate system can be analysed in this fashion.

The hardware setup used is pictured in Fig. 1. Images are captured on a copy stand with a high resolution monochrome television camera, and viewed on a television monitor. When the image is in satisfactory focus and fills the viewing area, it is transferred via an add-on board known as a frame-grabber to an IBM-compatible 80386 computer and stored on fixed or floppy disk for later processing. About 10 seconds are required to capture and store each image.

The Morphos program first takes the photographic image, stored as 262,144 pixels, or picture units, and determines the edge, or outline, as well as its centroid (center of gravity of the image). The program next takes the outline points and converts them to a waveform by determining the distance from the centroid to the edge along 512 equally spaced angles. The example picture in Fig. 2 is a leaflet of *Zamia furfuracea* L. f., showing the angles from the centroid to the outline. These 512 distances are used to calculate the Fourier transform, which consists of 512 'coefficients' or 'descriptors' that may be used in *PCA* or other multivariate statistical techniques. The transform is a unique set of values for each unique shape. This is known as 1:1 mapping.

An interesting property of the Fourier transform is that the first descriptor is most important in determining shape, and each subsequent descriptor is somewhat less important than the last. The ramification of this is that the shape can be reproduced from a small subset of coefficients from the beginning of the list of 512.

Approximately 30 - 50 of the first coefficients can be used to generate a shape which is indistinguishable from the original outline.

The second but perhaps the most important property of the transform is that the coefficients can be mathematically averaged, and the average used to reproduce an average shape. Figure 3 shows five leaflet outlines of *Z. furfuracea* and their averaging shape reconstructed from average Fourier descriptors. The method is promising not only for the comparison of different species, but for fine distinctions between leaflets of the same leaf, leaves from the same or different plants, or populations. The implications for possible verification of hybrid taxa are very encouraging. The Morphos program is continually being improved and expanded to allow more sophisticated operations useful for taxonomic analyses not only in the Cycadales, but in other biological groups in which outlines have significance for identification purposes.

Principal Component Analysis

Principal component analysis (*PCA*) is a phonetic analytical tool to help summarise or portray data so that similarities or differences between taxa are visually more obvious. The data matrix is mathematically manipulated to reduce variation in a particular study group to two or three dimensions. This allows visual presentation of the data. The data points are displayed as points or stick-and-ball objects in a two- or three-dimensional diagram. These diagrams are typically drawn by hand from computer-generated data or very roughly plotted by computer.

A good example of taxonomic use of *PCA* was the reinstatement of *Z. spartea* A.D.C. a narrow-leaflet taxon related to *Z. loddigesii* Miq. from Oaxaca, Mexico. *PCA* demonstrated (Schutzman, 1982) that a distinct morphological discontinuity, or gap, existed between it and *Z. loddigesii*, into which it had been sunk in synonymy by Schuster (1932).

PCA is the subject of another computer tool designed and written to aid biosystematic studies.

A program to more accurately and quickly display the results of *PCA*, called *PCAPLOT*, has been designed and written in our laboratory and continues to be improved upon. The examples in Figs. 4 - 6 are *PCA* scattergrams produced with *PCAPLOT* from Fourier transform data for three sample *Zamia* populations. Figure 4 illustrates the average shapes of (a) *Z. loddigesii*, (b) a naturally occurring *Z. loddigesii* x *Z. furfuracea* hybrid, and (c) *Z. furfuracea* together with their symbols as used in Figs. 5 and 6. Figure 5 is a three-dimensional *PCA* scattergram of the 15 leaflet outlines and 3 average leaflet outlines. Figure 6 illustrates the two-dimensional *PCA* scattergram, showing the position of the average leaflet by a star in each population. In each cluster of points, the average shape is found close to the centroid, illustrating that the mathematical average of leaflet shapes is picked up by *PCA*'s attempt to summarise variation in a few dimensions. The *PCAPLOT* program plots the data with either two or three axes, as shown by Figs. 5 and 6. If three axes are selected, the angle of view and height of the viewer relative to the scattergram are chosen interactively, and a two-point perspective view is calculated and displayed by the program. Identification numbers for the taxa, as well as zoom capabilities, are two additional useful features of the program., Future versions will place the symbols on each individual point or stick in the diagrams, as these now must be added manually.

These two computer programs, and the types of analyses they aid, are allowing us to perform more detailed and thorough investigation of taxonomic problems in the cycadales than have previously been possible.

Phylogenetic Analysis

While phonetic methods are typically used at or below the species level, phylogenetic studies are used at or above the species level. They attempt to produce hypotheses of evolutionary relationships of the taxa. Cladistic analysis, one type of phylogenetic study, is used to produce an evolutionary hypothesis in the form of a branching diagram called a cladogram, and is normally calculated by a computer because of the large number of calculations involved in even the smallest of studies.

The computer algorithm, or programmed procedures, looks for the cladogram which involves the fewest number of changes in characteristics needed to produce all the variation in the study group. This strategy for estimating phylogeny is called parsimony analysis, and the phrase is generally used interchangeably with cladistic analysis. This, however, is somewhat erroneous, since not all phylogenetic methods are designed around the search for parsimonious cladograms.

We have been assessing relationships among the cycad genera for some time now, first with the *PAUP* (vers. 2.4.1) and *CLADOS* (Beta test version) computer programs, and more recently, with J.S. Farris' Hennig86 program (1987). Our results differ somewhat from those reached by Crane (1988) and Stevenson (1990). This brings to light an unfortunate aspect about cladistic analyses: they are very sensitive to differing interpretation of characters. A careful study of characters and use of as many carefully chosen characters as possible should be helpful in stabilising results. Those who utilise cladistic analyses to create classifications are familiar with the concept of 'grouping by synapomorphy'. In the past, classifications have grouped organisms by both primitive and advanced features, but grouping organisms by primitive features often results in para- and polyphyletic groups. Now, it is generally accepted that shared derived features, called 'synapomorphies', are the most useful in producing classifications, since, generalising a bit, an advanced (or derived) feature is carried along by lineages for some time, even as they split up into many taxa, so the common possession of an advanced feature can identify taxa derived from a common ancestor (or 'stem lineage' - i.e., a monophyletic group.

An example of a synapomorphic feature is the cataphyll dimorphism seen in the cycad genera *Zamia* and *Ceratozamia*. This is important in supporting our argument that these two genera are sister groups, i.e., that they share a most recent common ancestor.

The coding of features for cladistic analysis is, as was mentioned above, highly sensitive to differing interpretation of characters.

One reason why our cladograms, one of which is pictured in Fig. 7, are so different from those obtained by Crane (1988) and Stevenson (1990 and this issue) is our different interpretation of some cycad features. An example of this is found in the genera *Stangeria* and *Bowenia*, which both Crane and Stevenson have claimed lack cataphylls on vegetative axes. We have good evidence from our plants that these structures are

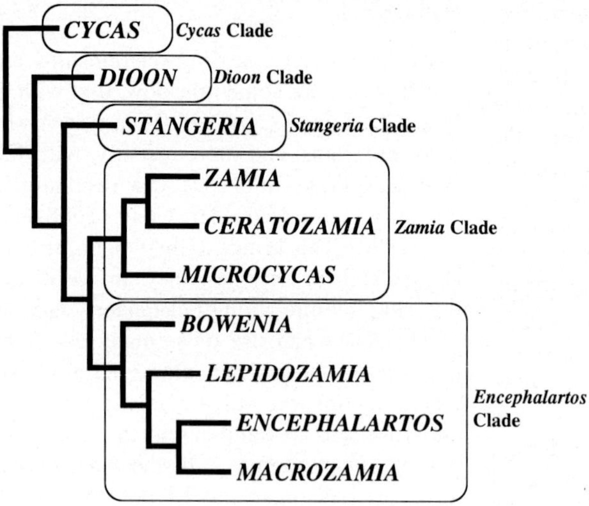

7

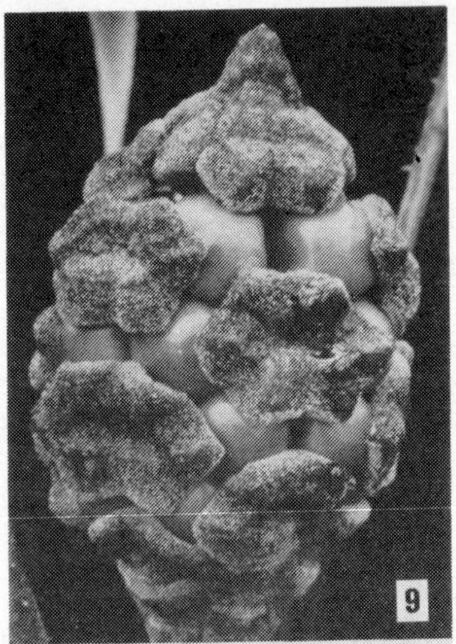

9

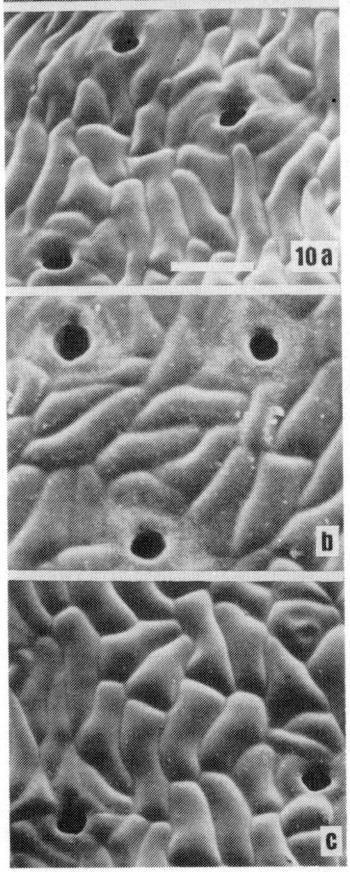

10 a b c

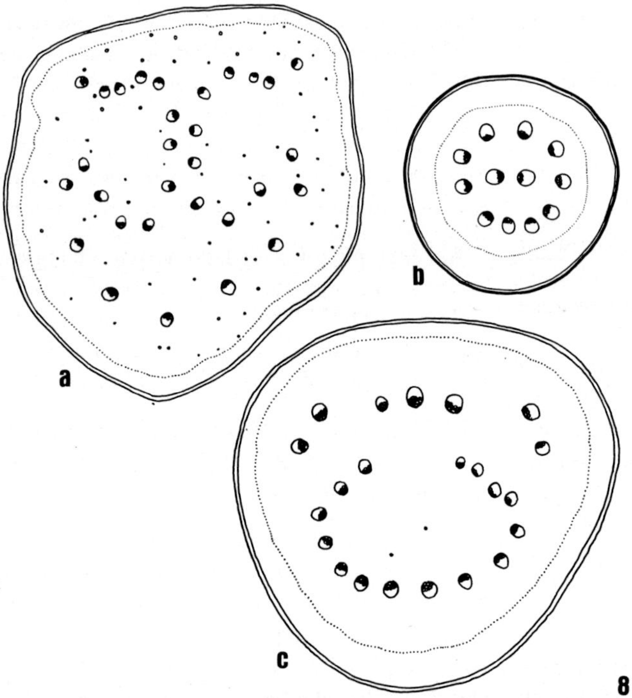

a b c

8

present in all stages of the life cycles of these two genera. So, while the lack of cataphylls to Crane and Stevenson is seen as a synapomorphy, we view that as a misinterpretation, and to our way of thinking *Stangeria* and *Bowenia* are not different from the remaining cycads in this respect. A.P. Vovides is collaborating with us on a forthcoming phyllotactic study of *Stangeria* cataphylls and leaves.

Another difference in interpretation concerns the petiole vasculature of the cycads. Most cycads have what is commonly known as the 'inverted omega' patterns of petiolar vascular bundles. In the analyses of Crane (1988) and Stevenson (1990a), *Bowenia* and *Stangeria* are coded identically as having a circular pattern of vascular bundles in their petioles. We view the *Stangeria* and *Bowenia* patterns of vasculature to be as different as night and day (Fig. 8). The *Stangeria* patterns seem more closely allied to the inverted omega pattern, while *Bowenia* only possesses a circle (or often two concentric circles) of vascular bundles. Once again, whereas Crane's (1988) and Stevenson's (1990) coding of characters assess the petiole vasculature of the *Bowenia* and *Stangeria* to be synapomorphous, our interpretation is that they do not share a common patterns of petiolar anatomy.

In our cladistic analysis using the Hennig86 program, we chose the genus *Cycas* as the outgroup (or sister taxon) to the remaining cycad genera, whereas in the works of Crane (1988) and Stevenson (1990), 'other seed plants' were used as an outgroup. The rationale for our decision to proceed in this manner is too lengthy to take up here, and the details of our analysis and the resulting classification are forthcoming (Schutzman and Dehgan manuscript, in prep). Our two cladograms place *Dioon* as very primitive and nearest *Cycas* and *Stangeria*.

This agrees with data from transmission and scanning electron microscopic studies of pollen by Dehgan and Dehgan (1988). Moretti (1990) recognised from karyotypic evidence that *Dioon* was distinct from all other New World Zamiaceae but placed it closer to the Encephalarteae *sensu* Johnson (1959) (*Encephalartos, Macrozamia*, and *Lepidozamia*). Additionally, *Bowenia* in both trees is basal to the Encephalarteae, and quite distant from *Stangeria*, a significant difference between the two works of Crane and Stevenson and our own.

Currently, four families are recognised in cycadales, the Boweniaceae, Cycadaceae and Stangeriaceae, which are monogeneric, and the Zamiaceae, comprising the remaining seven genera. Based on our cladograms, five monophyletic families can be recognised in the Cycadales using the Annotated Linnaean Hierarchy convention as advocated by E.O. Wiley (1979): the *Cycas, Dioon*, and *Stangeria* clades would be considered monogeneric families, the *Zamia* clade would include *Zamia, Ceratozamia*, and *Microcycas*, and the *Encephalartos* clade would include *Bowenia, Lepidozamia, Macrozamia* and *Encephalartos*. These groupings are superimposed on the cladogram pictured in Fig. 7. Subfamilies may have to be recognised in families described from the *Encephalartos* and *Zamia* clades. In no way do any of our works support a close relationship between *Bowenia* and *Stangeria*, which has in this issue been proposed by Stevenson.

←

Figures 7 - 10: Relationships of Cycad genera.

Fig. 7: One of two cladograms obtained from cladistic analysis of the cycadales using morphological data. Enclosed clades represent families in a forthcoming new classification of the cycadales.

Fig. 8: Petiole vasculature of cycads: (a) Invested omega pattern shared by all genera except *Stangeria* and *Bowenia*; (b) *Bowenia* pattern; (c) *Stangeria* pattern.

Fig. 9: Megasporangiate strobilus of *Zamia herrerae* Calderon & Standley showing raised bumps similar to those found on those of *Chigua restrepoi* Stevenson.

Fig. 10: Scanning electron micrographs of cycad abaxial leaf epidermis (scale = 50 μm: (a) *Zamia fischeri* Miq.; (b) *Chigua restrepoi* Stevenson; (c) undescribed *Zamia* sp. from NW Colombia.

This arrangement is in agreement with the ideas proposed by Dehgan and Dehgan (1988).

Another useful and quite necessary application of cladistic analysis should be the validation of new genera to prevent the recognition of paraphyletic and even polyphyletic groups, and bring the data matrix used under the close scrutiny of character analysis. Phonetic distinctions are only a first step in recognising a new taxon above the species level. An example of this is the recently described genus *Chigua* from Colombia (Stevenson, 1990b). Two major features are used to separate this genus from *Zamia*. The megasporophylls have six prominent bumps on the points of the hexagonal face, and the longitudinal axis of leaflets have a thickened vein described as a midrib. This first of these two features is routinely found in *Z. herrerae* Calderon & Standley, a Salvadoran and Mexican species (Fig. 9), and also occasionally in individuals of *Z. furfuracea*.

Electrophoretic work by Walters and Decker-Walters (1991) supports a close relationship of *Chigua* to certain species of *Zamia*. The 'midvein', which both Crane (1988) and Stevenson (1990A) code as a homology to the structures seen in leaflets of in *Cycas* and *Stangeria*, is reduced in the apical portion of leaflets until it is indistinguishable from other dichotomous parallel veins in the apex.

We feel that this is a feature unlike the complex midvein of *Stangeria* with its multiple vascular bundles, or that of *Cycas*, with a single bundle. The three structures should definitely not be considered homologous. These data, together with epidermal comparisons of *Chigua* to several *Zamia* species (Fig. 10 a - c), suggest that the new genus should be treated as part of *Zamia*, albeit exhibiting a few interesting apomorphic features. At best, the two species of *Chigua* should be considered distinctive species of *Zamia*. We believe that the use of characters such as these to describe a new genus would make the remainder of *Zamia* paraphyletic.

Acknowledgment

Experiment Station Journal Series No. R-014240. We thank Fe Almira and Ian Breheny for technical assistance.

References

***Bookstein, F.L., Strauss, R.E., Humphries, J.M., Chernoff, B., Elder, R.L., Smith, G.R.* 1982.** A comment upon the uses of fourier methods in systematics. Syst. Zool. 31: 85 - 92.

***Crane P,* 1988.** Major clades and relationships in the 'higher' gymnosperms. In Origin and Evolution of Gymnosperms, Chapter 3, C. Beck, ed. Columbia Univ. Press, N.Y.

***Dehgan, B., Dehgan, N.B.* 1988**. Comparative pollen morphology and taxonomic affinities in cycadales. Amer. J. Bot. 75: 1501 - 1516.

***Ehrlich, R., Pharr, Jr., Healy- Williams, N.* 1983.** Comments on the validity of fourier descriptors in Systematics. A reply to Bookstein et. al. Syst. Zool. 32: 202 - 206.

***Farris, J.S.* 1988.** Hennig86, Version 1.5. State University of New York at Stonybrook.

***Johnson, L.A.S.* 1959.** The families of cycads and the Zamiaceae of Australia. Proc. Linn. Soc. N.S.W. 84: 64 - 117.

***Moretti, A.* 1990.** Karyotypic data on North and Central American Zamiaceae (cycadales) and their phylogenetic implications. Amer. J. Bot 77: 1016 - 1029.

***Schuster, J.* 1932.** Cycadaceae. In: *Das Pflanzenreich* Part IV.I (Vol. 99) J. Cramer. Berlin.

***Schutzman, B.* 1982.** Preliminary systematic and taximetric studies of Mesoamerican *Zamia* L. (Zamiaceae). Unpub. M.S. Thesis, University of Florida, Institute of Food and Agricultural Sciences.

***Schutzman, B., Vovides, A.P.* 1985.** Phonetic and other systematic studies of the *Zamia loddigesii/Z. furfuracea* complex. Abstract of paper presented at 36th annual AIBS meetings, University of Florida, Gainesville, Aug. 11 - 15.

***Stevenson, D.W.* 1990a.** Morphology and systematics of the cycadales. Mem. N.Y. Bot. Gard. 57: 8 - 55.

***Stevenson, D.W.* 1990b.** *Chigua*, a new genus in the Zamiaceae with comments on its biogeographic significance. Mem. N.Y. Bot. Gard. 57: 169 - 172.

***Walters, T., Decker-Walters, T.S.* 1991.** The application of isozyme electrophoresis for examining taxon relationships within the cycadales. Can. J. bot. In Press.

***Wiley, E.O.* 1979.** An annotated Linnaean hierarchy, with comments on natural taxa and competing systems. Syst. Zool. 28: 308 - 337.

In Search of *Cycas micholitzii*

Ian S. Turner
Springs Farm, PO Box 2162, Harara, Zimbabwe

Abstract

Cycas micholitzii, from southern China and possibly Vietnam and Laos, is a tropical species with the unusual feature of leaf branching. *Cycas panzihuanensis*, from Sechuan Province, is an attractive plant locally propagated fairly extensively. *Cycas pectinata*, from Yunnan Province, is being transplanted for propagation to the botanic garden near Xishangbanna. A plant called *C. yunnanensis*, similar to *C. rumphii*, is also found in cultivation at Panzhihua. In Thailand, near Bangkok, a highly pubescent plant was found and may be an undescribed *Cycas*.

Cycas micholitzii is the odd man out in *Cycas* because of its beautiful multiple branching leaves. It grows in an area in south west China near the border with Vietnam and most likely in Vietnam and Laos also. I went to China in the hope of being able to see this plant growing in the wild and maybe collect a couple of specimens, but since the area where it grows is restricted because of military activity, I was unable to go there. However, I arrived in Kunming in the Yunnan Province of China in January 1989 and, since it was winter, found it very cold indeed. I had a contact at the botanic gardens in Kunming so the following day went to look for him but found he had left for Beijing. Some of the very helpful staff there showed me their *Cycas* collection, which included two plants of *Cycas micholitzii*. One was growing in a pot in the open and the other was in a pot in a greenhouse. The one outside was obviously suffering (like me) because of the cold weather; its leaves had not grown out to their full extent. Where these plants grow near Vietnam, the weather is much warmer. The plant in the greenhouse looked quite different to the one outside, so much so in fact that some of the people there though it was a different species. The Kunming Botanic Gardens, which is in fact the Kunming Institute of Botany, is very extensive and there are some very nice greenhouses with a varied collection of plants; cacti, orchids, bromeliads, ferns, etc., all to be found there, and outside there are some beautiful *Camellias* to mention only one of the many shrubs and trees to be seen.

Another *Cycas* which I was interested to see was *Cycas panzhihuaensis*. These plants grow near the town of Panzhihua, formerly known as Dukou. Panzhihua is a small mining town in the south of Sechuan Province. I have been in contact with the director of the botanic gardens in Panzhihua, Comrade Zhou, and he kindly came with an interpreter, Comrade Wang, to meet me in Kunming. The following day we travelled from Kunming to Panzhihua by road, which is about an 8 hour drive. The whole countryside in that area is very mountainous so the road twists and turns through the mountains and it was never possible to go very fast. The roads are quiet, with no private cars on the roads because no-one owns a car. There are quite a few transport lorries and government vehicles and two-wheeled tractors pulling trailers and, in some areas, lots of ponies pulling carts. Since it was winter when

Cycas panzhihuaensis

no rain falls, the whole area was very dry indeed, but in most of the valleys there was some irrigation and lots of vegetables and broad beans were being grown. The Chinese certainly work very hard in their fields, and the cabbages, etc. they were growing were about the best I have ever seen.

Cycas panzhihuaensis

My first sight of *Cycas panzhihuaensis* was at the hotel in the town of Panzhihua where I stayed my first night. There were some very nice specimens growing in the garden there. I only stayed in the hotel one night because Comrade Zhou very kindly let me stay in the good accommodation he has in his Botanic Garden. There are of course very many fine specimens of *Cycas panzhihuaensis*. This is a very beautiful cycad, similar in some respects to *Cycas taiwaniana*, but instead of the dark green leaves of *C. taiwaniana*, this species has leaves with a bluish-grey sheen to them.

The following day we went into the mountains where *Cycas panzhihuaensis* grows wild. We followed the road taken by the mine vehicles. The mine is in fact taking away whole sides of the mountains and the excavation work there is really on a grand scale. Seemingly quite a few cycads were lost when part of a hillside was taken. We left our car and carried on by foot past the mining operations and came to a place which has been declared a reserve for the cycads. There are very many of these beautiful plants growing there, some up to 8 feet tall; one plant looked as if it was growing out of a rock. A seed must have fallen into a crevice many years ago and found a crack to send its roots down to the soil; now the plant, which is about 5 feet tall, had filled the crevice and is just sitting on top of the bare rock enjoying the sunshine in the cool mountain air.

Every year, people from the Botanic garden collect all the seeds produced by the plants and grow them in the gardens so there are now very many plants in the nursery area of the gardens. This is a very good way to preserve a species. It can be said that there is no natural regeneration, but if the seeds were left in the wild, only a very small percentage would grow into mature plants. Now in cultivation, virtually all the seeds which germinate will grow. If at some time it is felt that the wild population is dwindling, it can be boosted by the cultivated plants. As I see it, with all the destruction of habitats these days for farming, road building, downs, etc. the only way a species is going to be preserved is by cultivating it.

The Panzhihua Botanic Garden has one specimen of *Cycas micholitzii*, but it is too cold in that area for it to grow to its full extent. They also have specimens of *Cycas pectinata* and what looked to me like *Cycas szechuanensis*. I had taken some *Encephalartos* species with me so was able to do a very worthwhile exchange.

Cycas micholitzii

Comrade Zhou told me about some other cycads growing at the steel works office which are thought to be *Cycas hainanensis*. There are two plants with very long leaves growing in big pots and each had a sucker. I was very grateful to the mine director for allowing me to have the suckers. There were some other cycads growing around the office and beds of flowering plants and many kinds of pot plants decorated the area, together with water fountains; it all looked very nice indeed.

Still with *Cycas micholitzii* in mind, we decided to go to a place called Xishangbanna which is in the far south of Yunnan. There is a tropical botanic garden there and I thought I would have a better chance of finding some of these plants because it is not too far from their habitat.

Comrade Zhou was unable to accompany us but he sent his assistant, Comrade Yung, who is quite knowledgeable about *Cycas* and is in fact the co-author with Comrade Zhou of *Cycas panzhihuaensis*, so we set off with Comrade Wang. It was a two and a half day journey to get to Xishangbanna by car. Still the roads passed over many mountain chains and at times we were driving on roads high above the valleys which were filled with clouds. The scenery was very beautiful.

The botanic garden at Xishangbanna is not only a botanic garden but a research centre also. Many kinds of fruit trees are being grown and experiments are carried out with other plants of agricultural value and an area has been set aside for the planting of endangered species, so the whole area covers many acres. One area in the garden was being landscaped and a lot of *Cycas pectinata* plants were being transplanted from the nearby forest to make quite a spectacular exhibit. One plant in particular stood out; it must have been at least 15 feet high and had many long branches. How it was transplanted to its new site remains a mystery. It certainly must have taken a lot of manpower!

The garden also has a nice collection of palms and some good specimens of *Roystonea, Arecastrum, Neodypsis, Phoenix* and *Arenga*, to name just a few, were all growing well. Unfortunately, no *Cycas micholitzii* were to be found but I was able to exchange a couple of the *Encephalartos* I had with me for some plants of a *Cycas* which looked different to the *Cycas pectinata* which grows in that area. They could be *Cycas yunnanensis*. *Cycas micholitzii* seems to be quite rare in cultivation and I was told that it is hard to find in its habitat. It is a pity that a botanic garden does not get some plants of these species together so that seeds can be produced. Here in Zimbabwe I am fortunate to have male and female *Cycas micholitzii* var. *simplicipinna* and have produced a good number of seedlings. So we get back to the conservation by propagation in cultivation again.

The Xishangbanna gardens are about 40 miles from the town of Xishangbanna, so after staying one night in the very good accommodation at the garden, we carried on to see the town which also has a small garden of its own. Xishangbanna is a town where most of the tourists from the West like to go. There is a very good hotel complex there with lots of small double-storey buildings where one can rent rooms. It is all set in a landscaped area with big trees, palms and flowering plants. They have bicycles for hire so it was quite usual to see the tourists pedalling their way around town.

The people who live in that south-west part of China are quite different from those from further north; they are more like the people from Thailand and Burma. There are some restaurants where one can get the delicious traditional food cooked in that area and, although I must admit sometimes it is hard to tell what one is eating, it all tasted very good indeed.

We left Xishangbanna and headed back to Kunming. although it is quite a long drive, there are always things of interest to see on the way. In Kunming, I was sorry to say goodbye to Comrade Wang and Comrade Yung, they had been good travelling companions and always did their best to make the trip go well.

I had to wait three days in Kunming until I was due to fly out to Hong Kong so I went back to the gardens there to see if my friend had returned from Beijing and was lucky to find that he had but at that time he was in town at a meeting. The following morning he came to see me at my hotel and we had a long talk about the *Cycas* at the gardens, not forgetting the small *Cycas micholitzii* I had seen there. I told him that I had some species of *Encephalartos* for exchange so he said he would go and talk it over with some of the heads of the departments and he would telephone me. I was very pleased to get a telephone call to say "come tomorrow morning", because they were interested in an exchange. I arrived bright and early at the garden and just about froze while waiting at a gate to be let into the greenhouse area. It certainly was very cold. I was glad to be given some hot green tea to warm me up when I got into the chief's office. I showed them the plants I had with me and they brought along some interesting looking *Cycas* species but it did not include a *Cycas micholitzii*. I explained to them that I had come from very far away Africa and the main object of my trip was to try to get a plant of that species and it would indeed be hard for me to have to return without one. After a long discussion between themselves and more green tea, I was very pleased to hear that although *Cycas micholitzii* is a very rare plant, since they have two in the garden they would let me have the small one from the greenhouse. So with that plant and another which looked like *Cycas szechuanensis* and a couple more *Cycas pectinata*, I returned to my hotel and had another drink (not green tea this time) to warm me up and celebrate my good fortune.

I had heard about a street market in Kunming where plants, birds and small animals are sold, so I decided to go to see what was on offer and was happy to find a man selling some small *Cycas* plants which looked like *Cycas yunnanensis*. I must admit that I had not heard about *Cycas yunnanensis* before I went to China but now, having seen these plants, they do look different to any other *Cycas* I know. Also on the market were some beautiful orchids and parakeets and some very nice monkeys and lots of very colourful goldfish. So it was quite an interesting place to spend some time seeing the items for sale. I got all my plants packed up and left the following morning aboard the very good airline China Aviation.

It was 5 weeks before the plants were unpacked and I was pleased to find most of them had new roots and some even had leaves growing inside the parcel. They continue to grow well and now, three months later, I have some of them planted out in the garden. The others are all growing well in pots. It seems important with small *Cycas* plants not to let them dry out; each plant had the old leaves cut off and was thoroughly washed then wrapped in newspaper and sealed in a plastic bag. There is enough moisture on the plant to keep the newspaper slightly damp inside the bag and new growth will start. If the plants are packed completely dry, it takes a long time for new roots to form and in the meantime the plant is drying out even more and usually deteriorates to such an extent that new growth is impossible. In the good climate of Zimbabwe, these plants will grow fast and I am looking forward to seeing seeds on some of them in the not too distant future.

I had a very pleasant stay in China and found the people very helpful. I would like to go back again one day to try to locate *Cycas micholitzii* growing wild.

Taxonomy of *Encephalartos* (Zamiaceae): Taxonomically Useful External Characteristics

P. Vorster
Botany Department, University of Stellenbosch,
7600 Stellenbosch, South Africa

Abstract

Macroscopical external characteristics of *Encephalartos* which have taxonomical value are enumerated and assessed. These include vegetative features, as well as characteristics of the male and female cones. These characteristics are very useful for circumscribing species, but not necessarily for determining phylogenetic relationships.

Introduction

Ever since the first species in the current genus *Encephalartos* was described in 1775, the definition of good taxonomical criteria has been a problem. Traditionally, species have been distinguished from each other on account of vegetative characteristics. While this worked reasonably well in the early years, we have now reached the stage where some species are vegetatively either indistinguishable from each other, or where the differences are so cryptic that these cannot unequivocably be described in words. Strangely enough, the morphology of the cones hardly featured in the taxonomy of *Encephalartos* until recently. The reason for this may be that taxonomists worked in distant countries, or at least far from the natural habitat, and had at their disposal only dried specimens or cultivated plants which seldom, if ever, flowered. Cycad cones are notoriously difficult to preserve in such a way that their finer characteristics remain visible.

Moreover, coning tends to be erratic, especially under artificial conditions, and when cultivated plants do cone, there are often no cones of related species available to highlight the differences. Indeed, when R.A. Dyer published his account of *Encephalartos* in 1965, the cones of *E. paucidentatus* were still unknown, even though this species had been described as early as 1926. Similarly, when Johnson published his monograph on the Australian Zamiaceae in 1959, the cones of several species had never been recorded. Over the years, descriptions of new species became more comprehensive, but were it not for accompanying illustrations and supporting type specimens, these would have been of little use. It is one thing to describe a species in detail, but quite another matter to distinguish it.

In his later years, R.A. Dyer did make a start in using cone characteristics for diagnostic purposes. Thus, in his paper on the *E. manikensis* group (Dyer and Verdoorn, 1969), he used both male and female cones; as well as in the defining of *E. heenanii* (Dyer, 1972) where the male and female cones are externally so strikingly similar. When I took over from Dyer, I decided to use the cones as my point of departure. The key structures affecting genetical isolation and thus speciation are the sexual organs, and I felt that these should give a much more true view of the differences and relationships between species.

Taxonomically Useful External Characteristics

In the following review, only external macroscopical characteristics are treated. The reason for this is that we still know very little about microscopical structures, anatomy, cytology and chemical constituents; and their taxonomical implications.

Vegetative characteristics

Amongst vegetative characteristics, the first to strike the observer is the stately trunk with regular 'chain-mail' pattern which is characteristics of so many species, but underground and therefore invisible in others. In species like *E. transvenosus* and *E. paucidentatus*, the stems are tall and columnar, habitually reaching a length of 6 metres, but occasionally twice that length. In most other arborescent species, the normal maximum trunk length is 2 to 4 metres, probably because the trunk in most species is not thick enough (i.e. sturdy enough) to withstand storm damage when growing taller than that. More difficult to understand is why species like *E. lehmannii*, *E. horridus*, *E. lanatus* and *E. cycadifolius* rarely have trunks exceeding 1 metre in length, even though these trunks seem to be perfectly strong enough to grow taller. These short trunks do not seem to be drawn continuously underground by contractile roots as in the case of species like *E. villosus* where the trunk is completely underground. Whereas, it is understood that the stems of species like *E. villosus, E. umbeluziensis, E. caffer, E. ngoyanus, E. humilis, E. ferox* and perhaps *E. cupidus* are drawn underground by contractile roots, there is no evidence that these underground stems ever attain the lengths of the exposed stems in other species. Apparently, it is not a case of their being short-lived, as in strongly suckering specimens there is no sign of the older stems dying off. The underground or exposed nature of the stem has been used to define species, but for identification it should of course only be used in population context where one can be sure that plants with unexposed trunks are not juveniles. To my mind, it is entirely senseless to use this characteristic for phylogenetic purposes, because there are at least two instances where plants without exposed stems are obviously closely related to others with well-developed trunks, literally dwarf siblings of the larger ones: here the obvious relationship between *E. humilis* and *E. laevifolius*, and *E. cupidus* and *E. dyerianua*, come to mind.

A perceptive observer can distinguish some species by the ornamentation of the trunk, i.e. leaf base remains. This relies on the size of the leaf bases, the width of the zones occupied by scale leaves, the size and shape of the scale leaves or their remains, and the presence or absence of a woolly indumentum in the crown or even on the remains of scale leaves or leaf bases. Species of which the trunks have a very distinctive appearance include *E. friderici-guilielmii, E. ferox, E. laevifolius, E. lehmannii*, and *E .inopinus*. Unfortunately, this appearance is subtle and very difficult to describe in words, so that it has hitherto not been used for taxonomical purposes.

The second feature which strikes the observer immediately is the appearance of the foliage: how long are the fronds, are they straight or recurved or incurved, are they green or glaucous, many or few? Frond length is used for descriptive purposes mostly, as this property can easily be influenced by local environmental factors. Yet there is a significant difference between the frond lengths of apparently closely related species such as *E. humilis* and *E. laevifolius*, or *E. ngoyanus* and *E. villosus*, and this feature has been used to key out species. The carriage of the fronds is often characteristic: in *E. transvenosus* they are straight and stiff; in *E. paucidentatus, E. heenanii, E. ghellinckii* and *E. cycadifolius* they are incurved so that the plant has a wineglass shape; in many species they are recurved to a greater or lesser extent; in *E. eugene-maraisii* and *E. lanatus* the tip is characteristically curved; and in *E. dolomiticus* they are reputed to be somewhat spirally twisted. Most species have more or less bright green fronds, but a number have strongly glaucous fronds which are characteristic for those species, and this characteristic has been used for taxonomical purposes. It is, however, an extremely dangerous feature to use without a very good understanding of the species concerned - for instance, in both *E. longifolius* and *E. laevifolius* bright green as well as glaucous populations occur. The number of fronds carried is also significant in its extreme manifestation: in *E. villosus* only three to five fronds are produced at a time and the total carried seldom exceeds ten, whereas in *E. transvenosus* up to 40 may be produced at a time and a trunk may carry more than 100 fronds.

The structures which have perhaps most widely been used for taxonomical purposes, are the leaflets. The orientation of the leaflets is highly characteristic; in most species they are ascending, but in *E. transvenosus* in the same plane, and descending (drooping) in *E .inopinus, E .paucidentatus, E. heenanii* and *E. ghellinckii.* In *E. caffer* and to some extent *E. friderici-guilielmii* the leaflets are ascending, but adjacent leaflets on the same side of the rachis are not in the same plane so that the frond has a plumose appearance. When seen in plane view, the leaflets of most species are directed towards the apex of the frond - this is especially noticeable in tropical African species like *E. hildebrandtii* and also in *E. lehmannii* as anyone would know who ever attempted to hand-pollinate this species or to cut off a cone. In *E. transvenosus* the leaflets in the centre of the frond spread almost at right angles as seen in plane view; while in *E. inopinus* the lower leaflets point towards the frond base, the centre leaflets spread at right angles, and the distal leaflets point towards the frond apex. A very stable characteristic is whether the proximal leaflets are reduced to prickles, and whether or not these extent to the very base of the frond or leave an unarmed petiole. This characteristic distinguishes *E. villosus* from *E. umbeluziensis*, and *E. dyerianus* from allied species, but its employment to distinguish *E. natalensis* from *E. altensteinii* may be open to criticism. Both character states occur in the same population of *E. inopinus*, but this situation is exceptional. The shape and dimensions of the leaflets to some extent go together. To anyone acquainted with the species, almost each species has a characteristic leaflet shape, but surprisingly this is by no means easy to describe in words. On the one end of the scale are the species with linear, very narrow leaflets, usually not much more than 5 mm wide, namely *E. cycadifolius, E. ghellinckii, E. friderici-guilielmii, E. lanatus, E. laevifolius* and *E. humilis*. At the other end of the scale is *E. ferox* with leaflets as wide as 80 mm. In *E. caffer* the leaflets are seldom longer than 100 mm, while in *E. villosus* and *E. paucidentatus* they are as long as 250 mm. In many species the lengths of the leaflets, not to mention the width, are intermediate between these extremes; and while the dimensions constitute good descriptive data, they are not really accurate for taxonomic purposes. The leaflet shape varies considerably between species, and in its extreme manifestations constitute useful taxonomic features. Thus, there is the group with linear leaflets already alluded to; there are the very broad leaflets rounded at both ends in *E. ferox*; and there are the conspicuously ovate leaflets of *E. woodii* and to a lesser extent *E. natalensis* (some forms), *E. transvenosus* and *E. longifolius*. In most other species the leaflets are narrowly elliptic and tapering to both extremities, rather than ovate. In the species with linear leaflets, the leaflets are straight, as they are also in *E. ferox*, but in many other species they are falcate to a greater or lesser extent, noticeably so in *E. paucidentatus* and *E. villosus*. In most species, but especially those with lobed or dentate margins, the leaflets are asymmetric (oblique). Lobed leaflets are usually constant and species-specific, as in *E. horridus, E. latifrons* and *E. arenarius*. In *E. trispinosus* it is however not uncommon to find unlobed leaflets, but that species is suspected of being involved in so much hybridising that its pure state is difficult to assess. Entire leaflets is also a constant characteristic - examples are *E. cycadifolius, E. lanatus, E. lehmannii, E. princeps, E. inopinus* and *E .paucidentatus* and *E. heenanii* when mature. On the other hand, *E. ferox* always has strongly dentate leaflets. One should however be extremely cautious when using leaflet dentation for taxonomic purposes. The leaflets of practically all species have at least some degree of dentation in the juvenile stage. Not only young plants, but also the first fronds produced after a fire or transplantation often have leaflets with much stronger dentation than before the event, and eventually leaflets with the normal degree of dentation get produced again. Lastly, even in species which do have dentate leaflets, individuals with completely entire margins to the leaflets are known. Examples are *E. transvenosus, E. villosus, E. lebomboensis, E. natalensis* and *E. altensteinii*. On the other hand, a few dentate or even lobed leaflets are sometimes found on mature plants of species associated with entire leaflets, such as *E. longifolius* and *E. lehmannii*. While one cannot be sure, one does get the impression that this is an atavism from past hybridisation, perhaps a long time ago if one takes present day distribution into account. Two more taxonomically useful leaflet characteristics are plane (flat) or recurved leaflet margins, and veination. In several species the leaflet margin is slightly thickened, but in *E. ghellinckii* it is revolute. The extent to which veins are visible on the abaxial leaflet surface varies from species to species, but is difficult to express in words and thus is not a characteristic which I would like to use. On the other hand, the veins are usually flush with the abaxial leaflet surface, but in *E. paucidentatus* and *E. heenanii* are conspicuously raised so that the abaxial surface is corrugated. The number of veins on the leaflets have been used to differentiate between *E. lanatus*, *E. laevifolius* and *E. humilis*, and this is something which deserves more attention. A characteristic which holds promise, perhaps even for phylogenetic purposes, but has never been studied, is whether the leaflets are succubous (the upper leaflets overlapping the lower in dorsal view) or incubous (the lower leaflets overlapping the upper in dorsal view). This feature is held to be taxonomically very important in the Hepaticae. True, Dyer has drawn attention to the angle of insertion of the leaflets in *E. princeps* as opposed to *E. lehmannii*, as well as in the *E. manikensis / pterogonus / concinnus / chimanimaniensis* group, but that has nothing to do with succubous or incubous.

Sexual characteristics

Female cones - As far as the cones are concerned, the females have up to now proved to be the most useful for taxonomic purposes, and there is the additional advantage that they are carried much longer in the mature state. The number of female cones carried per stem may have significance when distinguishing species. Thus, *E. lehmannii, E. trispinosus, E. arenarius* and *E. longifolius* invariably carry solitary cones. It is noteworthy that *E. lehmannii*'s sibling species *E. princeps* and *E. arenarius*' presumed sibling species *E. latifrons* bear multiple cones. *E. eugene-maraisii* seems never to carry more than two cones. The problem with this characteristic is that its application necessitates a profound knowledge of the species to be defined, and that it is of little if any use for identification purposes. For instance, a single cone on a plant of *E. natalensis*, a species which normally carries multiple cones, does not mean anything. *Encephalartos dolomiticus* was described as having solitary cones, yet cultivated plants carry up to three cones. Most female cones are sessile, but in *E. villosus* and *E. horridus*, for example, they are shortly stalked. The shape of the female cone is less easy to define, yet it is possible to draw a line between ovoid cones (e.g. *E. eugene-maraisii*) and cylindrical cones (e.g. *E. middelburgensis*). The colour of the female cone can present a dramatically distinct appearance; here one only has to think of the contrast between the brilliant salmon-red cones of *E. ferox* and the green cones of *E. arenarius*. It can be misleading though. In species like *E. villosus* and *E. umbeluziensis* the cones are dark green up to the stage where they are ready to be fertilised, after which they gradually turn bright orange-yellow. In species like *E. lehmannii, E. horridus, E. eugene-maraisii* and, to some extent, *E. middelburgensis*, the russet-brown colour is not due to inherent pigmentation, but to a microscopically fine indumentum which rubs off towards maturity to reveal the greenish underlayer. The presence and degree of pubescence on the cone scale faces (bullae) can be taxonomically significant. Thus, the cones of species like *E. ferox, E. dyerianus* and *E. villosus* are entirely glabrous; those of for instance *E. lebomboensis, E. longifolius* and *E. eugene-maraisii* have a microscopically fine indumentum; in *E. altensteinii*, *E. paucidenatus*, *E. transvenosus* and *E. laevifolius*, the female cones have a clearly visible hair covering (especially when young); while in for example *E. heenanii*, *E. aemulans* and *E. lanatus* the cones are covered in a very dense and shaggy trichome mat. The facets of the exposed bullae of the megasporophylls have been described and named by Melville (1957). I believe that basic pattern holds true for all species, but that the borders between the facets are not equally well expressed so that the bullae of some species may appear to have a reduced number of facets, or seem not to be differentiated into facets at all. The expression of the bulla facets is probably not a reliable taxonomic characteristic. The degree of projection or elevation of the central facet may however be diagnostic. In *E. lebomboensis* the bullae may be quite flat, but in *E. altensteinii, E. natalensis* and *E. transvenosus* the central facet is conspicuously elevated. A conspicuous and useful feature is the surface sculpture of the bullae. Thus, in *E. princeps, E. latifrons* and *E. aemulans* the bullae are tuberculate, in *E. trispinosus* and *E. horridus* numerous warty ridges radiate from the central facet, while in *E. villosus, E. inopinus* and *E. ferox* the facets are virtually smooth. The diameter of the central facet in relation to the diameter of the bulla does not at this stage appear to be a reliable characteristic.

Male cones - The male cones have not been found to be as taxonomically useful as the female cones. Most of the characteristics of the female cones listed above also hold true for the male cones, but there seems to be much less variation between the male cones of different species. One real practical problem is that the male cone rather rapidly changes in appearance as it matures, releases the pollen, and dries up; and only at the time of pollen release does the male cone present its mature appearance, if that is the correct word. In most species, the male cones are clearly stalked, sometimes very markedly so, like in the case of *E. villosus*, while in species like *E. lanatus* and *E. aemulans* they are sessile, and in some forms of *E. lebomboensis* the proximal part of the cone appears to be submerged in the apex of the trunk as if it never fully emerged. In shape the male cones of most species differ sharply from the female cones, being narrowly ovoid, but in *E. aemulans* the submature male cones are ovoid and externally look just like female cones. In *E. heenanii* the male and female cones are externally indistinguishable.

Microsporophylls have the same type of indumentum as the macrosporophylls, as well as the same range of pigmentation, but in all the material seen the surface textures are smooth, except sometimes for a tuberculate central facet. Where a considerable range of variation does occur is in the degree of projection of the central facet of the bulla. This may vary from slightly elevated in *E. dyeri, E. aemulans* and typical *E. lebomboensis*, to strongly projected in *E. lehmannii* and *E. arenarius*, to draw out into a drooping beak in *E. natalensis* and *E. eugene-maraisii*. In *E. villosus* the central facet is not raised, but it is enlarged to constitute almost the entire externally visible part of the microsporophyll, and the lower margin is often pectinately incised. A feature still to be studied is the distribution of sporangia on the abaxial surface of the microsporophyll.

Table 1 summarises external taxonomically useful vegetative and sexual characteristics and character states.

Conclusion

From the above, it should be clear that *Encephalartos* is blessed with a diverse range of variable characteristics which could be employed for taxonomical purposes. In isolation, these may be misleading, but in combination they constitute a powerful tool for defining species. As the cones are the organs for sexual reproduction, and as in most plants the reproductive organs play a decisive part in 'species recognition' as far as pollination by animal vectors is concerned, it is tempting to use the cones exclusively for taxonomical purposes. Whether this is feasible is another question; one only has to think of the very similar cones in *E. altensteinii, E. natalensis* and *E. transvenosus* to realise what problems are involved. To date, the characteristics of the cones, either alone or in conjunction with vegetative features, have allowed us to circumscribe *E. arenarius* as distinct from *E. latifrons, E. dy-*

eri, E. dolomiticus, and *E. middelburgensis* as distinct from *E. eugene-maraisii*; and *E. ferox, E. aemulans* and *E. heenanii* as distinct species. Yet sometimes the study of cones seems to add to the confusion rather than clarify the matter. In *E. lebomboensis* one finds populations in which the bullae of the female cones are quite flat, and the male cones sessile and slightly submerged in the apex of the stem, with bullae only slightly raised. In other populations the central facets of the megastobilate bullae are conspicuously raised, and the male cones are stalked with central facets of the bullae drawn out into drooping beaks. Yet we cannot find the slightest reason to separate these taxonomically. On the other extreme we are working on what we are convinced is a new species, which can only be distinguished by vegetative characteristics from what we presume to be its nearest relatives.

In conclusion, the characteristics listed above enable us to circumscribe species with confidence; but our understanding of cone morphology does not yet allow us in all cases to visualise phylogenetic groupings. The reason for this should probably be sought in an understanding of the functional value of these features, which surely play a part in preventing undue interbreeding where different species occur sympatrically.

As a postscript, it may be mentioned that chromosome studies, which are currently being undertaken, and chemical investigations (Osborne, Grobbelaar, and Vorster, 1988) have hitherto not been of any help in determining phylogenetic relationships, but that these avenues may eventually yield valuable data, especially if understood in functional context, and when used in conjunction with morphological data and geographical considerations.

References

***Dyer, R.A.* 1965.** The cycads of southern Africa. Bothalia 8: 405-514.

***Dyer, B.A.* 1972.** A new species of Encephalartos from Swaziland [*E. heenanii*]. Bothalia 10: 539-546.

***Verdoorn, I.C.* 1969.** *Encephalartos manikensis* and its near allies. Kirkia 7: 147-158.

***Johnson, L.A.S.* 1959.** The families of cycads and the Zamiaceae of Australia. Proceedings of the Linnean Society of New South Wales 84: 64-117.

***Melville, R.* 1957.** *Encephalartos* in central Africa. Kew Bulletin 1957: 237-257.

***Osborne, R., Grobbelaar, N, and Vorster, P.* 1988.** South African cycad research: progress and prospects. South African Journal of Science 84: 891-896.

***Robbertse, P.J., Vorster, P., and Van Der Westhuizen, S.* 1989.** *Encephalartos middelburgensis* (Zamiaceae): a new species from the Transvaal. South African Journal of Botany 55: 122-126.

***Vorster, P.* 1990.** *Encephalartos aemulans* (Zamiaceae), a new species from northern Natal. South African Journal of Botany 56: 239-243.

Field Studies of the Genus *Cycas* in Australia, Thailand and China

L.M. Whitelock
Cycad Gardens
Los Angeles, California, 10041 U.S.A.

Abstract

In Australia species of *Cycas* are sometimes difficult to separate because of overly precise descriptions coupled with vague locality data. Field studies have shown *C. media* to be a complex and highly variable species ranging over a huge area. Confusion has existed with *C. normanbyana, C. cairnsiana* and *C. kennedyana* on the Queensland coast. Species in the Northern Territory and Western Australia are less difficult because of their widely distinct and restricted distributions. In China and Thailand much of the native vegetation has been cleared for agriculture making field studies difficult; however, preliminary studies in Thailand indicate that several new species may be present.

Introduction

It seems paradoxical that *Cycas* having been the first cycad genus described (Linneaus, 1753), should at this time, 237 years later, have the reputation of being the most taxonomically confused. Over the 237 years of its existence there have been 75 species, 2 subspecies, 7 varieties and 4 forms described, disputed, confused, and finally submerged into synonymy with some of the original and more widely accepted species (e.g., *C. circinalis* and *C. rumphii*). At the present time approximately 26 species are recognised as being valid. Almost half of these (11) are endemic to Australia. There is a monumental job awaiting some taxonomist to untangle the species definitions within the *Cycas*. This undertaking is made all the more difficult given the huge area covered and the difficulties of both terrain and political unrest, both of which can impede travel.

Cycas has the widest distribution of all the cycad genera. It is found as far east as Fiji and Tonga, then west through New Caledonia, Vanuatu, Solomon Islands, Marianas Island, New Guinea, Australia, Southern Japan, the Ryukyu Islands, Philippines, Southern China, Republic of China (Taiwan), Vietnam, Kampuchea, Thailand, Laos, Indonesia, Malaysia, Burma, India, Sri Lanka (Ceylon), Madagascar, Comoros Islands and along the east coast of Africa (probably introduced by man). This gives an east-west range of about 12,862km (8,000 mi.). *Cycas* is found from

Table 1 Genus Encephalartos: Macroscopical external characteristics and character-states of taxonomical value		
Structure	**Characteristic**	**Character State**
A. Vegetative Characteristics		
TRUNK	Development	Well developed above ground. Mostly underground.
FROND	Length	
	Number	Less than 10. More than 10.
	Orientation	Straight. Incurved. Recurved. Spirally Twisted.
	Colour	Green. Glaucous.
	Petiole	Present. Absent.
LEAFLET	Orientation in transverse section of frond.	Ascending: - In same plane. - In different planes. - Spreading. Descending
	Orientation in plane view of frond.	Directed towards apex. Spreading. Lower leaflets directed towards base of frond.
	Orientation towards each other.	Succubous. Incubous.
	Reduction of lower leaflets to prickles.	No.

Table 1 (continued)

Genus Encephalartos: Macroscopical external characteristics and character-states of taxonomical value

Structure	Characteristic	Character State
LEAFLET (continued)	Shape	Linear, less than 5mm wide: - Margin revolute. - Margin thickened. - Margin plane. Not linear, 10-80mm wide: - Elliptical, broad ends rounded. - Elliptical, narrow ends acute. Ovate.
	Size: Width Length	
	Margin	Entire. Dentate. Lobed.
	Veins: Number	
	Veins: Protruding abaxially.	Yes. No.
B. Sexual Characteristics		
FEMALE CONE	Number	One. Two More than two.
	Presence of Stalk	Yes. No.
	Shape	Cylindric. Ovoid.
	Colour	Red. Yellow. Brown. Green.

Table 1 (continued) Genus Encephalartos: Macroscopical external characteristics and character-states of taxonomical value		
Structure	**Characteristic**	**Character State**
FEMLAE CONE	Indumentum	Present, microscopic Present, macroscopic: - Dense, long, shaggy - sparse, short Absent
	Expression of facets of bulla.	Yes. No.
	Elevation of central facet of bulla.	Yes. No.
	Surface Sculpture of bulla.	Tuberculate. Radiating warty ridges. Smooth.
MALE CONE	Number	One. Two. More than two.
	Similarity to female	Yes. No.
	Presence of stalk	Yes. No.
	Shape	Narrowly ovoid. Ovoid.
	Colour	Red. Yellow. Brown. Green.
	Indumentum	Present, microscopic. Present, macroscopic: - Dense, long and shaggy. - Sparse, short. Absent

Table 1 (continued) Genus Encephalartos: Macroscopical external characteristics and character-states of taxonomical value		
Structure	**Characteristic**	**Character State**
MALE CONE (continued)	Elevation of central facet of bulla.	Slightly: - Centered, sharply - Sharply defined. - Not centered. - Poorly defined and often Markedly: - Straight. - Drooping.

approximately 8º north of the Tropic of Cancer, to about 2º south of the Tropic of Capricorn, giving an overall north-south range of about 6,333km (3,800 mi.)!

Cycas occurs on the fringes of primary rain forest in some of its South Pacific range, but is more frequent in open situations on limestone outcrops near the shoreline. In mainland China, Viet Nam, Thailand, India and Burma, *Cycas* is found in tropical deciduous forest and on the borders of bamboo forests. They are also frequent on steep sided limestone hills rising above the surrounding plains. In Australia, *Cycas* can be found in scattered colonies in extremely dry and hot desert habitats in the northern portions of Western Australia and the Northern Territory.

Diagnostic Features

This variation of habitats has led to numerous adaptations of growth habits in *Cycas*, especially in the stem. Stems vary from dwarf subterranean, to tree-like proportions with many branches, bases 1 - 1.5m (3 - 5ft.) in diameter, and 9.2m (30ft.) or more in height. In many of the tropical species the usually persistent leaf bases are soon shed to form a smooth, skin-like surface. In drier or fire-prone areas the stem surface often cracks and fissures to form a hard tree-like bark, losing all remnants of the armor of old leaf bases. In some species the stems exhibit unusual prominent raised rings and in others the bases are swollen.

Unlike the wide variation shown in some cycad genera (such as *Encephalartos* and *Zamia*) the leaves of most *Cycas* species are quite similar. There is of course variation in length and width of both leaves and leaflets, and in the shape of the leaflets, such as straight, falcate, etc. Leaflet insertion into the rachis is also variable and leaves may be flat, or keeled to such a degree as to have the leaflets almost touching those form the opposite side of the rachis. Insertion may be at right angles to the rachis or angled forward to about 45º. Colour ranges from dark green to silver grey with many variations between the two extremes. Some species have leaves that are heavily and persistently tomentose. Leaflets in all but one species are simple and linear with one unbranched midrib, which easily separates them from all other cycad genera with the possible exception of Stangeria and *Chigua*, but these two genera have midribs with lateral branches. The exception in *Cycas* in regard to the simple leaflets is *C. micholitzii*. This species has the remarkable feature of dichotomous branching of the leaflets. Leaflets in the adult specimens may branch as many as three times giving the overall outline of the leaflet a fan like appearance. Leaflet width varies from species to species. Very narrow leaflets usually have revolute margins while those species with broad leaflets normally have flat margins. *Cycas* female 'cones' in *Cycas* easily separate them from all other genera. This is the only genus that does not form mature cone like structure encasing the seeds. In *Cycas* the individual female sporophylls are more like modified leaves and carry as many as five pairs of seeds attached along their margins. Each one of these sporophylls ends in a spatulate 'sterile blade' which may or may not be armed with marginal spines. In the study of *Cycas* species and their relationships, the morphology of the sporophyll is one of the best diagnostic features. The immature female sporophyll aggregates of *Cycas* may be divided into two types (see Fig. 1). The first, found mainly in the orient, India and the Philippines (see Fig. 2), produces a cone like structure with the sporophylls upright and tightly pressed together so as to form a compact dome in the center of the leaf crown. This 'cone' type remains in place until a new set of leaves forces its way through the center. Even as the sporophylls are forced outward and downward they persist in their compact alignment. I will refer to this type as the 'closed cone'. The second type with its distribution mainly in the South Pacific, Australia, the Indian Ocean, Madagascar and Africa, produce long and sometimes pendant sporophylls. From their inception they tend to be open and lax in their arrangement. This type I will refer to as the 'open cone'. In this group the seeds are always exposed in contrast to being tightly held within the 'dome' of the closed-cone group. These two groups also differ markedly in the size, shape and construction of their seeds. Seeds of the closed-cone group are generally small in size. Those of the open cones are usually somewhat larger, and some, especially those of the South Pacific and Indian Ocean islands, contain a layer of porous material between the sclerotesta and the endosperm that causes the seed to float. This floatation layer no doubt enables seed dispersal by ocean currents. The sarcotesta or fleshy seed coat is generally somber coloured in shades of brown, yellow-brown or orange-brown. A few species have a bright orange-red sarcotesta, and this colour may play a part in seed dispersal by attracting predators. The shape of the sclerotesta is also variable. The outline varies from globose to long ovate. All *Cycas* seeds are slightly compressed longitudinally and it is the only genus in which the sclerotesta does not have a built-in exit for the mergence of the radicle. In *Cycas* the sclerotesta splits at the midpoint of the broad end enabling the radicle to emerge. This splitting is caused by the swelling of the female gametophyte as it absorbs water through the sclerotesta. The surface of the sclerotesta is also variable, ranging form smooth, to fissured, to ridged (*C. wadei*) and finally to matted with persistent fibers (*C. beddomei*). So variable is the size, shape and surface of the sclerotesta that it can be a useful diagnostic feature for taxonomic treatments.

The male cones in *Cycas* relate more closely with cones of the other cycad genera. They are usually erect, ovoid or long ovoid cylindric, and covered with a fine tomentum which is commonly a golden-brown in colour. The sporophylls are wedge shaped, flat and generally taper abruptly to a sharp spine that is directed towards the cone apex. In some cases a well defined spine is absent, and in its place is only a rudimentary appendage.

Figure 1: *Cycas* sp., female cone,.Bynoe Harbour, N.T.

Australian Cycas

The Australian *Cycas* are probably the best known for any of the larger regions studied. As with other countries, most of the taxonomy of the genus was done early in the history of the country. However, the genus has been addressed in recent years with two new species having been named (*C. calcicola* and *C. pruinosa*). The recent interest in cycads has brought about considerable field work both by botanists and hobbyists. This has led to the discovery of numerous previously unknown populations. Two of these populations (Pine Creek and Bynoe Harbour, N.T.) are soon to be described, others will undoubtedly follow. (Fig. 1).

In Australia the *Cycas* is confined to the tropics and subtropics, occurring along the northern areas of Western Australia, the Northern Territory, and Queensland, then continuing down the Pacific coast as far at Mt. Perry. At present there are 11 valid *Cycas* species in Australia, four in Western Australia, three in the Northern Territory and four in Queensland.

Cycas kennedyana

The southern-most *Cycas* is *C. kennedyana* described by F. Mueller in 1882. It ranges from Mt. Perry,

Figure 2: *Cycas* sp., female cone, North of Bangkok, Thailand.

Queensland in the south to a little above Rockhampton in the north. This species is easily recognised by its slender trunk and its glaucous-green, strongly keeled and arching leaves. This cycad frequents the foothills of the coastal range at altitudes of about 150 - 300m. The habitat is typical of all the Queensland *Cycas* which inhabit open woodland on steep, dry, grassy hillsides on shallow rocky soil.

Cycas media

Also from Queensland and the most complex species of the Australian *Cycas* is *C. media*, described by Brown (1810). Colonies of plants attributed to this species range from Rockhampton in the south to the top of the Cape York Peninsula. This is a variable species morphologically, and each population investigated seemed a bit different. It is generally found on the Pacific slope of the coastal range, but numerous populations have been located well inland. The stems are usually less than 5m tall and seldom branched. The leaves are green, either flat or keeled, short or long, straight or arching. The petiole may or may not be armed, and is usually semi-terete. A great deal of field work will be needed to properly document this species.

Cycas normanbyana

Cycas normanbyana of Queensland and described by F. Mueller (1858) has been an enigma for years. Several botanists believed it to be from the Normanby Range near Port Dennison while other held it was from the Normanby River area of N. Queensland. However, when one reads Mueller's original description it clearly states that the type locality was the mountains at the mouth of the Burdekin River. The estuary of the Burdekin River is quite wide and within it there are several mouths of the Burdekin! To settle this question an expedition was mounted to survey the entire area of the Burdekin River estuary by helicopter. There are several small ranges of hills somewhat inland from the coast, all proved to be devoid of *Cycas*. At the coast there were two areas abruptly rising from the otherwise flat countryside. To the north is a large steep-sided rocky hill rising about 100m above the flats. This was a likely habitat, but also without a single cycad. To the south, Cape Upstart rises like an island, surrounded on the east by the Pacific Ocean, and on the west by the flat estuary of the Burdekin. It was on the western exposure of this promontory that *Cycas* was found in large numbers. They grow from sea level on beach sand, and continue up the ravines almost to the highest point. The cycads found here fit the description of *C. normanbyana* except for one detail. The difference is that these plants have six seeds per sporophyll as opposed to the two in the original description. However, the illustration used in the original description does not show a complete sporophyll. The drawing depicts the sporophyll ending immediately below the two seeds. It seems quite possible that an incomplete specimen was instrumental in depicting this species as producing only two seeds per sporophyll. Because the cycads on Cape Upstart agree in all other respects to the original description, and agree with the type locality, we must accept them as *C. normanbyana*. This species is very robust in growth and size. Stem reaches 6m tall and 25cm in diameter. Leaves are dark green, slightly keeled, 1.2 - 1.5m long, and 25 - 35cm wide. The petiole is unarmed, or armed in the upper third or half immediately below the leaflets.

Cycas cairnsiana

Cycas cairnsiana described by F. Mueller (1858) is the fourth and last of the described species from Queensland. It occurs in scattered colonies in an arid habitat between Irvingbank and Chillagoe in Northern

Queensland. These populations, growing generally in grassland, are sometimes quite extensive. The plants are small by comparison to other species and the height of the stem seldom exceeds 2m. The leaves are secund, stiff and arching, glaucous when fresh, then turning green with age.

The other two populations of *Cycas* from Queensland deserve additional study and appear to be good species. One is from the Mt. Surprise area and is found growing on steep hillsides in an arid and rocky habitat at an elevation of 500m. This is a dwarf plant with stems seldom 2m tall. The leaves average 86cm long, are strongly keeled and arching, and maintain a blue-gray colour even into old age. The leaflets average 18.6cm long and 6.3mm wide, are linear in outline with an abruptly acuminate apex and strongly revolute margins. The petioles are 16cm long and may be slightly to entirely armed. the megasporophylls average 18.6cm long, and usually carry 3 to 6 seeds, 4 being usual. The sterile blade average 4.2cm long and 2.4cm wide, with an average of five short flattened divisions per side. The seed is small sub-ovoid, and measures 37mm long and 28mm in diameter with the sarcotesta intact. The seed have a glaucous covering which gives them a powder-blue colour, completely masking the underlying yellow-brown colour of the ripe seeds. Male cones are golden-brown and average 23cm long and 7.5cm in diameter. These cycads seem to be most closely related to *C. cairnsiana*.

The other Queensland population has been commonly referred to as the 'Glen Idle Blue'. This is a much larger cycad than the Mt. Surprise Blue producing stems up to 10m tall, although the average plants range from 1-2m in stem height. The leaves are persistent blue-grey in colour and average 124cm long, have a medium keel, and are slightly arched. Leaflets average 20cm long and 5mm wide, the margins are strongly revolute. The petioles average 27cm long and may be unarmed or fully armed with 3mm long spines. Female sporophylls are glaucous, average 26cm long and carry from 3 to 6 seeds, but usually 4. The blade averages 5cm long and 2cm wide, will 11 divisions along each side and ending in a terminal spineless projection. The seeds are sub-globose and average 4cm in diameter. The ripe sarcotesta is dark brown, but the heavy glaucous bloom makes them appear purplish in colour. Male cones are rufous brown, ovoid in shape, and average 19cm long and 9cm in diameter. These plants grow in steep canyons on shallow rocky red soil, or under eucalypt woodland with ironbark predominating. Their altitudinal range is 600 - 750m.

Cycas angulata

In the Northern Territory there are at least five good species, three of them described and two undescribed. (Fig's. 3, 4). *Cycas angulata* was described by Brown (1810) and is quite common near the mouth of the Wearyan River where it empties into the Gulf of Carpentaria. It has also been reported on islands in the Gulf of Carpentaria and a somewhat doubtful report from Lawn Hill Station in Queensland. *Cycas angulata* is without a doubt the largest of the *Cycas* species in Australia. Stems commonly reach heights of 10m, and branching is not uncommon. Leaves measure 125cm long by about 27cm wide. Leaflets are 16cm long and 4 - 5mm wide with flat margins. Colour of new leaves is glaucous green turning green with age. Petioles are 25cm long and may be partially armed in juvenile specimens, but are usually unarmed in mature plants. Male cones are red-brown when emergent. Female sporophylls are long and terminate in a narrow blade with a long narrow apical spine. Each sporophyll carries from 4 to 10 seeds with 6 or 8 being more common. The habitat is open flat eucalypt woodland in sand or sand and clay at an elevation usually less than 100m.

Cycas armstrongii

Cycas armstrongii, named by Miquel in 1868, is the commonest cycad of the Northern Territory with an extensive range. It can be seen growing along the runway when one arrives at the Darwin airport. It is also a common 'weed' in lawns and vacant lots in this city. The range extends from Darwin south as far as Emerald Springs and east into the Kakadu area. *Cycas* armstrongii is one of the daintiest of the *Cycas* with trunks usually less than 7cm in diameter, and attaining an average height of 2m. The leaves are bright green, slightly keeled and less than 1m long by 25cm wide. Petioles are usually 18cm long and may or may not be

Figure 3: *Cycas* sp., Dutton River, N.T.

Figure 4: *Cycas* sp., Pine Tree, N.T.

armed. Male cones are small, about 12 - 20cm long, oval in shape and rust brown in colour. The female sporophylls are short with a triangular blade and usually hold four seeds. The habitat is open eucalypt woodland usually on sandy soils. The common name for this cycad is 'fire fern'. Because of its size, it is quite fern-like and is usually the first plant to leaf out after the annual grass fires. *Cycas armstrongii* has been observed to hybridize with both *C. calcicola* at Lichfield Park and with *Cycas* sp. on the Bynoe Peninsula.

Cycas calcicola

Cycas calcicola is of recent origin having been described by Maconochie (1978) from plants found growing on limestone just north of Katherine, N.T. The habitat is rocky, dry and the plants grow in open areas on limestone outcrops. The most distinctive feature of this cycad is the persistent silver tomentum on both sides of the leaves. In older leaves the upper surface becomes almost glabrous, but the tomentose condition persists on their underside. Stems are small, the tallest attaining about 3m. The leaves are flattened and about 85cm long. The leaflets are narrow, 2 - 3mm wide, with revolute margins, and are not decurrent on the rachis. Male cones are narrow-ovoid about 20cm long and 5cm in diameter. Female sporophylls 15cm long with 2 - 6 seeds just below the rhombic blade that ends in a weakly developed spine 10 - 16mm long. The ripe seeds are brown in colour with a glaucous surface, globular-ovoid, about 35mm long and 27mm in diameter. In the Lichfield Park area, colonies of *Cycas* identified as *C. calcicola* display a number of dissimilar features and should be more critically examined.

Other Northern Territory Cycads

There is a colony of specifically distinct *Cycas* to be found to the east of Pine Creek, N.T. (Fig. 4). They are closely allied to *C. armstrongii* as to shape, but are a bit heavier in the stem. Stems attain a height of over 4m, but the average size is about 1m tall and 13cm in diameter. The leaves are slightly keeled and a beautiful emerald green with a slight glaucous cast. They average 75cm long and 15cm wide. The leaflets are stiff, crowded or overlapping, 8cm long and 5 - 9mm wide, and are generally a lighter colour on the underside. Petioles are armed in juvenile plants and unarmed in mature plants. Male cones are narrow-ovoid 22cm long and 9cm in diameter. Female sporophylls are densely brown tomentose, about 24cm in length, and carry 1 - 4 seeds. The sterile blade is deltoid in shape, has a long terminal spine, but no marginal spines. The habitat is open eucalypt woodland generally on granitic soils.

On the Cox Peninsula, Bynoe Peninsula and Fog Bay areas there occurs another *Cycas* that is quite distinct, (Fig. 1). This cycad produces a stem to about 5m tall and 9cm in diameter. The leaves are more or less flat, a grey-green in colour, 125cm long and 21cm wide. The leaflets are 10cm long and 8mm wide, with slightly revolute to flat margins, and are held at about 90° angle to the rachis. Male cones are long-ovoid and rusty brown in colour. Female sporophylls have a sterile blade 5.5cm long and 2.5cm wide. The habitat is open eucalypt woodland on red soils.

Several other sightings of *Cycas* have been reported and may indicate additional new species. Most of these reports deal with Arnhem Land, one in particular on the Goyder River.

Western Australia Cycas

Western Australia is the habitat of four recognised species of Cycas. *Cycas pruinosa* was described by Maconochie (1978) from populations located near Kununurra and Lake Argyle. This is a small cycad with stems seldom exceeding 2m tall and 30cm in diameter. The leaves are grey-green, strongly keeled, straight to slightly curved, 90 - 100cm long and 16 - 26cm wide. Leaflets have revolute margins, are glabrous, angled forward at about 45º, 11 - 20cm long and 2 - 4mm wide and are not decurrent on the rachis. Petiole quadrangular, partially armed, about 10 - 15cm long. Male cone 40 - 50cm long and 9cm in diameter, the sporophylls with an 8 - 15mm long appendage directed towards the cone apex. Female sporophyll with 2 to 4 seeds, 30cm long with a sterile blade 15cm long, its margins pinnate-dentate with 20 - 30 spines up to 25mm long and 3 - 4 mm wide, the terminal spine 30 - 70mm long. Seed globular-ovoid, 40mm long, 25mm in diameter, surface pruinose. The habitat is arid, rocky, and with a few low-growing shrubs and spinifex, generally on steep slopes. In some areas two colour phases occur, blue-grey and dark-green.

Cycas basaltica

Cycas basaltica is described by Gardner (1923) from populations in the Kimberleys of Western Australia. This is a very distinctive species with stems from 1 - 3m tall. The emergent leaves are tomentose and glaucous-grey, 1 - 1.5m long, flat and appear to be somewhat deciduous in habit. Male cones long-conical, tan-tomentose, with a rudimentary appendage on each sporophyll face. Female sporophyll densely red-brown tomentose, with four seeds, blade narrow and entire. Habitat is on basalt rock in open woodland.

Cycas lane-poolei

Cycas lane-poolei was described by Gardner (1923) and has since been considered in synonymy with *C. armstrongii* by some botanists. It does have similarities with *C. armstrongii* in its growth habits and deciduous leaves. *Cycas lane-poolei*, however, differs in its large size and height of trunk, its glaucous leaves, larger seeds and almost compete lack of tomentum in the crown. This taxa is quite restricted in its range on the Mitchell Plateau as well as being geographically distinct from *C. armstrongii* by 400km!

Cycas furfuracea

Cycas furfuracea described by Fitzgerald (1918) is restricted to the Kimberley and King Leopold ranges of W.A. This is a large handsome cycad with extremely glaucous leaves. The stem reaches a height of 4 - 5m and diameter of 25cm. The leaves are straight, strongly keeled, and 120cm long and 35cm wide (flattened). The male cone is long-conical, 30cm long and 8cm in diameter. Female sporophylls are densely light-brown tomentose, carry 4 - 6 seeds and are 40cm long. The sterile blade is long and narrow with about seven segments one each side. The habitat is open eucalypt woodland on steep hillsides in shallow rocky soil.

Chinese Cycas

One month was spent in China during which all of the major botanic gardens were visited and their cycad collections studies. This work was hampered in some cases by the language barrier and others by lack of sufficient time for proper study, but observations were sufficient to form an opinion about the present taxonomy of the Chinese *Cycas*. Major differences were noted in the morphology of a number of the plants identified and belonging to some well-accepted species (i.e., *C. revoluta*, *C. taiwaniana* and *C. rumphii*, (Fig. 5). These differences might well lead to the description of several new species of *Cycas* from mainland China. Unfortunately, it was not possible to study any cycads in habitat. The following notes were taken from cultivated plants, some of which may not be typical of the same plants in habitat.

Cycas rumphii

Plants labeled *C. rumphii* and producing long broad leaves were grown at Beijing Botanic Gardens and at the South-China Botanical Garden at Guangzhou but were definitely not that species. These were dwarf plants that produced cones when the trunk was only 12cm in diameter and the same height. The female 'cone' produced by these plants were very small (12 - 15cm in diameter) and of the 'closed-cone' type. These two features alone prove this plant distinct from *C. rumphii*. It is said to grow natively in Yunnan and Guangxi Provinces.

Cycas siamensis

Plants labeled *C. siamensis* and collected in southern Guangxi were observed growing at the Guilin Botanic Garden new Kunming. These differed from the Thai plants in having glossy-dark green leaves and very gradually acuminate leaflets with revolute margins. This cycad produced a 'closed-type cone' at a small size (10 - 12cm in diameter). Adult plants exfoliate the old leaf-bases to produce a light-tan skin like epidermis. In larger plants a swollen base is formed as in *C. siamensis* of Thailand.

Cycas taiwaniana

Plants labeled *C. taiwaniana* and in cultivation at the South-China Botanic Garden differed form the type species from the Republic of China in two ways. The trunks were much stouter (from cultivation?) and exfoliated the old leaf-bases to form a skin like epidermis in the older and lower portions. The female sporophylls were almost glabrous in contrast to the densely tomentose counterparts from R.O.C. As in *C. taiwaniana*, females of this species have a 'closed cone'.

Cycas yunnanensis

Specimens of a cycad cultivated (and recently introduced) at the Kunming Botanic Gardens, which was jokingly referred to as *C. 'yunnanesis'* appeared quite distinct. This plant was said to be native to an area about 300km south of Kunming in Yunnan Province. The stems were up to 1m tall and 17.5 - 25cm in diameter. The leaves are dark green (not distinctly lighter below) and measure 140cm long and 27.5cm wide. Leaflets are 15cm long and 6mm wide, inserted into the rachis at about a 45º angle. The margins are not revolute. Petiole was 26cm long and 12mm in diameter, slightly four-angled and armed with 3mm long spines in only the upper half. Female 'cone' was the 'closed type', the sporophylls densely red-brown tomentose and carrying four seeds. Dry measurements of the sporophyll were 12cm long and 62mm wide at the blade. The blade is round in outline, bearing 22 - 24 divisions ending in long fine spines. Male cone not seen.

Figure 5: *Cycas circinalis*, female 'cone', Beijing Botanical Garden, China.

Cycas hainanensis

At the South-China Botanic Garden was a cycad collected on Hainan Island, but supposedly is not *C. hainanensis*. It has a stem 1m tall and long straight upright leaves. Unfortunately, there was no mature *C. hainanensis* to contrast with it.

Another cycad at this institution was labeled *C. taiwaniana* and had been collected on an island off the coast of Fujian Province. It differed from the Taiwan plants in having six seeds on each female sporophyll. Additional study is indicated.

Figure 6: *Cycas* sp., in habitat, Rataburi, Thailand.

Figure 7: *Cycas* sp., Tok Gnai Island, Thailand.

Cycas micholitzii

Several specimens of *C. micholitzii* were studied at the Guilin Botanic Garden. It was interesting to note that a very young individual had at least 50% of its leaflets undivided and the remainder only once-divided. This could indicate that seedling plants may not display the unique branching of the leaflets so prominent in mature plants.

There is no doubt that additional field work and critical study is needed for several of the *Cycas* of mainland China.

Thai Cycas

My time in Thailand was very limited and it was only possible to investigate eight populations of their native *Cycas*, (Fig's 6 - 12). All field work was carried on in southern Thailand. There was not sufficient time to travel north. Six of the populations were south of Bangkok, with one to the west, and one to the north.

Cycas siamensis

Our first stop was near Ratburi where we studied *C. siamensis*. This cycad has stems up to 2m tall crowned by numerous leaves. Some of these were maintaining to heads of leaves and did not appear deciduous as they sometimes do in cultivation. The habitat was dry, hilly country covered with vines and low shrubs and a few scattered *Ficus* sp. Leaves numerous, straight, flat, 140cm long and 30cm wide. Leaflets angled forward about 80º, slightly revolute margins, decurrent at reduced base, 22cm long and 8mm wide, medium-green above and yellow-green below. Male cone spindle-shaped, golden-brown tomentose, 34cm long and 9cm wide in diameter; sporophylls with an 18mm long spine held perpendicular to cone axis. Female 'cone' closed type, about 15cm in diameter, individual sporophyll with 2 - 3 seeds which are a light yellow-orange when ripe. Found at elevations from 150 - 600m.

Figure 8: *Cycas* sp., habitat, Tok Gnai Island, Thailand.

Cycas sp

Another colony investigated was about 10km from the Kampuchea border, (Fig. 6). The cycads grew on limestone cones rising about the surrounding flat plain at an elevation of 450m. The trunks attained 2 - 3m in height and about 15cm in diameter. They were almost impossible to climb to because of the vertical cliff face on which they grew. Leaves were 170cm long and 38cm wide. Male cones were 25cm long and 10cm in diameter. Female 'cones' were the closed type with 2 - 4 seeds and a sterile blade that was more or less round except for a long terminal spine. This plant did not fit any known species description.

To travel to extreme southern Thailand entailed catching a flight from Bangkok to Hat Yai which is situated on the Malay Peninsula. From there we drove to Satun where a boat was hired to take us down the coast and up a local river. Along the river were sheer cliffs upon which cycads could be seen, (Fig. 7). Some specimens were found on steep banks close to the river but to get there one had to wade waist keep through water, mud and mangroves to attain dry land. Once again we could not identify the species, but they may be closely related to *C. pectinata*. Stems were up to 7.5m long and 25cm in diameter. Leaves 125cm long and 35cm wide with the petiole 40cm long and armed 25% - 33% down from the lowest leaflets. Leaflets 20cm long and 9mm wide, margins not revolute. Immature female 'cones' were the closed type, light-brown tomentose, 25cm in diameter and 16cm high. Sporophylls with two seeds, 14cm long, the blade 7cm long including a long terminal spine. Male cones spindle shaped and averaged 34cm long and 8cm in diameter, each sporophyll with a 6mm long upright appendage. The habitat was almost always on vertical cliffs with the stems pendant or forming an S shape. The elevation ranged from 0 to 300m.

The next day we drove to Pak Meng beach in Trank Province. A boat was hired to take us to Tok Gnai Island about 3km offshore, (Fig. 8). This is an uninhabited island with vertical sides some of which were 300m high. *Cycas* plants were hanging from various cracks in the cliff, some with 10m long trunks. These plants did not differ much from the ones at Satun, so I will not detail them.

At Aw Manao in Naratiwat Province we found an interesting *Cycas* with an open-type female 'cone'. The largest stems were 4m tall and 11cm in diameter, their lower portions having lost all the old leaf bases and formed a smooth light-tan skin. The leaves were 140cm long and 46cm wide, the petiole 31cm long and armed in the upper 25%. Leaflets were slightly falcate, their upper surface slightly rounded, revolute margins, 30cm long and 13mm wide. Female sporophyll with two seeds, 30cm long, the sterile blade narrow and unarmed. Male cones not seen.

After a flight back to Bangkok we prepared for our last two trips. The first took us north and west towards Burma. After a seven hour drive we changed to a large

Figure 9: *Cycas* sp., large plant in habitat, Thailand near Burma border.

Figure 10: *Cycas* sp., male cone, same population as in Fig. 9.

4-wheel-drive truck to take us the last 90km. It was a rough trip, but we rode on top of the cab where we could get a good view of the country. It seems like the road was either going straight up or straight down, making our seating uncertain. After about 3 hours we arrived at a government forestry camp where we spent the night. Armed guards reminded us of just how close we were to the Burma border. The area seems untouched by man except for the road and camp. The trees are full of orchids and *Platycerium* ferns. The *Cycas* here is frequent to common and some specimens have grown to immense proportions. One specimen we photographed is over 10m tall with several branches. Its base is over 1m in diameter, (Fig. 9).

Leaves measure 190cm long and 50cm wide. The petiole is 53cm long and armed in the upper third. The male cones are ovoid and huge, measuring 35cm long and 20cm in diameter. They were dehiscing large amounts of pollen and a small beetle was found in these cones by the thousands! The cones are light brown in colour and each sporophyll has an upright spine 25mm long. The female "cones" are the closed type and measure 40cm in diameter by 23cm high. The sporophyll is 16cm long and 10cm wide and heavily armed with spines. Ripe seed is orange and measures 5cm long and 4cm in diameter. The seed is so tightly packed in the cone that its final shape can be triangular,

Figure 11: *Cycas* sp., same population as Fig's. 9 & 10, plant with female 'cone'.

Figure 12: *Cycas* sp., in habitat, near north of Bangkok, near Nakorn Swan, Thailand.

square or with many flat planes. Definitely not the regular ovoid or globose seed that is generally the case with *Cycas*. The habitat is open woodland adjacent to bamboo forests on steep hillsides. The elevation ranges from 460 - 1,000m.

The last population investigated was 240km due north of Bangkok some 240km near Nakorn Swan. This was the first glaucous-leaved *Cycas* we had seen on this trip and it was a beautiful plant, (Fig's. 10 - 12). The stem is 1 - 5m tall and 15 - 20cm in diameter, the lower portions forming a 'tortoise shell' formation of cracks, possibly due to yearly fires. Leaves are tomentose when emergent, glaucous and glabrous on the upper side when mature, 150cm long and 35cm wide. The petiole is 32cm long and armed 25% - 50% of its length nearest the leaflets. Leaflets are 17cm long and 9mm wide, have revolute margins, and are inserted at 90º to the rachis. Male cones are conical, 22cm long and 9cm in diameter, with a 5mm erect spine on each sporophyll face. Female 'cones' are the closed type, but had been some damaged by fire that specifics could not be noted.

The above field work has shown that a considerable amount of study remains to be done on the genus *Cycas*. Even in well-botanised countries such as Australia, new species are still being found. I hope that collectors and hobbyists will take the time to report new discoveries to the proper institutions. With such assistance taxonomists will find their job a much easier one!

Acknowledgements

A great deal of gratitude is owed to several people whose assistance made most of this study possible. In Australia, Pat and David Coutts were responsible for my having been able to study many of the *Cycas* of Queensland and the Northern Territory. Their assistance, generosity and friendship will never be forgotten. Joe Perner was my able guide in locating the *Cycas* of Western Australia. H.M. 'Monty' Anderson introduced me to many of the *Cycas* of the Northern Territory. In Thailand, Kampon Tansasha was my host, guide, translator and friend. Without his help little could have been accomplished there.

References

***Brown, R.* 1810.** Prodromus Florae Novae Hollandiae et Insulae Van-diemen. Johnson, London.

***Fitzgerald, W.V.* 1918.** The Botany of the Kimberleys, north-west Australia. J. & Proc. Roy. Soc. W. Australia 3: 102 - 224.

***Gardner, C.A.* 1923.** Botanical notes. Bull. Forest. Dept. W. Australia. No. 32, pp. 30 - 32.

***Linnaeus, C.* 1753.** Species Plantarum.

***Maconochie, J.R.* 1978.** Two new *Cycas* from the Northern Territory. Herb. N.T., Alice Springs.

***Miquel, F.A.W.* 1868.** Nouveaux materiaux pour servis a las Connaissance des Cycadees. Arch. neerl. Sci. exact. Natu. 3: 193 - 254.

***Mueller, F.V.* 1958 - 77.** Fragmenta Phytogeographie Australiae. 1: 81 - 86, 2: 179; 3: 38; 8: 169, 173; 10: 53; 11: 125.

***Mueller, F.V.* 1882.** Melbourne Chemist (*C. kennedyana*).

Part Six

Conservation

The Genus *Cycas* and Cycad Conservation in Australia Guide Lines

L.P. Butt
Associated Societies for Growing Australian Plants,
25 Ortive Street, Yeronga, Queensland, Australia

Abstract

Australian *Cycas*, once thought to consist only of *C. media*, now includes at least 11 species. As these become better known, efforts both to conserve them and to propagate them for horticultural purposes are bearing fruit.

Introduction

Quite a number of articles have now been written about that most ancient and very primitive group of plants, the Cycadales. Giddy's book 'Cycads of South Africa' came at precisely the right time to arouse fresh interest among nurserymen regarding these plants (Giddy, 1974). Until then the so-called 'Sago-palm' *Cycas revoluta*, was about the only species available in Australia and then only in selected nurseries. However, earlier than this Australian botanists like L.A.S. Johnson and J. Maconochie had been doing valuable work in classifying the Australian genera. Johnson's botanical papers (Johnson, 1959) opened a new door into our understanding of what valuable plants Australian cycads are, and the late John Maconochie's work on little-known species of *Cycas* was cut short by his untimely death in 1984.

Why, do you ask, are these plants so important when it is known that certain species have been declared noxious weeds because of their poisonous seed properties (toxic to cattle)? A good question, but the answers are very clear. Fossil records in our rocks show us that ancestors of today's cycads existed some 300 million years ago and were the primary form of vegetative life, existing as forests of seed ferns, the nearest relatives of our present-day cycads. What emerges is the fact that these plants of long ago, technically known as Cycadofilicales, were probably an offshoot of certain even-older fern-like families that had developed the ability to reproduce by seeds rather than spores.

After these plants died out there is a big gap of 100 million years before we find much further evidence of them. In the age of reptiles they appear again as two orders, the Bennetitales and Cycadales. These plants differed from their seed-fern ancestors in that nearly all had developed cones - only one genus of the cycadales retains its coneless characteristic and this has come down to us today in females of the family Cycadaceae to which belong the several species in the genus *Cycas*. This leads us to think that the much-maligned *Cycas* is the most primitive member of a very primitive order but further study is needed to link it positively with the considerably more ancient and now extinct order, the Cycadofilicales.

Species of *Cycas* have some superficial features in common with both tree ferns and certain palms but this concept changes dramatically when one considers their method of developing seed. Diamond or wedge-shaped appendages known as megasporophylls develop in the crown of the female plant and the seeds are borne in marginal notches around this spear-head spike. In some species the megasporophylls are augmented by hundreds of soft furry bracts which completely surround the seed as in *Cycas revoluta* (Assam and China). Others either sit on the plant's crown or hang pendently from around the crown as in *Cycas media* (north Queensland). *Cycas* seeds are generally ovate or globular and about the size of a walnut or bunya nut, and all are enclosed in a red to orange-yellow fleshy casing which is poisonous to cattle and will cause gastric troubles in man. However, this 'fruit' is much sought after by marsupials and rodents with no apparent ill effect to the creature.

Most of the literature published to date about Australian cycads has been about species of *Macrozamia, Lepidozamia and Bowenia* (Johnson, 1959; Butt, 1984), but this is changing. Studies into this genus by Maconochie involved the species of the Northern Territory and West Australia mainly because they were more accessible to him, stationed as he was at Alice Springs.

Cycas

The genus *Cycas* is distributed throughout Southeast Asia with a few in Madagascar. Australia, however, is richly endowed with species although only one, *Cycas media*, was formerly listed for the continent. Now it is known that there are eleven species with distinct possibilities of natural hybrids (Butt, 1990).

The genus is scattered in our tropic areas from northwestern Australia, right across the top of our Northern Territory to the eastern coastline of Queensland about as far as Gladstone. From Darwin, *Cycas* also occurs south into the centre of Australia.

Cycas media, the best-known member of the genus in Australia, is now generally accepted to grow north of Sarina and up into the coastal open forests past Cairns. It grows in extensive colonies and can sometimes be found in rainforest. New fronds after fire are a beautiful iridescent green colour and although the caudex is listed as 3 - 4 metres high, very old specimens of at least 10 metres have been found. Some have branching caudex heads. This species has been successfully grown from seed and can be found as a potted plant in certain nurseries as far south as Sydney. It is very hardy and frost resistant.

Cycas kennedyana - Around Rockhampton and the hill countries north to Bowen there occurs a species that has long been accepted as *Cycas media*, but it now emerges that this is more probably *C. kennedyana*. The cycad rarely grows more than 3 metres high, and has grey-green fronds which are blue-grey upon emerging. The seed heads hang pendant and the caudex (trunk) is generally only 20 - 36cm thick. This species is popular with collectors, has to some extent been grown from seeds, and will grow in cultivation in areas of tropic-subtropic to temperate habitat.

Cycas cairnsiana; C. normanbyana - Cycas cairnsiana is a very fine species also; it is often confused with *C. normanbyana* which I previously believed grew near Proserpine. Latest information to hand is that specimens thought to have been *C. normanbyana* probably are *C. cairnsiana*. Both species have thick strong arching fronds and bear orange-coloured, centrally disposed male cones. The strand of *C. normanbyana* in the area of Gloucester Park near Proserpine probably were planted there originally as the male cones are golden yellow and not orange as in *cairnsiana*. Both of these species carry only two seeds per megasporophyll, held vertically for some time before arching over. According to some authorities the spe-

Figure 1 : *Cycas armstrongii*, old leaf and fruit.

cies *normanbyana* occurs only on the range country of central queensland. The species will grow in cultivation in the same conditions as *C. media*.

Cycas armstrongii; C. calcicola - The Northern Territory has two outstanding species, *C. armstrongii* being the most prevalent (Fig. 1). When in full frond this narrow trunked cycad grows in close clusters and looks extremely palmlike. However, in hot dry times it becomes deciduous and defoliates. This species is synonymous with the west Australian *C. lane-poolei*. *Cycas calcicola*, another resident of the top end of the Northern Territory, grows mostly on limestone areas but often has been found growing with *C. armstrongii*. New fronds are silvery-fawn making this a most attractive species. It grows thickly in the Finnis River and Litchfield districts.

Cycas angulata - A little-known but very prominent species named *C. angulata* occurs in the extreme top end of the western side of the Gulf of Carpentaria and also the islands around the coast of Darwin. This stout-trunked species is quite distinct, having a great crown of green fonds. The trunk can grow to five metres tall and is sometimes branched. Not known to be in cultivation for any length of time but its requirements are the same as for *C. armstrongii* and *C. calcicola*.

Cycas pruinosa; C. furfuracea - The Kimberley Ranges area near the Ord River harbours a really fine *Cycas*, having a stout trunk and crowns of blue-grey fronds. *Cycas pruinosa* grows on rocky slopes in the area and produces possibly as much seed as *C. media*

Figure 2 : Len Butt in field dress with *M. moorei*.

or *C. kennedyana*. The ripe seeds of *C. pruinosa* are brown.

Another species, *C. furfuracea* comes from the King Leopold Ranges of the Kimberleys. It is rather smaller in height than *C. pruinosa* and holds its fronds more erect. The undersides of pinnae are heavily clothed with soft woolly scales. The ripe seeds are yellow.

Horticultural Practices

Apart from their botanical interest, species of *Cycas*, like all the palm-like members of the cycadales have very good landscape potential and also make very fine tub or potted plant specimens in situations around pools, on patios to augment palms and tree ferns on banks and terraces and also as front lawn decorations. *Cycas* will soon become as popular as the zamiads in cultivation, their crowns of graceful, arching fronds are always eye catching. Most species are hardy and can withstand drought.

All members of the genus *Cycas* will tolerate full sunlight, but if they have been previously growing in shade they should be put into full sunlight gradually. Generally they will get used to most soil types as long as drainage is adequate. All seeds are large and woody and can be sown horizontally just beneath the soil surface or halfway under as in coconut planting. They are notoriously slow in growth but not nearly to the extent generally supposed. Black and brown scale seem to be the main pests and should be watched carefully and treated with white oil and Rogor.

Conservation of Cycads

A great many interesting facts are emerging regarding the real accessibility of the various cycads to livestock and just how bad the poison really is. In the case of *Macrozamia moorei*, this plant could be a real test case in the fight to convince our authorities that our precious ancient plants are only dangerous under certain conditions for the general structure and growth of the species make it highly improbable that it would be a menace to well-managed and cared for stock (Figs. 2, 3). The large female cones are heavily armed with green spines at all stages from infancy to maturity and are generally surrounded by larger thorny spines that form on the petiole from the base of the frond to the commencement of the pinnae on the rachis. As the old fronds contain only minimal amounts of toxins the poison danger would seem to exist only at a time when natural wildfire or grazier-managed fire had destroyed these frond and spines giving access to wandering beasts to the exposed seeds and toxin-filled soft new fronds. In most situations only scrubber (wild) cattle are the beasts most likely to be affected. After having this explained, many farmers have fenced off areas heavily wooded with macrozamias, stopped regular burning and rescued affected beasts before they died of starvation. As the poisons are in the creatures joints and do not affect the meat, cows have been know to still bear good calves.

Like other cycads, the fronds of macrozamias have a high commercial value if collected and sold to the florist trade for manufacture of wreaths and as material for floral artists. The landscaping potential of just about all our native cycads is good as in their own environmental zone they fit perfectly and easily augment the known palms of the area. Possibly this will be the salvation of many species but it has its drawbacks. First of all, certain species have been pillaged for commercial gain even before they can be scientifically assessed. Although the taking of seeds is necessary for the survival of these plants, the taking of mature specimens is not.

Figure 3 : *Macrozamia moorei*, 8 metres high. Arcadia Valley, Queensland.

References

***Butt, L.* 1984.** Cycads of Australia. Australian Plants 13(101) : 3 - 35.

***Butt, L.* 1990.** An introduction to the genus *Cycas* in Australia. Palms and Cycads 27 (April - June) : 1 - 56.

***Giddy, C.* 1974.** Cycads of South Africa. C. Struik (Pty.) Ltd., Cape Town.

***Johnson, L.A.S.* 1959** The families of cycads and the Zamiaceae of Australia. Proc. Linn. Soc. N.S.W. 84(1) : 64 - 117.

Horticultural Practices and Conservation of Cycads

Bijan Dehgan and Fe Almira
Horticultural Systematics Laboratory,
Department of Environmental Horticulture
University of Florida, Gainesville, FL 32611 U.S.A.

Abstract

It has previously been suggested that if the demand for cycads is met through propagation and large scale commercial production, the decimation of wild populations by unscrupulous collectors will no longer occur. This statement appears naive at first, because production is extremely slow and costly, and the economics of removing large specimens from the wild are enticing when compared with the projected costs of long-term cultivation. Thus, a major challenge to conservationists is to produce plants which are horticulturally superior to any that could be removed from their native habitat. The first step in effective cycad cultivation is successful seed germination. Seeds of taxa such as *Bowenia*, *Ceratozamia*, *Dioon*, *Lepidozamia*, and *Mazrozamia* germinate rapidly and uniformly if collected when the embryo is fully developed and when the sarcotesta is removed. Others such as *Zamia* and, in particular, *Cycas*, suffer from morphophysiological complex dormancy involving germination inhibitors in the sarcotesta, a hard sclerotesta, unidentified physiological dormancy, and often embryo immaturity. Treatment of *Zamia* seeds with H_2SO_4 followed by GA_3 usually results in 100% germination. The embryo in *Cycas* is in the early stages of development when seeds first appear mature, but reaches maturity after several months of cool storage. However, germination following storage may be hastened by short treatment periods with H_2SO_4 followed by GA_3. The problem of slow growth may be at least partially resolved by use of well-drained soil mixes with Ph of 6.5 to 7, avoidance of under or over-irrigation, and periodic application of dilute liquid and micronutrient-amended slow-release fertilizers. Plant growth may be further hastened by application of growth regulators. Removal of the single primary root near its junction with the stem and indole-β-butyric acid (IBA) dip results in multiple 'primary' roots and numerous laterals. Multiple shoots, and thus greater number of propagules, are also possible with application of growth regulators.

Introduction

Ideally, cycads should be protected through preservation of their habitats, Unfortunately, this is not always possible. The demand for these horticulturally and botanically valuable plants, and financial motives of unscrupulous collectors often exceed protective ability of the local governments. Cycads are valued for a number of reasons. Avid collectors grow these plants for the simple love and pleasure of it, while others are interested in them for purely botanical reasons (Osborne, 1990). Neither of these two groups pose any threat to the existence of cycads. The greatest danger to cycad conservation is posed by a third group whose only interest is financial reward, irrespective of the damage. This may be illustrated by a recent purchase of 5,000 mature plants of *Zamia floridana* (= *Z. integrifolia*, *Z. pumila*) by a hapless nurseryman who was cognizant of their commercial worth for landscaping but had no experience with growing them. Despite professional guidance, more than 50% of the plants were lost in the first month because of excessive irrigation. The remaining individuals perished after a hard freeze. Similar, and perhaps more devastating, cases have been reported in Mexico (Tang, 1983, 1985, and Vovides, in this issue of the Proc.) and in South Africa (Osborne, 1990; Giddy, in this issue of the proc.).

What if it was possible to grow 5,000 plants of the same size in a few year? Would that satiate the demand and prevent wholesale destruction of cycads? This is not a multiple choice question; the answer, of course, is yes. It can be done with proper horticultural practices and some patience. As was suggested earlier (Dehgan, 1983), if the demand for cycads could be met through propagation and large scale commercial production, there would be no need for the decimation of wild populations. Such a supposition may at first seem simple minded. After all, it takes 100 years or longer for the trunk of a specimen of *Dioon* to reach a metre in height, posing a severe problem with time and patience. Moreover, the cost of digging up 5,000 plants from the wild is considerably less than growing them in a nursery for even one year. Thus, a paradox is created, The unique features of cycads, and consequently their beauty as landscape and foliage plants (Dehgan and McConnell, 1985), prompt many consumers to pay exorbitant sums to acquire them. In turn this creates a situation where conservation takes a back seat to financial gains.

One does not have to be a genius to realise that if a certain commodity is in demand, it would made economical sense to produce that commodity. Thus, the primary objective of this paper is to briefly review publications of the recent past on the subject and provide some research-supported practical suggestions on commercial production of cycads. A review of earlier literature on propagation and growth of cycads was presented by Dehgan (1983).

Seed Propagation

Long strides have been made towards rapid and uniform seed germination of cycad genera in the past decade. Generally, there does not appear to be any difficulty with seed germination of *Bowenia*, *Ceratozamia*, *Dioon* (Fig. 1b), *Lepidozamia*, *Stangeria*, and *Macrozamia*, as long as the embryo is fully developed (mature seed) and the sarcotesta (the fleshy coat) is removed before planting (unpublished data). *Microcycas calocoma* seeds have been successfully germinated by the staff of Fairchild Tropical Garden and reported to have a short period of maturation time, occasionally germinating as the cone falls apart (Hubbuch, 1987). In contrast, despite normal appearance, *Encephalartos* spp. seem to have the peculiar difficulty of not being properly pollinated and/or fertilized when under cultivation (Hubbuch, 1987 and pers. obs.). Thus, we have no experience with seed germination of this genus. However, seeds of some species collected from the wild do germinate without difficulty, while germination of others is erratic (Giddy, pers. comm.). Recently it was reported that thousands of plants of *Encephalartos transvenosus* (the Madjadji Cycad) have been produced and distributed to the public (Osborne, 1990).

The major difficulty seems to be with the commonly cultivated genera. *Zamia* and especially *Cycas* suffer from what is referred to by Nikolaeva (1977) as **'morphophysiological complex dormancy'**, in some species involving actual physiological dormancy, embryo

a
b
c
d
e
f
g

immaturity, chemical inhibitors in the sarcotesta, and the thick impermeable sclerotesta. *Zamia floridana*, for example, does not have embryo dormancy if seeds are collected after the cones fall apart. Removal of the sarcotesta and treatment of seeds with H_2SO_4 for about one hour results in excellent germination in 20 to 30 days.

However, if seeds with H_2SO_4 followed by 24 hour GA_3 soak at 1,000 ppm, very uniform germination of up to 100% (Fig. 1a) may then occur in about 2 weeks (Dehgan and Johnson, 1983). In *Z. furfuracea*, the optimal periods of treatment are about 20 minutes in H_2SO_4 and 24 hours in GA_3, respectively.

Some *Cycas* species are particularly difficult to germinate because at the time when seeds appear mature, they are actually in early stages of embryo development. Storage in a warm environment usually results in relatively rapid embryo development but also rapid loss of viability. In contrast, cold storage results in slow development of the embryo but minimal loss of viability. Best results are obtained when seeds are collected as they appear fully developed, stored in a cool (40º-50ºF) location for about 2 - 3 months, and the fleshy coat removed when ready to plant. Subsequent treatment with 20 - 30 minutes of H_2SO_4 and 24 hours of GA_3 gives excellent results (Dehgan and Schutzman, 1989). Cycad seeds are recalcitrant; they cannot be rehydrated once their water content falls below a critical minimum. This is often indicated by the complete separation of the female gametophyte from the sclerotesta (Fig. 1c), resulting in rattling of seeds and their floatation in water (Dehgan and Yuen, 1973). Floatation of seeds in water, however, is not always indicative of seed inviability. For example, the presence of spongy tissue in seeds of *Cycas circinalis* and related taxa results in buoyancy irrespective of viability (Dehgan and Yuen, 1983). In general, all mature cycad seeds appear to respond positively to a GA_3 soak as indicated by our studies and others (Ellstrand, et al Ornduff and Clegg, 1990; Butt, 1990).

Vegetative Propagation of Cycads

Despite attempts by many eminent scientists and some limited success in shoot or root differentiation, tissue culture of cycads has remained unsuccessful (see P.M. Resslar's paper in this issue). Cutting propagation of cycads is not possible because, unlike other gymnosperms, they do not possess lateral buds and therefore do not generate lateral shoots or branches. However, another but less productive method of vegetative propagation of cycads involves removal of rooting of 'side shoots', which are adventitious in origin (Stevenson, 1990) and commonly referred to as 'pups' or 'bulbs'.

A detailed discussion of this propagation method has been presented by Giddy (1974). Experimental methods that would result in an increase in the number of such shoots would also assist in vegetative propagation. Our most recent work (Dehgan and Almira, unpublished data), which involves localised application of growth regulators, has resulted in induction of stem branching and production of numerous side shoots. This method has proven so successful with

←

Figure 1: Propagation and Growth of cycads.

A. Uniform germination of *Zamia floridana* following treatment with H_2SO_4 and GA_3 treatment.

B. Uniform germination of *Dioon califanoi* without any treatment.

C. Early stages of female gametophyte separation (arrow) from the sclerotesta in seed of *Cycas revoluta*.

D. Growth regulator induced multiple stem branching in *Z.floridana*.

E. Growth regulator induced multiple steam branching in *C. revoluta*.

F. Growth regulator (BA + BA) induced multiple root branching in *Z. floridana*.

G. Growth regulator (BA) induced multiple root branching in *D. califanoi*.

Z. floridana (Fig. 1d) that the number of divisions may be counted only with extreme difficulty. Similar effects have been observed in *C. revoluta* (Fig. 1e). Because of the unusually rapid development of the caudex, young plants appear very large, reaching a size usually expected of 50 or more year-old wild individuals. Indeed, the results of this research may be the most significant contribution to horticultural production, and consequently conservation, of cycads. Due to circumstances beyond our control the results of this work cannot be presented at this time but will be published in detail elsewhere.

Growth Stimulation

It is common consensus that most cycad species develop much too slowly for large scale production to be economically feasible. This holds true, for most part, because growth in cycads is usually limited to annual production of only one set of leaves. There are but two exceptions to this statement. Under favorable conditions, seedlings may produce four or more flushes of growth, but with only one or a few leaves at a time. Also, mature plants exposed to severe cold or sudden shock, such as transplanting, often produce an additional set of leaves (pers. obs.). Thus, since stem growth is primarily dependent on leaf production, and in turn leaf development and elongation appear to be directly related to healthy root growth, it is logical that stimulating good root development would have a direct impact on the total plant growth.

The results of our experiments (Dehgan and Johnson, 1987) with *Z. floridana* (Fig. 1f) and more recently (Dehgan and Almira, unpublished data) with *Dioon califanoi* (Fig. 1g) show that when the primary root in *Zamia* is cut at or near its junction with the stem, and the cut end dipped in 2,000 ppm IBA for 5 seconds, usually two roots will regenerate. If the stem is dipped in IBA for 5 seconds, followed by 200 ppm BA (Benzyladenine) for 5 seconds, then 3 or 4 roots replace the original one.

The only problem is that BA-treated seedlings produce short, stubby roots that require a relatively long time before they continue normal growth. In *Dioon califanoi* IBA alone results in production of 3 or 4 roots. It is a matter of simple logic that if one root absorbs a certain amount of water and nutrients, two or more roots can accomplish the same task more efficiently. Consequently, the plants grow more quickly and, if fertilized and irrigated properly, they usually appear superior to those in the wild.

In terms of large scale production in containers, one can only discuss watering and fertilization in relation to the soil mix in which the plants grow. It has been pointed out that growth of cycads is negatively affected by severe reduction in soil moisture (Dehgan, 1983). Under cultivation, however, frequent irrigation is often a problem, resulting in reduced soil oxygen levels which directly affect nitrogen uptake and utilisation. By experimenting with various container media the following soilless mixes were recommended specifically for cycads based on their natural habitat preference:

1. Plants of the more mesic habitats, such as most species of *Zamia*, some *Ceratozamia*, *Bowenia*, *Cycas*, *Macrozamia*, *Lepidozamia*, and *Stangeria* require a well-drained soil mix which has an optimal water and nutrient holding capacity. The following mix has proven superior to commercial soil mixes:

   1 part by volume Metro-mix 500 or a similar soilless mix

   1 part by volume sharp sand

   1 part by volume perlite

   1 part by volume pine bark

   2.25 kg/M^3 (5lb/yd^3) of dolomite and 1.25 kg/M^3 (3lb/yd^3) perk micronutrients.

2. For plants of more xeric habitats, such as *Z. furfuracea*, *Encephalartos* spp., *Dioon* spp. and some *Ceratozamia*, the perlite is replaced by gravel-size porous fired clay or a similar product.

Both mixes have excellent drainage but significant water and nutrient holding capacity. The major problem with these mixes is their excessive weight. The unavailability of the specific components also may be a problem but substitutes are possible.

Experience and some habitat studies indicate that media with low Ph (acid soils) usually result in poor growth. For example, recent work of a graduate student at the University of Florida (Durando, 1989) indicated that such problems as 'frizzle top' and micronutrient deficiencies are indirectly caused by low soil pH. For best results the soil pH should be close to neutral or between 6.5 - 7.0, hence the reason for addition of dolomite.

Osmocote or other slow release fertilizers may be added to the mix or as a top dressing. Periodic additions of liquid fertilizers are also helpful. We use and recommend monthly application of 300 ppm Peters (20 - 20 - 20 or other formulations) water-soluble fertilizers. However, recent development of micronutrient amended slow release fertilizers have made possible maintenance of a reasonable level of fertility in the soil for relatively long period of time and avoidance of too-frequent fertilization (Dehgan and Almira, unpublished data). A direct correlation between fertilizer amounts and shade levels have been shown by Keever and Cobb (1986). They observed a higher rate of growth and production of a larger number of leaves in *Z. furfuracea* with 47% shade (as opposed to 72%) and an increase in weekly application of liquid fertilizer concentration from 200 ppm to 300 ppm.

Conclusions

The recent escalation of interest in cycads by botanical scientists, professional nurserymen, and amateur collectors has contributed to both a greater appreciation of and endangerment to cycads. Much new botanical information has resulted from the works of many, but there is a dearth of horticultural information that would contribute to production of superior plants. The outstanding private collections in several countries are of considerable significance in terms of germplasm availability, but the demand for plants has far exceeded the supply, resulting in decimation of many cycad populations. As conservationists and lovers of cycads we cannot afford to allow wholesale destruction of these magnificent ancient plants to continue. It behooves us to spend as much time, funds, and resources as necessary to produce superior plants in cultivation, thus making collection of wild plants superfluous. Such an endeavor is possible only through horticultural research. To our knowledge, we are the only academic institution conducting horticultural research on cycads. This is evidenced by the absence of other such reports in this conference proceedings. It is a certainty that, despite efforts by some governmental agencies, decimation of world populations will continue as long as the demand for these valuable plants is not met through horticultural production. Regrettably, the current available funding for horticultural research is much too limited to permit decisive undertakings. If cycads are to survive for future generations, considerable funds must be allocated to research on commercial production.

Acknowledgment

Experiment Station Journal Series No. R-O1426. We thank Drs., Alan Meerow and Michael Kane for review of the manuscript. We also thank Mr. Bart Schutzman for technical assistance.

References

***Butt, L.P.* 1990** An introduction to the genus *Cycas* in Australia. Publication Fund, Palm & Cycad Societies of Australia.

***Dehgan, B.* 1983**. Propagation and growth of cycads - a conservation strategy. Proc. Fla. State Hort. Soc. 96: 137 - 139.

***Jonson, C.R.* 1983.** Improved seed germination of *Zamia floridana* (*sensu lato*) with H_2SO_4 and GA_3. Sci. Hort. 19: 357 - 361.

***Dehgan, B., Jonson, C.R.* 1987**. Root branching in *Zamia floridana*: Effect of growth regulators and anatomical features. J. Amer. Soc. Hort. Sci. 112: 1041 - 1044.

***Dehgan, B., McConnell, D.W.* 1984.** Cycads with potential as foliage plants. Foliage Digest 7: 1 - 3.

***Dehgan, B., Schutzman, B.* 1983.** Effect of H_2SO_4 and GA_3 on seed germination of *Zamia furfuracea*. HortScience 18: 371 - 372.

***Dehgan, B., Schutzman, B.* 1989.** Embryo development and germination of *Cycas* seeds. J. Amer. Soc. Hort. Sci. 114: 125 - 129.

***Dehgan, B., Yuen, C.K.K.H.* 1983.** Seed morphology in relation to dispersal, evolution, and propagation of *Cycas* L. Bot. Gaz. 144: 412 - 418.

***Durando, J.E.* 1988.** A study on the expression of manganese deficiency in *Cycas revoluta*. Thesis submitted to the Grad. School, Univ. Fla., Dept. Environ. Hort.

***Ellstrand, N.C., Ornduff, Clegg, J.M.* 1990.** Genetic structure of the Australian cycad, *Macrozamia communis* (Zamiaceae). Amer. J. Bot. 77: 677 - 681.

***Giddy, C.* 1974** Cycads of South Africa. Purnell and Sons Ltd, Cape Town, South Africa.

***Hubbuch, C.* 1987.** Cycads: Propagation and container culture. Fairchild Trop. Gard. Bull. 42: 5 - 8.

***Osborne, R.* 1990.** A conservation strategy for the South African cycads. So. Afr. J. Sci. 86: 220 - 223.

***Stevenson, D.W.* 1990.** Morphology and systematics of the Cycadales. Mem. New York Bot. Gard. 57: 8 - 55.

***Tang, W.* 1983.** Letters to the editor. The Cycad Newsletter 6: 5 - 6.

***Tang, W.* 1985.** Cycad trade and conservation. Encephalartos 4: 22 - 23.

Cycad Conservation Legislation - Does it Work in South Africa?

C. Giddy
PO Box 45, P.O. Umlaas Road 3730
Republic of South Africa

Abstract

A summary is made of the objectives of the current cycad legislation in South Africa and the following questions are addressed: Are these objectives valid? Does the present legislation achieve these objectives? Is the legislation enforceable? Can a first world conservation strategy be applied in a third world situation?

---

1. The Objectives of the Current Cycad Legislation in South Africa

The objectives of the South African conservation legislation as embodied in the Environment Conservation Act of 1989 are:

a) to protect the ecological processes, natural beauty as well as preserving the biotic diversity of the natural environment;

b) to promote the sustained utilisation of species and ecosystems and the effective application and re-use of natural resources;

c) to contribute to the establishment, maintenance and improvement of environments which contribute to a generally acceptable quality of life for the inhabitants of the Republic of South Africa.

Cycads were for the first time protected under the Transvaal Flora Protection ordinance of 1940, with the other provinces following suit in 1957 (Cape), 195 (Natal) and 1969 (the Orange Free State). In 1967 the Transvaal Nature Conservation Ordinance elevated all cycads to 'Specially Protected' plants. Natal in 1972, the Cape Province in 1974. Only the Orange Free never introduced the two-tier classification system.

In addition, under south Africa's unique system of independent homelands', the departments of Nature Conservation of Ciskei, Transkei, Kwazulu, Venda, Lebowa, Bophutatswana, Kangwane, and Qua Qua have all enacted their own conservation legislation. In addition to the 12 separate provincial Conservation ordinances, cycads are also protected at national level under the Forest Act of 1984 and the Environment Conservation Act of 1989 and the International CITES agreement.

It should thus appear that cycads are adequately protected in terms of the law. Yet the decrease of natural populations with several species on the brink of extinction indicates that legislation alone will not save the cycads. Both local and international transgressions of the legislation occur.

2. Are These Objectives Valid?

South Africa is a curious mixture of First World technology and wealth, with a large underlay of Third World poverty and lack of skills. Where does conservation slot in? Can a First World conservation strategy in fact be applied to a Third World situation?

The concept - conservation - has become a very important and emotional issue in the developed world because a large majority of the people in the developed world see conservation primarily, in their own selfish interest, as the need to maintain areas of pristine scenic beauty for the purpose of their recreational needs. People in the Third World are no different except that their need is for survival rather than pleasure. In biological terms, other definitions and perceptions apply. As conservationists in the broad sense, we are becoming increasingly aware of the political, social and economic parameters involved in conservation. In order to achieve sustainable conservation measures we must accept that, in addition to habitat protection and management, we need the active support of the local people because social and economic factors will ultimately decide the fate of most species.

The greater proportion of the rural Natal population rely in one way or another upon the natural resources for survival on a daily basis.

An environmental policy which fails to incorporate or tolerate the daily needs of the people cannot succeed. Conservation therefore cannot be divorced from the economy of the region in which it is practiced. It follows then that if changes to lifestyle and culture are to be made as a consequence of the conservation activities, then there must be tangible benefits accruing to the communities involved.

In practical terms and with reference to Kwazulu, the areas with which I am most familiar, these benefits must include the availability of natural resources on a sustained yield basis. they need access to those crops naturally harvested from the area, as well as a percentage of any revenue made from tourism attracted to the area. This means the local people must be able to continue using in moderation, grazing land, crop land, timber and reeds for housing, firewood for domestic cooking and medicinal herbs. Hunting and fishing rights need to be defined and regulated. Here the difficulties and Western perceptions begin to surface.

Photographing Tongan tribesmen setting their ancestral fish traps on Kosi Bay Nature Reserve enchants tourists, seeing an antelope in a snare invokes righteous indignation, both are time honoured traditions but the Western perception differs. It is equally impossible to explain to a Zulu child why he may sell wild gathered mushrooms and fruit along the roadside but not cycads.

The Kwazulu Bureau for Natural Resources which is the conservation authority controlling approximately half the land area of Natal has adopted what can be described as a Third World conservation approach. Not only do they allow the use of natural resources within their nature reserves but 25% of the gross tourist revenue is paid back to the tribal authorities of the area concerned. A councillor from that tribe attends all meetings in Ulundi at which conservation matters relative to this area are discussed. They are free to spend the money on community projects of their choice.

During the 1989 - 90 financial year US$325,000 was earned through tourism of which US$81,000 was paid back to the local tribal authorities.

In addition, fish for local consumption to the value of US$95,000 was caught in the Kosi Bay Nature Reserve and reeds for roofing purposes to the value of US$20,000 was taken from the Reserve. Likewise it is hoped that the Kwazulu cycad nursery at Eshowe will, in addition to providing employment and revenue, also help to stem the illegal sale of habitat collected plants.

One of the most emotive issues in conservation circles is the use of medical plants, It has been esti-

Figure 1 (left): Children in rural Kwazulu selling *E. ngoyanus* plants along the roadsides.

mated that more than 40,000 people rely on the herbal trade for a living. This includes the sangomas or practitioners, the herbal traders and the gatherers. Surveys show that more than 85% of black people consult herbalists. More than 400 different plant species are used and the utilisation of certain species has placed them at risk while other species face extinction in the near future. A recent study indicated that the utilisation level of *Stangeria eriopus* in the Durban area alone was in excess of 50,000 plants per annum, based on regular surveys of stocks at the medicine markets. These plants were 5 - 10 year-old field collected plants. Traders complain that large plants are no longer available, indicating that natural regeneration in certain areas has ceased. The tubers of *Stangeria* are used to ward off evil spirits. With *Encephalartos* species the bark is stripped, leading to the eventual death of the entire tree. While the overall demand for medicinal plants has increased with population growth, and an economic network of gathering, sale and resale has developed to satisfy the urban demand, the natural areas supporting collectors has decreased. Factors contributing to this are the spread of agriculture, particularly sugar cane, afforestation and urban development. Added to this is the disturbing fact that the amount of plants gathered, far exceeds the demand.

Gathering has become an economic activity for unskilled women in the depressed tribal areas where no other employment exists. Due to the oversupply, prices are low and this leads to even more over-collecting in order to make a living. At the retail level, 10 five-year-old *Stangeria* tubers are sold for 50c ($0.018) or less than 2c each.

Several management options to regulate medicinal plant gathering have been put forward:

a) do nothing and allow species diversity to be reduced and species to become extinct;

b) manage the situation purely through strict law enforcement,

Figure 2 (right): Medical plants are a vital commercial resource for many rural communities. Plant gathering has become an economic activity for unskilled women in depressed tribal areas where no other employment exists.

Figure 3: Due to loopholes in the conservation laws these field-collected cycads may be legally sold once they cross provincial borders.

Figure 4: Despite large scale commercial production of cycad seedlings, the easy availability of field-collected plants hampers the effectiveness of nursery grown plants as a conservation strategy.

or on the positive side;

c) provide alternative legal sources of supply through cultivation, either by the private sector or conservation authorities;

d) manage existing natural areas with the cooperation and aid of the tribal authorities, the local sangomas (traditional healers) and conservation authorities.

The Natal Parks Board and the Bureau for Natural Resources have attempted to address the conservation and educational issues through the various Traditional Healers Associations but the problem is a complex one in terms of social, political and economic factors. In the past, traditional conservation practices helped to regulate supply and demand but increasing population densities have caused the breakdown of traditional practices and today some species can only be saved by direct intervention.

3. Is the present Cycad Legislation Enforceable?

More effective law enforcement will depend on:

I. A uniform conservation policy to be enforced throughout South Africa and the neighbouring States. At present, it is impossible to 'launder' plants between provinces, eg., the Transvaal protects only Transvaal species and Cape and Natal plants may be freely sold there. Many commercial sales from the homelands are covered by dubious letters of donation which are legal in terms of the ordinance.

II. With a uniform ordinance, specialist investigation teams will be able to cross provincial boundaries. At present, investigations are often hampered by attempts to establish under whose jurisdiction the plants were removed, transported and eventually sold. In the past, lack of manpower, lack of funding, and lack of expertise (taxonomic knowledge) has prevented effective enforcement. Under a single conservation authority such duplication will be eliminated and manpower and funds put to optimum use.

III. Concomitant with this is the need for heavier fines to serve as a deterrent once the matter reaches the courts. Plant theft is not regarded in a serious light by many magistrates. In a recent court case which enjoyed wide publicity in the media, a fine of US$75 was imposed on each of four counts (illegal export, import, transportation, and possession) a total of US$300. The value of the plants was estimated at US$225,000. Regrettably, these fines are laid down in the relevant legislation but a review is at present in progress along the lines of the recently increased fines for the illegal possession of rhino horn and ivory which now allow for fines of US$25,000 for a first offence.

Stricter law enforcement alone will not save the Cycads. On the positive side:

IV. We need a national educational campaign through the media to counter the ignorance and make the public aware of the conservation status of these plants. We need to differentiate between the First and Third World perceptions of cycads. To the affluent whites they are no longer decorative garden plants but a 'good investment'; to the tribal blacks, from whose areas they are bought at ridiculously low prices, they are a source of income in an economically depressed rural area. They are merely a resource they can commercialise. In terms of the so called commercial or investment value, the media should bear some of the blame. Newspaper headlines usually state 'Rare/Valuable/Treasured' plants, or "Plants to the value of $10,000 recovered in road block" etc. Seldom does the media point out that they are rare and valuable to science.

Cycads have suffered the indignity of being sold at well publicised auctions where exorbitant prices are paid by commercial landscapers in excess of their true value, ie., a well grown commercially produced palm would have served the same decorative purpose. Such auctions put all cycads at risk, both in the habitat as well as those in public and private gardens. As cycad lovers we are not proud of the spate of cycad thefts and such auctions reinforce the commercial value of cycads, a factor in many of the thefts.

V. Natural areas with viable populations of cycads should be identified and declared conservation areas. Our Natural Heritage Program for Sites of Conservation Significance already makes provision for this. These comprise biological sites selected by the regional conservation authorities after field surveys and evaluation in terms of either specialised habitats, et., wetlands, swamp or dune forests or localities of endangered species of animals or plants. These areas are not required by the State but remain in private ownership, yet enjoy the full protection of the conservation authorities plus management advice and assistance when needed.

VI. Concomitant with this is the need to establish properly managed artificial breeding colonies for the most endangered species. Recent field surveys by the Transvaal Department of Nature Conservation have shown that some natural populations are already extinct in the habitat or on the verge of becoming so. They list *Encephalartos cupidus, E. dolomiticus, E. dyerianus, E. eugene-maraisii, E. heenani, E. inopinus, E. laevifolius and E. middelburgensis* in this category. In the Cape Province, *E. latifrons and E. caffer* and in Natal *E. cerinus and E. aemulans* should be afforded the same protection. These species, because of their rarity, are in need of the most stringent protection. Habitat degradation, forestry activities, commercial exploitation, as well as herbalists, are listed as some of the causal factors.

It has been suggested that the most endangered species should not be in private possession unless their legal possession can be authenticated. This would exclude spurious donation letters.

A recommendation has been made that their possession should terminate with the lifetime of the present owner and that they should not become an article of commerce with each successive owner uprooting and relocating a living plant to the detriment of its survival. The CITES regulations are a case in point, where prior to the granting of an import permit the recipient must satisfy the authorities of the state of import that the facilities available will guarantee the plant's survival. In South Africa, cycads from disparate climatic regions, rainfall and temperature ranges, as well as soil types, are all grown in the same garden under the pretext that the collector in so doing, affords them better protection than in the habitat.

Such scientific breeding colonies should, in the interests of long term management and permanency, be run by a body such as the National Botanical Institute. These breeding colonies will serve not only as a gene pool in terms of biodiversity but also allow for the reintroduction of species into their former habitats, providing all precautions as to the genetic purity and disease free stock are met.

VII. The large scale cultivation of commercially produced seedlings at reasonable prices to the public should be encouraged. The private sector in particular should be involved and the establishment of such commercial nurseries in the rural areas close to existing habitats will supply employment and a useful income to third world populations.

Although the IUCN Threatened Plant Unit has placed more than half the world's cycads on the list of threatened species, I believe that we can still stem the tide by taking action now.

References

***Cunningham, A.B.* 1986.** 'An investigation of the herbal medicine trade in Kwazulu.' Phase 1 Report. Institute of Natural Resources.

***Department of Environment Affairs*:** Sites of Conservation Significance Program.

***Fourie, S.* 1989.** 'Conservation Status Report: Transvaal *Encephalartos* Species'. Directorate Transvaal Nature and Environmental conservation.

***Giddy, C, Vorster, P.* 1989.** Recommendations concerning more efficient protection of endangered Plant Species Report submitted to Cycad Commission of Enquiry, 1989.

***Kwazulu Bureau of Natural Resources*:** Environmental Policy Statement. The Environment Conservation Act. Act No. 73, 1989.

***The South African Association of Botanists* 1988:** Recommendations on Cycad Conservation.

Conservation Of East Coast Queensland Cycads

Grant Hawley B.A.(Hons)
Werona Holdings Pty. Ltd.
P.O. Box 512, Aitkenvale, Australia 4814

Abstract

This paper explains why I believe Queensland needs an active cycad conservation movement. I believe we need:

1. Accurate detailed information on Queensland cycads to determine different species and varieties;

2. a coordinated system of administering export permit requirements that is intelligible and sensible to all participating parties;

3. protection for some representative populations of species in the wild;

4. a public education campaign which focuses on the horticultural and aesthetic aspects of cycads and not just their noxious/toxic aspects.

Introduction

I am strongly committed to rational conservation (ecological sustainability) and reasonable use of our environment. I am Townsville based and the proprietor of the plant nursery, Werona Holdings Pty. Ltd., specialising in palm and cycad seeds and seedlings for Australian and overseas horticultural markets. Our propagation nursery is not open for public inspection. We produce each two months a mail order seed and seedling catalogue for individual plant enthusiasts; wholesale and retail nurseries whether local or overseas. My profession ensures me constant access to a wide area of east coast Australia for seed collection, photography, information gathering, communication with other professional seed collectors, commercial seed distributors, government agencies, nurseries and the public.

My professional interest in horticulture also extends to my leisure time activities. I contribute the Thursday's gardening column 'Now in Flower' to the local newspaper, the Townsville Bulletin and have presided over, and am currently on the executive of the North Queensland Palm Society.

I represented the community viewpoint on the very jovial Palmetum Technical Committee and I am currently president of both the Friends of the Palmetum and the North Queensland chapter of the Queensland Nursery Industry Association. These learning experiences both professional and leisure time have enhanced my belief in the need for an active cycad-conservation movement in Queensland.

Need for Accurate Detailed Information

A review of the abstracts of this conference and information generally suggests to me that we know more about cycad chemical make up than we do about naming individual Australian cycad species.

Most Australians, particularly Queenslanders, use the generic terms 'zamias' 'pineapple palm' or 'cycad', to describe the various members of the Cycadaceae family; much the same way as we Australians generalise with our eucalypts - which are all 'gums'. I am going to use the term 'cycads' throughout the paper.

The most valuable tool for the plant enthusiast and the horticulturist interested in Queensland cycads is Len Butt's outstanding work in the now-out-of-print issue of 'Australian Plants' published by the Society for Growing Australian Plants (Butt, 1984). The journal of the Palm & Cycad Society of Australia carried a few articles, but there are few basic published works describing the cycads outside the geographic bounds of those already described by Len Butt in this valuable yet slim volume.

We enthusiasts are forever being tempted with 'a larger, more diverse publication' but its like awaiting the first coning of a cultivated *Cycas* sp. 'Newcastle Range' - maybe next year. Len Butt's original work contributed to the conservation and acceptance of cycads by describing both their visual and horticultural qualities as well as pointing out their toxic ones. It seems that most other, more-scientific works focus on the cycad's infamous nature.

Classification and naming of east-coast-Queensland cycads is difficult, other than by their geographical location. Len admits he did not go far enough in 1984 when describing the east coast cycads, and his statement has special significance to me:

"Cycas media is the common Cycas of the east coast of central and northern Queensland, occurring in rainforest and in eucalypt forest. It is variable in height and form, and can vary so much that maybe natural crossing with other Cycas does occur. Near Cardwell, North Queensland, I found at least three variants growing within a twenty metre circle, fire blackened, but with great healthy rejuvenation in the fronds. Colour variation of leaves was from powder blue, dark green to pea green. All leaf fronds were young, at the same state of rejuvenation". (Butt, 1984; p 13).

Seedling Classification

As there is such noticeable variation in adult and seedling cycads I have begun to classify, grow and sell through our sales list (Grant Hawley's Palm & Cycad Catalogue), east coast Queensland cycads (especially '*C. media*' and its glaucous derivative *C.sp 'Marlborough'*) according to the geographical location of wild seed sources. Seed of a majority of all species sold is collected under personal supervision. Initially, two beds of 2,000 seeds of two *C. media* collections were grown side by side in a shaded nursery situation, progeny in each bed of germinated seedlings showed similarities of leaf colour and width of leaf pinnae to each other, but not those in the opposing bed. These were sold in our catalogues in 1989 as *Cycas media* CMT1 and CMT2.

In the nursery we have now developed a 'natural growth habitat' cycad seedling bank - as opposed to a cycad pollen bank. Reminiscent of large reclining filing cabinets, 10m-long by 1m-wide outdoor raised nursery beds have been established containing the progeny of two geographically isolated collections of the glaucous *Cycas sp. 'Marlborough'*. These seedlings are sold in our catalogues as *Cycas 'Marlborough'* CM1 and CM2. Outdoor growth of the seedlings in natural open shade similar to their natural habitat have resulted in 2 to 3 fronds per plant in less than 3 months.

Simple spray irrigation and initially time-consuming weed-mat cover ensures far fewer labour problems in the long run. Once cortex growth has expanded to the maximum of the original weed mat cut to set the seeds, the weed mat will be lifted and mulch used to suppress weed growth and allow for unrestricted cycad growth.

Nursery beds of *Macrozamia moorii* plants now at age 2 years grow in a similar outdoor situation. A project is to establish nursery beds of all geographical derivatives of *C. media/C. sp. 'Marlborough'*, as well as type species such as the glaucous *C. sp. 'Mt Surprise'* and *C.sp 'Robinhood'* and the geographical variants of *C. cairnsiana*, and *C.sp. 'Petford'*.

We envisage growing these species both for their conservation value as the adults in the wild are fast disappearing as well as being large saleable items. They are lovely plants.

The points that I have been trying to make here are:

1. the difficulty for the non-botanist in distinguishing between local cycads by appearance outside their geographical location;

2. based on geographical location and my observations there seems to be regional variants;

3. juveniles grown *en masse* from wild collected seed are variable according to geographic location;

4. finally - a lay person's guide (which includes a glossary of botanical terms) to identify the different east coast cycads would be welcomed.

Need for a Coordinated Intelligible System of Permit Requirements and Administration

Since 19 January, 1990, seed of cycads and zamias has been removed from Appendix II of CITES. This means, as I understand current Australian CITES laws, that I can now export wild-collected seed without a CITES authority, but to export artificially propagated plants I need a CITES authority.

Cycad seeds are heavy, expensive to freight overseas and will germinate given a variety of amazing factors. One of the best inducements to germination is via economy airmail which may take 2 - 3 weeks to arrive at the destination. In many countries, no permit is required for seed imports, however permits may be required for plant imports.

Should the importer be sent cycad seeds not needing a CITES export authority and which germinate en route, arriving with shoots, stock may be impounded and/or destroyed as having no accompanying CITES authority. Until there is a universally accepted meaning of a seed, germinated seed, and seedling, the best logic is to add the CITES export certificate regardless.

Some countries stipulate maximum ages, and entire lengths, including leaves and roots for the import of plants. Phytosanitary plant health certificate requirements also vary. Updates and local interpretations of current CITES regulations of importing and exporting countries many times do not correspond.

From the examples I have given here I hope I have illustrated the need for a universal terminology and simple, consistent import and export CITES requirements.

We Need More Protection for Some Representative Populations of Cycad Species in the Wild

Wild cycads in Australia have survived since colonial days in mostly inhospitable or uneconomic areas of the landscape. The sorts of areas in which cycads are now found include:-

- National parks
- state forests
- road and railway verges
- rocky or otherwise unsuitable rural, mining and suburban land.

Cycad survival in all but national parks and state forests is at high risk, due to their destruction by:-

- road widening
- burying of once above-ground public utility facilities
- removal of native vegetation for forest rehabilitation

- altered rural use
- mineral exploration
- suburban expansion

Cycads are now also becoming scarce in the field due to:

a) their value in a pot (1 metre high *Cycas* in 8" pot in local east coast nurseries retails at $90),

b) they are free for the taking in the bush,

c) their instant money and landscape appeal to unscrupulous roadside entrepreneurs on public land,

d) access to all of the above simplified by actual field location data being published in journals and gardening magazines.

Disappearing populations of wild cycads, particularly on public lands, results in reduced access for our legal seed collections for nursery propagation. The fertile seed-producing adults of our seed-collection areas on public land have in some cases, been almost totally eradicated by the travelling pick and shovel mob. These also operate in state forests, and private lands. It is most disheartening to visit localities expecting seeding plants only to find freshly dug holes, and occasional scatterings on the ground of green, immature seeds.

Cycads on private land have caused difficulties for property owners protecting their livelihoods. Australia is dependent on a rural export economy. Much time and energy has been expended by government agencies in cycad and noxious weed eradication. The landholders predicament with the zamia palm is universal. (Hardy, 1990).

Sheep and cattle are experimental feeders. They may be severely affected with the 'staggers' when cycad fronds are consumed. Cycads are normally found in inhospitable country to which animals go during the wet season or when the valley grasses have burnt or are sour. Eating the odd soft new cycad frond, eucalypt and other unsavoury greenery is normal.

It would be extremely sad if Australian cycads ended up in the same predicament as the South African cycad, *Encephalartos*. While we are probably a long way from that situation, the sooner enforced sensible regulations for public lands and an education program are in place the sooner we are likely to change attitudes. Australia's economy is not subject to the same pressures as some of the third-world countries where cycads are threatened.

We should have a more rational, measured approach to their conservation.

A system of issued and documented plant-collection permits, utilised by New South Wales (NSW) Forestry departments for elk and stag horns and orchids, should be implemented by Queensland government agencies for cycads. A few successful prosecutions and a hefty fine might discourage souveniring of wild plants on public land and selling to the unknowing plant purchaser. Cycad populations on public lands should be protected from souveniring by placing these areas in published 'forest reserves' making it an offence to remove plants or seeds without an official permit.

If the cycad population is locked away in a National Park, access to seeds, like access to seeds of the famous foxtail palm *Wodyetia bifurcata*, may be limited to the blackmarket. For a reason which I can never comprehend, the retention of the strict national parks philosophy - 'thou shalt not take' - can have the effect of partially destroying the very plants laws are meant to preserve. If seed cannot be obtained legally through National Park government organisations, much the same way as through Forestry organisations, ie., under collection permit or payment of royalty, to my mind as a conservationist in Australia, it may ensure the eventual eradication of the species, due to its rare plant appeal. In my view Queensland legislative changes in relation to seed collection of plants within national parks should be considered to overcome this very obvious problem.

We Need a Public Education Campaign Which Focuses on the Horticultural and Aesthetic Aspects of Cycads and not Just Their Noxious/Toxic Aspects

Cycads have value botanically, scientifically, aesthetically and as small counters to the build-up of atmospheric carbon dioxide. Historically, the worldwide status of cycads seems to have been as poisonous, noxious and/or toxic weeds. Australian children probably all know that Captain Cook's men ate cycad seeds and suffered. At school in Sydney (NSW) in the 1960's we were all under the impression that cycads and cyanide were associated!

Cycads need to be popularised generally for their aesthetic values, much the same way as the Greening of Australia program popularised a new appreciation of growing trees on public lands, farms and in suburban gardens.

The world and Australia now need plants, we need no longer annihilate the cycads. Encouragement to keep and grow cycads is required and for people to do this of course requires a change in public attitudes. While it may take a long time it is necessary to start somewhere and providing accurate information is a good start.

Incentives to keep and grow cycads as natural flora might also be considered. Most Queensland rural landholdings are leasehold and administered by the State Department of Lands. They encourage improvements in the form of fencing and capital works and rate it accordingly. Perhaps the Queensland Government Lands Department could coordinate with the federal government's desire to grow a billion trees in Australia and have areas of leasehold land occupied by cycads de-rated to non-grazing use, the landholder fencing the area as an income taxation deduction and using the cycad-dominated land for native flora regrowth, thus contributing to Australia's billion trees program and the retention of native habitat.

Understandably, landholders are wary of allowing free public access to their properties: gates are left open, breeding cattle are disturbed. One possibility is that seed of a variety of cycad species growing on crown lands or in leasehold situations could be collected by qualified Department of Forestry seed collectors, just like seed of many other economically valuable plants. This is collected and sold on a cost-recovery basis. The more that is collected, the less costly the seed. We have used this service with the collection of *Cycas* sp. 'Newcastle Range':. It works.

Turning to another aspect of education, the concept for Townsville's world-famous Palmetum was established through the initiative of a small band of local palm lovers, principally Jim Darley and other members of what became the North Queensland Palm Society. This is not generally known. Without community involvement and the Australian Bicentennial gift, with which the Townsville City Council were able to do wonders, it is unlikely there would be a Palmetum as it is today. There is no more bicentenary money. However, there is a smatter of public interest, and many generous organisations.

A public cycad park, paralleling the interest created by the Palmetum, could be initiated. If a site or sites within each state were chosen having direct access by the travelling public, an education program paralleling the Palmetum could be initiated quickly. Education, not eradication, should begin at school time, and not just be concentrated on by we interested adults in our plant societies.

Conclusion

Perhaps the next time the International Cycad Conference is held in Australia we will be able to show botanical gardens fulfilling some of the conservational educational aspirations I have outlined in this paper.

I would also hope that over the next few years in Australia, as interest in the environment and plants continues to grow, that cycads will be one of the beneficiaries and that we will have more positive legislation for the conservation of cycads in the wild and some rationalisation of requirements of souveniring from public lands, import and export of these amazing plants.

Recommendations for Cycad Conservation in Queensland

1. define species and varieties accurately through scientific research;

2. coordinate sensible CITES import and export permits so they are intelligible to users;

3. protect representative populations through implementation of sensible legislation and management;

4. focus an education program on the aesthetic, horticultural and conservation aspects of cycads.

References

***Butt, L.* 1984.** Cycads of Australia, in 'Australian Plants', Society for Growing Australian Plants 13 (101) : 3 - 36.

***Hardy, I.* 1990.** Zamia poisoning of cattle. QDPI Farm Note F14/Feb. 1990.

A Conservation Strategy for the South African Cycads

Roy Osborne
University of Natal, King George V Ave,
Durban, 4001 South Africa

This paper is currently in print in
The South African Journal of Science
and is reproduced below by kind permission of
the Editor, Graham Baker

Introduction

The plants popularly known as cycads are members of the order Cycadales and were at least co-dominant in much of the Jurassic flora, some 160 million years ago. Since that time two contrasting processes have been operating: (I) The absolute number of plants in habitat has decreased through the combined effects of climatic changes, competition with the Angiosperms and, most recently, because of man's effect on the environment. (II) The diversity of the cycad flora has increased through speciation processes. Globally there are approximately 182 cycad species in 11 genera (Stevenson and Osborne, 1992), most species occupying specific ecological niches with highly restricted distribution patterns in tropics and the neotropics. South Africa, Mexico and Australia are the three main endemic centres in this regard, South Africa itself having 34 species in two genera (Table 1). Most of these species are classified as endangered, vulnerable or rare by the Threatened Plant Unit (TPU) of the International Union for Conservation of Nature and Natural Resources (IUCN) (Gilbert 1984) based on information provided by local sources (Hall et al. 1980).

Laymen and botanists alike have long been fascinated by cycads. The laymen because of their early origin, decorative appeal of the foliage and trunk forms, odd fruiting structures, slow growth rate, comparative scarcity and rather exaggerated claims of monetary value. The botanists find the attraction in their many unusual and several unique biological features such as the coralloid root systems with nitrogen-fixing cyanobacterial symbionts, the presence in the seeds of high quantities of the acutely neurotoxic and chronically carcinogenic substituted azoxy glycosides, but perhaps most of all because of their reproductive biology. In this strictly dioceious plant group one finds the contrast between the ancestral motile spermatozoid condition and the advanced physiology of an intensely thermogenic male cone which appears to direct specific insect pollinators.

The sporadic reports of sex changes in cycads continue to puzzle those who seek simple explanations in

Table 1

A list of Cycad Species found in South Africa
(Format: Taxon, author(s), date locality/ies

ENCEPHALARTOS
Lehmann 1834 (52 species in Africa - 33 in Southern Africa)

E. aemulans	Vorster	1990	Natal
E. altensteinii	Lehmann	1834	E. Cape, Transkei
E. arenarius	R.A. Dyer	1956	E. Cape
E. caffer	(Thunberg) Lehmann	1834	E. Cape
E. cerinus	Lavranos & Goode	1989	Natal
E. cupidus	R.A. Dyer	1971	E. Transvaal
E. cycadifolius	(Jacquin) Lehmann	1834	E. Cape
E. dolomiticus	Lavranos & Goode	1988	E. Transvaal
E. dyerianus	Lavranos & Goode	1988	E. Transvaal
E. eugene-maraisii	Verdoorn	1945	W. Transvaal
E. ferox	Bertolini fil.	1951	N. Natal, Zululand, Mocambique
E. fiderici-guilielmi	Lehmann	1834	E.Cape, Transkei
E. ghellinckii	Lemaire	1867	Natal, Transkei
E. heenanii	R.A. Dyer	1972	E. Transvaal, Swaziland
E. horridus	(Jacquin) Lehmann	1834	E. Cape
E. humilis	Verdoorn	1951	E. Transvaal
E. inopinus	R.A. Dyer	1964	E. Transvaal,,
E. laevifolius	Stapf & Burtt Davy	1926	E. Transvaal Swaziland
E. lanatus	Stapf & Burtt Davy	1926	Central Transvaal
E. latifrons	Lehmann	1837	E. Cape
E. lebomboensis	Verdoorn	1949	N. Natal, E. & S.E. Transvaal, Swaziland, Mocambique
E. lehmannii	Lehmann	1834	E. Cape

E. lehmannii	Lehmann	1834	E. Cape
E. longifolius	(Jacquin) Lehmann	1834	E. Cape
E. middelburgensis	Vorster, Robbertse & v.d. Westhuizen	1989	Central Transvaal
E. natalensis	R.A. Dyer & Verdoorn	1951	Natal
E. ngoyanus	Verdoorn	1949	N. Natal, Zululand, S.E. Transvaal, Swaziland
E. paucidentatus	Stapf & Butt Davy	1926	E. Transvaal, Swaziland
E. princeps	R.A. Dyer	1965	E. Cape, Transkei
E. transvenosus	Stapf & Butt Davy	1926	N. & E. Transvaal
E. trispinosus	(Hooker) R.A. Dyer	1965	E. Cape
E. umbeluziensis	R.A. Dyer	1951	Swaziland, Mocambique
E. villosus	Lemaire	1867	E. Cape, Transkei, Natal, Zululand, S.E. Transvaal, Swaziland
E. woodii	Sander	1908	(Natal), extinct in nature
STANGERIA T. Moore 1853 (1 species endemic in South Africa)			
S. eriopus	(Kunze) Baillon	1892	E. Cape, Natal, Transkei, Zululand

terms of genetic or hormonal gender control processes.

The Conservation Problem

Over the past few years there has been an escalation of media reports covering alleged illegal collecting and trade of South African habitat-collected cycad specimens (Armstrong 1989; Comrie-Grieg and Cooper 199). These activities persist despite legislation at the Provincial level (Nature Conservation Ordinances; Natal No. 15 of 1974, Cape No. 19 of 1974 as amended in 1985, Transvaal No. 12 of 1983; and relevant sections from the National States legislation which apply to the homeland areas), the National level (Forest Act, Act. No. 122 of 1984, and the Forests Acts of Swaziland) and the International level (CITES regulations, in which all South African cycads are listed on Appendix 1), all intended to prevent such operations. One of the underlying causes of such activities is the demand by the 'status symbol' cycad owner who at all costs wants a cycad, perhaps to decorate a swimming pool or to create a conversation piece in the atrium. Size is more important than species but the owner rarely has the knowledge to properly maintain the specimen. Similar to this class is the fanatical collector, a misguided 'cycad lover' who must have one of each species in his garden, whatever the cost. The person described above rarely resorts to the criminal activity of removing a plant from habitat directly. Instead, they use an informal word-of-mouth network system to make contact with a trader who will arrange 'the deal', taking a hefty profit for his risky expertise. The informal network extends still further when plants are shipped abroad, with the services of an exporter and importer adding to the complexity of arrangements. At the bottom of this chain is the unsophisticated tribesman who performs the physical plant

Figure 1: A view of part of the Lebowa Government Nursery near Duiwelskloof, Transvaal, South Africa, where some 54,000 plants of Encephalartos transvenosus have been propagated from seed over the past few years for distribution at very reasonable prices to organisations and members of the public. Propagation and distribution on a large scale have the effect of reducing pressure by illegal collectors on habitat plants. (Photo: R. Osborne)

removal for a relatively trivial fee and who is hardly concerned with, or even aware of, the illegality of his actions.

The problem outlined in rather simplistic terms above, is neither easily nor quickly solved. The various conservation authorities are unable to maintain constant vigilance over all cycad habitats. Loop-holes in the legislation are exploited with 'letters-of-donation' of questionable authenticity in circulation and 'laundering' of cycads occurs by export and re-import from one province to another. Recent cycad auctions have had the effect of inflating prices to wholly unrealistic levels and making the risk vs. return factor more favourable to the unscrupulous trader.

Plants are not only removed from natural habitats; all botanical gardens and many private collections have reported a dismaying frequency of thefts of valuable specimens. Successful prosecutions of offenders are uncommon and the maximum fines that can be imposed by the courts are inadequate to serve an effective deterrent.

The situation is not without hope. A number of formal and informal bodies have done excellent work; these include the Provincial Nature Conservation authorities in South Africa and the homeland areas, the Government Departments of Agriculture, Environment Affairs and Water Affairs, the recently-formed Endangered Species Unit of the South African Police, various Municipalities, and conservation-orientated Societies. Their efforts have been variously directed at law-enforcement, establishment of reserves, propagation of plants and promotion of public awareness. Until recently, there was little overall co-ordination of the existing programs which thus suffered from the absence of a consolidated long-term objective. In the text which follows a seven-point action plant is submitted which embodies an overall philosophy for cycad conservation in South Africa. This has evolved from guidelines suggested by the Transvaal Directorate of Nature Conservation which subsequently formed the basis of a document put forward by the South African Association of Botanists (SAAB 1988) and is indeed similar to that proposed on an international level (Tang 1985). Furthermore, the spirit of this document is consistent with international guidelines for the conservation of biodiversity in Africa (Stuart and Adams 199).

A Proposed Seven-Point Cycad Conservation Action Plan

1. The existing legislation urgently needs rationalisation and revision. A concise, effective and nationally-uniform set of laws, which totally prohibits removal of any cycads from habitat, is necessary. The only circumstances in which removal could be justified would be when operations such as road and dam construction threaten the plants directly; in this case provision should be made for the relocation of specimens but only to approved botanical gardens or reserves. The legislation must also provide clearly for the authorised sale, donation, import and export of cycads and an unambiguous distinction must be made between field-collected and nursery-propagated plants. Severe penalties must be imposed on those contravening the law.

The so-called 'Cactus Law' of the USA (an IUCN Bulletin) may be a useful model to follow in this regard: transgressors may be deemed unfit to possess endangered plants and their collections confiscated.

2. Urgent studies should be made on the possibility of proclaiming additional reserves where cycad habitats are in jeopardy. The emphasis here must lie on the conservation of cycad habitats, rather than on cycad species or even populations (Hall, 1984). The means to this is already provided for in terms of the new Environment Conservation Act (Act No. 73 of 1989). Some excellent work has already been done in this respect as evidenced by the data in Table 2 (which also illustrates how many different controlling authorities are involved). Attention must be given to the matter of security of plants in these reserves, possibly through controlled access by persons identified in entrance records. The noble idea of re-introducing nursery-raised cycads to habitat must be carefully evaluated with attention being paid to the genetic and phytosanitary quality of the material, its prospect of survival to reach sexual maturity and the likelihood of future regeneration from such plants. For example, re-introduction of *Encephalartos woodii* from the present male clone to its original Ngoye Forest habitat would be quite literally a fruitless exercise. Ecological factors such as the frequency of the burn cycles and the presence or absence of specific insect pollen vectors must be carefully assessed. Consideration should be given to making controlled habitat-collected seed and pollen available to approved growers.

3. Further work must be done in creating and properly managing viable cycad collections outside the habitat areas. The existing comprehensive collections at sites such as the National Botanical Gardens in Pretoria, those at Kirstenbosch, the Lowveld Botanical Garden at Nelspruit and the Durban Botanical Gardens form an ideal nucleus for expansion. The National Botanical Institute is going ahead with a bold plan to stock each garden under its control with plantings of the cycads representative of its area. The initiative here was taken by the Lowveld Garden where a 'mini-farm' of several hundred *E. lebomboensis* plants was planted some years ago. An essential part of the management of these collections must be controlled pollination, seed-harvesting and and propagation programs. Expansion of the stocks of cycads in these gardens should be made from the numbers of plants confiscated from illegal trafficking operations.

One hopes that the large existing private collections might ultimately be distributed to these gardens through bequests.

Table 2 Cycad-orientated Southern African Nature Reserves		
Cycad Species	**Name of Reserve**	**Controlling Authority**
E. altensteinii	Bathurst State Forest	Cape. Dir. Nat. & Env. Conservation
E. arenarius	Alexandria State Forest Woody Cape Nature Reserve	Cape. Dir. Nat. & Env. Conservation Cape. Dir. Nat. & Env. Conservation
E. caffer	Grahamstown Cycad Reserve Longmore/Otterford State Forest Suurberg National Park	Cape. Dir. Nat. & Env. Conservation Cape. Dir. Nat. & Env. Conservation National Parks Board
E. cupidus	Blyderivierspoort Nature Reserve	Transvaal. Dir. Nat. & Env. Conservation
E. dyerianus	Lillie Nature Reserve	Transvaal Dir. Nat. & Env. Conservation
E. ferox	Kosi Bay Nature Reserve Lake Sibaya Reserve	KwaZulu Bureau Natural Resources KwaZulu Bureau Natural Resources
E. fiderici-guilielmi	Lawrence de Lange Nature Reserve	Queenstown Municipality
E. ghellinckii	Cathedral Peak Nature Reserve Giant's Castle Game Reserve Royal National Park	Natal Parks Board Natal Parks Board Natal Parks Board
E. heenanii	Malolotja Nature Reserve Songimvelo Game Reserve (incorporating Ida Doyer Nature Reserve)	Swaziland National Trust Commission KwaNgwane National Parks Board
E. horridus	Uitenhage Nature Reserve	Uitenhage Municipality
E. laevifolius	Blyderivierspoort Nature Reserve Malolotja Nature Reserve Starvation Creek Nature Reserve	Transvaal Dir. Nat. & Env. Conservation Swaziland National Trust Commission Dept Environment Affairs (Forestry Branch)
E. lanatus	Botshabelo Reserve	Middelburg Municipality
E. latifrons	Bathurst State Forest Waters Meeting Nature Reserve	Cape. Dir. Nat. & Env. Conservation Cape. Dir. Nat. & Env. Conservation
E. lebomboensis	Pongola Nature Reserve	Natal Parks Board Transvaal Dir. Nat. & Env. Conservation
E. lehmannii	Suurberg National Park	National Parks Board

Table 2 (continued)

Cycad-orientated Southern African Nature Reserves

Cycad Species	Name of Reserve	Controlling Authority
E. longifolius	Baviaanskloof State Forest Groendal Wilderness Area Loerie Dam Nature Reserve Longmore/Otterford State Forest Suurberg National Park Van Stadens Wildflower Reserve	Cape Dir. Nat. & Env. Conservation Cape Dir. Nat. & Env. Conservation Dias Divisional Council Cape Dir. Nat. & Env. Conservation National Parks board Dias Divisional Council
E. natalensis	Karkloof Falls Nature Reserve Krantzkloof Nature Reserve Oribi Gorge Nature Reserve Soada Nature Reserve Umgeni Valley Nature Reserve Umtamvuna Nature Reserve	Privately owned Natal Parks Board Natal Parks Board Natal Parks board Wildlife Society of Southern Africa Natal Parks Board
E. ngoyanus	Ngoye Forest Reserve Pongola Nature Reserve	KwaZulu Bureau Natural Resources Natal Parks Board Transvaal Dir. Nat. & Env. Conservation
E. paucidentatus	Malolotja Nature Reserve Songimvelo Game Reserve	Swaziland National Trust commission KwaNgwane National Parks Board
E. transvenosus	Happy Rest Nature Reserve Lekgalameetse Nature Reserve Mphaphuli Cycad Reserve Modjadji Nature Reserve	Transvaal Dir. Nat. & Env. Conservation Venda Government Venda Government Lebowa Government
E. trispinosus	Bathurst State Forest	Cape. Dir. Nat. & Env. Conservation
E. umbeluziensis	Mlawula Nature Reserve	Swaziland National Trust commission
E. villosus	Krantzkloof Nature Reserve Oribi George Nature Reserve Pongola Nature Reserve Shongweni Dam Reserve Umtamvuna Nature Reserve	Natal Parks Board Natal Parks Board Transvaal Dir. Nat. & Env. Conservation Natal Parks Board Natal Parks Board
Stangeria eriopus	Dlinza Forest Nature Reserve	Natal Parks Board

4. The horticultural demand must be adequately catered for. Cycads are majestic plants and public and private plantings will continue and indeed, should be encouraged whenever nursery stock is available. The essence here is a reliable supply of high quality material at reasonable prices. In this respect good work has been done by a number of bodies. As one example, the Lebowa Government has in the last few years propagated some 54,000 seedlings of *E.transvenosus* for distribution from its nursery near Duiwelskoof in the

Northern Transvaal. Equally laudable is the massive cycad propagation venture at Hartbeeshoek on the outskirts of Pretoria by the Transvaal Provincial Authority, where 150,000 cycad seedlings were raised and distributed over the period 1973 - 1988. It is hoped that other bodies will follow these examples with the joint goals of providing employment and relieving pressure on habitat populations. Several cycad-specific private nurseries contribute further to the availability of plants; these deserve our support and encouragement. The retail prices of plants produced in government and quasi-government nurseries must be market related so as not to prejudice the interests of the private commercial growers.

5. A national register of (sexually) mature plants in public and private gardens should be created. The logistic difficulty of getting the data and periodically updating it should not undermine efforts in this direction; a partial list would be better than none. This register would carry details such as the number and gender of plants outside their habitats. In several species, e.g., *E. dyerianus, E. heenanii, E. inopinus, E. latifrons,* the number of plants in gardens far exceeds the number remaining in habitat. These thus constitute a vital gene bank for the supply of pollen, seeds and other propagules. The efforts of the Cycad Society of Southern Africa in its program to establish and maintain seed and pollen banks deserve support. The registration could be conducted under the auspices of either that Society or by the Cycad Specialist Group of the Species Survival Commission of the IUCN.

6. Education must be encouraged wherever possible. Most cycad enthusiasts are receptive to meaningful information, as witnessed by the success of *Encephalartos,* magazine of the Cycad Society of Southern Africa. In the broad sense, educational programs must cover an awareness of the legislation and its necessity and should extend to practical aspects on cycad identification and propagation techniques. A key component of the educational parcel must be to stress that a garden specimen can assume 'feature' qualities only a few years after the nursery seedlings stage; 'having a cycad' does not require that a mature plant is removed from habitat and a sense of pride in the 'home-grown' must be instilled. Educational programs could be mounted under the auspices of the National Botanical Institute, the Botanical Society of South Africa, the Wildlife Society of Southern Africa, the Cycad Society of Southern Africa, and by various Universities where cycad research is active. A more comprehensive treatment of our cycad heritage should be included in school biology syllabi.

7. Cycad-orientated research needs stimulation. Good work is presently in progress at some Universities (Osborne, Grobbelaar and Vorster, 1988) but a lack of funding is inhibiting progress. Numerous opportunities are present. Of direct application would be projects such as an extension of the earlier work into cycad tissue culture (Osborne and Van Staden, 1987) and other propagative techniques. A more thorough knowledge of cycad reproductive biology would be immensely useful, as would studies on the dynamics of habitat populations and investigations into the storage of genetic material. Such research projects would benefit from a degree of national co-ordination; perhaps under the auspices of the Foundation of Research Development.

Recent Development

Early circulation of a draft of this text revealed that there was much common ground between the proposed strategy and plans already under consideration by various official and other bodies. The proposals herein have the support of the Transvaal Directorate of Nature and Environmental Conservation and the Wildlife Society of Southern Africa. Recommendations made at a meeting of senior nature conservation officials at Port Elizabeth in November, 1989 mirror these proposals to a large extent and plans are already in hand to revise the present legislation and to review management policies.

This text was tabled at *CYCAD 90* (The Second International Conference on Cycad Biology, Queensland, Australia, July 1990) and its wider application, in an international sense, was noted. This report may thus form the basis for a future global strategy for cycad conservation.

Further Information

Readers who wish to obtain further information on South African cycads generally should consult the books by Cynthia Giddy and Douglas Goode as detailed in the references below. Cycad enthusiasts, growers, research workers and interested laymen are invited to join the Cycad Society of Southern Africa which publishes a quarterly magazine 'Encephalartos', enquiries in this connection should be addressed to Membership Officer Pieter Stroebel, P.O. Box 189, 6000 Port Elizabeth, South Africa.

Acknowledgment

A number of people contributed to the preparation of this text; in particular I would like to acknowledge the assistance of Mr. K. Cooper, Mr. S. Fourie, Mr. S.A. Gerber, Mrs. C. Giddy, Mr. P.D. Glavovic, Prof. N. Grobbelaar, Dr. J.H. Grobler, Mr. B.J. Huntley, Mr. J.F. Jooste, Mr. H.J. Kemp and Dr. P. Vorster.

References

***Armstrong, S.* 1989.** South Africa faces scandal over export of rare plants. New Scientist, 12 August 1989, p. 21.

***Comrie-Grieg, J. and Cooper, K.* 1990.** Cycad Conservation and the "Madeira connection". African Wildlife 44: 21 - 23.

***Dyer, R.A.* 1965.** The Cycads of Southern Africa. Bothalia 8: 405 - 515.

***Giddy, C.* 1984.** Cycads of South Africa. Second revised edition. C. Struick, Cape Town.

***Gilbert, C.* 1984** Cycads: Status, trade, exploitation and protection, 1977 - 1982. World Wildlife Fund, U.S.A..

***Goode, D.* 1989.** Cycads of Africa. Struik-Winchester, Cape Town.

***Hall, A.V.* 1984.** Conservation of threatened natural habitats. South African National Scientific Programs Report No. 92.

***Hall, A.V., De Winter, M., De Winter, B., Van Oosterhout, S.A.M.* 1980.** Threatened plants of Southern Africa. South African National Scientific Programs Report No. 45.

***Osborne, R., Grobbelaar, N., Vorster, P.* 1988** South African cycad research: progress and prospects. South African Journal of Science 84: 891 - 896.

***Osborne, R., Van Staden, J.* 1987** In vitro regeneration of Stangeria eriopus. HortScience 22: 1326.

***South African Association of Botanists* 1988.** Broodboombewaring in S.A.. Encephalartos 15: 1 - 18.

***Stevenson, D.W., Osborne, R.* 1992.** A world list of cycads. Proceedings of the Second International Conference on Cycad Biology, Queensland, Australia, July 1990.

***Stuart, S.N., Adams, R.J.* 1990.** Biodiversity in Africa and its Islands. IUCN document (In Press).

***Tang, W.* 1985.** Cycad trade and conservation. Encephalartos 4: 22 - 23.

The Cycad Society of Southern Africa

Roy Osborne
Department of Chemistry, University of Natal
King George V Ave., Durban 4001 South Africa

The Cycad Society of Southern Africa was founded in 1983 as a non-profit organisation devoted to promoting interest in indigenous and exotic cycads. The aims of the society are to encourage propagation of cycads, to disseminate information on cycads, to maintain cycad pollen and seed banks, to promote all aspects of cycad conservation and to interact with similar organisations on an international basis. Membership of the society has increased steadily to the present 1990 levels of approximately 650 Southern African members and 200 overseas members (Figure 1), the membership body comprising cycad hobbyists, conservationists, professional growers and research workers. The society issues a quarterly magazine, ENCEPHALARTOS, which has been well-received internationally. The society is managed by an executive committee elected for a two-year term of office. Regional branches arrange public lectures and field excursions.

Enquiries may be directed to:

The Membership Officer
Pieter Stroebel
P.O. Box 189
Fort Elizabeth
600 South Africa

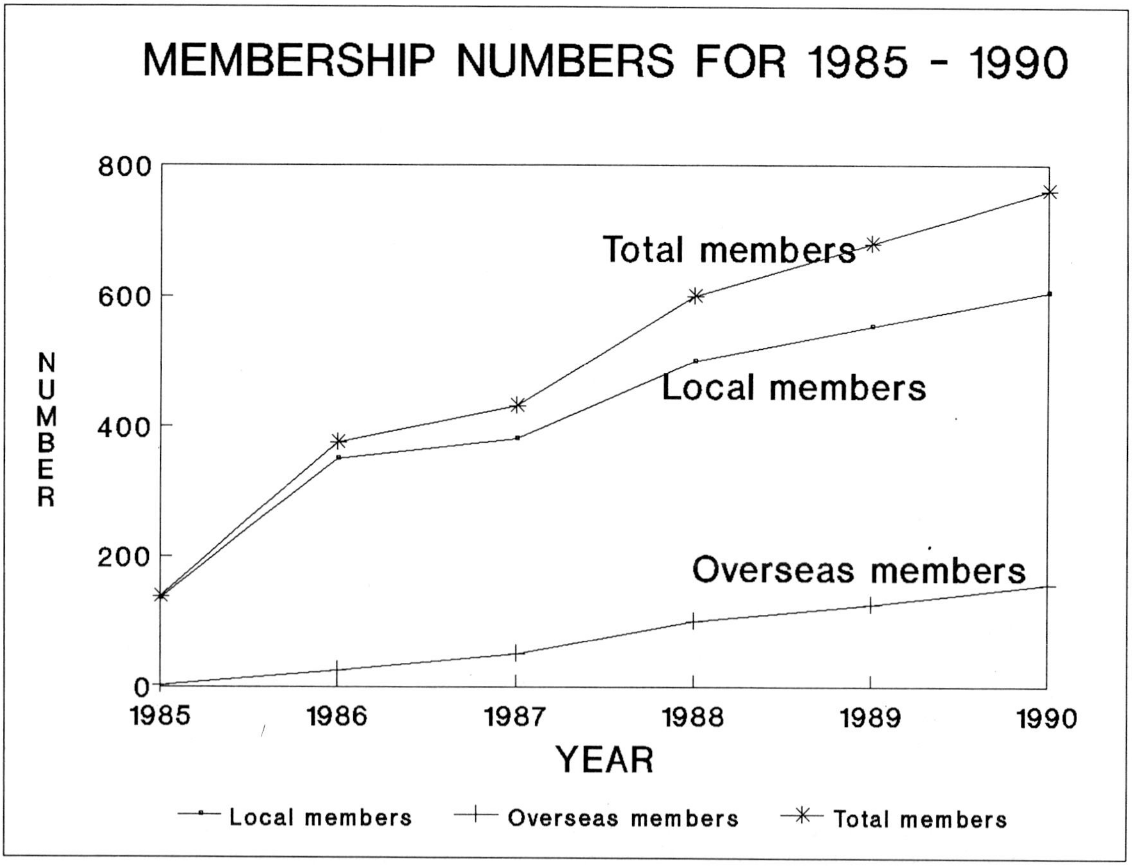

Figure 1. The Cycad Society of Southern Africa. Membership growth over the period 1985-1990.

Cycads in the Adelaide Botanic Garden

Roach, JF and John, C
Botanic Gardens of Adelaide
North Terrace, Adelaide 5000, South Australia

Adelaide is the capital city of the State of South Australia, situated on the central southern coast of the continent; latitude 34º 55' S, longitude 138º 35' E, altitude 50m above sea level. The climate is generally described as 'Mediterranean' with most of the rain falling from April to October. Average annual precipitation is 531mm while total annual evaporation is 1815.4mm. The Adelaide Botanic Garden, at is present site, has been open to the public since 1857.

Although there have always been cycads in this botanic garden, it was not until 1978 that a particular section of the garden was set aside for them. In 1984, an additional site was set aside to house the native species. The exotic collection is situated centrally within the gardens in an area where there are some overstorey trees which provide varying degrees of shade. The most shaded specimens are *Stangeria eriopus*, *Dioon edule* and *D. spinulosum*. Other successful species in cultivation include *Cycas revoluta*, *C. rumphii*, *C. taiwaniana*, *Encephalartos altensteinii*, *E. concinnus*, *E. lebomboensis*, *E. natalensis*, *E. poggei*, *E. princeps*, *E. trispinosus* and *E. villosus*. There is a gentle slope in most of this area which provides reasonable drainage, considering the soil type is red brown earth over clay marl, pH 8.5.

The Australian species are grown as a separate collection and form part of the understorey of an arboretum devoted to Australian species in the north-eastern corner of the garden. The basis for this collection was brought together from existing scattered plantings in 1984. The oldest of these, a *Macrozamia communis* has been in the Botanic Garden since 1958. Other successfully cultivated species include *Lepidozamia peroffskyana*, *Macrozamia fawcettii*, *M. lucida*, *M. miquelii*, *M. moorei* and *M. pauli-guilielmii*. Those plants which either need extra attention, have been in quarantine, or are not large enough to go out into the general collection, are held in a glass house or outdoor holding area. The latter area is provided with a minimum of 50% shading in summer.

Culture under glass provides protection from Adelaide's winters, which are often too wet and cold for many of the subtropical and tropical species, and can predispose them to rotting. Some subtropical Australian species that have not done well out-of-doors are *Bowenia spectabilis*, *Cycas calcicola*, *C. armstrongii*, *Lepidozamia hopei* and *Macrozamia heteromera*. These, together with a wide range of rainforest species from tropical Australasia and the near Pacific, are to be grown in the recently opened Bicentennial Conservatory where the temperature ranges between 16ºC and 33ºC and humidity ranges between 65 and 85%. The internal climatic conditions and free drainage growing medium (pH 6.5-7) provide excellent growing conditions for those species already planted: *Cycas circinalis*, *C .rumphii*, *C. normanbyana*, *C. calcicola*, *Lepidozamia hopei*, *Macrozamia platyrachis*.

Propagation of cycads at the Adelaide Botanic Garden is carried out under glass, with bottom heat maintained at 25ºC. Air temperature ranges from a minimum of 15ºC

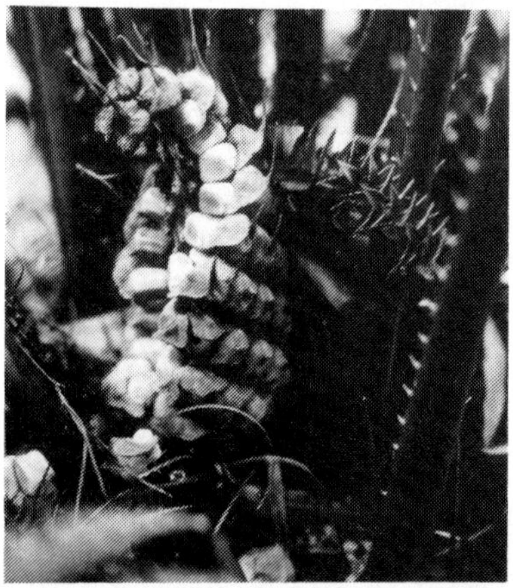

Figure 1 (top left): *Cycas circinalis* has responded well to conditions in the Bicentennial Conservatory.

Figure 2 (top right): *Cycas revoluta*: this specimen weighed 3.8 tonnes when transplanted.

Figure 3 (bottom left): *Macrozamia miquellii* in cone is growing as an understorey plant in the Australian Arboretum.

Figure 4 (bottom right): *Cycas thouarsii* is the fastest growing cycad in the collection. Received in May 1987 as seed. In June 1990 the longest leaf measured 100 cm and the caudex from ground level to apex measured 12 cm.

to a maximum of 28ºC; 75% shading is provided during the summer months.

Once leaves are established, young plants are transferred to a glasshouse where the temperature ranges between 15º and 33º C, and there is 60% shading during summer. Potted plants are grown in a free draining organic based potting mixture, with 2 parts by volume of coarse sand, 1 part peat, 2 parts composted pine bark (Pinus radiata) and 1 part 10 - 12 mm gravel; the final mixture has a pH of about 6. Plants are given a nutrient-balanced, foliar applied, fertiliser during the spring to autumn period.

In June 1990, the total collection numbered 52 species. Of these, 23 species are currently being grown outdoors. In general, most cycads do well in Adelaide as most species can be grown provided their drainage, in particular, and moisture requirements are met.

The World List of Cycads

Dennis W.M. Stevenson & Roy Osborne
New York Botanical Garden, Bronx, New York, 10458 U.S.A.
and
Department of Chemistry, University of Natal,
King George V Avenue, 4001 South Africa

This presentation is dedicated to the memory of the late
John Hendricks

Abstract

Our first 'World List of Cycads' was published in Encephalartos (Journal of the Cycad Society of Southern Africa) (Osborne and Hendricks, 1985), with some minor amendments in a supplementary list (Osborne and Hendricks, 1990). Further changes have made the previous lists obsolete and hence we have decided to present an updated list once again. The format too, is slightly different: in this list we give the names of taxa, the authors (without abbreviations) of the original descriptions and the date of their publications, followed by the countries or provinces where the plant is known to occur. The presentation is alphabetical by genera.

The total number of species in the list which follows is 182.

Note:

(i) In the localities for Zamia following, we have sometimes used provinces or countries; readers should thus note that Choco, for instance, is a province in the western coastal region of Colombia and that Tabasco, Queretaro, Veracruz, San Luis Potosi, Chiapas, Oaxaca and Yucutan are Mexican provinces.

(ii) The taxonomy of Zamia has always presented difficulties and is discussed by Eckenwalder (1980) and Stevenson & Sabato (1986). Further revision remains necessary.

BOWENIA Hooker ex Hooker f. (2 species) **Australia**			
B. serrulata	(W. Bull) Chamberlain	1912	Queensland
B. spectabilis	Hooker ex Hooker f.	1863	Queensland
CERATOZAMIA Brongniart (11 Species) **Mexico, Guatemala & Belize**			
C. eurphyllidia	V-Torres, Sabato & Stevenson	1986	Veracruz
C. hildae	G. Landry & M. Wilson	1979	San Luis Potosi, Queretaro
C. kuesteriana	Regel	1857	Tamaulipas
C. latifolia	Miquel	1848	San Luis Potosi, Queretaro & Hidalgo
C. matudae	Lundell	1939	Chiapas, Oaxaca & Guatemala
C. mexicana	Brongniart	1846	Veracruz, Hidalgo, Puebla
C. microstrobila	Vovides & Rees	1983	San Luis Potosi
C. miqueliana	H. Wendland	1854	Veracruz
C. norstogii	D. Stevenson	1982	Chiapas
C. robusta	Miquel	1847	Veracruz, Oaxaca, Campeche, Chiapas,Guatemala & Belize
C. zaragozae	Medellin-Leal	1963	San Luis Potosi
Note: No changes are expected in Ceratozamia. For a thorough treatment of the genus see Stevenson, Sabato and Vasquez Torres (1986).			
CHIGUA D. Stevenson (2 species) **South America**			
C. bernalii	D. Stevenson	1990	Colombia
C. restrepoi	D. Stevenson	1990	Colombia

CYCAS **Linnaeus (33 species)**			
Australasia			
C. armstrongii	Miquel	1868	Northern Territory
C. angulata	R. Brown	1810	N. Territory & Queensland
C. baguanheensis	L.K. Fu & S.Z. Cheng	1981	China
C. balansae	Warburg	1900	Vietnam
C. basaltica	C.A. Gardner	1923	Western Australia
C. beddomei	Dyer	1883	India
C. cairnsiana	F. Mueller	1876	Queensland
C. calcicola	J.R. Maconochie	1978	Northern Territory
C. chamberlainii	W.H. Brown & Keinholz	1925	Philippines
C. chevalieri	Leandri	1931	Vietnam
C.circinalis	Linnaeus	1763	widespread
C. conferta	Chirgwin	1990	Northern Territory
C. furfuracea	W.V. Fitzgerald	1918	Western Australia
C. guizhouensis	K. Lan & R. Zou	1983	China
C. hainanensis	C.J. Chen & C.Y. Chen	1975	Hainan Is., China
C. kennedyana	F. Mueller	1882	Queensland
C. media	R. Brown	1810	Queensland
C. micholitzii	Dyer	1905	
var. micholitzii			Vietnam, China
var. simplicipinna	Smitinand	1971	S.E. Asia
C. normanbyana	F. Mueller	1874	Queensland
C. panzihuaensis	L. Zhou & S.Y. Yang	1981	China
C. papuana	F. Mueller	1876	Papua New Guinea

C. pectinata	W. Griffith	1854	S.E. Asia
C. pruinosa	J.R. Maconochie	1978	Western Australia
C. revoluta	Thunberg	1782	Japan & Ryukyu Islands
C. riuminiana	Porte ex Regel	1863	Philippines
C. rumphii	Miquel	1839	S.E. Asia & Pacific Is
C. seemannii	A. Braun	1876	Fiji, Tonga & Vanuatu
C. scratchleyana	F. Mueller	1885	New Guinea
C. siamensis	Miquel	1863	S.E. Asia
C. szechuaensis	C. Chen & L.K. Fu	1975	China
C. taiwaniana	Carruthers	1893	Taiwan & S.E. China
C. thouarsii	Gaudichaud	1829	E. Africa & Indian Ocean Islands
C. wadei	Merrill	1936	Philippines
Note: We have ignored Schuster's (1932) work in view of the resulting confusion and considerable criticism of his treatment of the genus. Much field work needs to be done before the taxonomy of this genus could be considered satisfactory. Some useful information of the Australian representatives is provided in a popular booklet by Butt (1990). Work is presently in hand by Sabato and Stevenson which will address the issue of typification and valid publication of names.			
DIOON Lindley (10 species) **Mexico, Honduras, Nicaragua**			
D. califanoi	De Luca & Sabato	1979	Oaxaca
D. caputoi	De Luca, Sabato & V-Torres	1980	Puebla
D. edule	Lindley	1843	
var. edule			Veracruz, Hidalgo, Queretaro, San Luis Potosi, Tamaulipas
var. angustifolium	Miquel	1868	Tamaulipas, Nuevo Leon, San Luis Potosi
D. holmgrenii	De Luca, Sabato & V-Torres	1981	Oaxaca
D. mejiae	Standley & Williams	1951	N. Honduras, Nicaragua
D. merolae	De Luca, Sabato & V-Torres	1981	Chiapas

D. purpusii	Rose	1909	Oaxaca
D. rzedowskii	De Luca, Moretti, Sabato & V-Torres	1980	Oaxaca
D. spinulosum	Dyer	1884	Veracruz, Oaxaca
D. tomasellii	De Luca, Sabato & V-Torres	1984	var. tomasellii S.W. Coast, Mexico
Note: The taxonomy of Dioon is fairly complete and no significant changes are anticipated. A comprehensive review of the genus is given by Sabato and De Luca (1985).			
ENCEPHALARTOS Lehmann (52 species) **Africa**			
E. aemulans	Vorster	1990	Natal
E. altensteinii	Lehmann	1834	E. Cape, Transkei
E. arenarius	R.A. Dyer	1956	E. Cape
E. barteri	Carruthers ex Miquel	1868	
subsp. barteri			Nigeria, Ghana, Benin
subsp. allochrous	L.E. Newton	1978	Nigeria
E. bubalinus	Melville	1957	Tanzania, Kenya
E. caffer	(Thundberg) Lehmann	1834	E. Cape
E. cerinus	Lavranos & Goode	1989	Natal
E. chimanimaniensis	R.A. Dyer & Verdoorn	1969	Mocambique, Zimbabwe
E. concinnus	R.A. Dyer & Verdoorn	1969	Zimbabwe
E. cupidus	R.A.Dyer	1971	E. Cape
E. dolomiticus	Lavranos & Goode	1988	E. Transvaal
E. dyerianus	Lavranos & Goode	1988	E. Transvaal
E. eugene-maraisii	Verdoorn	1945	W. Transvaal
E. ferox	Bertolini fil.	1851	N. Natal, Zululand, Mocambique
E. fiderici-guilielmi	Lehmann	1834	E. Cape, Transkei

E. ghellinckii	Lemaire	1867	Natal, Transkei
E. gratus	Prain	1916	Malawi, Mocambique
E. heenanii	R.A. Dyer	1972	E. Transvaal, Swaziland
E. hildebrandtii	A. Braun & Bouche	1874	Kenya, Tanzania
E. horridus	(Jacquin) Lehmann	1834	E. Cape
E. humilis	Verdoorn	1951	E. Transvaal
E. inopinus	R.A. Dyer	1964	E. Transvaal
E. ituriensis	Bamps & Lisowski	1990	Zaire, Uganda
E. kisambo	Faden & Beentje	1989	Kenya
E. laevifolius	Stapf & Burtt Davy	1926	E. Transvaal, Swaziland
E. lanatus	Stapf & Burtt Davy	1926	Central Transvaal
E. latifrons	Lehmann	1837	E. Cape
E. laurentianus	De Wildeman	1903	Zaire, Angola
E. lebomboensis	Verdoorn	1949	N. Natal, S.E. Transvaal, Swaziland, Mocambique
E. lehmannii	Lehmann	1834	E. Cape
E. longifolius	(Jacquin) Lehmann	1834	E. Cape
E. manikensis	Gilliland (Gilliland)	1939	Zimbabwe, Mocambique
E. marunguensis	Devred	1958	Zaire
E. middelburgensis	Vorster, Robbertse & v.d. Westhuizen	1989	Central Transvaal
E. munchii	R.A. Dyer & Verdoorn	1969	Mozambique
E. natalensis	R.A. Dyer & Verdoorn	1951	Natal
E. ngoyanus	Verdoorn	1949	N. Natal, Zululand, S.E. Transvaal, Swaziland
E. paucidentatus	Stapf & Burtt Davy	1926	E. Transvaal, Swaziland
E. poggei	Ascherson	1878	Zaire

E. princeps	R.A. Dyer	1965	E. Cape, Transkei
E. pterogonus	R.A. Dyer & Verdoorn	1969	Mocambique
E. schmitzii	Malaisse	1969	Zaire, Zambia
E. sclavoi	De Luca, D. Stevenson & Moretti	1989	Tanzania
E. septentrionalis	Schweinfurth	1871	Uganda, Sudan
E. tegulaneus	Melville	1957	Kenya
E. transvenosus	Stapf & Burtt Davy	1926	N. Transvaal
E. trispinosus	(Hooker) R.A. Dyer	1965	E. Cape
E. turneri	Lavranos & Goode	1985	Mocambique
E. umbeluziensis	R.A. Dyer	1951	Swaziland, Mocambique
E. villosus	Lemaire	1867	E. Cape, Transkei, Natal, Zululand, S.E. Transvaal, Swaziland
E. woodii	Sander	1908	(Natal), extinct in nature
Note: The taxonomy of *Encephalartos* is near complete but several potentially new taxa are presently under investigation. The treatment of the ten-known South African taxa by Dyer (1965) remains valid and the popular works of Giddy (1984) and Goode (1989) serve as useful references.			
LEPIDOZAMIA Regal (2 species) **Australia**			
L. hopei	Regel	1876	Queensland
L. peroffskyana	Regal	1857	N.S.W. & Queensland
Note: The treatment by Johnson (1959) remains valid.			
MACROZAMIA Miquel (14 species) **Australia**			
Section Macrozamia Miquel			
M.communis	L.A.S. Johnson	1959	N.S.W.

M. diplomera	(F. Mueller) L.A.S. Johnson	1959	N.S.W.
M. macdonnellii	(F. Mueller ex Miquel) A.de Candolle	1868	Central Australia
M. miquelii	(F. Mueller) A.de Candolle	1868	N.S.W. & Queensland
M. moorei	F. Mueller	1881	N.S.W. & Queensland
M. riedlei	(Gaudichaud) C.A. Gardner	1930	Western Australia
Section *Parazamia* Miquel			
M. fawcetti	C. Moore	1884	N.S.W.
M. fearnsidei	D. Jones	1991	Queensland
M. heteromera	C. Moore	1884	N.S.W.
M. lomandroides	D. Jones	1991	Queensland
M. lucida	L.A.S. Johnson	1959	N.S.W. & Queensland
M. mountperriensis	F.M. Bailey	1886	Queensland
subsp. pauli-quilielmi			Queensland
subsp. flexuosa	(C. Moore) L.A.S. Johnson	1959	N.S.W.
subsp. plurinervia	L.A.S. Johnson	1959	Queensland
M. platyrachis	F.M. Bailey	1898	Queensland
M. secunda	C. Moore	1884	N.S.W.
M. spiralis	(Salisbury) Miquel	1842	N.S.W.
M. stenomera	L.A.S. Johnson	1959	N.S.W.
Note: The treatment by Johnson (1959) remains valid.			
MICROCYCAS (Miquel) A. De Candolle (1 species) **Cuba**			
M. calocoma	(Miquel) A. De Candolle	1868	W. Cuba

STANGERIA T. Moore 1853 (1 species) **South Africa**			
S. eriopus	(Kunze) Baillon	1892	E. Cape, Natal, Transkei, Zululand
ZAMIA Linnaeus (54 species) **South, Central & North America**			
Z. acuminata	Oersted ex Dyer	1884	Nicaragua, N. Panama, Costa Rica
Z. amazonia	D. Stevenson	1989	Brasil, Colombia, Venezuela
Z. amblyphyllidia	D. Stevenson	1987	Puerto Rico, Cuba, Jamaica
Z. amplifolia	Hort. ex Masters	1878	Choco
Z. angustifolia	Jacquin	1791	Bahamas, Cuba
Z. boliviana	(Brongniart) A. De Candolle	1868	N. Bolivia
Z. chigua	Seemann	1854	Choco, Panama
Z. cremnophila	Vovides, Schutzman & Dehgan	1988	Tabasco
Z. cunaria	D. Stevenson & Dressler	1990	Panama
Z. cupatiensis	Ducke	1992	Brazil, Colombia
Z. dressleri	D. Stevenson	1990	Panama
Z. fairchildiana	L.D. Gomez	1982	Costa Rica, Panama
Z. fischeri	Miquel	1854	Queretaro, San Luis Potosi, Veracruz
Z. furfuracea	Linnaeus fil. in Aiton	1789	Veracruz, Belize
Z. herrerae	Calderon & Standley	1924	El Salvador, Honduras, Guatemala, Chiapas
Z. inermis	Vovides, Rees & V-Torres	1983	Veracruz
Z. integrifolia	Linneaus fil. in Aiton	1789	Florida, Georgia, Bahamas, Cuba, Caiman Islands
Z. ipetiensis	D. Stevenson	1990	Panama
Z. jirijirimensis	R.E. Schultes	1953	E. Colombia
Z. lawsoniana	Dyer	1884	Oaxaca

Z. lecointei	Ducke	1915	Brazil. Venezuela
Z. lindenii	Regel ex Andre	1875	Ecuador
Z. lindleyi	Warszewicz ex A. Dietrich	1851	Panama (now extinct)
Z. loddigesii	Miquel	1843	Mexico
Z. manicata	Linden ex Regel	1876	N. Colombia, S. Panama
Z. montana	A. Braun	1875	N. Colombia
Z. monticola	Chamberlain	1926	Veracruz
Z. muricata	Willdenow	1806	N. Venezuela
Z. obidensis	Ducke	1922	Brazil
Z. obliqua	A. Braun	1875	Choco, S. Panama
Z. paucijuga	Weiland	1916	W. Mexico
Z. picta	Dyer	1884	W. Guatemala, Chiapas
Z. poeppigiana	Martius & Eichler	1863	Peru
Z. polymorpha	D. Stevenson	1990	Yucutan, Queretaro, Belize
Z. portoricensis	Urban	1899	Puerto Rico
Z. pseudomonticola	L.T. Gomez	1982	Costa Rica
Z. pseudoparasitica	Yates in Seemann	1854	N. Panama
Z. pumila	Linnaeus	1763	Dominican Republic, Puerto Rico, Cuba
Z. purpurea	Vovides, Rees & V-Torres	1983	Veracruz
Z. pygmaea	Sims	1815	W. Cuba, Isla de la Juventud
Z. regeliana	D. Stevenson	1990	Panama
Z. roezlii	Linden	1873	Choco
Z. skinneri	Warszewicz ex A. Dietrich	1851	Nicaragua, N. Panama, Costa Rica
Z. soconuscensis	Schutzman, Vovides & Dehgan	1988	Chiapas
Z.spartea	A. De Candolle	1868	Oaxaca

Z. splendens	Schutzman	1984	Chiapas
Z. standleyi	Schutzman	1989	Honduras
Z. sylvatica	Chamberlain	1926	Oaxaca
Z. tuerckheimii	J. Donnell Smith	1903	Guatemala
Z. ulei	U. Dammer	1907	W. Brazil, E. Peru
Z. variegata	Warszewicz	1854	Guatemala
Z. vasquezii	D. Stevenson	1990	Mexico
Z. verschaffeltii	Miquel	1869	Mexico
Z. wallisii	A. Braun	1875	Colombia

References

***Butt, L.* 1990.** An introduction to the genus Cycas in Australia. Publication Fund, Palm & Cycad Societies of Australia, 56 pp.

***Dyer, R.A.* 1965.** The cycads of Southern Africa. Bothalia 8: 405 - 515.

***Eckenwalder, J.* 1980.** Taxonomy of the West Indian cycads. J. Arnold Arb. 61: 701 - 722.

***Giddy, C.* 1984.** Cycads of South Africa. Second Revised Edition, C. Struik, 112 pp.

***Goode, D.* 1989.** Cycads of Africa. Struik Winchester, 256 pp.

***Johnson, L.A.S.* 1959.** The families of cycads and the Zamiaceae of Australia. Proc. Linn. Soc. New South Wales 85: 64 - 117.

***Osborne, R. and Hendricks, J.* 1985**. A world list of cycads. Encephalartos 3: 13 - 17.

***Osborne, R. and Hendricks, J.* 1985.** A world list of cycads - supplement. Encephalartos 5: 27.

***Sabato, S. and De Luca, P.* 1985.** Evolutionary trends in Dioon (Zamiaceae). Amer. J. Bot. 72: 1353 - 1363.

***Schuster, J.* 1932.** Cycadaceae.In A. Engler (ed.) Das Pflanzenreich. 99(IV.1): 1 - 168.

***Stevenson, D.W.* 1990.** Chiqua, a new genus in the Zamiaceae with comments on its biogeographic significance. Mem. New York. Bot. Gard. 57: 169 - 172.

***Stevenson, D.W., Osborne, R. and Hendricks, J.* 1990.** A world list of cycads. Mem. New York Bot. Gard. 57: 200 - 206.

***Stevenson, D.W. and Sabato, S.* 1986.** Typification of names in Zamia L. and Aulacophyllum Regel (Zamiaceae). Taxon 35: 134 - 144.

***Stevenson, D.W., Sabato, S. and Vazquez-Torres, M.* 1986.** A new species of Ceratozamia (Zamiaceae) from Mexico, with comments on vegetative morphology, habitats, and species distribution in Ceratozamia. Brittonia 38: 17 - 26.

Cycad Conservation In Mexico: *Dioon edule* Lindl - A Study Case

Andrew P Vovides
Instituto de Ecologia, A.C.,
Xalapa, Veracruz, Mexico

Abstract

Human pressure on populations of Mexican cycads are increasing, especially the disturbance of seed production by annual cropping of the cycad crowns for sale as ornamental plants by street vendors. Other populations are reduced to mutilated relict stands with nil regeneration.

Introduction

Mexico has a biota of world importance occupying 4th place world-wide for biological diversity, (Toledo, 1988). To give an example; estimates on the number of vascular plant species in Mexico are about 25,000 (Rzedowski, 1978). In the coastal state of Veracruz there are estimated to be approximately 9,000 species of vascular plants. Mexican vegetation types range from alpine and taiga in the high sierras, cactus-dessert, matorral, chaparral, tropical and subtropical thorn-forest, temperate woodlands, montane rain-forest or cloud-forest evergreen lowland tropical rain-forest.

In spite of efforts to conserve plant and wildlife within the 126 official protected areas of Mexico, much natural vegetation is disappearing, largely through human population growth, agricultural expansion very often on to marginal lands. This, and coupled with the ever present spectre of illegal plant traffic is taking heavy tolls on cycad populations (Vovides, 1986).

Native cycads in Mexico consist of about 35 species distributed amongst three genera; *Ceratozamia, Dioon,* and *Zamia*. With a conservative estimate, over 75% are endemic. They have been a neglected natural resource until recently (they are officially protected, but in spite of this, destruction and extraction continue unabated).

International Cycad Trade

Cycas commerce in Mexico has increased alarmingly over the recent years. Between 1977 and 1982 more than 130,000 cycads have been taken out of Mexico (Table 1) covered by 'bought' or falsified ministry of agriculture export permits. Gilbert (1984) showed that Mexico took first place world-wide for

Table 1

Total Number of Cycads Exported from Mexico during the Period 1977 - 1982
(Data from Gilbert, 1984)

1977	1978	1979	1980	1981	1982
3,481	-	6,503	12,391	102,126	8,299
TOTAL: 132,800					

cycad exports in 1981 showing a steady increase since 1979. In 1980 one single USA nursery imported 30,000 *Zamia furfuracea* plants per month to supply the landscape industry. In 1986 a single person from San Andres Tuxtla, Veracruz, obtained a permit to export two tons of *Z. furfuracea*. This species is endemic to the coastal area of the Los Tuxtla region of Veracruz, it has now been almost exterminated from most of its localities.

With this crisis in mind the Ministry of Ecology and Urban Development (SEDUE) of Mexico has financed a pilot study to obtain a cycad census for the state of Veracruz, this is summarised in Table 2. This inventory will be used to identify areas that need most protection since it includes data such as threats to the populations, land ownership, etc. This is aimed to concentrate rescue and conservation efforts in the more vulnerable areas. Though inventories can be criticised for being double edged swords, some idea of the status of cycads in the wild is needed in order to plan any conservation actions.

Table 2

Summary of the Cycad Inventory of Veracruz State, Mexico

(Based on Roy Osborne's format for the World Cycad Inventory)

*Veracruz Endemies

SPECIES	No. Estimated in Habitat				
	5-20	20-100	100-500	500-5000	5000+
Ceratozamia euryphyllidia*	+				
Ceratozamia mexicana y vars.					+
Ceratozamia miqueliana				+	
Dioon edule					+
Zamia fischeri			+		
Zamia furfuracea*				+	
Zamia inermis*				+	
Zamia loddigesii					+
Zamia purpurea				+	

In Mexico, cycads are not popular as ornamentals compared to other countries but their popularity is increasing. One can often see in Mexico city and other large cities, street salesmen touting foliage crowns of *Dioon edule* as 'palms'. These are the leaf crowns of large individuals that have been cut with a machete with a portion of the trunk apex, just enough to keep the leaves together, the whole thing being reminiscent of a giant Badminton shuttle-cock. The buyer is simply told to plant this in soil and "it will quickly take root". Of course, the starch-rich injured apex quickly rots and turns into a gelatinous mess. The rotting base is not noticed since this is usually completely buried in the soil to provide certain mechanical stability. Since *D. edule* leaves are very tough they stay green and rigid for over a month. Deterioration is not noticed until the leaflets begin to yellow and this often prompts the prospective cycad gardener to water more frequently than before, thus accelerating the rotting process in the cycad and the frustration in the gardener. Whole teams of collectors armed with cane-cutting knives, burlap sacs and pick-up trucks annually harvest natural *D. edule* populations decapitating large individuals.

Fortunately this does not kill the plants as it would a palm and eventually adventitious buds are produced which, if left unmolested, turn into branches or multiple crowns (Vovides, 1989). The adverse effect however, is the almost complete nullification of the population's seed production. The buds take some years to reach a minimum size to cone, by which time they are often harvested long before coning. This results in stagnant relict populations awaiting eventual deforestation. With this problem in mind an autecological study was undertaken on a natural stand of *D.edule* to learn a little more about its ability for survival in the wild (Vovides, 1990).

Ecology of *Dioon edule*

In Veracruz, *D. edule* is found mainly in tropical deciduous thorn forest and oak woodland, though relict stands can be found in grassland (after removal of the forest). Normally sporadic populations are formed, often locally abundant and very frequently in canyons, steep rocky slopes, and other isolated places with difficult access. A site was chosen about 30km southeast of Xalapa, off the mountain road that runs between Xalapa and Huatusco 600m above sea level. The site forms part of Cerro Papayal-Achichuca near the Rio de Los Pescados canyon, municipality of Coatepec, Veracruz. It is a calcareous hill that forms part of the Cretaceous formation of Chavarillo. It consists of metamorphic limestone with steep slopes and rocky outcrops. The soil is a black, shallow tropical rendzina with loose calcareous stones. The slope where the transect was laid out faces south and has an inclination of 43º (93%). The vegetation is mainly '*selva baja caducifolia*' (tropical deciduous thorn-forest) of Rzedowski (1978), with secondary succession stages in the more disturbed parts.

Four years of data from a 10m x 30m transect with a total of 139 individuals of *Dioon edule* were analysed. Survivorship curves, a life table, spatial distribution, and growth rate were investigated. The plants within the study transect were divided into eight height classes, and observations of leaf production for each height class over the 4-year period enabled an age of *Dioon edule* individuals to be estimated for each height class (Vovides and Peters, 1987).

A high mortality at the seedling and young stage of the life cycle was observed (plants up to 20cm tall), that comprise 78% of the total number of individuals in the transect. The remaining 22% are mature and juvenile plants.

Maximum trunk diameter is reached when the plant reached 40 - 50cm tall which also corresponds to the minimum height of plants observed with cones in this habitat. In heights below 50cm the leaf base increment is geometric while the plant is still increasing in diameter, and as a consequence, there is an increase in leaf production during diameter increment. After approximately 50cm height has been reached the number of leaves produced per flush tends towards a constant number, and the mean total number of leaf bases shows a linear relationship with height.

Chamberlain (1919) estimated the age of a *Dioon edule* plant 5 feet tall to be 970 years. In an attempt to improve on Chamberlain's method, age estimates were based on 3 to 5 plants per height class in the field (Vovides and Peters, 1987: Vovides 1990). In general, these estimates corroborate the age estimate found by Chamberlain (1919) with some very tall individuals being over 2,500 years old. The life table study showed that from the seed input over 250 years, only 109 have survived in the area of the transect. For the age class 750 - 1,000 years, the average expected life is estimated to be 625 years. The heaviest death rate occurs during the early part of the plant's life cycle, i.e., 88% of individuals aged between 0 and 250 years die (Vovides, 1990).

Estimate of fecundity of *Dioon edule* showed the interval in years between coning for males to be from 2.8 to 8.8 years and for females a much longer interval of 10 to 52 years was observed. During exceptionally favourable years this interval can be considerably shorter. Female cones contain between 80 to 250 seeds with a germination rate of 98% (Vovides, 1990).

Spatial distribution and pattern analysis showed *Dioon edule* to have a contagious or clumped distribution at block size 16 for total plants and block size 8 for plants 0-10cm tall which corresponded to approximately 8 and $4m^2$, respectively. The pattern for soil depth showed a similar trend but with lower variance. Although replicate transects would have provided results for a more detailed interpretation of spatial patterns, the steepness and limited area of the study site precluded the use of more than one transect. A skewed distribution of the plants toward the shallower soils is evident; 84% of the plants are on soil depths from 6 to 20cm and frequently on rocky outcrops.

The spatial distribution of the cycads on the shallower soils is probably due to competition for the deeper soils by the faster growing angiosperm species, thus, leaving the more inhospitable niches like these rocky outcrops and shallow soils open for the cycads and other xeric species that can colonise them. The dispersal of the cycad seeds by the rodent *Peromyscus mexicanus* that hoards seeds in small caverns and crevices in and around the rocky outcrops would also contribute to this distribution pattern. This rodent appears to be tolerant to the cycad toxins. In controlled experiments, *Peromyscus* ate ground-up *Dioon* seeds with impunity, whereas the *Rattus rattus* controls die within 24 hours after being fed the same diet (Gonzalez, 1990).

This habitat is subject to bush-fires during the dry season and though this does not affect established plants owing to their armour of persistent leaf bases, but can kill the less protected seedlings. Fire in the preceding year may induce females to cone during the following season (Grobbelaar, Meyer and Burchmore, 1989), probably due to the temporary release of minerals and nutrients. Endomycorrhiza has been detected in this population of *D. edule* (Vovides, 1988), and it has been widely reported that this kind of mycorrhizal association in angiosperms enhances the ability of the host plant to absorb nutrients and water (Barrow et al., 1977; Clarkson, 1974; Cox et al., 1980; Safir et al., 1972; Sanders and Tinker, 1971).

Both biotic and abiotic factors would necessarily have profound effects on the establishment of the cycads. Seedlings would be quickly eliminated during the dry season by fires and from areas where the soil is subject to drying and where quick anchoring by the seedling's vigorous taproot was not possible. The shallower soils and rocky outcrops are favourable to plant establishment by the presence of fissures where the taproots can obtain good anchorage and moisture from condensation, and also by seeds being 'forgotten' by rodents in shady crannies around the limestone rocks, thus contributing to the spatial distribution observed. These factors along with its xerophytic morphology may well give *Dioon edule* a selective advantage to survive on the poor soils of these exposed and harsh microhabitats. The banning of repeated decapitation of adult plants and human encroachment in this habitat would result in a higher seed production and thus survival., Owing to the high natural seedling mortality in the field it is thought that seeds could be harvested on a regular basis for propagation provided a minimum number of seedlings are artificially established in the field.

This could make a clear example case of conservation through propagation and resource management. The management of natural cycad populations may be an answer to their long-term survival in nature, but to achieve this a lot more needs to be known about cycad ecology and population dynamics.

Acknowledgments

I express my sincere gratitude to World Wildlife Fund - U.S., whose support through Grant No. 6543 made possible travel to Australia to attend Cycad '90; to Mr. and Mrs. Coutts for their generous hospitality in Townsville and the out-back station, Chudleigh Park, Qld, and to Stan and Jane Walkley during the post-Conference excursion in S.E. Queensland. Thanks is given to my technician, Senor Carlos Iglesias D., for assistance in the field in Veracruz to complete the Veracruz cycad inventory.

References

***Barrow, N.J., Malajczuk, N., Shaw, T.C.* 1977.** A direct test of the ability of vesicular-arbuscular mycorrhiza to help plants take up fixed soil phosphate. New Phytologist 78: 269 - 276.

***Chamberlain, C.J.* 1919.** The living cycads. Hafner Publishing, New York.

***Clarkson, D.T.* 1974** Ion transport and cell structure in plants. McGraw-Hill, London.

***Cox, G., Moran, K.J., Sanders, F., Nockolds, C., Tinker, P.B.* 1980.** Translocation and transfer of nutrients in vesicular-arbuscular mycorrhizas. III. Polyphosphate granules and phosphorus translocation. New Phytologist 84: 649 - 659.

***Gilbert, S.* 1984.** Cycads: status, trade, exploitation and protection 1977 -1982. Traffic (USA), World Wildlife Fund publication, Washington, DC.

***Gonzalez, A.* 1990** Algunas Interacciones entre *Dioon edule* (Zamiaceae) y *Peromyscus mexicanus* (Rodentia: Cricetidae)., La Ciencia y el Hombre (Universidad Veracruzana, Mexico) No. 5: 77 -92.

***Grobbelaar, N., Meyer, J.J.M., Burchmore, J.* 1989** Coning and sex ratio of *Encephalartos transvenosus* at the Modjadji Nature Reserve. South African Journal of Botany 55: 79 - 82.

***Rzedowski, J.* 1978.** Vegetation de Mexico. Limusa, Mexico City.

***Safir, G.R., Boyer, J.S., Gerdemann, J.W.* 1972.** Nutrient status and mycorrhizal enhancement of water transport in soybean. Plant Physiology 49: 700 - 703.

***Sanders, F.E.T., Tinker, P.B.* 1971.** Mechanism of absorption of phosphate from soil by *Edogone mycorrhizas*. Nature 223: 278 - 279.

***Toledo, V.M.* 1988.** Le diversidad biologica de Mexico. Ciencia y Desarrollo 14(81): 17 - 30.

***Vovides, A.P.* 1986.** Trade and habitat destruction threaten Mexican cycads. Traffic (USA) World Wildlife Fund 6: 13.

***Vovides, A.P., Peters, C.M.* 1987.** *Dioon edule*: La planta mas antigua de Mexico. Ciencia y Desarrollo 13 (73): 19 - 24.

***Vovides, A.P.* 1988.** Studies on Mexican cycads. Ph.D. thesis, University of Wales, Cardiff, U.K.

***Vovides, A.P.* 1989.** Problems of endangered species conservation in Mexico: Cycads an example. Encephalartos No. 20: 29 - 35.

***Vovides, A.P.* 1990.** Spatial distribution, survival, and fecundity of *Dioon edule* (Zamaiceae) in a tropical deciduous forest in Veracruz, Mexico. Amer. J. Bot. 77(12): 1532 - 1543.

Part Seven

Abstracts

The Improbable Story of Cycads as Human Food: Archaeological and Ethnographic Evidence of More Than 4,000 Years

Beaton, John M.
Department of Anthropology, University of California, Davis.
Davis, CA 95616

Among the many uses of cycads, their exploitation as a human food is reasonably well documented, although not often appreciated or understood. Among the many remarkable instances of man/plant relationships, the story of cycads-as-food is an extraordinary one. From the world perspective, there is evidence for the use of cycads as foods in the tropics and subtropics, but the role of cycads in these diets is highly varied. It is known that cycads can be a reasonably continuous and mundane factor in the diet, but this is rare. In many instances, their use is restricted to times of near-starvation, and for other peoples the eating of cycads is inextricably linked to ceremonial occasions. What binds all of these instances together is the clear evidence that cycads are highly toxic and are thus highly unlikely inclusions in a diet. The technology (leaching) which renders the toxin(s) harmless is comparatively simple but the methodology with which it is executed must be highly precise. A study of the prehistoric use of cycads, particularly the *Macrozamia* of Australia, indicates that the technology for the preparation of cycads is at least four thousand years old, and ranks arguably the oldest use of such technology known to man. It is an acute irony of the prehistory of man/plant relationships that this most toxic and dangerous plant may have been the first to be so tamed.

Elements of Cycad-Ingestion Toxicosis in Man and Other Mammals: Some Historical Considerations

Beaton, John M.
Department of Anthropology, University of California, Davis.
Davis, CA 95616

In recent years nutritionists, toxicologists, epidemiologists and others have become aware of the insidious nature of certain biochemical agents common, and possibly unique, to cycads. Two of these important analog compounds, termed cycasin and macrozamin, are known to be the proximal agents in a variety of acute and chronic disorders which affect many organs of mammalian physiology. While some observers consider that certain species of mammal, including some primates, regularly eat cycads, the overwhelming evidence from field and laboratory studies suggests that no mammal should be able to tolerate cycads as a significant component of its diet. It is proposed that the evolution of the cycad toxins was a Mesozoic phenomenon which predated the evolution of mammals and thus, for mammals, the cycads were never a useful food resource. Some implications for modern cycad distributions and livestock husbandry are noted.

The Cycad Collection of Ewanrigg Botanical Gardens in Zimbabwe

Chakavarika, Charles
Ewanrigg Botanical Gardens, PO Box 8119
Causeway, Harare, Zimbabwe.

The history of the gardens and the cycad collection is summarised and a report on the present status of the gardens' cycad collection is presented.

Insects as potential pollinators of *Encephalartos* spp. in South Africa with special reference to *E. villosus*.

(No Heading Provided)

Donaldson, John S.
National Botanic Institute, Kirstenbosch, P/B X7
Claremont 7735, South Africa.

Insects have been implicated in the pollination of some species of *Encephalartos*, most particularly *E. villosus*. However, little is known of the insects involved, their host specificity, their association with male and female cones or their ability to carry pollen. Insects collected from 17 species of *Encephalartos* included 11 species that were associated with cones of both sexes. Of the latter group, at least 6 species, all of which were beetles, were active when male cones were shedding pollen and when female cones were open to receive pollen. Three of these species were associated only with *E. friderici-guilelmii* indicating that the occurrence of entomophily needs to be investigated further in this species. The remaining insect species were less host specific being recorded from several cycad species, including *E. altensteinii* and *E. villosus*. Two insect species from *E. villosus*, namely *Antliarhinus zamiae* and *Porthetes hispidus*, were both capable of bearing substantial pollen loads and their behaviour indicated that they would come into close contact with the micropyle of their host plants.

Cultivation of Cycads From Seed, With Special Reference to *Encephalartos* spp

Giddy, Cynthia
PO Box 45, PO Umlaas Road, 3730
Natal, South Africa

A sequence of photographs is used to illustrate the salient points of cultivation of cycads from seed, based on the author's experience with *Encephalartos* spp. The aspects covered include techniques of pollination, harvesting, storage, testing and germination of seed and the first stage transfer of seedlings to nursery containers.

Encephalartos spp. in Habitat in Africa

Goode, Douglas
Durban Natural Science Museum
PO Box 2085, Durban, 4001, South Africa

A display is presented of photographs of *Encephalartos* species in habitat locations with special reference to the central African taxa. The display also includes a series of 'Protect the Cycads' posters with their conservation message for the Cape, Natal and Transvaal cycads. Finally, some examples of the art-work used in the author's book 'Cycads of Africa' (Struik Winchester, Cape Town, 256 pp., 1989) are shown and related to the concept of the book in its 21 year development.

DNA Variations in the Cycad Phycobionts

Grilli Caiola, Maria and Antonella Canini
Department of Biology, University 'Tor Vergata',
V.E. Carnevale, Rome, Italy

Previous studies on coralloid root gonidial zone of numerous cycads have evidenced the presence of a high number of heterocysts, very different vegetative cells and minute protoplasmatic structures containing thylakoidal systems.

To verify whether in so different cell types it was possible to establish DNA composition and content variations, a study was made on phycobionts of *Encephalartos gratus* Prain using specific fluorochromes for DNA. Parts of coralloid zone has been treated with acridine orange, DAPI and chromomycin in order to reveal total DNA, AT and GC bases respectively.

Measurements of fluorochrome fluorescence intensities were made on photographic prints of the specimens, taken in homogeneous conditions, by using an automatic image analysis system Quantimet 970 (Cambridge Instruments) equipped with QUIPS Version VO7.OO software.

Measurements of fluorescence intensity of different cells indicate the existence of an intense DNA variation from the apical to the basal zones both within the vegetative cells and heterocysts.

Senescence Patterns in the Female Cones of *Encephalartos*

Grobbelaar, N.
Department of Botany, University of Pretoria, South Africa

The cones of some species dry out without any of its components abscising. This is accompanied in some cases by the lengthwise splitting of the cone axis. In many cases, however, the cone scales abscise from the cone axis whereafter the seeds abscise from the cone scales.

In the latter cases, it is the swelling of the seeds which sets up a tension in the stalks of the cone scales which induces the formation of an abscission layer near the base of the cone scale stalks and the subsequent rupturing of the cone scales from the cone axis.

Specificity of the Cycad-Cyanobiont Symbiosis

Grobbelaar, N. and Marshall, J.
Department of Botany
University of Pretoria, South Africa

The cyanobionts in the coralloid roots of many cycad species were isolated and identified. Special attention was paid to the endophytes of *E. transvenosus*.

Although a *Calothrix* and an *Anabaena* species was each encountered once, all the other cyanobionts were *Nostoc* species. Several *Nostoc* species were encountered and all the cyanobionts were able to fix nitrogen. It was established that different root clusters on a single plant can house different cyanobiont species. In one case, two different cyanobiont species were encountered in a single root segment.

The cyanobionts isolated from *E. transvenosus* were compared morphologically, nutritionally and according to their own DNA and pigmentation.

Cycads of the Mesozoic: Once Again a Time of Diversity and Plenty

Hill, Chris. R.
The Natural History Museum
Cromwell Road, London SW7 5BD, UK

As the most primitive and in many respects the most bizarre seed plants surviving today, cycads both living and fossil have an inherent fascination for botanist and layperson alike. When it was realised in the early years of this century, however, that many of the fossils once classed as cycads were actually bennettitalean, the accepted description of the Mesozoic era as the 'Age of Cycads' began to appear inaccurate and misleading. Interest in fossil cycads subsequently waned. However, in recent years the balance has begun to be redressed, based on many significant discoveries of new fossil material worldwide, although the Gondwanan Mesozoic records mostly remain obscure. The new fossils discovered to date provide critical evidence of a wealth of diversity - beginning in the Permian - that had not previously been suspected, and they also indicate that extinct cycads continued to be remarkably diverse well into Cenozoic times.

Of special interest from my own work on the ancient rainforests of the British Jurassic, are two new genera of megastrobili which bridge the previously empty gap between *Cycas* and other Cycadales. Other fossils have contributed to a better understanding of the extinct paraphyletic family Nilssoniaceae and its pollen; whilst fossils collected last year in the Yorkshire Jurassic confirm for the first time that the plant which bore the net-veined leaves *Ctenis* is definitely cycadalean - even though the monocolpate pollen produced by its male strobili was remarkably like that of *Ginkgo*. In mankind's ongoing search for reserves of minerals and fossil fuels, clarification of the relationships, form and structure of such monocolpate pollen - from what we now know was a rapidly evolving group - has potential economic as well as scientific value, particularly for accurate biostratigraphic dating of Jurassic sediments.

Preliminary Observations on Insect/Plant Relationships in Allahabad Plants of *Cycas*

Pant, D.D. and Rita Singh
Department of Botany
University of Allahabad, Allahabad 211002, India.

The present study of the male cones of *Cycas circinalis* Linn. and young compact crowns of megasporophylls in female plants of *C. rumphii* Miq. and *C. revoluta* Thunb. growing in Allahabad has shown that the male cones alone are invaded by pollenivorous small beetles and other insects including ants. Once inside the cones the beetles were found to continue to live there by laying their eggs which hatched into larvae and gave rise to their next generation. Even though these investigations are in sharp contrast with those of Norstog *et al* and Tang who found snout weevils pollinating cones of *Zamia furfuracea* and *Z. pumila* in Miami, they find that their present observations can still be fitted in the generally accepted concept of evolution of insect pollination. The authors regard the insects invading the male cones of *C. circinalis* to be comparable with the insects which initiated insect/plant relationships by becoming pollenivorous before the relationship became reciprocal and the insects started pollinating the plants.

Chemotaxonomy of Cycads

Siniscalco Gigliani, Gesualdo
Dipartimento di Biologia vegetale
Facolta di Scienze, Universita di Napoli, Via Foria 223, 80139, Napoli, Italy.

The analysis of MAM glycosides and mucilages have proven to be a very important tool in understanding relationships among cycads and between cycads and other seed plants. In particular, the exclusive occurrence of MAM glycosides (cycasin and macrozamin) in cycads supports the view that cycads are a monophyletic group. Results support the view that Zamiaceae is a heterogeneous family. In particular, the tribe Zamieae which includes, according to Johnson (1959), *Bowenia*, *Ceratozamia*, *Microcycas* and *Zamia*, is heterogeneous in that *Bowenia* on the basis of macrozamin content appears well separated as shown by Stevenson (1981). The monosaccharide distribution patterns are characteristic at the generic level. Moreover, the mucilage composition of *Zamia* spp. does not support the subdivision of this genus into *Zamia* and *Aulacophyllum* as proposed by Regel (1876), because its pattern is identical in species attributed both to subgenus *Zamia* (i.e. *Z. furfuracea* and *Z. floridana*) and to the subgenus *Aulacophyllum* (i.e. Z. lindenii and *Z. wallisii*). Current researches are devoted to another MAN glycoside (neocyasin α) and to a free aminoacid (α-amino-β-methyl-amino-propionic acid) in order to verify whether the analysis of these compounds may have phylogenetic implications.

The Biogeography of the Neotropical Cycads and its Bearing on the Refugium Theory

Stevenson, D.
New York Botanical Garden, Bronx, New York 10458, USA.

A major tenet of the refugium theory of plant distribution and speciation patterns relates endemism and disjunction to past climatic changes in the neotropics. Briefly, the concept is that pleistocene glaciation patterns caused environmental changes, particularly with respect to precipitation decreases in the neotropics, and some areas remained as wet tropical refuges. These areas can be identified from the occurrence of high endemism, striking and similar disjunct distribution patterns, morphological variation of widespread polymorphic ochliospecies, and persistence of obviously xerophytic plants in rainforest as relicts of previous drying trends that have been reversed since the retreat of pleistocene glacial patterns. The areas of endemism are not areas of habitat diversity and are generally correlated with one or more of the other criteria listed. Because of a lack of information on the distribution and systematics of the neotropical cycads, they have not been considered in this context. However, given their known fossil record, it is clear that all neotropical genera except *Chigua* were present in what is now the new world long before pleistocene glaciation or the uplift of major new world mountain ranges. The present distribution of *Zamia* in South America and *Ceratozamia* in Mexico offer strong support for the Refugium Theory. With the exception of Atlantic coastal forests, each refugium in South America has an endemic species of *Zamia*. Moreover, *Zamia* in South America is virtually limited to refugia. It is interesting to note that some species of *Zamia* are polymorphic and disjunct. Some of these species are adapted to xerophytic conditions and occur on small isolated rocky mountains within the rainforest and are not found within large savannas. Additionally, there are several disjunct species pairs across the Andes, with one species of a pair on the Choco of Pacific Colombia and the other in the Amazon.

Ecological latitude of *Zamiaceae* in Mexico

Vazquez-Torres, M.
Centro de Investigaciones Biologicas
Universidad Veracruzana, Apdo. Postal 294, Xalapa, Veracruz, Mexico 91000.

The greatest number of living cycad species in the new world is found in the neotropics. In Mexico, they reach a major taxonomic representivity (approx. 35 species). This is probably due to the enormous geological, orographical, and climatic heterogeneity as well as the diversity of habitat. This contributes to the fact that the greatest taxonomic diversity of cycads occurs in Mexico. This paper concentrates on the distribution of *Ceratozamia* Brongn., *Dioon* Lindle., and *Zamia* L. that are principally located on the Sierra Madre Oriental and between the meridonals of the transisthmic mountains of Mexico. Only *D. edule* var. *angustifolium* Miq. is found to the north of the Tropic of Cancer in the Atlantic slope and *D. tomasellii* var. *sonorense* De Luca *et al* to the north of the Tropic of Cancer in the Pacific slope.

The water available in the habitats of cycads is very widely variable from very wet to xeric conditions as represented by mangroves and swamps to deserts, respectively. However, a majority of species are found in intermediate conditions. Altitudinal distribution varies from sea level to 2,200m, but most species are situated in the 500-1,000m interval. Substrate soils are also different for various species. These soils range from sandy to clay, volcanic and limestone. No species of neotropical cycads are known from gypsum and saline soils (desert alkaline soils).